WS00010330

This book is due for return on or before the last date shown below.

Don Gresswell Ltd., London, N21   Cat. No. 1208

D1356035

T

FUNCTIONAL ANALYSIS IN CLINICAL TREATMENT

FUNCTIONAL ANALYSIS IN CLINICAL TREATMENT

EDITOR

PETER STURMEY

AMSTERDAM · BOSTON · HEIDELBERG · LONDON
NEW YORK · OXFORD · PARIS · SAN DIEGO
SAN FRANCISCO · SINGAPORE · SYDNEY · TOKYO
Academic Press is an imprint of Elsevier

Academic Press is an imprint of Elsevier
30 Corporate Drive, Suite 400, Burlington, MA 01803, USA
525 B Street, Suite 1900, San Diego, California 92101-4495, USA
84 Theobald's Road, London WC1X 8RR, UK

This book is printed on acid-free paper. ∞

Library of Congress Cataloging-in-Publication Data
Application Submitted

British Library Cataloguing-in-Publication Data
A catalogue record for this book is available from the British Library.

ISBN 13: 978-0-12-372544-8
ISBN 10: 0-12-372544-5

For information on all Academic Press publications
visit our Web site at www.books.elsevier.com

Transferred to Digital Printing 2010

Working together to grow
libraries in developing countries

www.elsevier.com | www.bookaid.org | www.sabre.org

ELSEVIER   BOOK AID
            International   Sabre Foundation

To my colleagues at the Learning Processes Doctoral Program,
Queens College, City University of New York

CONTENTS

PREFACE

Behaviorism has long considered explanations and treatment of psychopathology. Miller and Dollard (1941) began to translate psychoanalytic into behavioral concepts. Skinner's *Walden Two* (1948), *Science and Human Behavior* (1953), and *Beyond Freedom and Dignity* (1971) developed that approach and addressed the conceptualization of psychopathology and its implications for case formulation and treatment. Skinner also expanded on earlier work by offering functional analytic approaches to explaining behavior change that might occur during classic psychotherapy and Rogerian therapy. Functional analytic approaches to psychopathology took off in the 1950s in areas such as mental retardation (Fuller, 1949; Lindsley, 1964a; Risley & Wolf, 1964, 1967; Wolf, Risley, Johnston, Harris, & Allen, 1967; Wolf, Risley, & Mees, 1964), schizophrenia (Ayllon & Michael, 1959; Lindsley, 1956, 1959, 1960, 1963; Lindsley & Skinner, 1954), tics (Barret, 1962), and geriatrics (Lindsley, 1964b) and influenced the development of behavior therapy (Kalish, 1981; Kanfer & Phillips, 1970; Ullman & Krasner, 1965; Wilson & Franks, 1982). Turkat's (1985) volume, *Behavioral Case Formulation*, was notable in illustrating the application of behavior analytic concepts to individual case formulation and in addressing its application to a wide range of clinical problems, such as fear, depression, personality disorders, substance abuse, and problems of later life. This early work formed the basis of subsequent development of functional approached to psychopathology.

Functional approaches to this psychopathology are characterized by focusing on current context; environmental variables that have a large impact on the presenting problem and that can be readily manipulated; operationalization of behavior and reliable measurement of its relationship to the environment; and case formulation and idiographic treatment that are grounded in behavioral concepts and aim to produce large, socially, and personally meaningful changes (Baer, Wolf, & Risley, 1968) that address goals that society recognizes as important, that use methods and achieve outcomes that are valued (Wolf, 1978). This volume shows that this approach has now been applied to the full range of psychopathology.

OVERVIEW OF THIS VOLUME

The first 3 chapters review basic philosophical underpinnings of functional analytic approaches. These first chapters describe the common learning processes that are involved in all behavior change processes and illustrate their application to case formulation and intervention with a wide range of populations. The next 17 chapters review the application of functional analysis to the major *Diagnostic and Statistical Manual of Mental Disorders* (4th edition; *DSM-IV;* American Psychiatric Association, 1994) categories of psychopathology using the following standard format: Each of these chapters briefly reviews the current diagnostic criteria and then moves on to describe a functional analytic model of that disorder, its functional assessment and analysis, and functional analytic-based interventions. Finally, each chapter illustrates these concepts and intervention methods with a case study, which includes functional assessment or analysis, development of a treatment plan, and evaluation of outcome. Cindy Anderson's final chapter provides an overview of these chapters and identifies common themes, emerging issues, and future directions.

THE CONTRIBUTORS

The contributors have made my job as editor an easy one. I selected them because they had established track records of empirical research in functional approaches to psychopathology and extensive professional experience in the hope that they would produce authoritative overviews of their areas of expertise and inform practitioners of how to conceptualize and treat psychopathology: I was not disappointed.

REFERENCES

American, Psychiatric Association. (1994). *Diagnostic and statistical manual of mental disorders* (4th ed.). Washington, D.C.: Author.

Ayllon, T., & Michael, J. (1959). The psychiatric nurse as a behavioral engineer. *Journal of the Experimental Analysis of Behavior, 2*, 323–333.

Baer, D. M., Wolf, M. M., & Risley, T. R. (1968). Some current dimensions of applied behavior analysis. *Journal of Applied Behavior Analysis, 1*, 91–97.

Barret, B. H. (1962). Reduction in rate of multiple tics by free operant conditioning methods. *Journal of Nervous and Mental Diseases, 135*, 187–195.

Fuller, P. R. (1949). Operant conditioning of a vegetative human organism. *American Journal of Psychology, 62*, 587–590.

Kalish, H. I. (1981). *From behavioral science to behavior modification.* New York: McGraw Hill.

Kanfer, F. H., & Phillips, J. S. (1970). *Learning foundations of behavior therapy.* New York: Wiley.

Lindsley, O. R. (1956). Feeding a kitten—A social reinforcer. In *Annual technical report #3, November, Contract N5-Ori-07662. Office of Naval Research*. Waltham, MA: Harvard Medical School, Behavior Research Laboratory.

Lindsley, O. R. (1959). Reduction in rate of vocal psychotic symptoms by differential positive reinforcement. *Journal of the Experimental Analysis of Behavior, 2*, 269.

Lindsley, O. R. (1960). Characteristics of the behavior of chronic psychotics as revealed by free-operant conditioning methods. *Diseases of the Nervous System* (Monograph Supplement), *21*, 66–78.

Lindsley, O. R. (1963). Direct measurement and functional definition of vocal hallucinatory symptoms. *Journal of Nervous and Mental Disease, 136*, 293–297.

Lindsley, O. R. (1964a). Direct measurement and prosthesis of retarded behavior. *Journal of Education, 147*, 62–81.

Lindsley, O. R. (1964b). Geriatric behavioral prosthesis. In R. Kastenbuam (Ed.), *New thoughts on old age (pp. 41–60)*. New York: Springer.

Lindsley, O. R., & Skinner, B. F. (1954). A method for the experimental analysis of the behavior of psychotic patients. *American Psychologist, 9*, 419–420.

Miller, N., & Dollard, J. (1941). *Social learning and imitation*. New Haven, NJ: Yale University Press.

Risley, T. R., & Wolf, M. M. (1964). Experimental manipulation of autistic behaviors and generalization into the home. In R. E. Ulrich, T. Stachnik, & J. Mabry (Eds.), *The control of human behavior* (pp. 193–198). Glenview, IL: Scott Foresman.

Risley, T. R., & Wolf, M. M. (1967). Establishing functional speech in echolalic children. *Behaviour Research and Therapy, 5*, 73–88.

Skinner, B. F. (1948). *Walden two*. New York: Macmillan.

Skinner, B. F. (1953). *Science and human behavior*. New York: The Free Press.

Skinner, B. F. (1971). *Beyond freedom and dignity*. New York: Knopf.

Turkat, I. (Ed.) (1985). *Behavioral case formulation*. New York: Plenum.

Ullmann, L. P., & Krasner, L. (Eds.). (1965). *Case studies in behavior modification*. New York: Holt, Rinehart & Winston.

Wilson, G. T., & Franks, C. M. (1982). *Contemporary behavior therapy. Conceptual and empirical foundations*. New York: Guilford Press.

Wolf, M. M. (1978). Social validity: The case for subjective measurement or how applied behavior analysis is finding its heart. *Journal of Applied Behavior Analysis, 11*, 203–214.

Wolf, M. M., Risley, T. R., Johnston, M., Harris, F., & Allen, E. (1967). Application of operant conditioning procedures to the behavior problems of an autistic child: A follow-up and extension. *Behaviour Research & Therapy, 5*, 103–111.

Wolf, M. M., Risley, T., & Mees, H. (1964). Application of operant conditioning procedures to the behavior problems of an autistic child. *Behaviour Research and Therapy, 1*, 305–312.

LIST OF CONTRIBUTORS

Laura Addison, Pediatric Feeding Disorders Program, Marcus Institute, The Kennedy Krieger Institute, Atlanta, GA

Cynthia M. Anderson, Educational and Community Supports, University of Oregon, Eugene, OR

Holly Bihler, Southern Illinois University, Carbondale, IL

Stacey Cherup, Gerontology Academic Program, University of Nevada, Reno, NV

Prudence Cuper, Department of Psychology, Duke University, Durham, NC

Robert Didden, Department of Special Education, Radboud University Nijmegen, The Netherlands

Mark R. Dixon, Southern Illinois University, Carbondale, IL

Erica Doran, Department of Psychology, Queens College, City University of New York, Flushing, NY

Claudia Drossel, Gerontology Academic Program, University of Nevada, Reno, NV

Richard F. Farmer, Oregon Research Institute, Eugene, OR

Jane E. Fisher, Gerontology Academic Program, University of Nevada, Reno, NV

Patrick C. Friman, Clinical Services, Flanagan's Boys' Home, Boys Town, NE

Sarah H. Heil, Department of Psychology, University of Vermont College of Medicine, Burlington, VT

Stephen T. Higgins, Department of Psychiatry, University of Vermont, Burlington, VT

Derek R. Hopko, Department of Psychology, University of Tennessee, Knoxville, TN

Sandra D. Hopko, Cariten Assist Employee Assistant Program, Knoxville, TN

Taylor E. Johnson, Southern Illinois University, Carbondale, IL

Craig H. Kennedy, Vanderbilt University, Nashville, TN

Janet D. Latner, Department of Psychology, University of Hawaii at Manoa, Honolulu, HI

C. W. Lejuez, Biology-Psychology, University of Maryland, College Park, MD

Thomas Lynch, Duke University Medical Center, Durham, NC

Michael Marroquin, Department of Psychology, Queens College, City University of New York, Flushing, NY

Rhonda Merwin, Duke University, Durham, NC

Prof. Ray Miltenberger, The Department of Child and Family Studies, University of South Florida, Tampa, FL

Nancy A. Neef, Ohio State University, Columbus, OH

John Northup, Ohio State University, Columbus, OH

Cathleen C. Piazza, Munroe Meyer Institute, University of Nebraska Medical Center, Omaha, NE

Joseph J. Plaud, Applied Behavioral Consultants, Inc., Whitensville, MA

Stacey C. Sigmon, Department of Psychology, University of Vermont College of Medicine, Burlington, VT

Peter Sturmey, Department of Psychology, Queens College, City University of New York, Flushing, NY

Robert Wahler, Department of Psychology, The University of Tennessee, Knoxville, TN

John Ward-Horner, Department of Psychology, Queens College, City University of New York, Flushing, NY

David A. Wilder, School of Psychology, Florida Institute of Technology, Melbourne, FL

W. Larry Williams, Department of Psychology, University of Nevada, Reno, NV

Stephen E. Wong, School of Social Work, Florida International University, Miami, FL

Doug W. Woods, Department of Psychology, University of Wisconsin-Milwaukee, Milwaukee, WI

Craig Yury, Gerontology Academic Program, University of Nevada, Reno, NV

1

STRUCTURAL AND FUNCTIONAL APPROACHES TO PSYCHOPATHOLOGY AND CASE FORMULATION

PETER STURMEY
JOHN WARD-HORNER
MICHAEL MARROQUIN

AND

ERICA DORAN
Queens College and The Graduate Center
City University of New York

This volume reviews functional analytic approaches to case formulation and treatment for all the major categories of psychopathology. The first three chapters review basic concepts and methods that these approaches use. In this chapter, we begin by contrasting structural and functional approaches to behavior, including psychopathology. We then define radical behaviorism by contrasting it with nonbehavioral approaches as well as other forms of behaviorism, such as earlier methodological behaviorism and more current forms of behavior and cognitive behavioral therapy. We go on to look at radical behavioral approaches to psychopathology and therapy, illustrating this with some of Skinner's work in this area. Since each chapter in this book that reviews particular forms of psychopathology concludes with an illustrative case study, we consider clinical case formulation, nonbehavioral and behavioral approaches to case formulation, and illustrate this with an example of a generic case formulation offered by Skinner.

STRUCTURAL AND FUNCTIONAL
APPROACHES TO BEHAVIOR

Structuralism and functionalism are two distinct approaches taken toward many intellectual endeavors, including anthropology, sociology, literature, linguistics, and indeed psychology. They differ in the status given to observations and the aim of the intellectual project. Structuralism takes observations to be tokens of underlying hidden structures that cannot be directly observed, but rather are inferred from superficial observations. These structures are seen as inherent and relatively unchanging attributes of the thing that is studied. For example, in literature, the surface words are seen as mere indications of one of a few underlying narrative structures. Similarly, in structural approaches to linguistics, spoken words are mere surface tokens of deep grammatical structures or the speaker's meaning. Hence, functionalism places great emphasis on the role of environmental variables and the relationship of the observed phenomenon to the environment. Biological evolution is one of the most characteristic examples of functionalism, which is closely related to behaviorism. Biological evolution sees surface observations—such as the structure of an organism, its organs, cells and physiology, and indeed its behavior—as the result of the selecting environment. Variations in the organism that are heritable and most adapted to the environment are likely to be differentially selected by the environment.

STRUCTURAL APPROACHES TO PSYCHOLOGY

Structuralism in psychology is concerned with the analysis of an experience, event, or idea down to its smallest part. Once the smallest parts have been found, they are classified into groups. Each molar experience is thought to have a scientific explanation. This explanation is in terms of the molecular parts that compose the molar experience. Molar experiences can be described as affective, sensational, and/or relational in nature. In this definition, the psychological area of interest is the private internal environment and how that environment produces surface behavior, focusing its concern on the "social or personal environment" (Calkins, 1906). This approach invites an analogy between psychology and chemistry in that the surface matter is analyzed into the component atoms that are then rationally organized to make up what is observed. Many early forms of structuralism depended on highly trained subjects to report their private experiences and perceptions as an attempt to observe and report the components of experience.

Fitremann (2006) described the structural approach to psychology and psychotherapy in which psychological events consist of a number of elements. The first of these elements is the "essential program," which is defined as necessary for survival and growth. This element is innate and described as a program searching for a particular stimulus, that stimulus then being used as a guide. The next element is "imprints," which is the element that is necessary to initialize a major program. The source of this element is an external specific trigger stimulus necessary for the onset of an essential program. The next element is the "structure," which is defined as a collection of the major programs at the stage of their present completion. The source of this element is internal and is described as a system of macrofunctions, linked with one another, in relation to the outer world. The last element is "contents," which is defined as input to the inner programs. Its source is both internal and external. This is described as the organization of stimuli from the external world for the development of programs within the organism.

Each of these elements interacts with the others and with the outside world. These interactions are private. Psychic functions are described as self-developing programs analogous to a computer system. Some components of a computer system, such as a printer, are external. Other components, such as a word processing program, are internal. The internal and external components require internal programming for them to function and interact with the outside world, such as the printing of a page or recording of data into a file. Thus, structural approaches are characterized by an organism that is involved in many internal and external processes that are driven by this internal program. The goal of structuralism is to define these processes and their relation to the internal and external world of the organism.

FUNCTIONAL APPROACHES TO PSYCHOLOGY

Functional approaches to psychology view observable behavior as something of interest and worthy of studying in and of itself. Observable behavior is viewed as an adaptation to the environment and thus useful to the organism. Functional approaches also seek to find lawful and reliable relationships between the environment and behavior. Such approaches are conceptually continuous with other functional approaches to behavior, such as biological evolution. For example, suppose we observe a lizard move from shade to the sun in the morning. Functionalist approaches to this problem—both evolutionary biology and behaviorism—would ask questions such as "What purpose does moving to the sun and shade serve for the organism?" If we observe a person speak, pick up a baby, or write a plan, a functional approach asks the same question.

To answer these questions, a functional approach would seek to discover reliable relationships between the environment and behavior. If we independently manipulated shade and temperature, where would the lizard move? If another person smiles or turns away, what happens to picking up the baby in the future? If we have a good, empirically based knowledge of the relationship between the environmental variables and the behavior of interest, could we turn the behavior on and off? Could we predict and control the behavior of interest? This is the goal of those endorsing the functional approach.

FUNCTIONAL AND STRUCTURAL APPROACHES CONTRASTED

From the preceding discussion, several differences between these approaches are immediately apparent. First, they differ as to what they consider to be important. Functional approaches focus on the observable and reliable, while structural approaches focus on nonobservable inferred constructs. Second, these approaches also differ in terms of causality. Structural approaches locate causes inside the person, while functional approaches locate the cause in the environment. For example, a structural linguist would say that a child does not speak because the child does not yet have the linguistic or cognitive structures to enable speaking to occur, while a functionalist might look at the child's learning history and current learning environment to determine reliable relationships between mutism and environmental variables. Third, these approaches also differ as to intervention. Because structural approaches locate the causes of behavior in the person's cognitive or linguistic structures, the intervention must be directed at stimulating, activating or repairing these structures. In contrast, functional interventions identify and alter the environmental variables that are reliably related to the behavior of interest.

These last two points have important implications for the nature of intervention. First, structuralists are to some extent pessimistic, as the hidden structures may be difficult to measure and modify. In contrast, functionalists tend to be optimistic about change, since it merely involves identifying reliable relationships between the environment and behavior. Even if these relationships are currently unknown, future observations and study could well reveal them. Second, structuralists are on weak ground in terms of explaining behavior because their explanations are necessarily circular. The presence of unobservable structures is inferred from behavior, yet these structures are then used to explain that same behavior. If behavior changes, structuralism infers that the structures changed and these changes in the structures caused the change. However, the structures can never be observed independent of the behavior, and hence such explanations can be characterized as explanatory fictions (Skinner, 1953, 1971, 1977).

RADICAL BEHAVIORISM

BEHAVIORISM

Behaviorism is the philosophy of science that underlies the practice of behavioral science, which is distinct from mainstream psychology. Chiesa (1994) noted that behaviorism is highly internally consistent. Its subject matter is carefully defined and its methods are universally agreed upon within the field, even though members of the field may study different kinds of behavior in very different contexts. Behaviorism also places greatest value on descriptive, observational data; remains very close to the data; and deemphasizes theory (Skinner, 1950). It is likely to observe a behavior and look for manipulable variables that influence that behavior. Behaviorism begins by asking the question "What would happen if . . .?" Thus, behaviorism deemphasizes theory. Based on many observations and generalizations, laws can be induced. For example, by observing operant extinction in a rat pressing a lever, an isolated neuron reinforced for firing with dopamine, and adjective use in undergraduates, one might induce some generalities about operant conditioning that might be more generally true.

Behaviorism operationally defines its terms. This is done in part because science requires such precision. Everyday language concerning behavior is imprecise and deceptive due largely to the extensive use of terms that have preexisting meanings and implications about the nature and causes of behavior. Everyday language is imbued with an existing system of thinking about behavior that often implies that people are the agents of behavior. Thus, if everyday terms are adopted, they first require careful examination as to their precise meaning. If they are not consonant with the science of behavior, their meaning should be specified or other terms should be used.

Behaviorism also adopts inductive, rather than classical theory-driven, hypothesis testing in science. Induction refers to science that gives prominence to the data. Induction is also a process of reasoning in which specific instances or observations are used to induce laws that are generally true and applicable to instances that have not yet been observed. Thus, induction assumes that we can generalize from instances to the general and that there is consistency between observations made in the past and the future.

Behaviorism is interested in variation in behavior. It seeks out the sources of variation in behavior of individual organisms. Rather than seeing variation as noise to be eliminated, variation is seen as the thing to be studied and explained.

Behaviorism also adopts an unusual view of causation. It replaces traditional notions of causality with two notions: functional relationships and selection. First, behaviorism seeks to discover lawful relationships between independent and dependent variables. When such relationships have been discovered, then it may be possible to predict and control behavior. Second, behaviorism looks to selection to explain variation in behavior. Selection of behavior can occur through natural selection or cultural selection, but most importantly for behaviorism, it occurs through selection of operant behavior during the life span of the organism. Of course, behavior can be a function of more than one independent variable, and one independent variable may systematically alter more than one behavior.

Three Sources of Behavior

Skinner (1953) described three sources of behavior: biological evolution, evolution of the operant during the organism's life span, and cultural evolution. The biological source of behavior is presented in the form of the evolution of reflex behavior and the capacity of the organism to respond to stimuli that may come to control the organism's behavior during its life span. Reflex behavior occurs when stimuli elicit behavior from the organism. For example, upon touching a hot stove, a child will quickly retract her hand. Such reflex behavior has survival value in that the behavior of removing one's hand from a hot surface prevents further injury. While some properties of the physical environment are enduring, such that natural selection can take place on such behavior, some aspects of the environment change rapidly. Biological selection occurs too slowly to keep up with such changes. The conditioning of reflex behavior to new environmental events is an adaptation to rapid changes in the environment.

The second source of behavior is the evolution of operant behavior over an organism's life span (Skinner, 1953). Some behavior becomes more or less probable than other behavior due to the consequences that follow behavior. This type of behavior operates on the environment to produce consequences, and the change in behavior results in a process of selection of a class of responses, a process called operant conditioning. Operant behavior differs from reflexive behavior in the way in which the two types of behaviors evolve and are conditioned. Reflexive behavior is of a physiological nature, inherited and unlearned. It is automatically elicited by events in an individual's environment. Respondent learning depends on stimulus pairings. In contrast, operant behavior is learned behavior, acquired through contact between emitted behavior and environmental events. Operant behavior is not inherited, but develops continuously over the course of an individual's life span.

The final source of behavior is cultural evolution (Skinner, 1953). Within any culture a number of controlling agencies—such as government, law, religion, ethical codes, education, economic control, and indeed psychotherapy—determine which behaviors are acceptable and thus, to a large extent, which behaviors will be reinforced. The behavior of members of a culture may be reinforced for following the rules of the controlling agencies, thereby increasing the likelihood that such behavior will recur in the future.

Cultural selection usually evolves gradually over generations. If cultural practices contribute to the survival of the culture, then these cultural practices may continue and propagate. For example, the controlling agencies of some cultures might prohibit fishing in certain waters due to pollution, thereby avoiding the costs of poor health and loss of production. Cultural practices become ossified and their original function may not be immediately apparent. For example, long after the pollution problem has been solved, the culture may continue to prohibit fishing even though the pollution problem no longer exists.

Behaviorism and Science

The philosophy of behaviorism underpins two related kinds of behavioral science: The Experimental Analysis of Behavior (EAB) and Applied Behavior Analysis (ABA). EAB studies basic learning processes in both nonhumans and humans. EAB is not concerned with the social significance of either the organisms or the responses it studies. Rather, it studies behavior in highly contrived environments in order to study learning in the most internally consistent and reliable manner possible. ABA is the application of the basic science to change behavior that is socially significant (Baer, Wolf, & Risley, 1968). ABA is concerned only with socially significant problems and strives to make large and socially significant changes in socially important behavior while simultaneously maintaining the conceptual rigor and scientific integrity of EAB.

BEHAVIORISM AND PSYCHOPATHOLOGY

STRUCTURALISM, FUNCTIONALISM, AND PSYCHOPATHOLOGY

Structuralist explanations of psychopathology abound. Indeed, the notion that unusual, personally distressful, or harmful behavior is somehow adaptive is hard to accept at first. Hence, structuralist explanations of psychopathology are palatable, as they avoid this apparently irrational conclusion. The ideas that someone has a mental illness, a neurochemical imbalance, brain damage, a defective cognitive structure, genetic disorder, or trauma-induced personality damage all seem reasonable and appealing explanations. These explanations may offer simple solutions, such as psychotropic medication, or at least appear to remove stigma and responsibility for behavior, which may appear humane. Many find such explanations intriguing, whereas behavioral explanations are often prosaic.

Yet there is evidence that quite extreme and unusual forms of behavior (Layng, Andonis, & Goldiamond, 1999; Schaefer, 1970), including human behavior (Ayllon, Hauton, & Hughes, 1965), can be acquired though shaping. The functional nature of psychopathology has been enshrined in the psychodynamic notion of secondary gains, in which the problem provides some additional benefit, such as special status, financial benefit, avoidance of work or responsibility, and avoidance of acknowledging the true reasons for the problem. Much is deemed to be psychopathological precisely because of its consequences: We notice and react differently to behavior deemed to be psychopathological in that we find it disturbing.

BEHAVIORISM AND PSYCHOPATHOLOGY

Behaviorism, by contrast, gives no special status to psychopathology. Behaviorism explains unusual—even the most unusual—behavior in the same terms and with the same variables as any other behavior. Psychopathology, like other behavior, is to be operationalized, its sources of variation are to be tracked down, and its functional relationships must be discovered. If one wishes to change the psychopathological behavior, then it must be treated through the same methods used in the modification of other behavior. In explaining psychopathology, the explanations of behavior and behavior change are the same as those used to explain any behavior change.

The EAB often studies arbitrary, conveniently measured responses in nonhuman animals that are cheap, whose learning history and environment can readily be controlled, and where the magnitude and generality of the behavior change may not be important. When intervening to modify behavior in applied settings, ABA has a much more difficult task. Baer et al. (1968) identified seven dimensions of ABA: ABA must be applied, behavioral, analytic, technological, conceptually systematic, effective, and have generality. Wolf (1978) added that ABA must also be socially valid. *Applied* means that society has defined the behavior, organism, and stimuli as important. *Behavioral* refers to a pragmatic emphasis on what organisms do, rather than their reports of their behavior. Behavior is a physical phenomenon that actually exists in time and space. Its physical parameters, such as frequency, duration, latency, and intensity, and its products can be measured reliably. Any change in the data must clearly be ascribed to the behavior of the organisms observed, rather than changes in the behavior of the observers. *Analytic* refers to the believable and reliable demonstration of a functional relationship between independent and dependent variables. ABA rejects statistical methods and group designs. These methods erroneously emphasize the averaged data of the nonexistent average subject at one point in time, rather than the variations in behavior over time of the actual person of concern. Statistics and group designs also erroneously give importance to a minimal standard of change that is merely greater than chance, instead of socially or personally important. Finally, they also make the serious error of inferring that an independent variable caused a change. This is done despite the fact that in many group experiments, some subjects' behavior may not have changed at all, while still other subjects' behavior changed significantly in the *opposite* direction! Researchers using group designs, however, still infer that the independent variable caused the change, even when that change may not have happened (Chiesa, 1994)! In contrast, ABA analyzes variations in the behavior of individual organisms and its relationship to explicitly manipulated independent variables using single-subject experimental designs, such as reversal and multiple baseline designs. By reliably demonstrating that the presentation and withdrawal of an independent variable reliably controls a dependent variable, we can conclude that a functional relationship exists (Sidman, 1960). ABA interventions are also *technological,* meaning that they are operationalized and reliably implemented to such a degree that another person could implement them accurately. ABA interventions are *conceptually systematic,* meaning that their methods are described with reference to behavior analytic concepts. Thus, describing a procedure as "attention extinction" is preferred to "planned ignoring," since the former states the learning that appears to be occurring and the latter merely describes an intervention technique. An ABA intervention is said to be *effective* when it produces behavior change that is of practical value. Clients and members of society, rather than behavior analysts, generally define this standard of change. Finally, *generality* refers to behavior change that occurs over time, settings, and a set of responses judged to be important. Generalization of behavior change must be planned and systematically implemented, rather than merely hoped for. Wolf (1978) added that ABA must also "find its heart" by demonstrating that its goals, procedures, and effects were judged to be socially valid. Thus, merely eliminating stuttering but leaving the person talking in an unusual monotone would be socially invalid. The outcome goals must be those that society recognizes as important. Additionally, the methods used must also be acceptable and valued. For example, society and clients usually view restrictive intervention methods as unacceptable.

BEHAVIORISM AND NONBEHAVIORAL APPROACHES TO PSYCHOPATHOLOGY

Behavioral approaches to psychopathology are quite distinct from nonbehavioral approaches in a number of ways. Obviously, behavioral approaches are very empirical, encompassing some of the most extensively studied and evaluated intervention procedures. Of course, other approaches to psychopathology may also be very empirical, but other features distinguish ABA from these approaches. ABA interventions involve careful preintervention assessments that operationalize and measure the psycho-pathological behavior, identify the controlling independent variables that inform and may be part of the intervention method, develop an operation-alized treatment plan, and evaluate its impact on the behavior of interest primarily through observational data. Each treatment plan is tailored to each individual based on the individually assessed controlling variables identified in the pretreatment assessment. Hence, the same topography may be treated in very different ways, depending on the functional rela-tionships identified, while very different target behaviors, populations, and intervention procedures may be regarded as conceptually identical.

Interventions informed by radical behaviorism can be distinguished from behavior therapy. Both approaches are learning-based interventions, and thus, they share some conceptual similarities, empirical rigor, and intervention methods. Behavior therapy tends tie certain intervention tech-niques to certain diagnoses, such as the use of flooding for phobias. It is also more likely than ABA to use group experimental designs and infer-ential statistics. Hence, behavior therapy is less likely to be interested in therapy drive by individual differences in function and more likely to give psychiatric diagnosis greater prominence. Similarly, interventions informed by radical behaviorism can be distinguished from cognitive behavior therapy and cognitive therapy. Again, both approaches share some simi-larities mentioned earlier. However, cognitive behavior therapies and cog-nitive therapies are often structural in that they give cognitions the status of causes of behavior and attempt to infer the status of these unobservable structures; hence, one of the aims of intervention is to remedy these broken structures, rather than focusing on the person's behavior in the natural environment.

Some of the preceding features of ABA can be hard for nonbehaviorists to accept. They may concede the usefulness of behavioral procedures, at least in certain circumstances and populations. They may concede that there is a greater quantity of data for a particular application, at least for the moment, and may be comforted by the possibility of good empirical studies that may yet arrive to support their viewpoint. Yet, many are still troubled by *behaviorism*. For example, the refusal to accept self-reports as sufficient measures of behavior change still seems odd to those who place greatest emphasis on the personal experience of psychopathology, rather than on observable behavior. There are, however, even more fundamental areas of disagreement. The use of single-subject experiments in place of group designs remains an insurmountable hurdle to some scientists. For example, Roth and Fonagy's (2005) otherwise excellent book on evidence-based practice in psychotherapy referred to single-subject experimental designs as "single-case studies" (p. 25) instead of the correct term "single-subject experiments." In so doing, they undermined the status of single-subject experiments and implied that such experiments could not be used to infer causality or have external validity. Roth and Fonagy also excluded these experiments from further consideration on the basis that their results cannot be generalized to "broader clinical populations" (p. 25). Thus, they confused the aims and methods of inductive behaviorism with typical hypothetico-deductive science and excluded much relevant evidence of the efficacy of behavioral interventions from their review. Other fundamental disagreements include (a) the notion that private events are behavior to be explained, rather than the cause of behavior, (b) that mental illnesses do not cause the unusual behavior, and (c) that the person is not an autonomous agent. Many find this model of people offensive. In a world where drug companies have successfully convinced many people that shoplifting, a bad night's sleep, and feeling sad are mental illnesses to be diagnosed and treated with drugs, where services have the explicit aim of promoting autonomy and self-determination, many of the philosophical underpinnings of behaviorism are unacceptable.

SKINNER AND THERAPY

Psychotherapy refers to a wide range of methods used to treat emotional or behavioral problems by psychological rather than physical means. In psychotherapy, a trained person establishes a professional relationship in order to remove or minimize symptoms, remedy maladaptive patterns of behavior, and promote positive personal development. There is a very wide range of therapies, some of which have been extensively evaluated (Roth & Fonagy, 2005), although the field is littered with fad, sometimes dangerous, and unevaluated therapies (Lillienfeld, Lynn, & Lohr, 2003; Jacobson, Foxx, & Mulick, 2005).

Skinner wrote extensively about the potential application of behaviorism to psychopathology long before there were data to support such speculations (Skinner, 1948, 1953, 1957, 1971), laying the foundation for ABA (Morris, Smith, & Altus, 2005). However, Skinner rarely endorsed any specific method of intervention, although he did advocate some general principles. Credit for coining the term *behavior therapy* is generally given to Ogden Lindsley, one of Skinner's students. Lindsley established the Behavior Research Laboratory in 1952, which is widely regarded as the first human operant laboratory, to analyze the behavior of persons with schizophrenia.

Skinner also engaged in public debate with Carl Rogers (1948, 1951) whose client-centered therapy placed much of the responsibility for the treatment process on the client, with the therapist apparently taking a nondirective role. The central hypothesis of nondirective, client-centered therapy is "that the individual has within him or her self vast resources for self-understanding, for altering her or his self-concept, attitudes, and self-directed behavior—and that these resources can be tapped if only a definable climate of facilitative psychological attitudes can be provided" (Rogers, 1986 quoted in Brodley, 1986). In this debate, Skinner (1953) agreed that it is important for a therapist to serve the role of non-punitive audience for the client, but believed the therapist's role does not end there. Skinner criticized Rogers's approach to therapy because of its position on the question of control of client behavior in therapy, noting that there are four possible solutions to the question: denying control, refusing control, diversifying control, and controlling control. Rogers (1948) argued that the therapist must refuse to control the client's behavior, for to do otherwise would lead to the "subtle control of persons and their values and goals by a group which has selected itself to do the controlling" (p. 212). Skinner (1953), however, pointed out that "... [t]o refuse to accept control is merely to leave control in other hands ... " (p. 439). While Rogers asserted that the individual holds the solutions to his or her problems inside himself or herself, Skinner argued that, depending on the prior and current environmental controls or influences that have produced that individual and his or her problems, no such acceptable solution may be within that individual. Thus, it is necessary for the therapist to exert control of client behavior, rather than merely remain a passive yet supportive listener, in order to enable the client to eventually alter his or her problematic behavior.

In Skinner's (1953) view, the control exerted over the individual by various external agents, such as parents, peers, employers, religious groups, and the government, restricts the individual's behavior. While these controls can often be appropriate and beneficial to both the individual and society, they can also result in harmful behavior, particularly where the control exerted is either excessive or inconsistent. Examples of such harmful behavior might include escape responses, revolt against the controlling agent, and passive resistance. Skinner also noted that aversive control can give rise to emotional by-products measurable by respondent behaviors, such as fear, anxiety, anger, rage, and depression, which may distort productive patterns of behavior and strengthen operant behavior that may lead to poor client outcomes. Such controlling agents may lead to maladaptive operant behaviors, such as the individual who turns to drugs, alcohol, or other ineffective and damaging forms of escape from the emotional by-products of control. The Skinnerian view of therapy is therefore inexorably tied to control: Psychotherapy is an important source of societal control of behavior, and the job of the therapist is to counteract the harmful effects of inappropriate forms of control by modifying the control of behavior.

In Skinner's (1953) own words, "behavior is the subject matter of therapy rather than the symptom" (p. 379). As noted earlier, Skinner avoided endorsing specific forms of intervention, such as the token economy. Surprisingly, for Skinner the goal of therapy and the therapist was to enable the client to exercise self-regulation, in order to control his or her behavioral responses to the aspects of the environment that previously evoked the maladaptive responding. Skinner used the term *self-control* to refer to a learning process whereby a person emits one behavior to change the probability of another. For example, one might put a recycle bag next to one's desk in order to make recycling paper more likely. Placing the recycle bag next to the desk is referred to as the *controlling response,* and recycling paper is referred to as the *controlled response.* The independent variables that control the controlling response are the ultimate cause of self-control, rather than the controlling, autonomous, free, and dignified self. Hence, the therapist's job is to discover and manipulate those independent variables that make the client's effective controlling responses more likely. An analogy might be drawn as follows. Many people experience personal difficulties that involve extinction and punishment and their negative emotional by-products. Most people, however, have an effective self-regulation repertoire that enables them to terminate these aversive situations effectively and quickly. If one is bored (one's current behavior is being extinguished) and irritated (one experiences the negative side effects of extinction), one finds other things to do (one engages in behavior that is likely to be reinforced), thereby removing the negative emotional side effects of extinction. Learning such self-regulatory repertoires may be at the heart of resilience, prevention of mental health problems, and relapse prevention. The job of the radical behaviorist therapist is to teach effective self-control by identifying and manipulating the variables that influence the controlling response. For example, a therapist might instruct their client to go look at the parrots in order to be happier. Hopefully, the therapist's instructions are antecedent stimuli that control the client's controlling response of going to the pet store, and that behavior in turn hopefully results in the private and public behaviors we call "happy" that are incompatible with the clients problem behavior.

Skinner (1953) recognized that merely identifying the environmental conditions controlling behavior would not necessarily result in the successful modification of the problematic behavior. In some circumstances, the response is due to the excessive control and punishing effects of the controlling agent. Here, the role of therapy is also to address these variables and modify their effects. Perhaps a fearful child should be assigned to a different classroom away from their punitive and fear-provoking teacher. At other times, the maladaptive response may be the result of inadequate controls in the individual's history, such that the client has not learned self-regulation, in which case the role of therapy should be to supply additional controlling variables to facilitate acquisition of self-control. Where the inappropriate response is the result of excessive reinforcement, such as in drug addiction, the therapist will need to assist in arranging new contingencies in which the problematic behavior will be extinguished and effective and acceptable forms of behavior will be promoted.

Skinner (1953) recognized that therapists should use constructivist techniques to assist their clients. He noted that it will often be the case that the client lacks an adequate repertoire of responses to a given stimulus or controlling agent. Thus, it falls to the therapist to employ constructive techniques to strengthen already-existing responses and to add new responses where such appropriate responses are absent. Skinner recognized that therapists cannot foresee all possible future circumstances, so therapists cannot teach an appropriate response for every condition encountered by an individual. Thus, the ultimate goal of therapy is to set up a repertoire of self-regulation in which the individual can respond appropriately to both current and novel circumstances as they arise. For example, a client who uses excessive alcohol intake as an escape response needs to have a repertoire of better ways of escaping from aversive stimuli in general, not just the current stimuli the therapist has identified as triggering this maladaptive responding. Perhaps the client will need to be taught how to avoid such future circumstances, or other techniques for lessening the circumstances' abilities to evoke the maladaptive responses. The client should thus leave the therapist armed with the repertoire necessary to respond to both foreseen and unforeseeable future situations, and able to use self-control to regulate his or her responding. Hence, from the beginning, Skinner identified and addressed the issue of generalization of behavior from therapy sessions to other settings. The various techniques the behavioral therapist uses to so arm the client with the ability to self-regulate will be addressed in Chapter 3, "Advanced Concepts and Methods of Intervention in Behavioral Approaches to Psychopathology."

CLINICAL CASE FORMULATION

GENERAL FEATURES

Many approaches to conceptualization and treatment of psychopathology use case formulation to distill the essential features of the individual's presenting problem to guide the development of an individualized treatment plan. Eells, Kendjelic, Turner and Lucas (1998) defined case formulation as "... a hypothesis about the causes, precipitants, and maintaining influences of a person's psychological, interpersonal, and behavioral problems. The function of a case formulation is to integrate rather than summarize descriptive information about the patient" (p. 146). Many types of case formulations are possible, but some commonalities exist between different approaches. Eells et al. (1998) described three shared features of all case formulations: "They emphasize levels of inference that can readily be supported by a patient's statement in therapy. The information they contain is based largely on clinical judgments rather than patient self-report. The case formulation is compartmentalized into preset components that are addressed individually in the formulation process and then assembled into a comprehensive formulation . . ." (p. 145). Thus, although the content and focus may vary with respect to the psychological constructs used, most case formulations include similar elements and serve similar purposes. All case formulations are abstractions of client problems. Precipitating events are often described as well as resulting symptoms and problems. A case formulation creates a description of how these events relate to one another and influence the client's problem (Eells et al., 1998).

The continuing use of case formulations and the expanding number of books in this area have focused attention onto the specific characteristics of case formulations (Bruch & Bond, 1998; Eells, 1997; Eells et al., 1998; McWilliams, 1999; Persons, 1989; Sturmey, 1996; Tarrier, Wells, & Haddock, 1998; Weerasekera, 1996). Researchers have developed an interest in what is required to produce an accurate and complete case formulation that can be used to guide treatment.

Bergner (1998) defined the optimal clinical case formulation as

> an empirically grounded, organization of all of the key facts around some factor that not only integrates all of the information obtained, but in doing so identifies the core state of affairs from which all of the client's difficulties issue. Further it would do so in such a way that this information becomes highly useable by the clinician and the client in matters such as their selection of a therapeutic focus; identification of an optimum therapeutic goal, and generation of effective forms of intervention. Most importantly, the existence of such a formulation would allow the clinician to focus therapeutically on that one factor whose improvement would have the greatest positive impact on the client's overall problem or problems. (p. 287)

Bergner (1998) further stated that the function of the assessor is to sort the data into causal categories, to bring to light relevant information or variables that the client may not have considered to be part of the problem, and to "identify the core state of affairs from which all of the client's difficulties issue" (p. 289). A summary of past variables that may have contributed to current problems is helpful. In order for a case formulation to be useful for the treatment of a client, a case formulation must contain information regarding current functioning and the relation of modifiable current variables to the client's problem. To this end, Bergner asserted "the best case formulation would be in terms that identify those factors that currently maintain the problem and that permit ready translation into effective therapeutic action" (p. 192).

Bergner (1998) listed a set of guidelines for case formulations. The first step is to determine the facts of the case. The therapist collects data, determines and organizes the relevance of various factors, and uses present factors to decide which past factors may have contributed to the client's dysfunction. Bergner described this method as a "streamlined approach that will reduce the collection of extraneous information" (p. 197). The second step is the development of an explanatory account. Here, the therapist organizes the facts into a cause-and-effect relationship with respect to the client's present symptoms. The therapist creates a molar account by focusing on broad patterns in the client's descriptions and ignores irrelevant details. Following this, the therapist fits these patterns into an established cause-and-effect relationship. The last step is to check, implement, and revise the formulation, as necessary, subjecting the preliminary formulation to a series of questions including: (a) Is the formulation consistent with the observed facts of the case? (b) Does it account for all of the facts? (c) Does it provide a good fit with the pattern or explanation alleged? (d) Is it useful in terms of an intervention that will provide a sound treatment? and (e) Does the case formulation provide answers to these questions so the assessor can then implement treatment? The therapist then adds the information collected during treatment to the formulation for reevaluation and revision (Bergner, 1998).

NONBEHAVIORAL APPROACHES

Case formulations can be made from many different theoretical perspectives. Much early work, such as Freud's and Luria's case descriptions, involved the analysis and treatment of individual cases. More recently, interest in case formulation has focused on cognitive-behavioral, cognitive, and eclectic case formulation, although Eells' (1997) volume included case formulations from a wide range of approaches. Calam's (1998) formulation of an eating disorder illustrates the cognitive-behavioral approach to case formulation. She wrote

> The foundations of [Tanya's] . . . eating disorder were to be found in early physical and emotional neglect in an environment pervaded by the threat of violence and loss. The extent to which pathological interactions in the family centered around food was probably an important vulnerability factor. The trigger to sensitivity about weight, the comment about her legs, came at a time of intense vulnerability . . . dieting fulfilled a need for a sense of control, self-starvation led to hunger, and thus to binge eating with vomiting as a strategy to prevent weight gain . . . She believed herself to be a bad person, deserving violence from her partner . . . Tanya had developed dissociative strategies to protect her from the intense fear that she experienced at home. . . . (p. 141)

What are some of the features of nonbehavioral approaches to case formulation? They are often structuralist in that part of the process is to use the surface presenting behavior and symptoms to uncover the hidden cause of psychopathology within the person, including hidden trauma, an arrested stage of emotional development, a psychiatric diagnosis, or putative biological damage related to the presenting problem, such as neurochemical imbalances or damage to neurological structures. Calam's (1998) formulation illustrates a mixture of both structuralist and functionalist elements. On the one hand, great store is set on structuralist explanations, such as a "need for a sense of control" as the cause of vomiting and on variables which, even if they are important cannot be manipulated, such as the person's history. On the other hand, there are also very functionalist elements. For example, self-starvation could be construed as reinforcer deprivation, which establishes food as a very powerful reinforcer, with binge eating followed by vomiting forming a response chain (with reinforcement taking place during a binge and after vomiting), although Calam's formulation does not use these behavior analytic terms. (Compare this formulation to that of Farmer and Latner presented in Chapter 18, Eating Disorders," in this volume.) The avoidance function of dissociative strategies also sounds very reminiscent of behavior maintained by negative reinforcement.

BEHAVIORAL APPROACHES TO CASE FORMULATION

Behavioral approaches to case formulation share some important features with other approaches. They are, for example, all relatively brief statements abstracted from the presenting problem that guide the development of an individual treatment plan. Behavioral approaches to case formulation differ from other approaches, however, in the kinds of variables that are deemed to be important, the nature of causality, the role of client history, the kinds of data that the therapist collects, the status given to self-reported data, and to some extent the kinds of interventions implied by the formulation.

Behavioral approaches to case formulation emphasize current behavior as the thing to be explained and modified. Note that in radical behaviorism, behavior includes both public behavior that more than one person can observe, as well as private events, such as thoughts, feelings, and emotions that are behaviors that only one person can observe. The inclusion of private events raises problems for some radical behaviorists because they cannot be observed reliably (Lamal, 1998), although some have attempted to conduct experimental analyses of private events (Friman, Hayes, & Wilson, 1998; Taylor & O'Reilly, 1997). Behavioral approaches also focus on current environmental independent variables that can be manipulated and that have a large impact on the behavior of interest (Haynes & O'Brien, 1990). Although the individual's learning history may be important, it is nevertheless deemphasized because of problems in obtaining accurate and reliable information and the inability to manipulate historical variables. Hence, behavioral case formulation involves rejecting large amounts of information, such as history and independent variables that cannot be manipulated or that do not have a large impact on the behavior of interest.

This results in more parsimonious and treatment-focused formulations (Sturmey, 1996).

Recall that in radical behaviorism, traditional notions of causality are replaced with functional relationships. Experimental functional analysis systematically manipulates some independent variable and observes the shape of the function that results. This approach led Baer et al. (1968) to define a functional analysis as follows: "An experimenter has achieved an analysis of a behavior when he can exercise control over it . . . [it is] an ability of the experimenter to turn the behavior on and off . . . " (p. 94). This will seem a tall order to many clinicians. Even in areas where ABA has had most impact on practice, most clinicians use descriptive methods to assess behavior, but do not directly manipulate it (Desrochers, Hile, & Williams-Moseley, 1997). Indeed, large differences are apparent in the different chapters of this volume regarding the use of pretreatment experimental assessment. In some areas, such as mental retardation and child disorders, these methods have been extensively developed. Other areas, such as depression and anxiety, continue to rely mostly on descriptive methods. The interventions that behavioral formulations imply are environmental ones. If current contingencies and antecedents are reliably related to the presenting problems, then intervention consists of modifying these relations, either by teaching the person to modify his or her own environment through self-management or by teaching others to modify the environment to make adaptive behavior more likely.

SKINNER'S CASE FORMULATION

In *Beyond Freedom and Dignity*, Skinner (1971) offered a generic case formulation, which is summarized in Table 1.1. It translates the everyday language of emotional and mental states into behavioral concepts and identifies the variables that control the client's behavior. He wrote, "What he [the client] tells us about his feelings may permit us to make some informed guesses about what is wrong with the contingencies, but we must go directly to the contingencies if we want to be sure, *and it is the contingencies that must be changed if his behavior is to be changed*" (p. 147, emphasis in original). Thus, Skinner's approach treats the client's self-reports cautiously and emphasizes the importance of observing whether the contingencies described can be observed.

Skinner did not go on to complete the case formulation exercise and translate the formulation into a treatment plan. However, it is easy to do so. Such a treatment plan is summarized in the third column of Table 1.1. Note that almost all the implied interventions involve reinstatement of schedules of reinforcement in order to reinstate effective behavior and remove the emotional side effects of extinction. This implied treatment plan can be compared to Hopko, Hopko, and LeJuez's account of

TABLE 1.1   A Summary of Skinner's Generic Case Formulation and Its Implied Treatment Plan

Everyday Language	Behavioral Process	Implication for Intervention
He lacks assurance, feels insecure, or is unsure of himself.	His behavior is weak and inappropriate.	Reinforce appropriate behavior and teach him to self-regulate so he can make reinforcement of his own behavior more likely.
He is dissatisfied or discouraged.	He is seldom reinforced, and as a result his behavior undergoes extinction.	Implement a reinforcement schedule.
He is frustrated.	Extinction is accompanied by emotional response.	Removal of extinction will remove accompanying emotional responses.
He feels uneasy or anxious.	His behavior frequently has unavoidable aversive consequences, which have emotional effects.	Teach him appropriate ways to avoid aversive consequences and or give him effective behavior.
There is nothing he wants to do or enjoys well; he has no feeling of craftsmanship, no sense of leading a purposeful life, no sense of accomplishment.	He is rarely reinforced for doing anything.	Implement reinforcement schedule.
He feels guilty or ashamed.	He has previously been punished for idleness or failure, which now evokes emotional responses.	Promote adaptive behavior to remove opportunity for idleness and failure.
He is disappointed in himself and disgusted with himself.	He is no longer reinforced by the admiration of others, and the extinction which follows has emotional effects.	Either reinstate behavior that others can admire or train others to reinforce whatever appropriate behavior he still has.
He becomes hypochondriacal.	He concludes that he is ill.	Teach him another rule.
He becomes neurotic.	He engages in a variety of ineffective modes of escape.	Teach him effective modes of escape.

contemporary behavioral approaches to case formulation and treatment of depression in Chapter 15, "Mood Disorders," in this volume. We can infer that the job of the therapist is not to directly implement the changes in reinforcement schedules unless absolutely necessary, but rather to shape the client's self-control skills so he can discriminate and state the problematic contingencies. Then, by discovering his own functional assessment, the client can learn to arrange his own environment to reinstate the contingencies of reinforcement that would remove his problem behavior and its associated negative emotional side effects.

REFERENCES

Ayllon, T., Hauton, E., & Hughes, H. B. (1965). Interpretation of symptoms: Fact or fiction? *Behaviour Research and Therapy, 3,* 1–7.

Baer, D. M., Wolf, M. M., & Risley, T. R. (1968). Some current dimensions of applied behavior analysis. *Journal of Applied Behavior Analysis, 1,* 91–97.

Bergner, R. M. (1998). Characteristics of optimal case formulations: The linchpin concept. *American Journal of Psychotherapy, 52,* 287–300.

Brodley, B. (1986, August). *Client-centered therapy—What it is? What it is not?* Paper presented at the First Annual Meeting of the Association for the Development of the Person-Centered Approach, Chicago.

Bruch, M., & Bond, F. W. (Eds.), (1998). *Beyond diagnosis. Case formulation approaches in CBT.* Chichester: Wiley UK.

Calam, R. (1998). Eating disorder, self-image disturbance and maltreatment. In N. Tarrier, A. Wells, & G. Haddock. *Treating complex cases. The cognitive behavioural therapy approach* (pp. 131–154). Chichester: Wiley UK.

Calkins, M. W. (1906). A reconciliation between structural and functional psychology. *Psychological Review, 13,* 61–81.

Chiesa, M. (1994). *Radical behaviorism: The philosophy and the science.* Boston: Authors Cooperative.

Desrochers, M. N., Hile, M. G., & Williams-Moseley, T. L. (1997). Survey of functional assessment procedures used with individuals who display mental retardation and severe problem behaviors. *American Journal on Mental Retardation, 101,* 535–546.

Eells, T. D. (1997). *Handbook of case formulation.* New York: Guilford.

Eells, T. D., Kendjelic, E. M., Turner, L. C., & Lucas, C. P. (2005). The quality of psychotherapy case formulations: A comparison of expert, experienced, and novice cognitive-behavioral and psychodynamic therapists. *Journal of Consulting and Clinical Psychology, 73,* 579–589.

Fitremann, J-M. (2006). *Basics of structural psychology.* Retrieved June 8, 2006, from http://www.structuralpsy.org/Pages/BasicsStructPsychology.html

Friman, P. C., Hayes, S. C., & Wilson, K. G. (1998). Why behavior analysts should study emotion: The example of anxiety. *Journal of Applied Behavior Analysis, 31,* 137–156.

Haynes, S. N., & O'Brien, W. H. (1990). Functional analysis in behavior therapy. *Clinical Psychology Review, 10,* 649–668.

Jacobson, J. W., Foxx, R. M., & Mulick, J. A. (2005). *Controversial therapy for developmental disabilities: Fads fashion and science in professional practice.* Mahwah, NJ: Lawrence Erlbaum.

Lamal, P. A. (1998). Advancing backwards. *Journal of Applied Behavior Analysis, 31,* 705–706.

Layng, T. V., Andonis, P. T., & Goldiamond, I. (1999). Animal models of psychopathology: The establishment, maintenance, attenuation, and persistence of head-banging by pigeons. *Journal of the Experimental Analysis of Behavior, 30*, 45–61.

Lillienfeld, S. O., Lynn, S. J., & Lohr, J. M. (Eds.). (2003). *Science and pseudoscience in clinical psychology*. New York: Guilford Press.

McWilliams, N. (1999). *Psychoanalytic case formulation*. New York: Guilford.

Morris, E. K., Smith, N. G., & Altus, D. E. (2005). Skinner's contributions to applied behavior analysis. *The Behavior Analyst, 28*, 99–132.

Persons, J. B. (1989). *Cognitive therapy in practice. A case formulation approach*. New York: Norton.

Rogers, C. (1948). Divergent trends in methods of improving adjustment. *Harvard Educational Review, 18*, 209–219.

Rogers, C. (1951). *Client-centered therapy. Its current practice, implications, and theory*. Boston: Houghton Mifflin.

Roth, A., & Fonagy, P. (2005). *What works for whom? A critical review of psychotherapy research*. New York: Guilford Press.

Schaefer, H. H. (1970). Self-injurious behavior: Shaping head-banging in monkeys. *Journal of Applied Behavior Analysis, 3*, 111–116.

Sidman, M. (1960). *Tactics of scientific research. Evaluating experimental data in psychology*. New York: Basic Books.

Skinner, B. F. (1948). *Walden two*. New York: Macmillan.

Skinner, B. F. (1950). Are theories of learning necessary? *Psychological Review, 57*, 193–216.

Skinner, B. F. (1953). *Science and human behavior*. New York: The Free Press.

Skinner, B. F. (1957). *Verbal behavior*. New York: Appleton-Century-Crofts.

Skinner, B. F. (1971). *Beyond freedom and dignity*. New York: Knopf.

Skinner, B. F. (1977). Why I am not a cognitive psychologist. *Behaviorism, 5*, 1–10.

Sturmey, P. (1996). *Functional analysis in clinical psychology*. Chichester: Wiley UK.

Tarrier, N., Wells, A., & Haddock, G. (1998). *Treating complex cases. The cognitive behavioural therapy approach*. Chichester: Wiley UK.

Taylor, I., & O'Reilly, M. F. (1997). Toward a functional analysis of private verbal self-regulation. *Journal of Applied Behavior Analysis, 30*, 43–58.

Weerasekera, P. (1996). *Multiperspective case formulation. A step toward treatment integration*. Malabar, FL: Krieger.

Wolf, M. M. (1978). Social validity: The case for subjective measurement or How applied behavior analysis is finding its heart. *Journal of Applied Behavior Analysis, 11*, 203–214.

2

OPERANT AND RESPONDENT BEHAVIOR

PETER STURMEY
JOHN WARD-HORNER
MICHAEL MARROQUIN

AND

ERICA DORAN

Queens College and The Graduate Center,
City University of New York

Behavioral approaches to psychopathology assume that the learning processes that underlie the acquisition and maintenance of psychopathology and its treatment are identical to those relatively few learning processes that underlie any behavior change process. Consonant with the notion of striving for parsimony and simplicity in science, including applied science, even many apparently complex examples of behavior change may be reducible to this limited range of learning processes. Biological and cultural evolution are important determinants of behavior (Skinner, 1953), but unfortunately, these variables are usually not readily identifiable and their relationship to current psychopathology cannot be assessed directly. Moreover, they cannot be readily manipulated during therapy. Hence, behavioral approaches focus instead on those variables in the current environment of which behavior is a function, that can be manipulated, and that have a large effect on behavior (Haynes & O'Brien, 1990).

The purpose of this chapter is to review two forms of learning that may underlie the conceptualization and treatment of psychopathology: respondent and operant behavior. Chapter 3, "Advanced Concepts and Methods of Intervention in Behavioral Approaches to Psychopathology," goes on to review stimulus equivalence and other derived relationships; rule-governed behavior; and complex behavior, such as modeling, chaining, and self-control, and their relationship to psychopathology.

Functional Analysis in
Clinical Treatment

RESPONDENT BEHAVIOR AND
PSYCHOPATHOLOGY

Respondent behaviors are elicited by stimuli and apparently occur automatically in the presence of these stimuli. They are elicited by antecedent stimuli and are relatively insensitive to their consequences. Respondent behaviors serve an adaptive role for the organism by regulating its physiology or otherwise contributing to the safety of the organism. For example, the eye blink conditioned reflex that occurs when an object nears the eye helps to protect the eye from damage. Similarly, a conditioned fear/anxiety response readies the organism for action. Skinner (1953) speculated that, under principles of natural selection, this reflex learning reflects those aspects of the environment that do not change from generation to generation, such as gravity or threats to the physical integrity of the organism.

Respondent behavior is acquired through respondent conditioning, which involves manipulating antecedent stimuli to affect behavior. Respondent conditioning is also known as classical or Pavlovian conditioning. Pavlov conducted physiological research on digestive secretions of dogs. He noted that, after repeated exposure to a researcher who often brought food to the dogs, the mere appearance of the researcher (a previously neutral stimulus) could elicit salivatory responses similar to those elicited by the presentation of food (an unconditioned stimulus). From this observation and further research, Pavlov articulated a general law of conditioning: "[A]fter repeated temporal association of two stimuli, the one that occurs first comes eventually to elicit the response that is normally elicited by the second stimulus" (Pavlov, cited in Leslie, 1996, p. 5). Thus, without any respondent conditioning, an unconditioned stimulus (US) initially elicits an unconditioned response (UR). During respondent conditioning, a neutral stimulus (NS) is paired with an unconditioned stimulus (US). After repeated pairings of the NS with the US, the neutral stimulus becomes a conditioned stimulus (CS), eliciting a conditioned response (CR) that is usually similar to the UR (Miltenberger, 2004).

There are four variations of respondent conditioning, differing from each other in terms of the timing of the presentation of the NS and the US: delay conditioning, trace conditioning, simultaneous conditioning, and backward conditioning. In a delay conditioning procedure, the NS is presented first, and then the US is presented after the start of the presentation of the NS, but before the NS terminates. For example, first a tone (NS) is presented, and second, while it is still ongoing, food (US) is also presented. In trace conditioning, the NS is again presented prior to the presentation of the US, but the NS is terminated before the presentation of the US. For example, a tone (NS) is presented and terminated, and then the food (US) is presented. In simultaneous conditioning, the NS and US are presented simultaneously. For example, a tone (NS) sounds at precisely the same time that the food (US) is presented. Finally, in backward conditioning, the US is presented first, followed by the presentation of the NS. For example, the food (US) is presented, followed by the presentation of a tone (NS).

Several other factors influence the acquisition of respondent behavior, including (1) the nature of the stimuli, (2) the temporal relationship between the stimuli, (3) the number of pairings of the stimuli, (4) prior exposure to the CS, and (5) the contingency between the stimuli (Leslie & O'Reilly, 1999). The nature of the stimuli includes, for example, its intensity; generally, the greater the stimulus intensity, the more effective it will be and the more salient a CS it will become. The temporal relationship between the CS and US also bears on the effectiveness of conditioning. Generally, conditioning is facilitated when there is a very short period of time between the presentation of the NS and the US. Likewise, while one pairing may at times be sufficient to establish respondent conditioning; generally, a greater number of pairings will produce stronger conditioning. The order and temporal relationship between the NS and the US play a significant role in whether and how fast respondent conditioning occurs. Respondent conditioning occurs most quickly if the US is presented immediately after the onset of the NS. In the case of backward conditioning, it is less likely that respondent conditioning will occur (Miltenberger, 2004). Last, the subject's learning history and prior exposure to the CS may also influence acquisition of the CR. A subject who has previously been exposed to the NS without the presentation of the US will likely be more resistant to respondent conditioning than a subject who lacks this prior experience.

Respondent conditioning can affect behavior in many ways. In respondent conditioning, the association between a stimulus and a response is unidirectional in that a stimulus elicits a response, but a response cannot operate to produce a consequential stimulus, since responding does not have an effect on the presentation of the US or CS. This is significant, since in many respondent conditioning situations, the presentation of stimuli is not under the organism's control. The CR is elicited by a naturally occurring CS. This inability to control the stimuli that elicits a response limits the application of respondent conditioning procedures in treatment. Even the therapist may have problems producing dark thunder clouds or tingling sensations in the client's head upon demand! In contrast to respondent procedures, operant procedures are response-dependent, and operant learning is therefore instrumental in the treatment of behaviors that have a direct effect on environmental change (Domjan, 2004).

Following respondent conditioning, an organism also responds to stimuli that are different from those involved in the original acquisition of a conditioned response. For example, after a person is stung by a bee, other flying insects, such as mosquitoes and flies, may also come to elicit avoidance behavior that is similar to the one that was evoked following the bee sting. This is known as stimulus generalization. Stimulus generalization of respondent behavior occurs with stimuli that are physically similar to those involved in the original acquisition of respondent behavior. For example, if acquisition of a dog phobia took place to a CS, such as loud barking and a large dog, generalization might occur to yapping and small dogs. Such mechanisms, however, do not account for all the generalization that might occur following respondent conditioning. For example, following an incident of severe embarrassment during public speaking, a person might avoid many social situations that are not physically similar to the stimuli involved in the original conditioning. Hence, other mechanisms might account for generalization of respondent behavior, such as higher-order conditioning, rule-governed behavior, and transfer of function within a stimulus class (see Chapter 3, "Advanced Concepts and Methods of Intervention in Behavioral Approaches to Psychopathology").

RESPONDENT BEHAVIOR AND PSYCHOPATHOLOGY

Respondent conditioning has been implicated in the acquisition of many forms of psychopathology, especially those related to fear or anxiety, as well as unusual reflex problems. Thus, many simple and complex phobias (Wu, Conger, & Dygdon, 2005), trauma- and stress-related conditions (such as post-traumatic stress disorder [see Chapter 16, "The Fear Factor: A Functional Perspective on Anxiety," by Friman]), morbid grieving, recurrent nightmares, some anger problems (Grodzinsky, & Tafrate, 2000), and psychosomatic problems (such as asthma and environmental sensitivities [De Peuter et al., 2005; DeVriese et al., 2000; Hermans et al., 2005; Van Den Bergh, 1999]) may be respondent behavior and treated accordingly. Similarly, abnormal gagging, coughing, and other problematic reflexes can also be conceptualized and treated within a respondent behavior framework (Foster, Owens, & Newton, 1985; Fulcher & Cellucci, 1997). Respondent conditioning models have also been applied to conditioned nausea that may occur during radiotherapy for cancer patients (Greene & Seime, 1987). Hence, respondent conditioning has been implicated in the acquisition of a wide range of psychopathology.

Application

There are relatively few applied examples of the acquisition of respondent behavior, since most clinical work is concerned with the removal or control of inappropriate respondent behavior, although Whitehead, Lurie, and Blackwell (1976) studied the acquisition of a conditioned blood pressure decrease in people with both normal and hypertensive blood pressure. One neglected potential application is the incorporation of conditioned positive emotional responses into treatment of some respondent-based psychopathology (see discussion of Paunovic [1999], below).

HIGHER-ORDER RESPONDENT CONDITIONING

Higher-order conditioning involves the pairing of a second NS with a CS, such that the second NS also becomes a CS. For example, the instances of conditioning discussed previously involved using a tone as an NS. After repeated pairings with the US (food), the tone would become a CS. If this CS were now paired with a novel NS, such as a flash of light, eventually the flash of light would also become a CS, taking on the same response-eliciting properties as the tone and the food (Miltenberger, 2004).

An understanding of the principles of higher-order conditioning is useful in many situations, such as accounting for generalization of respondent conditioning. For example, suppose a client demonstrates anxiety upon passing a newsstand, to the point at which he alters his daily routine to avoid newsstands, even though no traumatic experiences involving a newsstand or related stimuli have occurred. It is possible the client's response is due to higher-order conditioning. For example, one possibility is that for this client, like many people, the sound of a dentist's drill has become a CS. The sound of the drill itself does not result in pain (the US), but due to repeated pairings of the NS (the sound of the drill) with the US (pain), the NS is now a CS capable of evoking a similar response, such as muscle contraction, increased heart rate, feelings of stress, and fear/anxiety. Now, if the client repeatedly reads a magazine in the waiting room when he hears the CS, higher-order conditioning could result in the magazine also becoming a CS. Thus, when the client later passes a stack of magazines, the magazines may come to elicit the CR.

Higher-order conditioning may occur between various physical stimuli and verbal stimuli, such as spoken words and privately uttered words. Hence, words, such as *death* or *sunshine*, may all come to elicit a variety of conditioned responses through higher-order conditioning, as well as through other learning processes. Studies using self-report data have found that the severity of fear and phobias can be predicted by aversive events that might enhance initial conditioning, such as accidents, injuries, and embarrassment related to the fear, and by positive events, such as positive or pleasurable events related to the fear (Dygdon, Conger, & Strahan, 2004; Wu et al., 2005). These studies suggest that stimulus pairings occurring after an initial conditioning event that may be higher-order conditioning are be important in the acquisition of phobias. Laws and Marshall (1990) considered the role that higher-order conditioning might play in sexual deviancy. The authors examined sexual patterns as behaviors that are acquired and established through both Pavlovian and operant conditioning, learned through observation and modeling, as well as direct experience and reinforcement. The authors asserted that the fantasizing engaged in by individuals during masturbatory behavior (and reinforced by orgasm) leads to the higher-order conditioning of the deviant behavior, refining that behavior, and increasing the likelihood that it will occur in the future. These studies further support the proposition that higher-order conditioning can be important in the acquisition of psychopathology.

As with acquisition of simple respondent behavior, there is little interest in the application of higher-order conditioning in treatment of psychopathology, although such applications could be possible and important.

RESPONDENT EXTINCTION

If the CS is repeatedly presented without being paired with the US, then the presentation of the CS will also eventually no longer elicit the CR. This is called respondent extinction. Once a CR has been subjected to respondent extinction, however, it is still possible that a later presentation of the CS will elicit the CR. This is known as spontaneous recovery. The later-elicited CR is generally of a smaller magnitude than the original CR and will again cease if the CS is presented in the absence of the US (Miltenberger, 2004). Spontaneous extinction may occur in applications with humans; therefore, clinicians should be vigilant for this phenomenon. The likelihood that respondent extinction will occur increases with the number of presentations of the CS without the US. (See discussion of transfer of extinction function across members of a response class in Chapter 3, "Advanced Concepts and Methods of Intervention in Behavioral Approaches to Psychopathology.")

Applications of Respondent Extinction

Respondent extinction may underlie a variety of treatments of fear and anxiety, including trauma-related disorders that involve repeated presentation of the CS without the US. They include flooding, implosion, systematic desensitization, various forms of self-managed exposure, systematic desensitization, assertiveness training for people who are socially anxious, Wolpe's (1958) various reciprocal inhibition procedures, Masters and Johnson's (1966) sensate focus for sexual problems, anger management involving exposure to anger-eliciting CS, rehearsal relief for traumatic nightmares, and exposure therapy for morbid grieving drug-related problems. Skinner (1953) speculated that respondent extinction was one of the learning mechanisms which accounted for the behavior changes that sometimes occur in classical psychotherapy. During therapy, the client is encouraged to talk about emotional topics, which were perhaps previously paired with distressing situations, which the client rarely discussed. The therapist also may repeatedly present various CS by bringing up emotionally difficult topics. Thus, the client is repeatedly exposed to CS by speaking about the emotional topics with the therapist, by the therapist's comments in regards to the topics, or by the client's private recollections of the of the topics in the absence of the situation in which the descriptive thoughts or words were previously conditioned—public and perhaps private conditioned verbal stimuli—in the absence of the unconditioned stimuli. Hence, respondent extinction may occur during psychotherapy.

Respondent extinction may also take place in various forms of behavior therapy. In behavior therapy, the procedures that the therapist uses to present the CS without the US vary considerably in their nature, speed, and intensity. For example, rehearsal relief for nightmares may involve presenting the CS by repeatedly writing and/or drawing the content of the recurrent nightmare, whereas flooding for agoraphobia may involve exposure to a maximally fear-evoking CS, such as a trip to a crowded store. Whatever the form of the procedures, they all seem to involve repeated exposure to the CS without the US.

Application: Respondent Extinction and Higher-Order Conditioning

Hermans, Craske, Mineka, and Lovibond (2006) and Hermans et al. (2005) provided reviews of the application of respondent extinction to human behavior. One interesting example is Paunovic's (1999) application of respondent extinction and respondent conditioning for the treatment of post-traumatic stress disorder (PTSD) and depression. The participant was a 34-year-old man who had a long history of trauma. When he was 20 years old, he fought in a war in which he killed enemies and was exposed to many dead bodies. His best friend and brother were both killed during the war. He also suffered non-war-related traumas. His girlfriend, whom he planned to marry, was severely burned in an accident. Approximately two months prior to his treatment, five individuals attacked the patient, and he suffered from broken ribs and a concussion. His PTSD symptoms became so severe that his girlfriend left him and he lost his job.

The treatment plan consisted of 17, 90-min sessions, broken into a number of different treatment phases. One phase, exposure countercondi-tioning, consisted of using higher-order respondent conditioning to condition the press of a fingertip to several happy events. The conditioning consisted of having the patient imagine happy events (CS), which in turn elicited happy feelings (CR). Next, higher-order conditioning was conducted by squeezing the top of the patient's finger when he was experiencing the happy events. Thus, through the pairing of the pleasure-evoking stimuli with the press of a fingertip, a press to the fingertip also elicited a pleasurable CR. After using higher-order conditioning to condition a number of happy events to the press of a finger, the participant was asked to imagine traumatic events or was presented with traumatic stimuli (CS). The traumatic images and stimuli produced feelings (CR) similar to those experienced during the traumatic event. During exposure to the traumatic CS, the incompatible happy feelings were assumed to be elicited by pressing the patient's finger. Exposures to the traumatic CS were short, 2–3 min sessions that continued until the CS no longer produced traumatic CRs.

Clinical interviews and self-rating scales that were administered before, during, and after treatment demonstrated that the exposure counter-conditioning treatment was effective at eliminating the client's PTSD symptoms.

OPERANT BEHAVIOR

Operant behavior has been defined as "[b]ehavior that acts on the environment to produce an immediate consequence and, in turn, is strengthened by that consequence" (Miltenberger, 2004). As such, it refers to behavior that "operates" on the environment, hence the name "operant." It has also been described as behavior that is said to be voluntary or purposeful (Leslie, 1996), in contrast to behavior that is reflexive or otherwise apparently outside the subject's control. An operant response is not elicited by an antecedent stimulus; rather, operant conditioning relies on the reinforcing or punishing effects of a consequence to strengthen or weaken operant behavior. Operant behavior is sensitive to contingencies. For example, if a consequence that was previously delivered contingently is now delivered at a different rate or is delivered independent of the behavior, the behavior frequency should decrease. Contingency is not a dichotomousbut a dimensional variable, since the degree of contingency that exists can be parametrically manipulated (Hammond, 1980). Operant behavior is controlled by many variables, including the schedule of reinforcement, the response effort required, the density of reinforcement available for the operant class of interest compared to other operant classes, and deprivation of the reinforcer that maintains the operant behavior. These variables and their role in operant conditioning are explored further in the following sections.

ACQUISITION AND MODIFICATION OF EXISTING OPERANT BEHAVIOR

Operant behavior is a function of four classes of consequences: positive reinforcement, negative reinforcement, positive punishment, and negative punishment. The term "positive" refers to the presentation of a stimulus following the behavior, while the term "negative" refers to the removal of the consequential stimulus. Reinforcement refers to the process whereby an operant behavior is strengthened by the manipulation of consequences, while punishment refers to the process whereby an operant behavior is weakened by the manipulation of conseqeunces.

As stated previously, several other factors also influence the acquisition and maintenance of operant behavior. These factors relate both to differences in the consequences and differences in the subjects, and include the immediacy and consistency of the consequence, the magnitude of the consequence, individual differences related to the person's learning history, and reinforcer deprivation (Miltenberger, 2004, p. 71). To be effective, the consequence must be delivered immediately after the response is emitted. The longer the time between the response and the consequence, the weaker the contiguity between the two, and the less likely operant conditioning will occur. Moreover, the consequence must be contingent upon the response. That is, the consequence should reliably follow the performance of the behavior or, as may be the case, the absence of the particular behavior, such as when a subject is presented with a reinforcer for omitting a response for a particular time interval. A response that is regularly followed by an immediate consequence is more likely to be acquired. If the consequence is delayed or occurs only occasionally following the response during acquisition, the consequence is less likely to have an effect on the response (Miltenberger, 2004, pp. 70–73).

Certain variables related to the individual also affect the acquisition of operant behavior. For example, certain consequences may function as aversive stimuli to some individuals but function as appetitive stimuli to others and the same stimulus may function as a reinforcer or a punisher at different points in the same person's life, depending upon their learning history. Thus, the presentation of a particular stimulus contingent on one individual's behavior may increase that behavior. The same procedure may have the opposite effect with another person or within the same person at different times.

The schedule of reinforcement also affects the acquisition and maintenance of an operant response. The consequence may be delivered on a continuous reinforcement schedule, such that every occurrence of the response being conditioned results in the delivery (or removal) of the stimulus. Alternately, the individual's behavior may be reinforced intermittently, whereby only some of the occurrences are reinforced. The intermittent delivery of the consequence may be based on either a certain number of responses (ratio) or the passing of a certain amount of time (interval), and may be fixed or variable. For example, if a reinforcer is delivered after every five responses, this would be a fixed ratio schedule. If a reinforcer is delivered, on average, after every five responses, this would be a variable ratio schedule. Each of these schedules—fixed ratio, variable ratio, fixed interval, and variable interval—is naturally encountered in a variety of situations. Experimental manipulations of schedule have shown that each schedule has a different characteristic impact on the acquisition and maintenance of operant behavior.

APPLICATION OF OPERANT INTERVENTIONS

Many interventions are based on operant models of psychopathology. Indeed, there are too many to discuss in detail here. However, operant-based interventions are exemplified by identifying reinforcers, teaching new behaviors, increasing existing adaptive behaviors with simple reinforcement, increasing existing adaptive behaviors using concurrent schedules, decreasing undesirable behaviors with reinforcement and punishment, and using reinforcer deprivation.

Identifying Reinforcers

The therapist must identify reinforcers that not only will be effective but also can be readily manipulated. An accurate functional assessment of a client's problem may identify reinforcers maintaining the problematic behavior that may also be suitable for inclusion in an intervention. Verbally competent humans often have a wide range of reinforcers and may report reinforcers in a largely accurate manner. Simple interviews are often adequate to identify reinforcing consequences in some circumstances, although the therapist should be cautious about using such self-report data in many circumstances, as the accuracy of client verbal behavior is readily influenced by many variables during therapy.

A wide range of psychometric instruments has been developed to assess potential reinforcers. Cautela and Kastenbaum (1967) developed the *Reinforcer Survey Schedule* for use in adult mental health settings. There are now at least 23 variants of this scale, designed to be used with children, adolescents, seniors, psychiatric patients, children with autism, people with mental retardation, people with visual impairments and in such varied contexts as marriage, social behavior, sex, work, parent-child relationships, and school (Cautela & Lynch, 1983, Table 1). These scales have served as models for the subsequent development and elaboration of a wide range of similar psychometric tools. These tools often, but not always, identify functional reinforcers that may be incorporated into treatment plans.

In some cases, identification of reinforcers may be quite challenging, such as when working with individuals with poor verbal skills, young children, or with disorders that are characterized by few effective reinforcers, such as depression and chronic schizophrenia, or that have one powerful but problematic reinforcer, such as in addictions and obsessions. A variety of stimulus preference assessment technologies address this issue. For example, Northup, George, Jones, Broussard, and Vollmer (1996) identified reinforcers for four children with attention deficit hyperactivity disorder (ADHD). They used a preference questionnaire, along with verbal and pictorial choice methods. They presented the questionnaire verbally to the children, who then rated the items on a 3-point scale. In the verbal and pictorial methods, they asked the child to select between two alternatives presented simultaneously. They then validated these preference assessment methods by asking the child to complete academic tasks for tokens that he or she could then trade in for items identified in the various preference methods. The survey method accurately identified reinforcers on only

55% of occasions, whereas verbal and choice methods accurately identified reinforcers on 70% and 80% of occasions. Hence, direct assessments tend to be more accurate than survey methods. Potential reinforcers can also be identified by observation of approach and avoidance behavior in the natural environment.

Therapists can also use the Premack Principle to identify reinforcers. High probability behaviors may function as reinforcers if they are made contingent upon low probability behaviors (see Mitchell & Stoffelmayr [1973] for an example of the Premack Principle as applied to very inactive people with schizophrenia). Progressive ratio schedules have also been used to identify reinforcers. In progressive ratio schedules, participants must emit more and more responses for access to stimuli until they no longer respond. The number of responses emitted is a parametric measure of the stimulus' strength as a reinforcer (Roane, Lerman, & Vorndran, 2001).

When one is assessing stimuli as potential reinforcers, it is important to recognize the dynamic nature of reinforcement. When a primary reinforcer is delivered after a behavior, not only will reinforcement take place, but some degree of satiation may also occur. This small amount of satiation may weaken the effectiveness of that stimulus as a reinforcer in the future. A stimulus might function as a reinforcer for one response, but not another. For example, a stimulus might function as a reinforcer for merely reaching to pick up the reinforcer, but not for completing an effortful or aversive task associated with therapy. Whether and to what extent a stimulus functions as a reinforcer may also depend on the recent deprivation of that stimulus. For example, activity might function as a powerful reinforcer after a period of sitting, but not after a period of activity (Michael, 1982, 2000). Finally, the effectiveness of a stimulus as a reinforcer may depend on the availability of other reinforcers: In a reinforcer-deprived environment, a stimulus might function as a reinforcer; whereas in a reinforcer-rich environment, the same stimulus might be ineffective. Likewise, one might be able to enhance the effectiveness of a stimulus as a reinforcer by reducing the overall availability of reinforcers. Cautela (1984) referred to this latter idea when describing the "general level of reinforcement" in a client's life. Hence, in order to circumvent the problem of reinforcer satiation, therapists should (a) prefer to use generalized reinforcers, such as tokens, points, or money, to reduce the likelihood of satiation; (b) establish secondary reinforcers; (c) vary the reinforcers used; and (d) employ client choice of stimuli on a regular basis.

Application: Increasing Existing Adaptive Behavior

A common clinical problem occurs when an important behavior, although in the client's repertoire, is emitted at an inappropriate frequency. For example, a shy person may interact with others, but only rarely.

Alternatively, a person might complain about minor health problems constantly, but rarely talk about other topics. In such situations, it may be that there is an inappropriate schedule of reinforcement supporting the inappropriate frequency of the target behavior. Alternatively, the schedule of reinforcement for other behaviors may be inappropriate. For example, perhaps the schedule of reinforcement maintaining healthy behavior is too weak, resulting in higher rates of unhealthy behavior.

In such situations, one possible approach is to enhance the schedule of reinforcement for the desirable target behavior. An example of this approach, applied to a very challenging problem that has not been dealt with effectively by other methods, comes from Henry and Horne (2000). These researchers attempted to increase speaking in five elderly people with severe dementia. The target behavior was responding correctly to questions such as "What is this?" and "Can you give me the cup?" Baseline data showed that these behaviors were present at variable but consistently low frequency. The addition of a simple reinforcement schedule resulted in modest, although variable, increase in speech. These results are promising given the severity of the problem addressed and the absence of effective alternate methods of intervention.

Application: Decreasing Undesirable Behavior With Reinforcement

The acquisition of operant behavior has a broad impact on the entire behavioral repertoire. Not only does the reinforced operant increase; other responses decrease and the entire behavioral repertoire becomes predictable and organized in a way that may be very useful (Leslie & O'Reilly, 1999). These properties of operant behavior have obvious implications for treatment of psychopathology; often the problem referred is one in which the person's behavioral repertoire is "out of control," disorganized, or fragmented. For some people, the problem may be the absence of appropriate behavior due to skills deficits, or the lack of environmental support for the existing appropriate behavior. In either case, the establishment and support of some operant behavioral repertoire may address many of these problems.

Dixon et al. (2004) used differential reinforcement of alternative behavior (DRA) to reduce the inappropriate verbal behavior of four adults aged 20–62 years with acquired brain injuries from traumatic accidents. Inappropriate verbal behaviors included profanity, aggressive remarks, suicidal utterances, and sexual utterances. Prior to treatment, analog baseline sessions were conducted in order to determine the function of the inappropriate verbal behavior. For two participants, escape from demands maintained the inappropriate vocalizations, and for the other two participants attention maintained the inappropriate vocalizations. For the participants whose utterances were maintained by escape from demand, treatment consisted of preventing escape for inappropriate utterances, ignoring inappropriate utterances, and allowing escape contingently upon appropriate verbal utterances. For the participants whose utterances were maintained by attention, treatment consisted of withholding attention for inappropriate utterances and providing contingent attention for appropriate utterances. In all cases, the DRA treatment decreased the frequency of inappropriate utterances. For two participants, the frequency of appropriate behavior also increased.

A second example comes from Stitzer, Bigelow, Liebson and Hawthorne (1982), who examined the effect of contingent reinforcement in the form of money, take-home methadone, or selecting a methadone dose to take home in 10 men addicted to opiates. These reinforcers were made contingent upon clear urine screens indicating no recent illicit drug use. This resulted in a significant reduction in illicit drug use, and 8 of 10 participants remained drug free for at least 2.5 consecutive weeks during the 12-week treatment. Simple reinforcement of the absence of the target behavior or other specific behavior is a commonly used intervention method that can be applied to a very wide range of problems.

CONCURRENT OPERANT BEHAVIOR

So far we have described operant behavior as if there were only a single class of operants and a single contingency. This is obviously an oversimplification, since more than one concurrent schedule of reinforcement is the rule, rather than exception. Herrnstein (1970) hypothesized that an organism allocates its responses between multiple schedules of reinforcement depending on the reinforcer schedule, reinforcer magnitude, reinforcer delay, and response effort on each schedule. For example, if the response effort increases, reinforcer magnitude or quality decreases, or reinforcer delay increases, then the organism may allocate more responding to the second schedule.

This approach has been extensively used to model interventions for treatment of psychopathology. To simplify, one envisions one schedule maintaining the problematic behavior and a second schedule maintaining the healthy behavior. The aim of intervention is to weaken the schedule maintaining the problem behavior by either reducing reinforcer quality, increasing the effort necessary to engage in the problem behavior, or increasing the time to reinforcement or some combination of these methods, while making the opposite changes to strengthen the desirable behavior. For example, interventions for ADHD have construed ADHD as a disorder in which two classes of behavior are on two concurrent schedules. Impulsive behavior is behavior maintained by immediately available, albeit small and low-quality, reinforcers, whereas desirable, self-controlled behavior is maintained by larger, better quality reinforcers, but that are delayed (see Chapter 5, "Attention Deficit Hyperactivity Disorder," by Neef and

Northup). Likewise, behavioral interventions for depression construe depressed behavior as being low-effort behaviors maintained by escape from aversive stimuli and healthy, nondepressed behavior as being effortful behaviors maintained on too lean a schedule of reinforcement. Hence, behavioral interventions for depression focus on weakening environmental support for depressed behavior (such as by training family members to ignore complaining) and by strengthening healthy, nondepressed behaviors (using such techniques as shaping and establishing new contingencies of reinforcement to maintain these alternate behaviors; see Chapter 15, "Mood Disorders," by Hopko, Hopko and LeJuez). Both of these problems are topographically very different, but both can be modeled in terms of concurrent operant behavior.

Application

Dixon and Falcomata (2004) provide an application of concurrent schedules to a common problem—noncompliance with rehabilitation exercises. A 31-year-old man with acquired brain damage participated; he had problems in carrying out rehabilitation exercises which had resulted in deterioration of his neck muscles to the point that he rarely kept his head up. The experimenters construed this problem in terms of two concurrent schedules. In one schedule, avoidance of physical therapy resulted in immediate removal of demands. In the second schedule, compliance with therapy resulted in effort and some unspecified, delayed reinforcer after holding his head up. Intervention consisted of teaching the client to select one of two schedules—one involving an immediate, small quantity of a reinforcer for not participating and another involving a delayed, large quantity of a reinforcer for completing the exercises. Initially, both consequences were delivered without a delay. Over time, access to the larger reinforcer was progressively delayed.

During baseline, the client selected the immediate small reinforcer more frequently than the larger delayed reinforcer. Following training on selecting the delayed larger reinforcer, he progressively held his head up for longer periods of time. At the end of training, the client chose the delayed larger reinforcer over the immediate smaller reinforcer on almost all occasions. Thus, concurrent schedules can be used to model a variety of clinical problems and to design interventions to increase desirable behavior.

STIMULUS CONTROL OF OPERANT BEHAVIOR

Antecedent stimuli that have been associated with reinforcement or punishment come to exert control over operant behavior. Both nonsocial stimuli (such as lights, dial tones, written instructions) and social antecedent stimuli (such as verbal instruction or even the presence of another person) can all come to control operant behavior, if the organism has a learning history with these stimuli. For example, the smell of smoke or the act of entering a car in the morning may make it more probable that a person will engage in smoking. For some people, the presence of certain other people associated with a history of punishment may occasion social withdrawal.

Stimulus control is established by a consistent history of consequences associated with two sets of stimuli. One stimulus—the S^D—is associated with a schedule of reinforcement, and a second stimulus—the S-delta—is associated with extinction. Hence, responding is more likely to occur in the presence of the S^D than the S-delta. Thus, establishing or modifying stimulus control consists of consistent reinforcement and nonreinforcement in the presence of two sets of stimuli.

Many forms of clinical intervention do not involve teaching new behavior but rather involve bringing existing behavior under appropriate stimulus control. For example, one might teach a person with an eating disorder to eat three healthy meals a day at set times and places. Here, since eating is already in the client's behavioral repertoire, some programmatic reinforcement is delivered by the therapist or family members associated with the new S^Ds. Likewise, reinforcement is withheld for eating at inappropriate times.

Application: Assessing Stimulus Control

In order to modify the stimulus control of operant behavior, one first needs to identify the antecedent stimuli that control behavior. Moore, Edwards, Wilczynski, and Olmi (2001) manipulated task demands and social attention given to four typically developing children aged 4–9 years in order to determine which events were associated with problem behavior. All participants engaged in problem behavior such as tantrums when completing homework. A multielement design was used to examine the children's behavior under five different conditions. The researchers used a free play condition as a baseline, in which each child played alone in a room. For the high demand condition, the parents instructed their children to engage in a low probability task that was successfully completed less than 25% of the time. For the low demand condition, the parents instructed their children to engage in a high probability task that was successfully completed more than 85% of the time. The parents provided a lot of attention noncontingently in both the high and low social attention conditions. The authors collected data on child-specific problem behavior, compliance with parental instructions, and social initiations.

For two of the children, the problem behaviors and low levels of compliance occurred almost exclusively during the high demand condition, suggesting that task demand evoked problem behaviors and increased noncompliance. For the other two children, the problem behavior occurred much more often in the high attention condition, suggesting that social attention alone evoked problem behaviors. Furthermore, these two children's social initiations remained low during the low attention condition, which also supported the conclusion of attention-maintained problem behavior. This study shows that stimulus control can be assessed by systematic manipulation of antecedents.

Application: Acquisition of Stimulus Control

O'Reilly, Green, and Braunling-McMorrow (1990) incorporated the acquisition of stimulus control in promoting safe behavior in four clients with acquired brain injuries who showed a lot of unsafe behavior in their home. They identified hazardous situations and appropriate safe behavior in these situations. For example, if soap was left in the bath tub, then the client should pick up the soap and place it in the sink. O'Reilly et al. established stimulus control of client safe behavior by (a) giving them a checklist with a task analysis of safe behavior, (b) having them read the checklist out loud until no errors occurred, (c) asking the clients to read each step as they went along and check it off as they completed it, (d) reminding clients to check off a step if they failed to do so, (e) asking the clients to complete the items on the checklist while they were out of the room, and (f) then evaluating the state of the room and giving the clients feedback on accurate completion of the task analysis. This intervention was effective in establishing stimulus control by the checklists of client safe behavior in these clients. Additionally, after training was completed, clients exhibited safe behavior without the checklist, indicating that transfer of stimulus control from checklists to the natural environment had also occurred.

Application: Modifying Presentation of Antecedents

Interventions based on stimulus control interventions can also systematically present or remove discriminative stimuli. This approach has been successfully used in the treatment of obesity by eliminating discriminative stimuli to eat, by removing readily available and/or visible junk food from the home, and presenting antecedent stimuli for appropriate eating, such as having a variety of nonfattening foods readily available. Stuart (1967) and Loro, Fisher, and Levenkron (1979) found this approach to be effective in producing weight loss. Thus, modification of stimulus control might be one element of effective treatment of obesity as well as other forms of psychopathology.

TRANSFER OF STIMULUS CONTROL

On some occasions, the behavior of interest is already under stimulus control, but the stimulus control is inappropriate. For example, a person with an eating disorder may be very likely to eat only high calorie food after 4 p.m, but not low calorie food or an food at other times. Here, the stimulus control of eating is problematically narrow. Alternatively, an obese person may eat any kind of food 10 times a day, at inappropriate times (such as late at night), and/or in inappropriate settings (such as in the car and at the workplace). Here, the eating is under problematically broad stimulus control.

Transferring stimulus control from existing stimuli to other stimuli is an important aspect of employing antecedent control procedures. Transfer of stimulus control can be accomplished through a number of procedures that progressively modify the existing antecedents controlling behavior, including prompt fading, prompt delay, and stimulus fading. Prompt fading consists of the gradual removal of a prompt. For example, a client might initially write an instruction and post it in a prominent position. Later, the reminder might be progressively faded by making it smaller or placing it in a less conspicuous spot. A prompt delay procedure consists of gradually lengthening the amount of time that elapses between the presentation of the discriminative stimulus and the prompt. For example, the number of seconds between a verbal and a physical antecedent might gradually be increased. A stimulus prompt involves initially changing the discriminative stimulus so that the learner will be more likely to make the correct response (Miltenberger, 2004).

Application: Modifying Stimulus Control of Elective Mutism

Wulbert, Nyman, Snow, and Owen (1973) provided an ingenious example of modifying stimulus control of behavior in a 6-year-old girl with elective mutism. The girl, of average intelligence, did not speak or engage in motor activities while at school. While at home, however, she engaged in conversation and other activities with her mother. The researchers compared the efficacy of two treatment packages to increase verbal and motor behavior. The experimental treatment consisted of stimulus fading coupled with positive reinforcement and timeout. During this treatment, the mother sat with her child in a room and presented tasks to her child. The fading procedure consisted of gradually increasing the presence of an experimenter, while the girl's mother delivered praise for verbal and motor compliance. A timeout procedure was later added to this treatment. For the control treatment, the mother was not present, and the experimenter delivered task instructions, timeouts, and praise; fading was not included. The fading procedure, along with praise and timeouts, were effective. Here, verbal and motor compliance increased in the presence of adults who previously had been associated with no verbal or motor behavior. Furthermore, two other experimenters were also faded into the sessions, and the procedure was later used successfully at the girl's school with teachers and students. As the procedure was applied to more individuals, the fading steps progressed more quickly, and the girl eventually began participating in school activities without prompts.

Lack of Generalization: An Example of Inappropriate Stimulus Control

Generalization of responses to relevant stimuli or environmental events other than those directly addressed in intervention is another important aspect of the use of antecedent control procedures. Generalization is a common problem in all forms of therapy. For example, a client who has learned a cognitive strategy in a treatment session located in a psychologist's office should use that strategy when faced with a novel situation in his or her home. Similarly, if a client relapses as a medication is tapered, one might construe this as a problem of generalization across drug doses.

Behavioral approaches construe the lack of generalization as a problem of inappropriately narrow stimulus control: The person shows the desirable behavior when presented with some, but not all, relevant S^Ds. For example, a shy person might respond assertively while role-playing, but not in the natural environment; an angry person may be able to relax in the presence of minor provocations, but not in the presence of direct insults. Stokes and Bear (1977) identified a number of strategies that may be used to increase the likelihood that a skill learned in one situation will generalize to other similar situations. Some of these strategies include using common stimuli, using multiple exemplars, and training responses to contact the natural reinforcement contingency. In using common stimuli, the therapist includes relevant stimuli from the natural environment in the training environment. Using multiple exemplars consists of training a number of different examples of the stimuli to which a response should generalize. In conducting any intervention, it is important to select skills that will be maintained in the natural reinforcement contingencies in the relevant environments.

Application: Promoting Generalization of Staff Behavior

An interesting example of the generalization problem comes from the area of staff training, where the staff trainers seek to promote generalization of staff behavior relating to behavioral interventions. Often, we train staff in workshops or role-play but are perplexed that the skills that they appear to understand in one context are not used in another.

Ducharme and Feldman (1992) examined the effectiveness of four interventions on the generalization of staff teaching skills. They trained direct-care staff to use a 10-step teaching procedure to teach clients self-care skills. Staff training included written instructions, single case training, common stimuli training, and general case training. The written instructions condition consisted of providing staff members with written instructions on teaching clients. In the single case training condition, they used role-play to teach the staff how to employ the 10-step teaching procedure correctly with one client in the program. The common stimuli condition was the same as the single case training condition, with the exception that *in vivo* training was conducted with a client. In general case training, they used role-play with 12 client programs, which they selected because (a) the programs were representative of the variety of stimulus situations (SDs) that might be encountered when teaching clients, and (b) the programs allowed the staff to practice all response variations that might be encountered when teaching any client. The authors conducted probe sessions on untrained programs after each training session to assess generalization across clients, settings, and programs. Written instructions had no effect on staff behavior. The single case and common stimuli training did increase the correct use of teaching skills, but only modestly and not across all programs. After staff received general case training, however, they consistently taught novel programs which they had not been trained to conduct directly 85% correct or better.

OPERANT EXTINCTION

Operant extinction occurs when the contingency between the operant and its reinforcer is broken. This may occur by removing the consequence maintaining the behavior or by breaking the contingency by continuing to provide the consequence independently of the response. Operant extinction is associated with phenomena that are broadly the opposite of the phenomena observed during the acquisition of operant behavior. Thus, not only does the operant become less frequent, but other behavior increases in frequency, the behavioral repertoire becomes less predictable and novel behavior temporarily emerges. Additionally, operant extinction is often accompanied by "emotional side effects," such as aggression and crying.

In order to intervene using extinction, several conditions must be met. First, it is necessary to identify the consequence maintaining the behavior. For example, when one uses extinction for a child who tantrums when playing with her sister, it would be necessary to determine before intervention whether the reinforcer maintaining the tantrums was perhaps access to toys, access to a parent, or access to a preferred location. If this is not done accurately at the outset, not only is intervention based on extinction not possible, but the target behavior may be inadvertently strengthened (Iwata, Pace, Cowdery, & Miltenberger, 1994). Second, it must be possible to manipulate the consequence. This can be difficult, particularly when socially mediated behavior is involved. For example, if peers in a classroom cannot be taught to stop taunting the client (a fellow student) and this is the consequence maintaining some target behavior in the client, then extinction is again not possible. Third, it is necessary to withhold the consequence completely to avoid inadvertently shaping higher rates of the target behavior. This means that other people need to be able to tolerate the extinction burst. Again, if one is dealing with a socially mediated behavior, then one has to be able to teach all participants to conduct the extinction procedure accurately.

Extinction alone is not only a difficult procedure to implement and sometimes plagued with undesirable side effects, but it is also an inefficient form of behavior change. Combining extinction with reinforcement of other behavior is a much more efficient way to reduce an undesirable behavior. If the procedure is well designed, it may also result in an increase in appropriate, functionally equivalent behavior or other desirable behavior. Hence, practitioners generally should not use extinction alone, but should combine it with a reinforcement procedure.

Application: Extinction and Stimulus Control

France and Hudson (1990) used extinction and stimulus control to decrease the frequency of night waking in seven typically developing 8- to 20-month-old infants. Night waking consisted of "any noise from the child, sustained for more than 1 min, heard between the time of sleep onset (first substantial period of quiet) and an agreed upon waking time" (France & Hudson, p. 92). During baseline, the experimenters instructed the parents to continue to react to night wakings as they had in the past. During intervention, the parents received instructions on night waking, inadvertent reinforcement of night waking in the form of parental attention, extinction, and stimulus control. The parents were instructed to ignore night wakings (extinction), except if the child was ill, and to implement a consistent bedtime routine (establish stimulus control). In a nonconcurrent multiple baseline design, the experimenters collected data on the frequency and duration of night wakings and completed a sleep disturbance scale. For all but one infant, the frequency of night wakings decreased. For all infants, there was a decrease in the duration of night waking, and the parents reported less sleep disturbance on the rating scale. Further, reductions in the frequency and duration of night wakings were maintained based on information obtained at the 3-month and 2-year follow-ups.

Application: Extinction and DRA

Munford and Pally (1979) demonstrated the effectiveness of extinction and differential reinforcement in the treatment of psychogenic vomiting. The patient was an 11-year-old boy of average intelligence. Vomiting first began to occur when the boy was 3 years old. Until a few months before treatment, the vomiting had occurred intermittently. With the death of the boy's sister, however, the frequency of vomiting increased from 2 to 10 times daily. Previous medical attempts to treat the vomiting were unsuccessful, and medical examinations were unable to determine a physical cause of the problem. During a 1-week baseline, a caregiver was asked to record the frequency of vomiting and was also asked to identify possible antecedent and consequent events. During this period, the boy vomited 29 times. Attention in the form of sympathy and help cleaning up was identified as the consequence that often followed vomiting. No antecedents were identified that regularly preceded vomiting. Therefore, treatment consisted of (a) training the boy's caregivers to withhold attention when the boy vomited, (b) having the boy clean up by himself, and (c) training the caregivers to provide attention only after the boy completed daily chores and homework assignments.

During the first week of treatment, vomiting occurred 20 times. This number dropped to near-zero levels in the following 2 weeks. Anecdotal data provided by the caregiver also suggested that the boy was doing much better academically. At the 12-week and 1-year follow-ups, no occurrences of vomiting were reported.

SHAPING

Shaping consists of two processes that progressively differentiate existing responses: extinction and differential reinforcement. In shaping, naturally occurring variations in a response that more closely approximate the target response are differentially reinforced, while other previously reinforced response variations are placed on extinction. This extinction of some response variations may induce response variability, which in turn may produce novel forms of behavior that may also be differentially reinforced, if they also approximate to the target behavior. As the existing behavioral repertoire approximates more closely to the terminal response, the criterion for differential reinforcement is adjusted so that it moves closer to the target behavior. This again involves extinction of previously reinforced responses, which in turn induces further response variability. Shaping has been used effectively to modify the form, intensity, and numerous other response parameters.

Shaping has also been used to teach new behaviors and reinstate a very wide range of motor, verbal and other skills in varied populations. Naturally occurring shaping has also been implicated in the development of a wide variety of psychopathologies, including self-injury (Layng, Andronis, & Goldiamond, 1999; Schaefer, 1970) and sleep disturbance in infants (Blampied & France, 1993).

Shaping is in many ways an undesirable method of intervention: It can be an excruciatingly slow method of changing behavior. If behavior can be changed quickly—with instructions or modeling, for example—then these methods are obviously preferred. Shaping also requires that the person doing the shaping can discriminate and differentially reinforce appropriate response variations and can modify the criterion for shaping very precisely. Finally, because shaping necessarily involves extinction of behavior, the emotional side effects of extinction must be considered. Nevertheless, if a response is not in a client's repertoire and cannot be taught using more efficient methods, then shaping may be an appropriate method of intervention.

Application

Isaacs, Thomas, and Goldiamond (1966) conducted the classic application of shaping to psychopathology when they reinstated speech in a mute patient diagnosed with catatonic schizophrenia. They first observed that the patient made eye movements in the presence of a stick of gum. Hence, prior to intervention, some operant behavior (eye movements) probably had a history of reinforcement with gum. This current operant behavior was the first approximation to speech, even though it was topographically very dissimilar to the final goal behavior. During the initial six treatment sessions, eye movements were reinforced in the presence of gum. In the next step, gum was withheld until slight movements of the lips occurred. Thus, previously reinforced behavior was placed on extinction, response variability may have been induced, and novel behavior that more closely approximated speech was differentially reinforced. In subsequent steps, shaping was used to establish any vocalization, then sounds closer to the word "gum," then approximations of "gum, please," and finally answering a personal question.

Shaping is readily applicable to a wide range of psychopathology. For example, Waranch, Iwata, Wohl, and Nidiffer (1981) shaped approach responses in an adult with mental retardation and a severe phobia of mannequins. By using shaping in conjunction with *in vivo* desensitization, they successfully treated the individual's phobia. Dahlquist (1990) used shaping, in conjunction with other operant conditioning techniques, to successfully treat a typically developing teenager exhibiting persistent vomiting behavior. By employing these techniques, Dahlquist gradually expanded the period of time between ingestion and vomiting to the point at which vomiting no longer occurred.

PUNISHMENT

Punishment refers to a reduction in the future probability of a behavior following the application of an aversive stimulus or removal of an appetitive stimulus contingent upon the target behavior. Punishers may be primary or secondary. Many high-intensity stimuli—loud noises, bright lights, extreme heat and cold—often appear to function as inherent punishers. Indeed, because such stimuli threaten the integrity of the organism, there may be a considerable evolutionary advantage to organisms that avoid such stimuli efficiently. Many stimuli, however, function as punishers only after some learning has occurred. Society is full of such secondary punishers: The parent who raises his or her hand in warning, for example, is likely using secondary punishers. Indeed, many parents go out of their way to ensure that certain things they say are established as secondary punishers by systematically pairing warnings with loss of reinforcers. A wide range of stimuli have been employed in punishment procedures to address successfully a very wide range of problematic behaviors.

The programmatic use of punishment is perhaps one of the most controversial intervention approaches. Punishment has often been used excessively in the past, and such practices continue today. Punishment is especially controversial when used with populations with limited ability to consent and in restrictive settings and where alternative, less restrictive interventions may also be effective. It raises many complex ethical issues. However, punishment is a natural component of the physical environment: We learn to walk more carefully on a slippery surface to avoid the aversive consequence of losing balance and falling. Punishment is also a very common part of the social environment for all humans: The polite "excuse me" or "wait" may well be an effective secondary punisher. Indeed, one of the most common forms of self-management—the alarm clock—involves self-punishment for remaining in bed. Punishment may also be a component of many nonbehavioral therapies, such as when a therapist looks a little peeved and suggests that his or her client try the assignment one more time. In many circumstances, society resorts to punishment as an acceptable and effective way to control human behavior, such as fines. The use of punishment is also problematic, however, because its effectiveness is likely to be highly reinforcing to the change agent by terminating the undesirable and possibly aversive client behavior, which is aversive to the change agent. As such, any intervention that incorporates punishment must be very carefully constructed, implemented and monitored.

Behavioral interventions are often effective in removing or reducing the use of restrictive procedures and replacing them with equally effective, but more acceptable, reinforcement-based interventions (Sturmey & McGlynn, 2002). Indeed, the development of interventions based on function analysis has led to the reduction in the proportion of research based on punishment procedures (Pelios, Morren, Tesch, & Axelrod, 1999). Clearly, an important ethical imperative for those who engage in behavioral interventions is to remove punishment-based interventions and replace them with other interventions that are not only effective but also more socially and ethically acceptable.

Application: Behavior Reduction with Punishment

Phillips, Phillips, Fixsen, and Wolf (1971) provided evidence of how the removal of tokens (negative punishment via response cost) can effectively reduce behavior. The participants were boys in the social welfare system who were identified as the most likely to become delinquent youths based on current aberrant behavior. At the time of the study, a token economy, in which the boys could earn or lose points by engaging in specified adaptive or maladaptive behavior, was already in place. Depending on the number of tokens earned, the boys were able to exchange their tokens for various privileges. In the experiment, the token system was extended to tardiness by deducting points for each minute a boy was late for dinner. A baseline measure of tardiness was recorded prior to the implementation of two treatment phases. During the first treatment, the boys were provided with a 5-min warning prior to dinner, and a bell was rung to indicate when it was time to eat. One hundred points were lost for each minute the boys were late. The second treatment phase was the same as the first, except that the boys were only given warnings that they would lose tokens if they were late, but the warnings were never followed by the actual loss of tokens. It was demonstrated that tardiness was reduced only when followed by a loss of points and that verbal warnings had no effect on decreasing tardiness; in fact, the amount of tardiness returned to baseline levels during this phase.

Application: Removing Unnecessary Punishment Procedures

Excessive use of restraint, seclusion, and emergency psychotropic medication remains problematic not only in institutional but also in community settings. A wide range of populations are at risk for these restrictive procedures, including people with mental retardation, psychiatric disorders, and children and adolescents in various residential and forensic settings. The failure to implement simple positive and proactive procedures and policies that facilitate or encourage their use may well contribute to the excessive use of these procedures (Donat, 1998; Sturmey & McGlynn, 2002).

Donat (1998) addressed this issue in a psychiatric hospital. Intervention consisted of requiring behavioral consultation, functional assessment, and development of a behavioral plan following restraint use in high-risk clients. This procedure resulted in a 62% reduction in restraint use in 53 high-risk clients.

DEPRIVATION, SATIATION, AND SETTING EVENTS

Behavioral interventions have often emphasized environmental events that occur close in time to the behavior of interest; these interventions are easiest to study and these variables are easier to manipulate reliably. However, events more distant in time that influence the behavior may also be of importance. One class of events is deprivation of the reinforcer that maintains the behavior of interest. Reinforcer deprivation has two effects (Michael, 1982, 2000). First, it makes the reinforcer temporarily more powerful. Second, deprivation also evokes behaviors related to that reinforcer. For example, if one has not interacted with other people for a long time (and the other people are effective reinforcers), then two kinds of changes occur. First, one will work very hard to access other people—for example, making multiple phone calls until one gets to talk to someone. Second, a wide range of the behaviors related to that reinforcer become more frequent; for example, one might start looking for other people or respond more rapidly at the sound of another person's voice. The converse of deprivation—that is, when large quantities of a reinforcer have been delivered recently—is satiation. Generally, satiation has the opposite effect of deprivation: Behaviors related to the reinforcer reduce in frequency, and the stimulus temporarily ceases to function as a reinforcer. Although satiation may appear to be a different operation from deprivation, it may be more parsimonious to consider it as merely zero deprivation. Thus, satiation and deprivation are merely two ends of one continuum.

Interventions based on deprivation and satiation include first identifying the reinforcer maintaining the behavior of interest and manipulating its presence or absence before the behavior of interest occurs. If one wishes to increase a desirable behavior to compete with an undesirable behavior, for example, one might deliberately deprive the person of the reinforcer maintaining that behavior. To increase cooperative toy play in children to compete with fighting behavior, one might thus identify the most preferred toys, ensure that they were not available for several hours, and use them in play sessions when fighting was most likely to occur. Satiation can also be used to decrease an undesirable behavior. If the reinforcer maintaining an undesirable behavior is known and can be manipulated, then that reinforcer could be delivered at fixed times throughout the day, or immediately prior to a time when the undesirable behavior is very likely to occur. For example, if one knows that binge eating is maintained in part by access to sympathy from friends, and hence, is more likely following periods of no social contact, one might include frequent social activities, especially prior to times when binge eating is most likely, as part of a function-based intervention.

Application

Buchanan and Fisher (2002) evaluated the effects of noncontingent reinforcement on disruptive vocalization in two adults, aged 82 and 89 years old, with dementia. Functional analyses identified the consequence maintaining vocalization was attention and sensory consequences for one person and attention only for the second. Treatment consisted of delivering these consequences noncontingently. The noncontingent delivery of the consequences that were apparently maintaining the disruptive vocalizations resulted in a significant reduction in the target behavior.

CHAPTER SUMMARY

Important aspects of psychopathological behavior may be respondent or operant behavior. Correct identification and analysis of the target behavior, whether it is respondent or operant, and its controlling variables can be used to design individually based interventions. The next chapter looks at more complex forms of learning and their relationship to psychopathology.

REFERENCES

Blampied, N. M., & France, K. G. (1993). A behavioral model of infant sleep disturbance. *Journal of Applied Behavior Analysis, 26*, 477–492.

Buchanan, J. A., & Fisher, J. E. (2002). Functional assessment and noncontingent reinforcement in the treatment of disruptive vocalization in elderly dementia patients. *Journal of Applied Behavior Analysis, 35*, 99–103.

Cautela, J. R. (1984). General level of reinforcement. *Journal of Behavior Therapy and Experimental Psychiatry, 15*, 109–114.

Cautela, J. R., & Kastenbaum, R. (1967). A reinforcement survey schedule for use in therapy, training, and research. *Psychological Reports, 20*, 1115–1130.

Cautela, J. R., & Lynch, E. (1983). Reinforcement survey schedules: Scoring, administration, and completed research. *Psychological Reports, 53*, 447–465.

Dahlquist, L. (1990). The treatment of persistent vomiting through shaping and contingency management. *Journal of Behavior Therapy and Experimental Psychiatry, 21*, 77–80.

De Peuter, S., Van Diest, I., Lemaigre, V., Li, W., Verleden, G., Demedts, M., et al. (2005). Can subjective asthma symptoms be learned? *Psychosomatic Medicine, 67*, 454–461.

DeVriese, S., Winters, W., Stegen, K., Van Diest, I., Veulemans, H., Nemery, B., et al. (2000). Generalization of acquired somatic symptoms in response to odors: A Pavlovian perspective on Multiple Chemical Sensitivity. *Psychosomatic Medicine, 62*, 751–759.

Dixon, M. R., & Falcomata, T. S. (2004). Preference for progressive delays and concurrent physical therapy exercise in an adult with acquired brain injury. *Journal of Applied Behavior Analysis, 37*, 101–105.

Dixon, M. R., Guercio, J., Falcomata, T., Horner, M. J., Root, S., Newell, C., et al. (2004). Exploring the utility of functional analysis methodology to assess and treat problematic verbal behavior in persons with acquired brain injury. *Behavioral Interventions, 19*, 91–102.

Domjan, M. (2004). *The essentials of learning and conditioning.* Belmont, CA: Thomson Wadsworth.

Donat, D. C. (1998). Impact of a mandatory behavioral consultation on seclusion/restraint utilization in a psychiatric hospital. *Journal of Behavior Therapy and Experimental Psychiatry, 29*, 13–19.

Ducharme, J. M., & Feldman, M. A. (1992). Comparison of staff training strategies to promote generalized teaching skills. *Journal of Applied Behavior Analysis, 25*, 165–179.

Dygdon, J. A., Conger, A. J., & Strahan, E. Y. (2004). Multi-modal classical conditioning of fear: Contributions of direct, observational and verbal experiences. *Psychological Reports, 95*, 133–153.

Foster, M. A., Owens, R. G., & Newton, A. V. (1985). Functional analysis of the gag reflex. *British Dental Journal, 158*, 369–370.

France, K. G., & Hudson, S. M. (1990). Behavior management of infant sleep disturbance. *Journal of Applied Behavior Analysis, 23*, 91–98.

Fulcher, R., & Celucci, T. (1997). Case formualtion and behavioral treatment of chronic cough. *Journal of Behavior Therapy and Experimental Psychiatry, 28*, 291–296.

Greene, P. G., & Seime, R. J. (1987). Stimulus control of anticipatory nausea in cancer chemotherapy. *Journal of Behavior Therapy and Experimental Psychiatry, 18,* 61–64.

Grodzinsky, G. R., & Tafrate, R. C. (2000). Imaginal exposure for anger reduction in adult outpatients: A pilot study. *Journal of Behavior Therapy and Experimental Psychiatry, 31,* 259–279.

Hammond, L. J. (1980). The effects of contingency upon the appetitive conditioning of free-operant behavior. *Journal of Experimental Analysis of Behavior, 34,* 297–304.

Haynes, S. N., & O'Brien, W. H. (1990). Functional analysis in behavior therapy. *Clinical Psychology Review, 10,* 649–668.

Henry, L. M., & Horne, P. J. (2000). Partial remediation of speaker and listener behaviors in people with severe dementia. *Journal of Applied Behavior Analysis, 33,* 631–634.

Hermans, D., Craske, M. G., Mineka, S., & Lovibond, P. F. (2006). Extinction in human fear conditioning. *Biological Psychiatry, 15,* 361–368.

Hermans, D., Dirikx, T., Vansteenwegen, D., Baeyens, F., Van den Bergh, O., & Eelen, P. (2005). Reinstatement of fear responses in human aversive conditioning. *Behaviour Research and Therapy, 43,* 533–551.

Herrnstein, R. J. (1970). On the law of effect. *Journal of the Experimental Analysis of Behavior, 13,* 243–266.

Isaacs, W., Thomas, J., & Goldiamond, I. (1966). Application of operant conditioning to reinstate verbal behavior in psychotics. In R. Ulirich, T. Stachnick, & J. Mabry (Eds.), *Control of human behavior, Volume 1* (pp. 199–202). Glenview, Illinois: Scott Foresman.

Iwata, B. A., Pace, G. M., Cowdery, G. E., & Miltenberger, R. G. (1994). What makes extinction work: An analysis of procedural form and function. *Journal of Applied Behavior Analysis, 27,* 131–144.

Laws, D. R., & Marshall, W. L. (1990). A conditioning theory of the etiology and maintenance of deviant sexual preference and behavior. In W. L. Marshall, D. R. Laws, & H. E. E. Barbaree (Eds.), *Handbook of sexual assault. Issues, theories, and treatment of the offender* (pp. 209–229). New York: Plenum Press.

Layng, T. V., Andronis, P. T., & Goldiamond, I. (1999). Animal models of psychopathology: The establishment, maintenance, attenuation, and persistence of head-banging by pigeons. *Journal of Behavior Therapy and Experimental Psychiatry, 30,* 45–61.

Leslie, J. C. (1996). *Principles of behavioral analysis.* Amsterdam, The Netherlands: Harwood Academic Publishers.

Leslie, J. C., & O'Reilly, M. F. (1999). *Behavior analysis: Foundations and applications to psychology.* Amsterdam, The Netherlands: Harwood Academic Publishers.

Loro, A. D. Jr., Fisher, E. B. Jr., & Levenkron, J. C. (1979). Comparison of established and innovative weight-reduction treatment procedures. *Journal of Applied Behavior Analysis, 12,* 141–155.

Masters, W. H., & Johnson, V. E. (1966). *Human sexual response.* Boston: Little & Brown.

Michael, J. (1982). Distinguishing between discriminative and motivational functions of stimuli. *Journal of the Experimental Analysis of Behavior, 37,* 149–155.

Michael, J. (2000). Implications and refinements of the establishing operation concept. *Journal of Applied Behavior Analysis, 33,* 401–410.

Miltenberger, R. G. (2004). Respondent conditioning. In V. Knight (Ed.), *Behavior modification principles and procedures* (3rd ed., pp. 136–142). Belmont, Ca: Wadsworth Publisher. Location: Publisher.

Mitchell, W. S., & Stoffelmayr, B. E. (1973). Application of the Premack principle to the behavioral control of extremely inactive schizophrenics. *Journal of Applied Behavior Analysis, 6,* 419–423.

Moore, J. W., Edwards, R. P., Wilczynski, S. M., & Olmi, D. J. (2001). Using antecedent manipulations to distinguish between task and social variables associated with problem behavior exhibited by children of typical development. *Behavior Modification, 25,* 287–304.

Munford, P. R., & Pally, R. (1979). Outpatient contingency management of operant vomiting. *Journal of Behavior Therapy and Experimental Psychiatry, 10,* 135–137.

Northup, J., George, T., Jones, K., Broussard, C., & Vollmer, T. R. (1996). A comparison of reinforcer assessment methods: The utility of verbal and pictorial choice procedures. *Journal of Applied Behavior Analysis, 29,* 201–212.

O'Reilly, M. F., Green, G., & Braunling-McMorrow, D. (1990). Self-administered written prompts to teach home accident prevention skills to adults with brain injuries. *Journal of Applied Behavior Analysis, 23,* 431–446.

Paunovic, N. (1999). Exposure counterconditioning (EC) as a treatment for severe PTSD and depression with an illustrative case. *Journal of Behavior Therapy and Experimental Psychiatry, 30,* 105–117.

Pelios, L., Morren, J., Tesch, D., & Axelrod, S. (1999). The impact of functional analysis methodology on treatment choice for self-injurious and aggressive behavior. *Journal of Applied Behavior Analysis, 32,* 185–195.

Phillips, E. L., Phillips, E. A., Fixsen, D. L., & Wolf, M. M. (1971). Achievement place: Modification of the behaviors of pre-delinquent boys within a token economy. *Journal of Applied Behavior Analysis, 4,* 45–59.

Roane, H. S., Lerman, D. C., & Vorndran, C. M. (2001). Assessing reinforcers under progressive schedule requirements. *Journal of Applied Behavior Analysis, 34,* 145–167.

Schaefer, H. H. (1970). Self-injurious behavior: Shaping head-banging in monkeys. *Journal of Applied Behavior Analysis, 3,* 111–116.

Skinner, B. F. (1953). *Science and human behavior.* New York: The Macmillan Company.

Stitzer, M. L., Bigelow, G. E., Liebson, I. A., & Hawthorne, J. W. (1982). Contingent reinforcement for benzodiazepine-free urines: Evaluation of a drug abuse treatment intervention. *Journal of Applied Behavior Analysis, 15,* 493–503.

Stokes, T. F., & Baer, D. M. (1977). An implicit technology of generalization. *Journal of Applied Behavior Analysis, 10,* 349–367.

Stuart, R. B. (1967) Behavioural control of overeating. *Behaviour Research and Therapy, 5,* 357–365.

Sturmey, P., & McGlynn, A. (2002). Restraint reduction. In D. Allen (Ed.), *Ethical approaches to physical interventions. Responding to challenging behavior in people with intellectual disabilities* (pp. 203–218). Plymstock: BILD Publications.

Van den Bergh, O., Stegen, K., Van Diest, I., Raes, C., Stulens, P., Eelen, P., et al. (1999). Acquisition and extinction of somatic complaints in response to odors: A paradigm relevant to investigate Multiple Chemical Sensitivity. *Occupational and Environmental Medicine, 56,* 295–301.

Waranch, H. R., Iwata, B. A., Wohl, M. K., & Nidiffer, F. D. (1981). Treatment of a retarded adult's mannequin phobia through in vivo desensitization and shaping approach responses. *Journal of Behavior Therapy and Experimental Psychiatry, 12,* 359–362.

Whitehead, W. E., Lurie, E., & Blackwell, B. (1976). Classical conditioning of decreases in human systolic blood pressure. *Journal of Applied Behavior Analysis, 9,* 153–157.

Wolpe, J. (1958). *Psychotherapy by reciprocal inhibition.* Stanford, CT: Stanford University Press.

Wu, N. Y., Conger, A. J., & Dygdon, J. A. (2005). Predicting fear of heights, snakes and public speaking from multimodal classical conditioning. *Medical Science Monitor, 12,* 159–167.

Wulbert, M., Nyman, B. A., Snow, D., & Owen, Y. (1973). The efficacy of stimulus fading and contingency management in the treatment of elective mutism: A case study. *Journal of Applied Behavior Analysis, 6,* 435–441.

3

ADVANCED CONCEPTS AND METHODS OF INTERVENTION IN BEHAVIORAL APPROACHES TO PSYCHOPATHOLOGY

PETER STURMEY
JOHN WARD-HORNER
MICHAEL MARROQUIN

AND

ERICA DORAN
*Queens College and The Graduate Center,
City University of New York*

The procedures described in Chapter 2, "Operant and Respondent Behavior," involve the modification of a single respondent or operant response class. These processes do involve some complexity, such as when they result in the broad reorganization of the behavioral repertoire. However, the behavior changes described in Chapter 2 are both relatively simple and account only for some learning related to psychopathology. This chapter goes on to discuss other types of learning including rule-governed behavior, stimulus equivalence and derived relations, and other forms of complex behavior, such as modeling, chaining, and self-control.

RULE-GOVERNED BEHAVIOR

Chapter 2 described how antecedent stimuli come to control operant behavior. One form of antecedent control of operant behavior is known as rule-governed behavior. Rule-governed behavior is a form of verbal behavior in which a person states a rule which alters the future probability of other behavior and is hence a discriminative stimulus for some other behavior. For example, a person might state "a drink might do me good," which might translate as a statement describing likely contingencies, such as "drinking will be consequated by reduction in tension." Rules may be learned from other people, as when a parent instructs a child to "be polite to strangers" and then socially reinforces following that rule. People may also learn to state their own rules, as when someone says to himself or herself "that candy will make me fat" without instruction from other people.

Such rule-following behavior may be efficient in that it is not necessary to come in contact with contingencies to learn. In so doing, it may result in behavior that is relatively insensitive to contingencies. Hence, if we have learned a general repertoire of rule-following behavior, we may learn in a very efficient manner. For example, following the rule "don't drive down the freeway tonight" that someone else states, even though we do not come into contact with the contingencies that may or may not exist on the freeway, may be a very effective form of learning.

Application

Taylor and O'Reilly (1997) provided an example of the application of rule-governed behavior that is especially interesting, since it attempted to implement and analyze private rule-governed behavior. The participants were four adults with mild mental retardation living in community settings. The behavior of interest was shopping in a supermarket, which consisted of 21 steps. Taylor and O'Reilly took data in actual supermarkets and in a classroom simulation setting. Initially, they taught participants to use overt self-instruction consisting of stating the problem, stating the correct response, describing the response, and self-acknowledgment. Training also occurred in the actual supermarket using picture prompts and overt self-instruction. Once the participants had mastered overt self-instruction, the experiments taught the participants to state the rules covertly using instructions and modeling. Taylor and O'Reilly observed that this intervention was effective in teaching the participants to self-instruct privately. The participants shopped independently without any external prompts and without overtly stating the rules. As evidence that the procedure was indeed rule-governed behavior, rather than merely chaining or some other behavioral process, Taylor and O'Reilly showed that when private rule stating was presumably blocked by having the client repeat random numbers, the participants no longer shopped independently. Thus, they concluded that private rule-governed behavior did indeed control the overt shopping behavior.

STIMULUS EQUIVALENCE, DERIVED RELATIONS, AND PSYCHOPATHOLOGY

STIMULUS EQUIVALENCE

Earlier discussions of stimulus control have been limited to antecedents that are, for the most part, physically similar, and accounts of stimulus generalization have used the physical similarity between training and novel stimuli to explain stimulus generalization. In reality, however, behavior is often under the stimulus control of stimuli that are physically very different. Hence, more generalization has to be accounted for than is described by the physical similarity of stimuli. For example, a person with a phobia might act fearfully when criticized by many different people, but the person may also act fearfully when hearing, seeing, or reading about criticism. In this example, all the different antecedent stimuli are functionally equivalent, since they all evoke fearful behavior, even though the stimuli are physically quite different. The person did not learn many of these fearful responses directly. So, how can behavior analysis begin to address this issue?

Stimulus equivalence is a behavioral account of this phenomenon. A set of stimuli is said to be equivalent when it shows the four properties of reflexivity, symmetry, transitivity, and equivalence. For example, suppose we consider three stimuli: A, B, and C. These stimuli would be said to be equivalent only if (a) the participant matches each stimulus to itself (reflexivity); (b) after learning that A is the same as B, the participant responds as if B is the same as A (symmetry); (c) after further learning B is the same as C, the participant responds as if A is the same as C (transitivity); and finally, (d) the participant responds as if C is the same as A (equivalence; Sidman, 1994).

In experimental studies stimulus equivalence is prototypically taught using matching to sample (MTS) training. In MTS training, several classes of stimuli are presented, often on a computer screen—for example, dogs (Class 1), trees (Class 2), and tables (Class 3). Each class contains stimuli in different formats, such as photos, line drawings, and written words. Hence, for each class there are several stimuli, such as a photo of a dog (A1), a line drawing of a dog (A2), and the written word *dog* (A3). During training, a sample stimulus, such as a photo of a dog (A1), is presented followed by the presentation of two or more comparison stimuli, such as the words *dog* (A3), *tree* (B3), and *table* (C3). The subject then selects one of the comparison stimuli by pressing a key. Feedback is given for correct (A3) and incorrect (B3) responses, such as presenting the words *right* or *wrong*.

The learning that takes place in stimulus equivalence training has several interesting features. First, many relationships that are not directly taught emerge. For example, after learning the relationships A1–B1 and B1–C1, a participant may also learn the untrained relations A1–C1 and C1–A1. Second, if a new relationship is taught—say the word *chien* (D1) to a member of the class C1, then many more untrained relationships may emerge, such as D1–B1, D1–A1, A1–D1, and A1–D1 (Fields & Verhave, 1987). Equivalence classes may have a substantial number of members. Thus, many relationships may exist between the stimuli within a class. Further, relationships can be taught between existing equivalence classes that can establish new and even larger classes with many stimulus relationships. For example, a child might be taught that all primates, birds, and reptiles are vertebrates. Such teaching not only establishes relationships within the three classes but also establishes many novel relationships among all the members of the three classes.

The resulting behavior that emerges after stimulus equivalence training corresponds in many ways to what, in everyday language, would be said to be "understanding an idea." A person who points to a dog when someone says "dog" and who points to a photograph of a novel dog when someone says "chien" after being taught that *chien* is the same as *dog* appears to understand the concept "dog."

Application

There have been a few applications of stimulus equivalence learning to psychopathology. Cowley, Green, and Braunling-McMorrow (1992) used equivalence training to teach three men with acquired brain injuries to name their therapists' faces. Two of the three had received extensive rehabilitation, and staff continued to prompt all three to use their names. The stimuli were spoken names (A), color photographs (B), hand-written names (C), and color photographs of nameplates from office doors (D). The relationships were all taught using MTS teaching. To address the social validity of these procedures, the experimenters measured progress during training and also asked the men to find the therapists in natural settings.

The experimenters found that the participants largely, though imperfectly, formed the equivalence class in that novel, untrained relationships did indeed emerge. There was also some evidence that the effects of training generalized to the natural environment, since some participants learned to find therapists in the natural environment when they were asked to find their therapist. Stimulus equivalence training has also been used to teach many complex concepts, such as those involving money, reading, numbers, color, and other concepts (Sidman, 1994).

TRANSFER OF FUNCTION WITHIN
A STIMULUS CLASS

Stimuli may have a number of functions. For example, stimuli can function as consequences when they are used as reinforcers or punishers or discriminative stimuli. One of the most interesting recent lines of research related to stimulus equivalence is that of so-called transfer of function within a stimulus class. For example, in a study on avoidance responding, Auguston and Dougher (1997) provided evidence showing the way equivalence classes may explain how previously neutral stimuli may come to evoke avoidance behavior. Eight college students participated. During the first phase, participants underwent equivalence class training and testing in a typical MTS procedure, consisting of the presenting of a sample stimulus followed by the presentation of three comparison stimuli. Three telegraph keys were located beneath the computer screen correlated with the comparison stimulus. Participants responded to the comparison stimuli by pressing one of the three telegraph keys. In the second phase, a classical conditioning procedure was implemented in which one stimulus (CS+) from class 1 was paired with shock, and one stimulus (CS−) from class 2 was not paired with shock. In the third phase, avoidance training occurred. This phase was the same as the second phase except that participants could press the correct telegraph key to escape shock. The fourth phase tested for transfer of the avoidance response to the other members of each of the two classes that were not directly taught. During stimulus presentations, presses to the correct key provided a measure of avoidance. All of the participants met criterion during training, and all participants also showed avoidance responding in the presence of class 1 stimuli and a virtual absence of responding in the presence of class 2 stimuli. These results demonstrate the transfer of avoidance responding to other stimuli of the same class.

Research on transfer of stimulus functions has important implications for clinical work. First, it may be a model for the acquisition of various forms of psychopathology. For example, suppose a client already has an equivalence class of stimuli that we might loosely call "threatening people." When encountering a new person, the client learns that this new person is equivalent to other stimuli in this existing class; then the functions that were already attached to the existing members of the class may transfer to the new stimulus. Thus, now the new person has become a member of the existing class of threatening people and elicits anxiety and avoidance responses, even though no direct conditioning has taken place. Similarly, if the findings reported by Auguston and Dougher (1997) are generalizable to clinical phobias, this might offer a model for the successful treatment and/or spread of phobic and trauma-related behavior. For example, transfer of extinction might occur when respondent extinction takes place during desensitization to other stimuli that are members of an equivalence class (e.g., to the actual objects and people that formerly also elicited fear).

OTHER DERIVED RELATIONSHIPS

Equivalence is one kind of relationship that exists between stimuli. Other relationships are also possible, such as opposites, bigger-smaller, faster-slower, before-after, smaller-larger, and others. When people derive relationships among stimuli that are all equivalent, they learn that all members of the stimulus class are equivalent or, more loosely, the same. However, when relationships are derived among stimuli that are not equivalent, the derived relationships are more complex. For example, if an anxious person has already learned that "A is more fearful than B" and "B is more fearful than C," and now learns that "D is more fearful than A" and "E is as fearful as D," then the person will now learn multiple new derived relationships that are not equivalent.

Application

Derived relationships have considerable potential in explaining the acquisition of psychopathology, especially how psychopathology arises without direct learning. For example, transfer of respondent extinction functions between members of an equivalence class could be offered as an explanation for how exposure therapy operates, when respondent extinction takes place with only a limited sample of members of the equivalence class. While some research on transfer of function related to clinical phenomena, such as Auguston and Dougher's (1997) work on transfer of avoidance behavior, has been conducted, to date there have been no direct applications of this work to clinical phenomena.

BEHAVIORAL ACCOUNTS OF NOVEL BEHAVIOR

Psychopathology often involves much novel behavior that is not directly learned. A simple phobia may clearly involve conditioning (such as a fear of dogs following receipt of a painful dog bite), but even here we must still explain how the person behaves fearfully to novel stimuli (such as dogs physically dissimilar to the dog that rendered the bite). In many cases, there is no clear conditioning event, and many disorders, such as generalized anxiety disorders and depression, are diffuse in that they involve behavior in many apparently unrelated environments. Most behavior analytic accounts of novel behavior that is not directly taught appeal to one or two learning processes, such as stimulus generalization and reinforcement of novel responding. Derived relationships and transfer of function, however, may explain much novel complex behavior related to psychopathology.

COMPLEX BEHAVIOR

MODELING

Modeling, also called the modeling prompt and observational learning, is another form of learning based on stimulus control of operant behavior. Modeling is defined as "a type of prompt in which the trainer demonstrates the target behavior for the learner" (Miltenberger, 2004, p. 562). It has been used to teach a very wide range of behavior, but especially social and language behavior. Modeling is one specific instance of various forms of behavior which are under the stimulus control of another person. In the case of modeling, the behavior of the learner must match the partner's behavior. Other examples of where another person's behavior is the discriminative stimulus for another person's behavior include situations such as when the behavior of the partner may be a discriminative stimulus for an opposite or a mirror response from the learner, as in the case of ballroom dancing, fencing, and other sports (Skinner, 1953).

Modeling involves both respondent conditioning and operant components. Respondent conditioning is involved when the emotional responses of the model affect how the learner responds to a given stimulus, such as when a child sees another child happily playing with a toy. The responses emitted by the model, such as laughter, function as conditioned stimuli for a conditioned emotional response in the learner, such as happiness. This emotional response is described as a vicarious emotional response. This can also be demonstrated in the case of phobias, in which the classically conditioned emotional response of fear may be acquired, such as when a child observes another person react fearfully to a stimulus. The fear of the parent may vicariously produce fear in the child. If all of this occurs in the presence of some neutral stimulus, that stimulus may become a conditioned stimulus for the emotional response of fear (Powell, Symbaluk, & MacDonald, 2005).

Modeling also involved operant conditioning, specifically in how the observer comes to imitate the model's behavior. For modeling to be effective, the observer must be able to attend to and imitate the person modeling the behavior. When modeling is used correctly, the behavior exhibited by the model should act as a discriminative stimulus for imitation by the observer. In the operant conditioning paradigm, the consequences of the organism emitting a given behavior can either be punishing or reinforcing, and as a result, the operant behavior will either decrease or increase in frequency, respectively. Through modeling, however, the observer does not have to experience the consequence directly. If an observer sees the model receiving a punishing consequence, the observer is less likely to engage in the observed behavior. Likewise, if the observer sees the model being reinforced, the observer is more likely to engage in the observed behavior.

When the student has imitated a given skill, an opportunity to use reinforcement or punishment procedures has been created. If the modeled behavior was imitated correctly, the learner can be presented with a reinforcer to increase the future probability of that behavior occurring. If some approximation of the modeled behavior is imitated correctly, reinforcement can be used in a shaping procedure to produce a target behavior not currently in the learner's repertoire (Powell et al., 2005).

Considering the operant conditioning component involved in the acquisition of modeled behavior, these same operant factors play a role in the maintenance of behaviors learned in this manner. The consequences that the learner receives as a result of imitating a behavior that has been modeled influence the future probability of that correct imitation. Reinforcing consequences will increase the future probability of a behavior occurring, while punishing consequences will decrease the future probability of that behavior occurring.

The reinforcement and punishment history of the learner also plays a role in the maintenance of modeled behavior. If a learner has been punished in the past for a given behavior, it may be difficult to maintain a new, similar behavior that was taught using modeling, particularly if the learner is still punished for the previous behavior. This might be seen, for example, when attempting to teach a child how to push other children on a swing at a playground. If this child has a history of being punished for a similar behavior, such as aggressively pushing a sibling, it may be difficult to teach or maintain appropriate swing pushing on a playground.

Applied Intervention Using Modeling

LeBlanc et al. (2003) used video modeling to teach perspective-taking skills. Perspective-taking skills are behaviors that lead us to conclude that the person appears to understand that another person's beliefs about events may differ from reality and guide future behavior. Three boys with autism, aged 7 through 13 years participated. This is an interesting example of ABA because previously nonbehavioral researchers had attempted to explain such phenomena in terms of cognitive variables (Baron-Cohen, 1997) and had not been very successful at intervening to teach children with autism theory of mind (Drew et al., 2002).

LeBlanc et al. (2003) used the Sally Anne false-belief tasks (Baron-Cohen, Leslie, & Firth, 1985), in which a participant has to state the behavior of a puppet that has a limited exposure to certain events that have occurred and of which the child participant has had full exposure. If the participant predicts that the puppet will behave according to only the events to which the puppet has been exposed (in contrast to the full range of experiences to which the child has been exposed), then the child has demonstrated perspective taking. After being given a pretest with a Sally Anne task, participants were shown a video of a person correctly completing a Sally Anne task, with the video camera zooming in on specific cues. Subjects were then asked to answer perspective-taking skills questions. Correct answers resulted in reinforcement with verbal praise and preferred edibles. Incorrect answers resulted in a replay of the correct responses via videotape. The participants were then tested using the Sally Anne tasks. While all three participants failed the Sally Anne task in the pretest, after training using video modeling, all three participants were able to pass the various perspective-taking tasks presented.

While the tasks were successfully taught using video modeling, Leblanc et al. (2003) noted that they did not have a behavioral explanation of why video modeling was successful. The behavioral explanation could be conceptualized using the operant model discussed earlier. Perspective-taking behavior could have been operantly taught through observing the model first engaging in the particular skills needed for perspective taking, such as focusing on specific cues, and then receiving reinforcement. When the participant was given the opportunity to imitate the behavior, the participant's behavior was also reinforced, which increased the likelihood of that response class occurring in the future. The continued reinforcement of the response classes emitted by the participants could have aided generalization when variants of the tasks were presented. The observed behavior of the model may also have functioned as a discriminative stimulus for imitating that behavior, specifically focusing attention on certain contextual cues.

Modeling has been extensively used in a wide range of clinical applications in areas of social and language skills training for people with severe psychiatric disorders, developmental disabilities, and assertiveness training in a variety of contexts. It is also often combined with various exposure therapies for anxiety disorders, in which the therapist models behavior for the client to imitate.

CHAINING

Chaining is an important procedure in teaching complex behavior to individuals (Leslie & O'Reilly, 1999). A behavioral chain is a complex unit of behavior that consists of a number of individual responses that are emitted in a specific sequence (Miltenberger, 2004). The first response in a behavioral chain is initiated in the presence of a discriminative stimulus. The completion of each response functions as, or acts to produce, a subsequent discriminative stimulus that sets the occasion for the next response in the chain. Further, the completion of each response also functions as a conditioned reinforcer for the previous step of the chain (Leslie & O'Reilly, 1999). For example, dirty silverware may be a discriminative stimulus that sets the occasion to clean a fork. The first response in the chain is to put soap on the fork. The presence of soap on the fork would then serve as a conditioned reinforcer for having put soap on the fork, and it would also function as a discriminative stimulus for turning on the water. In this way, a chain of dishwashing responses is emitted, and the terminal response is reinforced by a clean fork.

Prior to implementation of a chaining procedure, it is necessary to first perform a task analysis. A task analysis consists of analyzing a complex behavioral unit into several smaller stimulus-response units, which are referred to as links. In the preceding example, the four links may be putting soap on the fork (R1), turning on the water (R2), rinsing the fork (R3), and turning off the water (R4). The discriminative stimuli for each response, respectively, would be the dirty fork (for R1), soap on the fork (for R2), running water (for R3), and a clean fork (for R4).

Chaining may be used to teach a complex behavior using three main methods: backward chaining, forward chaining, and total task presentation (Miltenberger, 2004). In backward chaining, the last response is taught first, and reinforcement is provided after successful attempts. After the learner masters the last response in the chain, the second-to-last response is taught. In this way, each response is successively added onto the chain, and the learner completes the chain on every single trial. For example, applying backward chaining to washing a fork would require that the learner be first taught to turn off the water. Next, the learner would be taught to rinse the soap. To teach the second-to-last response, the instructor would provide prompts to rinse the soap. The learner would then need to independently complete the last response in the chain to obtain reinforcement.

In forward chaining, the first rather than the last response in the chain is taught first. In this procedure, the discriminative stimulus is presented, and reinforcement is provided for the successful completion of the first response. After the first response is mastered, training is conducted for the second response. This consists of presenting the discriminative stimulus for the first response, allowing the first response to occur independently, and teaching and reinforcing approximations of the second response. For example, the learner is presented with a dirty fork (discriminative stimulus) and is then taught to put soap on the fork. Teaching the second response (turning on the water) would consist of presenting the dirty fork, allowing the learner to independently complete the first response, and then prompting and reinforcing turning on the water.

In contrast to forward and backward chaining, total task presentation consists of teaching the entire complex behavior in one trial (Miltenberger, 2004). With this procedure, the learner is prompted through the entire chain, and reinforcement is provided when the learner completes the chain. The learner would be prompted at the outset to perform the full sequence of putting soap on the dirty fork, turning on the water, washing the fork, and turning off the water in a single trial. Reinforcement would be provided when the learner completes the entire sequence of the chain.

In comparing the three methods, backward chaining has the advantage of allowing the learner to earn the natural reinforcer at the end of the chain (Miltenberger, 2004). This procedure is especially useful when teaching learners that have difficulty learning complex behavior. In contrast, forward chaining has the advantage of providing extra practice for the responses that occur in the beginning of the chain (Leslie & O'Reilly, 1999). However, disadvantages of forward chaining are that arbitrary reinforcers are used to teach the earlier responses in the chain (Miltenberger, 2004, p. 224) and that earlier responding is placed on extinction as chaining progresses. Finally, total task presentation has the advantage of practicing all of the responses each time the complex behavior is initiated. The disadvantage of this technique, however, is that reinforcement is delayed until the entire sequence is practiced, which makes teaching the response chain more difficult. The relative merits of the different forms of chaining are not always clear, and the practitioner may have to resolve them on a case-by-case basis.

Application of Chaining

Wong and Woolsey (1989) provided an example of the application of chaining. In a multiple baseline design across responses, four patients with schizophrenia were taught a series of conversation statements when approached by a staff member. Specifically, the participants were taught to greet a therapist ("Hi"), address a therapist by name ("John"), respond to a question ("I am doing well"), ask a personal question (e.g., "How is it going?"), and ask a question in response to the topic of conversation. The training procedure consisted of using forward chaining to teach one conversational statement at a time, and prompting, feedback, and instruction were used to help initiate responses. Therefore, the procedure consisted of teaching the patients to first say "hello" when greeted by a staff member. Reinforcement, in the form of tangible items and praise, was provided after each correct response. After mastering the first response, participants were taught the second response (addressing the therapist by name). Teaching the second response consisted of the patient saying "hi" to a staff member greeting and then implementing the training procedure to teach the patient to address the staff member by name. If the previously learned response was incorrect, reinforcement could not be earned for the response being trained. Thus, the patient was required to (a) correctly, and independently, emit the previously learned conversational statements and (b) respond correctly to the prompts provided for the second response for reinforcement to be delivered.

Three of the four patients successfully acquired the entire series of conversational statements. The patient who did not acquire the chain did emit some of the conversational responses. However, this patient did not maintain the skill after training was removed. The lack of success with this patient was attributed to ineffectiveness of the purported reinforcer.

SELF-CONTROL

Skinner (1953) provided an analysis of self-control, without reference to a controlling or motivating self, that proved to be an important model for intervention in the area of mental health. Skinner proposed that self-control consists of two responses: a controlling response and a controlled response. The controlling response changes the probability of the controlled response. Hence, the variables of which the controlling response is a function are the focus of intervention. For example, a controlling response might be to put one's gym bag next to the front door at night. If this alters the probability of going to the gym, then going to the gym is a controlled response. Variables that alter the probability of placing the gym bag next to the door, such as antecedents and contingencies that reinforce doing so, then become the focus of intervention. Skinner himself used this extensively to manage his own behavior to become productive and happy throughout his long career (Epstein, 1997; Skinner, 1987; Skinner & Vaughn, 1983).

Recall from Chapter 1, "Structural and Functional Approaches to Psychopathology and Case Formulation," that Skinner speculated that self-control strategies were the most important method for intervening in mental health because they may result in greater generalization of behavior change. Indeed, he hinted that some mental health problems, such as suicide, might be construed as examples of inappropriate forms of self-control. Suicidal behavior reduces the probability of all future behavior to zero and thereby terminates all aversive stimuli. If other effective controlling responses could be taught that served the same function—teaching more effective ways of avoiding aversive stimuli—then suicidal behavior might be reduced. This suggests that teaching forms of self-control might be an important strategy to teach resilience to psychopathology. Examples of common self-control strategies include (a) presentation of stimuli to increase the probability of a desired behavior, such as carrying healthy snacks to work or placing reminders to engage in preventative actions; (b) accurate self-reinforcement for appropriate behavior, such as self-praise contingent upon completing a desired task; (c) removal of stimuli that make an undesirable behavior more likely, such as throwing junk food out of the house to prevent unhealthy snacking; (d) self-restraint, such as sitting on one's own hands to reduce the probability of nail biting; and (e) use of drugs to change the probability of a behavior of interest, such as refraining from caffeine consumption in the evening to increase the probability of sleeping and so on (Skinner, 1953).

Application

James (1981) reported an innovative application of self-control when he taught a man who stuttered to self-punish stuttering using time out. (Self-punishment sounds odd, of course, but then remember that most people self-control using self-punishment every day using an alarm clock to punish staying in bed.) In the first study, James taught the client to discriminate accurately stuttering from nonstuttering and to press a buzzer and stop talking for at least 2 s contingent upon each occurrence of stuttering. They construed not talking as a form of time out from a high probability behavior. After several sessions, James then instructed the client to simply stop talking without using the buzzer. Later, they instructed the client to use this procedure at home, on the phone, and in shops. Self-punishment effectively reduced disfluent speech in both laboratory and community settings, and the effects were maintained at 6- and 12-month follow-ups.

In a second study with the same participant, a response cost procedure was evaluated in which failure to use the time-out procedure would result in a small loss of money. James showed that the addition of the response cost procedure resulted in much greater accuracy of the implementation of timeout. This in turn had the effect of reducing disfluent speech. Note that this study follows Skinner's analysis of self-control exactly. The variables controlling the controlling response—the response cost to increase correct use of timeout—influenced both the controlled response *and* resulted in a change in the controlled response when stuttering was reduced.

SUMMARY

Rule-governed behavior, stimulus equivalence, modeling, chaining, and self-control account for other forms of behavior and may form the basis for a variety of other interventions. Familiarity with these other learning processes permits the clinician a greater understanding of the learning that may take place during the development of psychopathology and gives the clinician a wider range of methods of intervention.

REFERENCES

Auguston, E. M., & Dougher, M. J. (1997). The transfer of avoidance functions through stimulus equivalence classes. *Journal of Behavior Therapy and Experimental Psychiatry, 28,* 181–193.

Baron-Cohen, S. (1997). *Mind blindness. An essay on theory of mind and autism.* Cambridge, MA: MIT Press.

Baron-Cohen, S., Leslie, A. M., & Frith, U. (1985). Does the autistic child have a "theory of mind"? *Cognition, 21,* 37–46.

Cowley, B. J., Green, G., & Braunling-McMorrow, D. (1992). Using stimulus equivalence procedures to teach name-face matching to adults with brain injuries. *Journal of Applied Behavior Analysis, 25*, 461–475.

Drew, A., Baird, G., Baron-Cohen, S., Cox, A., Slonims, V., Wheelwright, S., et al. (2002). A pilot randomized control trial of a parent training intervention for preschool children with autism. Preliminary findings and methodological challenges. *European Child and Adolescent Psychiatry, 11*, 266–272.

Epstein, R. (1997). Skinner as self-manager. *Journal of Applied Behavior Analysis, 30*, 545–568.

Fields, L., & Verhave, T. (1987). The structure of equivalence classes. *Journal of the Experimental Analysis of Behavior, 48*, 317–332.

James, J. E. (1981). Behavioral self-control of stuttering using time-out from speaking. *Journal of Applied Behavior Analysis, 14*, 25–37.

LeBlanc L. A., Coates A. M., Daneshvar, S., Charlop-Christy, M. H., Morris, C., & Lancaster, B. M. (2003). Using video modeling and reinforcement to teach perspective-taking skills to children with autism. *Journal of Applied Behavior Analysis, 36*, 253–257.

Leslie, J. C., & O'Reilly, M. F. (1999). *Behavior analysis: Foundations and applications to psychology.* Amsterdam, The Netherlands: Harwood Academic Publishers.

Miltenberger, R. G. (2004). *Behavior modification principles and procedures* (3rd ed.). Pacific Grove, CA: Wadsworth.

Powell, R. A., Symbaluk, D. G., & MacDonald, S. E. (2005). Observational learning, language, and rule-governed behavior. In V. Knight (Ed.), *Introduction to learning and behavior* (pp. 358–472). Belmont, CA: Thomson Wadsworth.

Sidman, M. (1994). *Equivalence relations and behavior: A research story.* Boston: Authors Cooperative.

Skinner, B. F. (1953). *Science and human behavior.* New York: Macmillan.

Skinner, B. F. (1987). A thinking aid. *Journal of Applied Behavior Analysis, 20*, 379–380.

Skinner B. F., & Vaughn, M. E. (1983). *Enjoy old age. A program of self management.* New York: Norton.

Taylor, I., & O'Reilly, M. F. (1997). Toward a functional analysis of private verbal self-regulation. *Journal of Applied Behavior Analysis, 30*, 43–58.

Wong, S. E., & Woolsey, J. E. (1989). Re-establishing conversational skills in overtly psychotic, chronic schizophrenic patients. Discrete trials training on the psychiatric ward. *Behavior Modification, 13*, 415–431.

4

FUNCTIONAL ANALYSIS METHODOLOGY IN DEVELOPMENTAL DISABILITIES

ROBERT DIDDEN

*Department of Special Education of the Radboud University
Nijmegen, The Netherlands*

Since the first publications more than 30 years ago, the functional analysis approach has significantly contributed to improving the lives of individuals with developmental disabilities, including autism. Important gains in both conceptual and applied work have been made in the teaching of skills and the assessment and remediation of problem behavior (Carr, Innis, Blakeley-Smith, & Vasdev, 2004). Individuals with developmental disabilities are at increased risk for the development of problem behaviors and associated psychopathology, such as self-injurious (Rojahn, 1994), aggressive (Gardner & Cole, 1993), stereotypic (Rojahn & Sisson, 1990), and other problem behaviors (Emerson et al., 2001). Without adequate treatment or support, such behaviors persist (Emerson et al., 2001). Risk factors include poor self-help and communication skills, deficits in social and problem-solving skills, lack of self-help skills, punitive parenting practices, restricted access to materials and activities, certain genetic disorders and physical and neurological conditions, and psychiatric disorders (Dekker, Koot, van der Ende, & Verhulst, 2002; Emerson et al., 2001; McClintoch, Hall, & Oliver, 2003). Such factors are more prevalent in people with disabilities than in typically developing people.

Clinicians working in such diverse settings as special schools, residential facilities, clinical treatment or outpatient settings, and community agencies are confronted with a wide range of problem behaviors and psychopathology. Functional analysis methodology provides an empirically validated framework for its assessment and treatment. In this chapter, we (a) describe basic assumptions of this approach, (b) review methods for conducting functional analysis in clinical practice, (c) review function-based treatments, and (d) present a brief example of its application.

BASIC ASSUMPTIONS AND PRINCIPLES

Functional analysis is a methodology for systematically investigating relationships between problem behavior and environmental events. Its purpose is to identify variables controlling behavior(s) and to generate hypotheses about its function(s). A treatment is then selected that matches this function (Gardner, Graeber-Whalen, & Ford, 2001; Hanley, Iwata, & McCord, 2003; Sigafoos, Arthur, & O'Reilly, 2003).

The first studies demonstrating behavior-environment relationships in individuals with developmental disabilities were published in the early 1960s (Berkson & Mason, 1963; Lovaas, Freitag, Gold, & Kassorla, 1965). These early reports showed that problem behavior is not a characteristic feature of a person, but reflects a response to environmental conditions. Problem behavior in individuals with developmental disabilities is conceptualized as a learned response that is evoked and maintained by environmental conditions and is influenced by establishing operations, antecedents, and consequences (Carr et al., 2004). The four general classes of consequences are (a) positive social reinforcement, (b) negative social reinforcement, (c) positive automatic reinforcement, and (d) negative automatic reinforcement. For example, negative automatic reinforcement occurs when a target behavior produces an alleviation or reduction in an internal aversive stimulus, such as physical discomfort or tension, that is attenuated or terminated contingent upon removing someone from an uncomfortable wheelchair (Wilder & Carr, 1998). Individuals differ with respect to which stimuli function as reinforcers (Fisher, Hinnes, & Piazza, 1996; Taylor & Carr, 1992).

It is the function of the behavior, not its topography, that guides treatment selection. Problem behavior is not conceptualized as a symptom of an underlying pathology or personal trait (e.g., personality disorder, genetic disorder, depression, attachment disorder) or developmental stage, but as a response that is lawfully related to environmental conditions (see Sturmey [1996] for a discussion of structuralist and functionalist approaches to human behavior). For example, even problem behaviors more or less characteristic for a specific genetic disorder, such as self-injury in Lesch-Nyhan and Cornelia de Lange syndromes, may show considerable variability across environmental conditions (Didden, Korzilius, & Curfs, In press;

Duker, 1975; Moss et al., 2005) and may be treated with interventions based on functions rather than diagnosis or topography.

ESTABLISHING OPERATIONS

During the past decade, increased attention has been given to the concept of establishing operations or EOs. EOs are factors, such as reinforcer satiation and deprivation, that alter the relationship between antecedent events, the subsequent behavior, and its maintaining consequences (McGill, 1999). For example, social deprivation may be an EO that increases the reinforcing value of attention and that increases the likelihood of behavior that has previously resulted in attention. EOs may be highly idiosyncratic (Adelinis, Piazza, Fisher, & Hanley, 1997). The three classes of EOs are (a) physical, (b) biological, and (c) social. For example, Carr, Smith, Giacin, Whelan, and Pancari (2003) showed that menstrual discomfort was an EO for problem behaviors in three women who were living in a residential facility. Other studies have shown that environmental complexity, sleep deprivation, and ear infection may function as EOs for problem behavior (Duker & Rasing, 1989; O'Reilly, 1997; O'Reilly & Lancioni, 2000). McAtee, Carr, and Schulte (2004) developed the *Contextual Assessment Inventory*, which explores and identifies possible EOs. Clinicians may then continue with direct observation to investigate relationships between EOs and problem behavior further (Bodfish & Konarski, 1992; Horner, Vaughn, Day, & Ard, 1996).

FUNCTIONAL ANALYSIS METHODOLOGY

Functional analysis methodology includes methods for the assessment of functional properties of problem behavior. A distinction is made between descriptive and experimental methods. Descriptive or nonexperimental methods are also referred to as functional assessment. Experimental methods or functional analysis refers to procedures that systematically manipulate environmental conditions to assess effects on the rates of problem behavior.

DESCRIPTIVE ANALYSIS

Descriptive analysis involves methods of both indirect and direct observation of the target behavior and environmental events. Such methods are typically implemented in naturally occurring applied settings.

Indirect Observation

"Indirect" means that these methods do not require direct observation of the person exhibiting the problem behavior and include interviews and rating scales. These methods rely on reports by informants who are in daily contact with the person, such as parents, caregivers, and teachers, or on reports by individuals themselves. In the scientific literature, several instruments have been described such as the *Functional Analysis Interview (FAI)*, *Motivation Assessment Scale (MAS)* (Durand & Crimmins, 1988), and *Questions About Behavioral Function (QABF)*. In this section I will briefly review two of these instruments.

Functional Analysis Interview

O'Neill et al. (1997) developed the FAI which is a protocol for conducting an extended clinical interview with informants. Informants are asked open-ended questions about features of the problem behavior (e.g., duration, topography), immediate and distal events that most likely evoke that behavior, and consequences that follow the behavior and that may maintain it. Such a protocol may be especially useful in case of low-frequency/high-intensity behaviors and takes about 45 to 90 min to complete. It provides the clinician with a comprehensive set of data that can be used to generate hypotheses on behavioral function. The FAI may be easily adapted for use with other populations, such as individuals with mild developmental disability or learning disabilities. Only a few studies have been published that report data on reliability and validity of the FAI (Cunningham & O'Neill, 2000; Yarbrough & Carr, 2000).

Questions About Behavioral Function

Rating scales such as the QABF (Matson & Vollmer, 1995) were developed as an alternative to analog baselines (see "Multiple Experimental Analyses" below). The QBAF is a 25-item rating scale, and each item is rated on a four-point Likert-type scale. The QABF consists of five subscales addressing five maintaining variables: (a) nonsocial (automatic) reinforcement, (b) tangible reinforcement, (c) attention, (d) escape, and (e) physical discomfort. Items are included that describe social avoidance and physical discomfort, making the QABF more comprehensive than other scales. Each subscale contains five items, and subscale as well as total scale scores are calculated. The subscale with the highest score indicates the most likely function.

The QABF has sound psychometric properties. Paclawskyj, Matson, Rush, Smalls, and Vollmer (2000) found acceptable to high inter-rater and test-retest correlations as well as high internal consistency for each subscale. Factor analysis yielded five factors corresponding to the QABF-subscales. The QABF also has convergent validity (Paclawskyj, Matson, Rush, Smalls, & Vollmer, 2001; Shogren & Rojahn, 2003). In an important experimental study, Matson, Bamburg, Cherry, and Paclawskyj (1999) demonstrated predictive validity of the QABF for treatment selection in an impressive sample of participants (i.e., $N = 180$) who showed self-injury, aggression, and stereotypic behaviors. Interventions based on functions identified with the QABF resulted in a substantially larger decrease in problem behaviors than non-function-based treatments.

Direct Observation

During direct observation, environment-behavior relationships are systematically recorded and analyzed. Methods of direct observation are the foundation of the functional analytic approach and date from the late 1960s (Bijou, Peterson, & Ault, 1968). This section will describe two methods: (a) scatter plot and (b) antecedent behavior consequent (ABC) analysis.

Scatterplot

The scatterplot is a tool for investigating temporal characteristics of problem behavior (Touchette, MacDonald, & Langer, 1985). The scatterplot is a recording sheet or grid in which time intervals are blocked within and across successive days. Length of intervals may vary between 5 and 30 min, or more depending on practical constraints and/or behavior frequency. Observers record the extent to which the target behavior and other events occur within these intervals. Scatterplot data may reveal patterns of responding in the behavior, and the target behavior may be related to specific activities, time of day, presence of particular individuals or settings, and combinations of these and other variables.

Touchette et al. (1985) presented data on patterns of self-injurious and aggressive behavior in three individuals with developmental disabilities. Scatterplot data collected for 7 days revealed that the rate of target behavior was relatively high when a particular staff member was present and during certain activities. For a third subject, the data could not be interpreted. Substantial reductions in the target behaviors were accomplished by rescheduling the staff member and by revising scheduling of activities. Kahng et al. (1998) evaluated the use of scatterplots in 20 individuals with severe to profound disability who showed self-injury and aggression. Visual analysis of the Scatterplots failed to reveal any reliable temporal response pattern.

Antecedent Behavior Consequent Analysis

Two methods of ABC analysis may be distinguished: (a) ABC charts and (b) recording sequences of behavior and environmental events. ABC charts allow the observer to write down narrative descriptions of a target behavior and record events that immediately precede and follow the behavior. The consistency with which specific events appear contiguous to the problem behavior is then analyzed. The *Detailed Behavior Report* (DBR) is an example of a chart for conducting an ABC analysis (Groden, 1989). The DBR also records more distal relevant events and conditions. Some preliminary data suggest that the DBR is a useful and reliable instrument for assessing behavioral function (Groden & Lantz, 2001).

Bijou et al. (1968) were among the first to describe a method for conducting an ABC analysis. This method may reveal sequences of behavior and associated events through time (Lerman & Iwata, 1993). Data may be expressed as percentages (e.g., number of intervals with a consequent event that followed the target behavior divided by the total number of intervals with that behavior). For example, Mace and Lalli (1991) conducted an ABC analysis of bizarre speech in a 46-year-old man with moderate developmental disability who lived in a group home. They found that, for example, during task demands, bizarre speech was on a mean of 65% of the occasions followed by discontinuation of the ongoing task by staff. This high percentage suggests that bizarre speech was intermittently maintained by escape from demands.

Emerson et al. (1996) used time-based lag-sequential analysis to identify antecedent and consequent events of aggressive and self-injurious behavior in three children with severe developmental disability. This technique involves calculation of conditional probability of the onset, occurrence, or termination of one event at specific points in time to the onset, occurrence, or termination of another event. For example, to test whether self-injurious behavior is escape-motivated, the conditional probability of self-injury while demands (i.e., antecedent event) are placed upon the individual is compared with the overall chance or unconditional probability of self-injurious behavior. Then the statistical significance of the difference between the conditional and unconditional probability of an event is tested using the z-statistic (for a description of sequential analysis of naturally occurring events, see Emerson et al., 1996).

Experimental Analyses

Experimental functional analyses require the experimental manipulation of antecedents and/or consequences that are hypothesized to influence a target behavior. Potential maintaining and eliciting events are manipulated to assess control over responding. Two types of experimental analyses are distinguished: (a) single experimental analysis and (b) multiple experimental analysis.

Single Experimental Analysis

During single experimental analysis, only one event is presented and withdrawn to assess its effect on the rate of the target behavior. A functional relationship is demonstrated if the frequency or duration of the target behavior (i.e., dependent variable) changes in relation to a change in the event (i.e., independent variable). For example, Carr and McDowell

(1980) demonstrated a functional relationship between self-injurious scratching in an 11-year-old boy with below average IQ and contingent attention from his parents in the home setting. Prior to experimental analysis, informal observations in the home setting indicated that self-scratching was often followed by attention in the form of reprimands and physical comfort from his parents. Data collected in an ABA design showed that the frequency of self-scratching varied as a function of parental contingent attention. Function-based treatment, consisting of response contingent time out and differentially reinforcing nonscratching, resulted in a marked and long-standing decrease in the number of wounds resulting from scratching.

Multiple Experimental Analysis

Multiple experimental analysis may be conceptualized as a series of single experimental analyses. Two types of multiple analyses have been described: (a) analog baselines and (b) structural analysis. Iwata, Dorsey, Slifer, Bauman, and Richman (1982) developed and validated functional analysis methodology using analog baselines. "Analog" means that test conditions may resemble conditions found in natural environments. Typically, the individual is observed during four conditions, each containing an EO, an antecedent event, and a consequent event hypothesized to maintain the problem behavior. These conditions refer to three classes of reinforcement: (a) attention, (b) escape, and (c) sensory. There is also a fourth control condition. A systematic relationship between responding and the four conditions indicates a behavioral function.

In a large sample of individuals with profound to mild developmental disability who showed self-injurious behavior ($N = 152$), Iwata and his colleagues (1994) found differentiated patterns of responding across conditions in most cases, indicating sensitivity of self-injurious behavior to different contingencies. This methodology has been extended and replicated in a large number of studies and in a wide range of problem behaviors (see Table 4.1). A variation of analog baselines is a brief functional analysis which uses only one or two repeated experimental and control conditions (Derby et al., 1992) and can thus identify functions more rapidly in applied settings, such as outpatient clinics.

Structural analysis differs from analog baselines in that multiple events are manipulated in the natural context of the individual. Methods of structural analysis have been developed as an alternative for direct observation in artificially created contexts. For example, Sigafoos and Meikle (1996) observed self-injurious and aggressive behavior in two 8-year-old boys with autism in a classroom. They instructed teachers to create situations (e.g., child is ignored, child is presented with an instructional demand) in which four functions of problem behavior were tested. Results suggested that problem behaviors were maintained by teacher attention and access to objects. Subsequent functional communication training (see "Functional Communication Training" below), consisting of teaching the children to verbally ask for help and request objects by pointing to a line drawing, resulted in near zero levels of problem behavior.

FUNCTION-BASED TREATMENTS

Within functional analysis methodology, selection of treatment is based on the identified function(s) of problem behavior. Function-based treatment can take the form of (a) modifying EOs, (b) modifying antecedent events, (c) removing or altering the reinforcing or consequent event, and/or (d) teaching the individual functional and adaptive skills that compete with or replace the problem behavior. In this section, we selectively review several important function-based approaches. Table 4.1 provides an overview of function-based treatment related to behavioral function and topography.

TABLE 4.1 Categorization of Function-Based Treatments

Behavioral Function	Function-Based Treatment	Behavioral Topography	Research Article
Attention/ tangibles	NCR	self-injury	Vollmer et al. (1993)
		bizarre speech	Mace & Lalli (1991)
		problematic speech	Carr & Britton (1999)
		aggression/	Hagopian et al. (1994)
		destruction	Hagopian et al. (1994)
		pica	Piazza et al. (1998)
	DRO/I/A	self-injury	Vollmer et al. (1993)
	FCT	aggression, self-injury	Carr & Durand (1985)
		various topographies	Hagopian et al. (1998)
		breath holding	Kern et al. (1995)
		various topographies	Derby et al. (1992)
	EXT	sleep disruption	Didden et al. (2002)
		self-injury	Duker (1975)
	TO	self-injury	Carr & McDowell (1980)
		self-injury	Hanley et al. (1997)
Automatic	NCR	self-injury, stereotypy	Sprague et al. (1997)
		rumination	Wilder et al. (1997)
		rumination	Rast & Johnson (1986)
		stereotypy	Higbee et al. (2005)
	EXT	eye poking	Kennedy & Souza (1995)
		eye poking	Lalli et al. (1996)
		hand mouthing	Lerman & Iwata (1996b)
		stereotypy	Rincover et al. (1979)

(Continues)

TABLE 4.1   (*Continued*)

Behavioral Function	Function-Based Treatment	Behavioral Topography	Research Article
	DRO/I/A	self-injury	Cowdery et al. (1990)
		stereotypy	Rincover et al. (1979)
	FCT	stereotypy	Durand & Carr (1987)
	Change in EO/	self-injury	Favell et al. (1982)
	antecedent	self-injury	Horner (1980)
Escape/ avoidance	Change in EO/ antecedent	self-injury	Smith et al. (1995)
		aggression	Adelinis & Hagopian (1999)
		aggression, self-injury	Horner et al. (1991)
		aggression	Piazza et al. (1997)
		stereotypy	Tustin (1995)
		destruction	Piazza et al. (1996)
		various topographies	Dunlap et al. (1991)
	NCR	self-injury	Vollmer et al. (1995)
	Choice-making	aggression	Foster-Johnson et al. (1994)
		disruption	Romaniuk et al. (2002)
	EXT	self-injury	Iwata et al. (1990)
		food refusal	Cooper et al. (1995)
		stereotypy	Durand & Carr (1987)
FCT		aggression, self-injury	Carr & Durand (1985)
		aggression, self-injury	Lalli et al. (1995)
		aggression, self-injury	Hagopian et al. (1998)
		vomiting	Lockwood et al. (1997)
	DRO/I/A	destruction	Piazza et al. (1996)
Escape/ Avoidance (increased arousal)	Relaxation	self-injury	Steen & Zuriff (1977)
		disruptive behavior	Mullins & Christian (2001)
	Change in EO	self-injury	O'Reilly (1997)
	Routines	self-injury	Horner et al. (1997)

Note: NCR = Noncontingent reinforcement; DRO/I/A = Differential reinforcement of other/incompatible/alternative behavior; FCT = Functional communication training; EXT = Extinction; TO = Time out; EO = Establishing operation.

MODIFYING EOS

An approach to modifying EOs is noncontingent reinforcement or NCR, in which the reinforcer is delivered systematically and in a response-independent format (for a review, see Carr et al., 2000). For example, NCR in escape-maintained problem behavior may consist of providing escape from a task at regular time intervals and irrespective of whether the individual exhibits problem behavior or not. NCR is started with relatively dense reinforcement schedules (i.e., short time intervals) that, when low frequencies of the target behaviors are established, are gradually thinned while maintaining low behavior frequencies. Prior to starting NCR, mean interresponse time of the occurrence of the target behavior is assessed. The reinforcer is then delivered at the end of a time interval, irrespective of whether the target behavior has occurred during or at the end of that interval or not. Occurrence of the target behavior during the interval does not lead to resetting that interval (such as in procedures of differential reinforcement of other behavior or DRO). If the function cannot be identified or if a maintaining reinforcer cannot be withheld (e.g., in case of automatic reinforcement), NCR with arbitrary reinforcers may be effective, provided stimuli are identified through preference assessment (Duker, Didden, & Sigafoos, 2004; Higbee, Chang, & Endicott, 2005) and function as reinforcer.

Other approaches based on manipulation of EOs include studies that have shown that changes in the classroom curriculum (Dunlap, Kern-Dunlap, Clarke, & Robbins, 1991) and providing opportunities for choice-making (Shogren, Faggella-Luby, Bae, & Wehmeyer, 2004) may be highly effective. Apparently, such interventions lead to a reduction in the aversive properties of instructional demands. A change in EOs may also be effective in case of automatically reinforced problem behavior. For example, Horner (1980) showed that the frequency of self-injurious and stereotypic behavior was markedly reduced by enriching the social and physical environment.

MODIFYING ANTECEDENTS

Interventions based on modifying antecedents have included removal, fading, and other manipulation of antecedents. For example, this approach may be indicated in the treatment of problem behavior that is escape-maintained. Escape from task demands may be treated by changing task characteristics, such as its level of difficulty, speed of presentation, novelty, and duration (Smith, Iwata, Goh, & Shore, 1995), and interspersing easy with difficult tasks (Horner, Day, Sprague, O'Brien, & Heathfield, 1991). For example, Pace, Iwata, Cowdery, Andree, and McIntyre (1993) initially eliminated all task demands and gradually increased the number of demands concurrent with low levels of self-injury.

EXTINCTION

Extinction (EXT) occurs when the contingency between a target behavior and its reinforcing consequence is interrupted. EXT involves withholding the consequent event maintaining the problem behavior upon its occurrence (Ducharme & Van Houten, 1994; Lerman & Iwata, 1996a) or making the consequent noncontingent on the target behavior. EXT may take several forms depending on the consequence maintaining the behavior. Attention extinction may occur if attention from caregivers or others is withheld contingent on the occurrence of the target behavior. Note that ignoring an individual's problem behavior will be effective in reducing that behavior only when that behavior is maintained by contingent social attention. Extinction of escape/avoidance behavior consists of preventing the individual from escaping from or avoiding the event that elicits negatively reinforced problem behavior. Finally, in case of automatic positive reinforcement, sensory extinction involves removing the sensory consequences resulting from problem behavior.

Extinction is associated with several negative side effects. A clinically important side effect of extinction is the occurrence of the so-called extinction burst, which is an increase, albeit temporary, in the frequency or intensity of the target behavior at the beginning of the treatment (see Didden, Curfs, van Driel, & de Moor [2002] for an extinction burst during treatment of attention-maintained sleep disruptive behavior). Extinction is also accompanied by increased behavioral variability, as seen by the emergence of novel behaviors or the reemergence of old behaviors. This can be the basis of reinforcement of appropriate behavior.

The extinction burst is not as frequent as is expected (Lerman & Iwata, 1996a). However, clinicians should be aware that an extinction burst is a signal of treatment effectiveness, not ineffectiveness. EXT in combination with differential reinforcement often results in a faster reduction in the target behavior than EXT used alone (Lerman & Iwata, 1996a) and is therefore preferable to EXT alone.

FUNCTIONAL COMMUNICATION TRAINING

Functional Communication Training (FCT) involves teaching an individual a specific communicative response that serves the same function as the problem behavior (Carr et al., 1994; Sigafoos & Meikle, 1996). FCT is also referred to as differential reinforcement of communicative behavior. In FCT the problem behavior and the new communicative response should be functionally equivalent and have the same function. Problem behavior and the new communicative response then belong to the same response class because they have identical consequences.

Several instructional procedures exist for teaching alternative and communicative responses to individuals with developmental disabilities (Carr et al., 1994; Duker et al., 2004). FCT is most often used for problem behaviors controlled by social contingencies. For example, FCT of attention-maintained problem behavior consists of teaching an individual communicative responses with which she or he is able to request attention. FCT for problem behavior that has the function to escape from demands may take the form of teaching an individual socially appropriate means of asking for a break and escape.

For FCT to be effective, it should be combined with extinction of the target behavior, and the new communicative response should require less response effort to perform than the target behavior (Hagopian, Fisher, Sullivan, Acquisto, & LeBlanc, 1998). Derby et al. (1997) reported impressive long-term effectiveness of FCT for a variety of problem behaviors. Two meta-analyses have shown that FCT is associated with relatively large effect sizes (Carr et al., 1999; Didden, Duker, & Korzilius, 1997).

COMPREHENSIVE FUNCTIONAL ANALYTIC MODEL

Clinicians may use the functional analytical approach in an attempt to explain and treat problem behaviors in individuals with developmental disabilities. Within this approach, variables affecting problem behavior are investigated. In case of individualized treatment plans, a comprehensive, integrative, or multicomponent model should broaden its focus beyond assessing relationships between problem behaviors and environmental conditions. When assessing behavioral function and designing a function-based treatment, presence of and relationship to a variety of factors, such as anxiety and depression (Friman, Hayes, & Wilson, 1998; Sturmey, 2002), physical discomfort (Bosch, van Dyke, Smith, & Poulton, 1997), hostile attributions and other cognitions (Fuchs & Benson, 1995; Pert, Jahoda, & Squire, 1999), medication side effects, system factors (McAfee, 1987), coping and problem-solving skills (Hartley & MacLean, 2005; O'Reilly et al., 2004), and caregivers' attributions of the cause of problem behavior (Hastings, 1997), should be considered. However, reliable measurement and assessment of such factors may be difficult or even impossible in individuals with profound to mild disabilities.

Recording and analyzing relationships between problem behavior and environmental conditions, however, remain the central feature of the functional analysis methodology, as problem behavior is conceptualized as the final outcome of an individual's learning history during interactions with environmental conditions. In daily clinical practice, clinicians should use the following seven steps: (1) identify problem, then select and reliably define target behavior(s); (2) design and use scatter plot and perform unstructured and/or structured ABC recordings; (3) interview caregivers and client, if possible, using the FAI and complete rating scales/checklists such as the QABF; (4) conduct single experimental analyses; (5) formulate hypothesis of function; (6) design and implement function-based treatment; and (7) evaluate treatment.

CASE STUDY

PARTICIPANT

Tim was a 17-year-old adolescent who functioned in the mild range of developmental disability. He had relatively good adaptive and communicative skills that were appropriate for his mental age. Because of his disruptive and aggressive behavior at home, he had been placed in a residential facility about a year ago living with seven other young men of his age. During the day he visited a day care center where he worked in a sheltered workshop-type setting. In the living room and at the workshop, he exhibited highly disruptive behaviors, such as elopement, yelling and shouting, and verbally aggressive behaviors toward peers and caregivers. His verbal aggression mainly consisted of threatening others, causing fear in his caregivers and peers. In an attempt to control verbal aggression, he regularly was sent outside to calm down, sometimes was sent home from work, and was also sent to his room contingent on verbal aggression when he was in the residence's living room. Although this behavior was not frequent, it was nevertheless highly intensive and severe. Given its adverse consequences, verbal aggression was targeted for functional assessment and treatment.

BASELINE

Prior to a change in management of Tim's verbal aggression, a baseline was taken for 2 weeks. During baseline, several caregivers in both settings completed the QABF. A structured functional analysis interview was conducted with a caregiver in each setting who worked with him at least 6 months. For 2 weeks, caregivers in both settings recorded antecedent and consequent events when the target behavior occurred and other relevant information, such as sleep and affect. Next, a scatter plot was used in which the occurrence of the target behavior was recorded within 2-hr intervals. During a short interview with Tim, he told the interviewer that he often felt tired when he was being verbally abusive. He admitted that his verbal aggression was a problem for others as well as for himself.

After 2 weeks, results of functional assessment were discussed with Tim and his caregivers. The highest score on the QABF was found on the escape/avoidance scale followed by attention and physical discomfort. The target behavior exclusively occurred during situations in which others were present and was usually triggered by a demand from a caregiver, being teased by peers, and when being criticized. In the afternoon he often showed signs of sleepiness. He almost always was reprimanded by caregivers when he was verbally abusive and was removed from the situation. Data from the scatter plot showed that the likelihood of occurrence of the target behavior was greatest on Mondays and lowest on weekends. Furthermore, verbal aggression was observed most often between 7 a.m. and 9 a.m., 1 p.m. and 3 p.m., and 6 p.m. and 7 p.m. These were periods in which more demands were placed on Tim (e.g., getting up, work) and in which he participated in a crowded group, such as mealtimes. Caregivers thought that his verbal aggression resulted from frustration and anger, that he was spoiled, and that he did not want to comply with instructions.

The functional assessment suggested that his verbal abuse was a type of escape-avoidance motivated behavior elicited by demands and crowded social situations in which he sometimes was teased. The behavior was most probably maintained by intermittent escape from demands and social situations. Antecedents included situations in which demands were placed on Tim or when he was criticized. Likely EOs included crowded situations as well as sleepiness. There were no symptoms of a psychiatric disorder or medical problems. He also slept for a normal amount of 9 to 10 hr each night. The mean baseline frequency of verbal abuse was 3 per day (range: 0–5).

TREATMENT

Treatment was initially implemented for 2 weeks. Modifying EOs consisted of presenting Tim with as many choices between work tasks, eating a meal with the group or in his room, shortening delays before giving him prompts and help during work, and preventing unwanted group interactions as much as possible. Alternate behavior was strengthened by giving Tim opportunities for a break upon a socially appropriate request and increased rates of social attention for socially appropriate behaviors and on-task behavior. In order to modify the contingencies maintaining his tantrums, he was not sent to his room, outside, or to his home after verbal abuse. When Tim was abusive, his caregiver would join Tim in another room, instruct him to relax, and, after he had calmed, discuss with him alternative and appropriate ways to cope with the situation that elicited the behavior. Tim was not verbally punished and his caregiver's tone was emotionally neutral. Negative self-statements (e.g., "I am a bad person") were redirected in positive self-statements (e.g., "I can manage it"). After Tim had calmed, both his caregiver and Tim reentered the room, and Tim was redirected back to his task. Following Tim's redirection, his caregiver paid extra attention to Tim for the remainder of the time period. Initial treatment resulted in a marked decrease of rate of verbal aggression to a mean of 0.8 per day (range: 0–2). This effect was maintained during observations scheduled 3 months following the start of treatment.

CONCLUSION

In this chapter, I briefly reviewed functional analysis methodology for individuals with developmental disabilities. Developments within the functional analytic approach have undoubtedly led to an increase in our understanding of what may cause problem behavior as well as to the design of effective and humane treatments. Several meta-analyses in this area have shown that behavioral treatments based on outcomes of a functional analysis have larger effect sizes in terms of a reduction in the target behavior than treatments that are not based on such an analysis (see Campbell, 2003; Carr et al., 1999; Didden et al., 1997; Didden, Korzilius, van Oorsouw, & Sturmey, 2006).

Function-based treatments are the first choice when behavioral function of problem behavior is reliably identified. Some limitations and strengths of various methods of descriptive and experimental analyses have been noted (Carr et al., 2004; Iwata, Kahng, Wallace, & Lindberg, 2000; Sturmey, 1995). First, the function is not identified in cases in which the problem behavior shows an undifferentiated pattern across conditions and situations. Furthermore, functional analysis may not be possible if the individual is prevented from exhibiting a target behavior (e.g., the individual is restrained to prevent life-threatening self-injury or aggression) or if an individual lives in a highly restricted environment (e.g., separation and isolation to prevent dangerous aggression). Non-function-based treatments, such as skills teaching and differential reinforcement of appropriate behavior, should then be considered in an attempt to change the individual's situation and enhance the individual's future perspectives. In some cases, punishment-based procedures should be considered. For example, Duker and Seys (2000) described highly effective treatment using aversive stimuli for chronic and life-threatening self-injurious behavior in individuals who were mechanically restrained. In these cases, function of the behavior could not be identified and other treatments, including pharmacological treatments, had failed.

Functional analysis methodology has been validated in children and adults with profound to moderate developmental disability and autism. In the scientific literature less attention has been paid to individuals living relatively independent in the community or at home with their family, those with dual diagnosis, and/or those showing low-frequency and high-intensity problem behaviors (e.g., fire setting, sexual offending). At present, the status of functional analysis methodology in these target groups remains unclear, although some studies have been published indicating that this approach is valid for use in community settings, such as group and family homes (Feldman, Condillac, Tough, Hunt, & Griffiths, 2002; O'Reilly, Lancioni, & Taylor, 1999). The functional analytic approach has been successfully adapted for use with individuals with emotional and behavioral disorders (Lewis, Scott, & Sugai, 1994; Newcomer & Lewis, 2004; Reed, Thomas, Sprague, & Horner, 1997). However, its relationship to treatment effectiveness in this group remains to be determined. A meta-analysis by Gresham et al. (2004) showed that effect sizes for non-function-based treatments did not significantly differ from those of function-based treatments in children and youngsters with average cognitive abilities.

Future studies should address verbal behaviors, including private verbal behavior such as thoughts and feelings (Friman et al., 1998). For example, depressive (verbal) behaviors (Sturmey, 2005), psychotic-type behavior (Mace & Lalli, 1991), and obsessive-compulsive-type behaviors (Didden et al., In press) should be further targeted for functional analysis, as the same learning processes that may cause aggression and self-injury may also underlie these behaviors. As a result, functional analysis methodology may be successfully integrated in cognitive behavioral treatment programs, such as anger management, that are increasingly being used in clinical and outpatient settings (e.g., Rose, West, & Clifford, 2000). For this purpose, existing functional analysis methodology should be further adapted and refined for use with individuals with verbal abilities and mild cognitive impairments.

REFERENCES

Adelinis, J., & Hagopian, L. (1999). The use of symmetrical 'Do' and 'Don't' requests to interrupt ongoing activities. *Journal of Applied Behavior Analysis, 32*, 519–523.

Adelinis, J., Piazza, C., Fisher, W., & Hanley, G. (1997). The establishing effects of client location on self-injurious behavior. *Research in Developmental Disabilities, 18*, 383–391.

Berkson, G., & Mason, W. (1963). Stereotyped movements of mental defectives: Situational effects. *American Journal of Mental Deficiency, 68*, 409–412.

Bijou, S., Peterson, R., & Ault, M. (1968). A method to integrate descriptive and experimental field studies at the level of data and empirical concepts. *Journal of Applied Behavior Analysis, 1*, 175–191.

Bodfish, J., & Konarski, E. (1992). Reducing problem behaviors in a residential unit using structured analysis and staff management procedures: A preliminary study. *Behavioral Residential Treatment, 7*, 225–234.

Bosch, J., van Dyke, D., Smith, S., & Poulton, S. (1997). Role of medical conditions in the exacerbation of self-injurious behavior. *Mental Retardation, 35*, 124–130.

Campbell, J. (2003). Efficacy of behavioral interventions for reducing problem behaviors in persons with autism: A quantitative synthesis of single-subject research. *Research in Developmental Disabilities, 24*, 120–138.

Carr, E., & Durand, M. (1985). Reducing behavior problems through functional communication training. *Journal of Applied Behavior Analysis, 18*, 111–126.

Carr, E., Horner, R., Turnbull, A., Marquis, J., McLaughlin, D., McAtee, M., et al. (1999). *Positive behavior support for people with developmental disabilities: A research synthesis*. Washington, DC: American Association on Mental Retardation.

Carr, E., Innis, J., Blakeley-Smith, A., & Vasdev, S. (2004). Challenging behavior: Research design and measurement issues. In E. Emerson, C. Hatton, T. Thompson, & T. Parmenter (Eds.), *International handbook of applied research in intellectual disabilities* (pp. 423–441). London: Wiley & Sons.

Carr, E., Levin, L., McConnachie, G., Carlson, J., Kemp, D., & Smith, C. (1994). *Communication-based intervention for problem behavior*. Baltimore: Paul H. Brookes.

Carr, E., & McDowell, J. (1980). Social control of self-injurious behavior of organic etiology. *Behavior Therapy, 11*, 402–409.

Carr, E., Smith, C., Giacin, T., Whelan, B., & Pancari, J. (2003). Menstrual discomfort as a biological setting event for severe problem behavior: Assessment and intervention. *American Journal on Mental Retardation, 108*, 117–133.

Carr, J., & Britton, L. (1999). Idiosyncratic effects of noncontingent reinforcement on problematic speech. *Behavioral Interventions, 14*, 37–43.

Carr, J., Coriaty, S., Wilder, D., Gaunt, B., Dozier, C., Britton, L., Avina, C., & Reed, C. (2000). A review of "noncontingent" reinforcement as treatment for the aberrant behavior of individuals with developmental disabilities. *Research in Developmental Disabilities, 21*, 377–391.

Cooper, L., Wacker, D., McComas, J., Brown, K., Peck, S., Richman, D., Drew, J., Frischmeyer, P., & Millard, T. (1995). Use of component analysis to identify active variables in treatment packages for children with feeding disorders. *Journal of Applied Behavior Analysis, 28*, 139–153.

Cowdery, G., Iwata, B., & Pace, G. (1990). Effects and side effects of DRO as treatment for self-injurious behavior. *Journal of Applied Behavior Analysis, 23*, 497–506.

Cunningham, E., & O'Neill, R. (2000). Comparison of results of functional assessment and analysis methods with young children with autism. *Education and Training in Mental Retardation and Developmental Disabilities, 35*, 406–414.

Dekker, M., Koot, H., van der Ende, J., & Verhulst, F. (2002). Emotional and behavioral problems in children and adolescents with and without intellectual disability. *Journal of Child Psychology and Psychiatry, 43*, 1087–1098.

Derby, K., Wacker, D., DeRaad, A., Ulrich, S., Asmus, J., Harding, J., et al. (1997). The long-term effects of functional communication training in home settings. *Journal of Applied Behavior Analysis, 30*, 507–531.

Derby, K., Wacker, D., Sasso, G., Steege, M., Northup, J., Gigrand, K., et al. (1992). Brief functional assessment techniques to evaluate aberrant behavior in an outpatient setting: A summary of 79 cases. *Journal of Applied Behavior Analysis, 25*, 713–721.

Didden, R., Curfs, L., van Driel, S., & de Moor, J. (2002). Sleep problems in children and young adults with developmental disabilities: Home-based functional assessment and treatment. *Journal of Behavior Therapy and Experimental Psychiatry, 33*, 49–58.

Didden, R., Duker, P., & Korzilius, H. (1997). Meta-analytic study on treatment effectiveness for problem behaviors with individuals who have mental retardation. *American Journal on Mental Retardation, 101*, 387–399.

Didden, R., Korzilius, H., & Curfs, L. (In press). Skin-picking in individuals with Prader-Willi syndrome: Prevalence, functional assessment, and its comorbidity with compulsive and self-injurious behaviors. *Journal of Applied Research in Intellectual Disabilities*.

Didden, R., Korzilius, H., van Oorsouw, W., & Sturmey, P. (2006). Behavioral treatment of problem behaviors in individuals with mild mental retardation: A meta-analysis of single case studies. *American Journal on Mental Retardation, 111*, 290–298.

Ducharme, J., & van Houten, R. (1994). Operant extinction in the treatment of severe maladaptive behavior: Adapting research to practice. *Behavior Modification, 18*, 139–179.

Duker, P. (1975). Behaviour therapy of self-biting in a Lesch-Nyhan patient. *Journal of Mental Deficiency Research, 19*, 11–19.

Duker, P., Didden, R., & Sigafoos, J. (2004). *One-to-one training: Instructional procedures for learners with developmental disabilities.* Austin, TX: Pro-Ed.

Duker, P., & Rasing, E. (1989). Effects of redesigning the physical environment on self-stimulation and on-task behavior in three autistic-type developmentally disabled individuals. *Journal of Autism and Developmental Disorders, 19*, 449–460.

Duker, P., & Seys, D. (2000). A quasi-experimental study on the effect of electrical aversion treatment on imposed mechanical restraint for severe self-injurious behavior. *Research in Developmental Disabilities, 21*, 235–242.

Dunlap, G., Kern-Dunlap, L., Clarke, S., & Robbins, F. (1991). Functional assessment, curricular revision, and severe behavior problems. *Journal of Applied Behavior Analysis*, *24*, 387–397.

Durand, M., & Carr, E. (1987). Social influences on 'self-stimulatory' behavior: Analysis and treatment implications. *Journal of Applied Behavior Analysis*, *20*, 119–132.

Durand, M., & Crimmins, D. (1988). Identifying variables maintaining self-injurious behavior. *Journal of Autism and Developmental Disorders*, *18*, 111–126.

Emerson, E., Kiernan, C., Alborz, A., Reeves, D., Mason, H., Swarbrick, R., Mason, L., & Hatton, C. (2001). The prevalence of challenging behaviors: A total population study. *Research in Developmental Disabilities*, *22*, 77–93.

Emerson, E., Reeves, D., Thompson, S., Henderson, D., Robertson, J., & Howard, D. (1996). Time-based lag sequential analysis and the functional assessment of challenging behaviour. *Journal of Intellectual Disability Research*, *40*, 260–274.

Favell, J., McGimsey, J., & Schell, R. (1982). Treatment of self-injury by providing alternate sensory activities. *Analysis and Intervention in Developmental Disabilities*, *2*, 83–104.

Feldman, M., Condillac, R., Tough, S., Hunt, S., & Griffiths, D. (2002). Effectiveness of community positive behavioral intervention for persons with developmental disabilities and severe behavior disorders. *Behavior Therapy*, *33*, 377–398.

Fisher, W., Hinnes, H., & Piazza, C. (1996). On the verbal content of verbal attention. *Journal of Applied Behavior Analysis*, *29*, 235–238.

Foster-Johnson, L., Ferro, J., & Dunlap, G. (1994). Preferred curricular activities and reduced problem behaviors in students with intellectual disabilities. *Journal of Applied Behavior Analysis*, *27*, 493–504.

Friman, P., Hayes, S., & Wilson, K. (1998). Why behavior analysts should study emotion: The example of anxiety. *Journal of Applied Behavior Analysis*, *31*, 137–156.

Fuchs, C., & Benson, B. (1995). Social information processing by aggressive and nonaggressive men with mental retardation. *American Journal on Mental Retardation*, *100*, 244–252.

Gardner, W., & Cole, C. (1993). Aggression and related conduct disorders: Definition, assessment and treatment. In R. Barrett & J. Matson (Eds.), *Psychopathology in the mentally retarded* (pp. 213–252). Needham Heights, MA: Allyn & Bacon.

Gardner, W., Graeber-Whalen, J., & Ford, D. (2001). Behavioral therapies. In A. Dosen & K. Day (Eds.), *Treating mental illness and behavior disorders in children and adults with mental retardation* (pp. 69–100). Washington, DC: American Psychiatric Press.

Gresham, F., McIntyre, L., Olson-Tinker, H., Dolstra, L., McLaughlin, V., & Van, M. (2004). Relevance of functional behavioral assessment research for school-based interventions and positive behavioral support. *Research in Developmental Disabilities*, *25*, 19–37.

Groden, G. (1989). A guide for conducting a comprehensive behavioral analysis of a target behavior. *Journal of Behavior Therapy and Experimental Psychiatry*, *20*, 163–169.

Groden, G., & Lantz, S. (2001). The reliability of the Detailed Behavior Report (DBR) in documenting functional assessment observations. *Behavioral Interventions*, *16*, 15–25.

Hagopian, L., Fisher, W., & Legacy, S. (1994). Schedule effects of noncontingent reinforcement on attention-maintained destructive behavior in identical quadruplets. *Journal of Applied Behavior Analysis*, *27*, 317–325.

Hagopian, L., Fisher, W., Sullivan, M., Acquisto, J., & LeBlanc, L. (1998). Effectiveness of functional communication training with and without extinction and punishment: A summary of 21 inpatient cases. *Journal of Applied Behavior Analysis*, *30*, 229–237.

Hanley, G., Iwata, B., & McCord, B. (2003). Functional analysis of problem behavior: A review. *Journal of Applied Behavior Analysis*, *36*, 147–185.

Hanley, G., Piazza, C., Fisher, W., & Adelinis, J. (1997). Stimulus control and resistance to extinction in attention-maintained self-injurious behavior. *Journal of Applied Behavior Analysis*, *18*, 251–260.

Hartley, S., & MacLean, W. (2005). Perceptions of stress and coping strategies among adults with mild mental retardation: Insight into psychological stress. *American Journal on Mental Retardation, 110,* 285–297.

Hastings, R. (1997). Staff beliefs about the challenging behaviors of children and adults with mental retardation. *Clinical Psychology Review, 17,* 775–790.

Higbee, T., Chang, S., & Endicott, K. (2005). Noncontingent access to preferred sensory stimuli as a treatment for automatically reinforced stereotypy. *Behavioral Interventions, 20,* 177–184.

Horner, R. (1980). The effects of an environmental "enrichment" program on the behavior of institutionalized profoundly retarded children. *Journal of Applied Behavior Analysis, 13,* 473–491.

Horner, R., Day, M., & Day, J. (1997). Using neutralizing routines to reduce problem behavior. *Journal of Applied Behavior Analysis, 30,* 601–614.

Horner, R., Day, M., Sprague, J., O'Brien, M., & Heathfield, L. (1991). Interspersed requests: A nonaversive procedure for reducing aggression and self-injury during instruction. *Journal of Applied Behavior Analysis, 24,* 265–278.

Horner, R., Vaughn, B., Day, H., & Ard, W. (1996). The relationship between setting events and problem behavior. In L. Koegel, R. Koegel, & G. Dunlap (Eds.), *Positive behavioral support* (pp. 381–402). Baltimore, MD: Paul H. Brookes.

Iwata, B., Dorsey, M., Slifer, K., Bauman, K., & Richman, G. (1982). Toward a functional analysis of self-injury. *Analysis and Intervention in Developmental Disabilities, 2,* 3–20.

Iwata, B., Kahng, S., Wallace, M., & Lindberg, J. (2000). The functional analysis model of behavioral assessment. In J. Austin & J. Carr (Eds.), *Handbook of applied behavior analysis* (pp. 61–89). Reno, NV: Context Press.

Iwata, B., Pace, G., Dorsey, M., et al. (1994). The functions of self-injurious behavior: An experimental-epidemiological analysis. *Journal of Applied Behavior Analysis, 27,* 215–240.

Iwata, B., Pace, G., Kalsher, M., Cowdery, G., & Cataldo, M. (1990). Experimental analysis and extinction of self-injurious escape behavior. *Journal of Applied Behavior Analysis, 23,* 11–27.

Kahng, S., Iwata, B., Fischer, S., Page, T., Treadwill, K., Williams, D., et al. (1998). Temporal distributions of problem behavior based on scatter plot analysis. *Journal of Applied Behavior Analysis, 31,* 593–604.

Kennedy, C., & Souza, G. (1995). Functional analysis and treatment of eye poking. *Journal of Applied Behavior Analysis, 28,* 27–37.

Kern, L., Mauk, J., Marder, T., & Mace, C. (1995). Functional analysis and intervention of breath holding. *Journal of Applied Behavior Analysis, 28,* 339–340.

Lalli, J., Casey, S., & Kates, K. (1995). Reducing escape behavior and increasing task completion with functional communication training, extinction, and response chaining. *Journal of Applied Behavior Analysis, 28,* 261–268.

Lalli, J., Livezey, K., & Kates, K. (1996). Functional analysis and treatment of eye poking. *Journal of Behavior Analysis, 29,* 129–132.

Lerman, D., & Iwata, B. (1993). Descriptive and experimental analyses of variables maintaining self-injurious behavior. *Journal of Applied Behavior Analysis, 26,* 293–319.

Lerman, D., & Iwata, B. (1996a). Developing a technology for the use of operant extinction in clinical settings: An examination of basic and applied research. *Journal of Applied Behavior Analysis, 29,* 345–382.

Lerman, D., & Iwata, B. (1996b). A methodology for distinguishing between extinction and punishment effects associated with response blocking. *Journal of Applied Behavior Analysis, 29,* 231–233.

Lewis, T. J., Scott, T. M., & Sugai, G. (1994). The Problem Behavior Questionnaire: A teacher-based instrument to develop functional hypotheses of problem behavior in general education classrooms. *Diagnostique, 19*, 103–115.

Lockwood, K., Maenpaa, M., & Williams, D. (1997). Long-term maintenance of a behavioral alternative to surgery for severe vomiting and weight loss. *Journal of Behavior Therapy and Experimental Psychiatry, 28*, 105–112.

Lovaas, O., Freitag, G., Gold, V., & Kassorla, I. (1965). Experimental studies in childhood schizophrenia: Analysis of destructive behavior. *Journal of Experimental Child Psychology, 2*, 67–84.

Mace, F., & Lalli, J. (1991). Linking descriptive and experimental analysis in the treatment of bizarre speech. *Journal of Applied Behavior Analysis, 24*, 553–562.

Matson, J., Bamburg, J., Cherry, K., & Paclawskyj, T. (1999). A validity study on the Questions About Behavioral Function (QABF) scale: Predicting treatment success for self-injury, aggression, and stereotypes. *Research in Developmental Disabilities, 20*, 163–176.

Matson, J., & Vollmer, T. (1995). *Questions About Behavioral Function (QABF)*. Baton Rouge, LA: Disability Consultants, LLC.

McAfee, J. (1987). Classroom density and the aggressive behavior of handicapped children. *Education and Treatment of Children, 10*, 134–145.

McAtee, M., Carr, E., & Schulte, C. (2004). A Contextual Assessment Inventory for problem behavior: Initial development. *Journal of Positive Behavior Interventions, 6*, 148–165.

McClintoch, K., Hall, S., & Oliver, C. (2003). Risk markers associated with challenging behavior in people with intellectual disability: A meta-analysis. *Journal of Intellectual Disability Research, 47*, 405–416.

McGill, P. (1999). Establishing operations: Implications for the assessment, treatment, and prevention of problem behavior. *Journal of Applied Behavior Analysis, 32*, 393–418.

Moss, J., Oliver, C., Hall, S., Arron, K., Sloneem, J., & Petty, J. (2005). The association between environmental events and self-injurious behaviour in Cornelia de Lange syndrome. *Journal of Intellectual Disability Research, 49*, 269–277.

Mullins, J., & Christian, L. (2001). The effects of progressive relaxation training on the disruptive behavior of a boy with autism. *Research in Developmental Disabilities, 22*, 449–462.

Newcomer, L., & Lewis, T. (2004). Functional behavioral assessment: An investigation of assessment reliability and effectiveness of function-based interventions. *Journal of Emotional and Behavioral Disorders, 12*, 168–181.

O'Neill, R., Horner, R., Albin, R., Storey, K., Sprague, J., & Newton, J. (1997). *Functional analysis of problem behavior: A practical guide* (2nd ed.). Pacific Grove, CA: Brooks/Cole.

O'Reilly, M. (1997). Functional analysis of episodic self-injury correlated with recurrent otitis media. *Journal of Applied Behavior Analysis, 30*, 165–167.

O'Reilly, M., & Lancioni, G. (2000). Response covariation of escape-maintained aberrant behavior correlated with sleep deprivation. *Research in Developmental Disabilities, 21*, 125–136.

O'Reilly, M., Lancioni, G., O'Donoghue, D., Sigafoos, J., Lacey, C., & Edrisinha, C. (2004). Teaching social skills to adults with intellectual disabilities: A comparison of external control and problem-solving interventions. *Research in Developmental Disabilities, 25*, 399–412.

O'Reilly, M., Lancioni, G., & Taylor, I. (1999). An empirical analysis of two forms of extinction to treat aggression. *Research in Developmental Disabilities, 20*, 315–325.

Pace, G., Iwata, B., Cowdery, G., Andree, P., & McIntyre, T. (1993). Stimulus (instructional) fading during extinction of self-injurious escape-behavior. *Journal of Applied Behavior Analysis, 26*, 205–212.

Paclawskyj, T., Matson, J., Rush, K., Smalls, Y., & Vollmer, T. (2000). Questions About Behavioral Function (QABF): A behavioral checklist for functional assessment of aberrant behavior. *Research in Developmental Disabilities, 21*, 223–229.

Paclawskyj, T., Matson, J., Rush, K., Smalls, Y., & Vollmer, T. (2001). Assessment of the convergent validity of the Questions About Behavioral Function scale with analogue functional analysis and the Motivation Assessment Scale. *Journal of Intellectual Disability Research, 45*, 484–494.

Pert, C., Jahoda, A., & Squire, J. (1999). Attribution of intent and role-taking: Cognitive factors as mediators of aggression with people who have mental retardation. *American Journal on Mental Retardation, 104*, 399–409.

Piazza, C., Contrucci, S., Hanley, G., & Fisher, W. (1997). Nondirective prompting and noncontingent reinforcement in the treatment of destructive behavior during hygiene routines. *Journal of Applied Behavior Analysis, 31*, 165–189.

Piazza, C., Fisher, W., Hanley, G., LeBlanc, L., Worsdell, A., Lindauer, S., & Keeney, K. (1998). Treatment of pica through multiple analyses of its reinforcing functions. *Journal of Applied Behavior Analysis, 31*, 165–189.

Piazza, C., Moes, D., & Fisher, W. (1996). Differential reinforcement of alternative behavior and demand fading in the treatment of escape-maintained destructive behavior. *Journal of Applied Behavior Analysis, 29*, 569–572.

Rast, J., & Johnston, J. (1986). Social versus dietary control of rumination by mentally retarded persons. *American Journal of Mental Deficiency, 90*, 464–467.

Reed, H., Thomas, E., Sprague, J., & Horner, R. (1997). The student guided functional assessment interview: An analysis of student and teacher agreement. *Journal of Behavioral Education, 7*, 33–49.

Rincover, A., Cook, R., Peoples, A., & Packard, D. (1979). Sensory extinction and sensory reinforcement principles of programming multiple adaptive behavior change. *Journal of Applied Behavior Analysis, 12*, 221–233.

Rojahn, J. (1994). Epidemiology and topographic taxonomy of self-injurious behavior. In D. Gray & T. Thompson (Eds.), *Destructive behaviors in developmental disabilities: Diagnosis and treatment* (pp. 49–67). Thousand Oaks, CA: Sage Publications.

Rojahn, J., & Sisson, L. (1990). Stereotyped behavior. In J. Matson (Ed.), *Handbook of behavior modification with the mentally retarded* (pp. 101–113). New York: Plenum Press.

Romaniuk, C., Miltenberger, R., Conyers, C., Jenner, N., Jurgens, M., & Ringenberg, C. (2002). The influence of activity choice on problem behaviors maintained by escape versus attention. *Journal of Applied Behavior Analysis, 35*, 349–362.

Rose, J., West, C., & Clifford, D. (2000). Group interventions for anger in people with intellectual disabilities. *Research in Developmental Disabilities, 21*, 171–181.

Shogren, K., Faggella-Luby, M., Bae, S., & Wehmeyer, M. (2004). The effect of choice-making as an intervention for problem behavior: A meta-analysis. *Journal of Positive Behavior Interventions, 6*, 228–237.

Shogren, K., & Rojahn, J. (2003). Convergent reliability and validity of the Questions About Behavioral Function and the Motivation Assessment Scale: A replication study. *Journal of Developmental and Physical Disabilities, 15*, 367–375.

Sigafoos, J., Arthur, M., & O'Reilly, M. (2003). *Challenging behavior and developmental disability*. London: Whurr.

Sigafoos, J., & Meikle, B. (1996). Functional communication training for the treatment of multiple determined challenging behavior in two boys with autism. *Behavior Modification, 20*, 60–84.

Smith, R., Iwata, B., Goh, H., & Shore, B. (1995). Analysis of establishing operations for self-injury maintained by escape. *Journal of Applied Behavior Analysis, 28*, 515–535.

Sprague, J., Holland, K., & Thomas, K. (1997). The effect of noncontingent sensory reinforcement, contingent sensory reinforcement, and response interruption on stereotypical
and self-injurious behavior. *Research in Developmental Disabilities, 18*, 61–77.

Steen, P., & Zuriff, G. (1977). The use of relaxation training in the treatment of self-
injurious behavior. *Journal of Behavior Therapy and Experimental Psychiatry, 8*,
47–48.

Sturmey, P. (1995). Analog baselines: A critical review of the methodology. *Research in
Developmental Disabilities, 16*, 269–284.

Sturmey, P. (1996). *Functional analysis in clinical psychology.* London: Wiley.

Sturmey, P. (2002). Mental retardation and concurrent psychiatric disorder: Assessment and
treatment. *Current Opinion in Psychiatry, 15*, 489–495.

Sturmey, P. (2005). Behavioural formulation and treatment of depression in people with
mental retardation: Formulations and interventions. In P. Sturmey (Ed.), *Mood disorders
in people with mental retardation* (pp. 293–315). Kingston, NY: NADD Press.

Taylor, J., & Carr, E. (1992). Severe problem behaviors related to social interaction. *Behavior Modification, 16*, 305–335.

Touchette, P., MacDonald, R., & Langer, S. (1985). A scatter plot for identifying stimulus
control of problem behavior. *Journal of Applied Behavior Analysis, 18*, 343–351.

Tustin, R. (1995). The effects of advance notice of activity transitions on stereotypic behavior. *Journal of Applied Behavior Analysis, 28*, 91–92.

Vollmer, T., Iwata, B., Zarcone, J., Smith, R., & Mazaleski, J. (1993). The role of attention
in the treatment of attention-maintained self-injurious behavior: Noncontingent reinforcement and differential reinforcement of other behavior. *Journal of Applied Behavior
Analysis, 26*, 9–21.

Vollmer, T., Marcus, B., & Ringdahl, J. (1995). Non-contingent escape as treatment for self-
injurious behavior maintained by negative reinforcement. *Journal of Applied Behavior
Analysis, 28*, 15–26.

Wilder, D., & Carr, J. (1998). Recent advances in the modification of establishing operations
to reduce aberrant behavior. *Behavioral Interventions, 13*, 43–59.

Wilder, D., Draper, R., Williams, W., & Higbee, T. (1997). A comparison of noncontingent
reinforcement, other competing stimulation, and liquid rescheduling for the treatment
of rumination. *Behavioral Interventions, 12*, 55–64.

Yarbrough, S., & Carr, E. (2000). Some relationships between informant assessment and
functional analysis of problem behavior. *American Journal on Mental Retardation, 105*,
130–151.

5

ATTENTION DEFICIT HYPERACTIVITY DISORDER

NANCY NEEF

The Ohio State University

AND

JOHN NORTHUP

University of Iowa

DIAGNOSIS AND RELATED CHARACTERISTICS

Attention Deficit Hyperactivity Disorder (ADHD) is a disorder of unknown etiology with defining characteristics of inattention, overactivity, and impulsivity (American Psychiatric Association [APA], 2000). It is the most prevalent disorder in the school-age population of children in the United States. Although the *DSM-IV* estimates that ADHD affects 3–5% of children, recent population studies have yielded much higher estimates, ranging from 5% to 16% of the school-age population (Rowland et al., 2001, 2002; Scahill & Schwab-Stone, 2000). ADHD is much more common in males than in females; the ratio is approximately 6:1 in clinic-referred samples and 3:1 in community-based samples (Barkley, 1998; Pastor & Reuben, 2002).

Diagnosis is typically made using *DSM-IV* criteria (APA, 2000). The *DSM-IV* uses descriptive criteria to group behaviors into diagnostic categories. A diagnosis of ADHD-predominantly inattentive type is made if at least six of nine listed behaviors indicative of inattention are present. Examples of symptoms include the following: often fails to give close attention to details or makes careless mistakes in schoolwork, work, or other activities; often has difficulty sustaining attention in tasks or play activities; and often does not seem to listen when spoken to directly. These

Functional Analysis in
Clinical Treatment

behaviors must persist for at least 6 months to a degree that is maladaptive and inconsistent with developmental level. A diagnosis of ADHD-predominantly hyperactive-impulsive type is made if at least six of nine listed behaviors indicative of hyperactivity-impulsivity are present. Examples of symptoms include the following: often fidgets with hands or feet or squirms in seat, and often talks excessively. A diagnosis of ADHD-combined type is made if six or more inattentive *and* hyperactive-impulsive symptoms are present. A diagnosis of any of the three types of ADHD also requires the presence of some of the listed symptoms before seven years of age and in two or more settings, as well as clear evidence of clinically significant impairment in social, academic, or occupational functioning. In addition, ADHD is diagnosed only if the symptoms do not occur exclusively during the course of a pervasive developmental disorder, schizophrenia, or other psychotic disorder and are not better accounted for by another mental disorder.

ADHD is chronic and persists over the life span (National Institutes of Health, 1998). Indeed, the results of prospective, longitudinal investigations indicate that for approximately 50% of children with ADHD, problems associated with impulsivity will continue into adulthood (Barkley, Fischer, Edelbrock, & Smallish, 1990; Weiss & Hechtman, 1993). ADHD frequently co-occurs with additional emotional, behavioral, and learning problems. Disruptive behavior disorders such as oppositional defiant disorder and conduct disorder are the most common, and are more frequently correlated with the hyperactive/impulsive type of ADHD. Internalizing disorders, such as depression, anxiety, and learning disabilities, are more frequently associated with the inattentive type of ADHD (Wolraich, Hannah, Baumgaertal, & Feurer, 1998). Approximately 26% of children classified as having a learning disability, 43% of children classified as having an emotional disability, and 40% of children classified as having another health impairment also have a diagnosis of ADHD (Forness, Kavale, Sweeney, & Crenshaw, 1999). Estimates from follow-up studies indicate that more than half of children with ADHD will experience school failure (Barkley et al., 1990). The clinical significance of this is shown by the observation that as many as 46% are suspended from school each year, and approximately one third drop out of school (Barkley, 1998).

PROBLEMS WITH STRUCTURAL APPROACHES

A categorical diagnosis of ADHD may be useful for a variety of reasons. For example, a diagnosis may determine an individual as legally eligible for specific services and accommodations and may also suggest a trial of a particular course of medication. However, a categorical diagnosis of ADHD is of limited utility for directing any treatment options other than medication. There is no commonly accepted measure for diagnosing

ADHD. Diagnosis is typically based on behavior rating scales and indirect measures which have been criticized because the descriptors typically do not have objective anchors, are not operationally defined, and involve subjective judgments that may be imprecise and inconsistent (Atkins & Pelham, 1991; Gulley & Northup, 1997; Kollins, Ehrhardt, & Poling, 2000; Stoner, Carey, Ikeda, & Shinn, 1994). In addition, they do not take into account the context for the behavior, such as the consequences for behaving one way relative to another. Perhaps their most serious limitation is that the topographical features that have traditionally been used to diagnose ADHD are of limited utility in providing insight into the functional properties of the problem behavior of these individuals (Critchfield & Kollins, 2001).

The development of functional analysis methodologies has contributed to the current emphasis on function rather than topography. This approach has been developed with other populations, principally severe problem behavior of individuals with developmental disabilities, but has not characterized approaches to the diagnosis and treatment of ADHD.

FUNCTIONAL ANALYTIC APPROACH TO ASSESSMENT

The primary goal of a functional analytic model of ADHD is to understand environmental conditions that may maintain or exacerbate specific problem behaviors associated with ADHD in order to develop effective, individually tailored interventions. Thus, the primary purpose of a functional analysis and assessment approach to ADHD is to identify specific environmental conditions that have a large effect on the target behaviors and that may be modified.

The importance of developing individualized treatments for children with a diagnosis of ADHD and associated behavior problems has been increasingly recognized in recent years. Individual differences in the type and severity of problem behaviors associated with ADHD are the rule rather than the exception. The inherent heterogeneity of a diagnosis of ADHD, as shown by the three subtypes and 21 listed "symptoms," virtually precludes the development of a single universal treatment to match the diagnosis. Rather, a functional model attempts to match appropriate environmental modifications to specific problem behaviors presented by any individual child. As DuPaul, Eckert, and McGoey (1997) wrote, "One size doesn't fit all."

Currently, ADHD is assumed to be of a neurobiological origin. A functional approach to ADHD is not necessarily inconsistent with this assumption, as a functional approach makes no assumptions concerning etiology. Rather, the model suggests that many problem behaviors associated with

ADHD may not be due to biological factors alone, but rather may be due to an interaction between a child's biological characteristics and specific environmental events (DuPaul et al., 1997; Northup et al., 1999).

A functional analysis of ADHD proceeds through a series of steps. First, the specific ADHD subtype must be identified, which may suggest a general class of treatments. For example, treatments for inattention may generally take a different form than those for impulsivity (as described later). Second, it is necessary to identify the specific behaviors that led to the diagnosis (e.g., often leaving seat, talking excessively), as well as their frequency and severity. Behaviors that have been most commonly included are out-of-seat, inappropriate vocalizations, playing with objects, and off-task. The first three behaviors are often combined and may form a response class that may be collectively referred to as disruptive behavior. However, each behavior is first analyzed separately, as it is possible that the different behaviors may serve different functions (Derby et al., 1994). For example, it is not uncommon for off-task to serve a different function than other more overtly disruptive behaviors (Northup & Gully, 2001). Third, it is essential to determine if any co-occurring, emotional, behavioral, or learning problems exist. Co-occurring behaviors, such as noncompliance, aggression, and social and academic problems, are often of greater immediate concern than the core symptoms that initially led to a diagnosis of ADHD.

The preceding steps translate a categorical diagnosis of ADHD into a functional model that can provide the foundation for the development of individualized treatments. After specific target behaviors have been identified, a functional analysis can proceed to identify when, where, and for whom these behaviors are of most concern—that is, the specific context in which the problem behaviors occur. Most problem behaviors associated with ADHD are by definition context specific. For example, young children are expected and encouraged to talk, run, and jump on the playground, but not in the classroom. One advantage of this approach is that the model assumes problem behaviors to be context specific and goes beyond addressing core symptoms as universal and chronic problems.

ASSESSMENT METHODS

The functional analysis model of assessment and treatment was developed in the area of developmental disabilities (Iwata, Dorsey, Slifer, Bauman, & Richman, 1982/1994) and has been extended to other populations and behavior problems; it remains an active and emerging area of research. Nevertheless, a number of studies have demonstrated the possibility of extending functional analysis procedures across diverse groups of children and to disruptive behavior problems such as those commonly associated with ADHD (Broussard & Northup, 1995, 1997; Umbreit, 1995).

As noted, the first step of a functional analysis of ADHD is to identify the specific context in which problem behavior or behaviors occur. This is of particular importance, as assessment should occur either directly in that context or under conditions as representative of that context as possible. For children with ADHD this is often but not always the classroom and during academic tasks. Thus, a descriptive analysis is highly recommended prior to a functional analysis for children with ADHD. In addition to identification of the specific context in which problem behaviors occur, a descriptive analysis may also identify specific forms of teacher attention, peer attention, and escape as well as other idiosyncratic variables unique to a particular classroom.

Although problem behaviors related to ADHD most often occur in the classroom, a descriptive analysis may also document problems occurring in other settings. For example, Northup, Gulley, Edwards, and Fountain (2001) documented significant social withdrawal during recess for one participant when she received stimulant medication for ADHD behaviors that were of concern only in another setting. For another student with a diagnosis of ADHD, Kodak, Grow, and Northup (2004) conducted a functional analysis of elopement that occurred only in the context of kickball. In subsequent sections we identify two approaches to conducting a functional analysis of behavior associated with ADHD: Traditional approaches based on Iwata et al.'s model and those based on reinforcer dimensions and temporal discounting.

TRADITIONAL FUNCTIONAL ANALYSIS PROCEDURES

Most functional analyses that have included children with ADHD have used procedures generally based on those described by Iwata et al. (1982/1994). For example, Umbreit (1995) conducted a brief classroom-based functional analysis for an 8-year-old boy with a diagnosis of ADHD. All assessment conditions were procedurally consistent with those of Iwata et al. and included contingent teacher and peer attention and escape from academic tasks. The authors concluded that the child's behavior was maintained by escape from academic tasks. Subsequently, a curriculum-based assessment was conducted to evaluate the effects of several antecedent events. The results of both assessments were combined to develop an intervention that significantly reduced the child's disruptive behavior in the classroom. Boyajian, DuPaul, Handler, Eckert, and McGoey (2001) conducted brief functional analyses in a preschool classroom for three children aged 4 and 5 years who were described as at risk for ADHD. This study demonstrated that useful functional analyses can be conducted for preschool children who are at risk for ADHD, but who have no other developmental delays. The authors noted the potential utility of functional analysis for early intervention for ADHD.

Procedural Variations

Iwata et al.'s (1982/1994) functional analysis procedures have proven useful for other populations. However, the unique characteristics associated with various other populations and problem areas necessitate some procedural variations. Thus, some functional analysis procedures that have been described for children with ADHD have varied from those of Iwata et al. in important ways.

Several studies involving children with ADHD have described functional analysis conditions in which all experimental manipulations were conducted in the context of a child being asked to work independently on an academic task. For example, Gunter, Jack, Shores, Carrell, and Flowers (1993) conducted classroom-based functional analyses. They compared the attention and ignore conditions while the participants were engaged in independent seatwork. Northup et al. (1995) conducted functional analyses in regular education classrooms for three students referred for ADHD-related disruptive behavior. Following a descriptive analysis, each student was exposed to three experimental conditions: teacher attention, peer attention, and escape. All conditions began with a request for the student to be seated and quietly complete an academic assignment. Disruptive behavior occurred most frequently during a different condition for each of the three students.

If experimental manipulations are conducted in the context of asking a student to complete an academic task, it is important to carefully identify task difficulty. The types of tasks and demands that are most likely to occasion aberrant behavior in the area of severe developmental disabilities may be overt and relatively discernible by direct observation, such as dressing or self-care. However, tasks like reading and math that may be aversive to typically developing children in educational settings are more subtle and are characterized by many relatively minor variations. For example, three-digit addition problems may be too difficult for a child, but two-digit addition problems may not be. Alternatively, academic tasks may be too easy for some children and may be aversive if they are experienced as boring. Thus, it is recommended that academic assessment information be obtained from a variety of sources, including teachers and permanent work products, and from other assessment methods concerning the difficulty level of various academic tasks. Curriculum Based Measurement (CBM; Shinn, 1995) is a well-researched procedure that is used to measure academic performance on materials taken directly from the student's curriculum. Basic CBM data are derived from brief (1- to 3-min) fluency and accuracy measures in reading, math, spelling, or written expression. CBM has the advantage of providing systematic procedures for conducting brief academic assessments that can be administered repeatedly. CBM has also been very useful, if not essential, for identifying the difficulty level of various academic tasks to be used in analog assessments for children in regular educational settings.

Sometimes all experimental conditions are conducted in the context of an academic task, and task difficulty may be manipulated as an antecedent manipulation. For example, an escape condition may be conducted under both an instructional level and a difficult task condition, while all other conditions are conducted using an instructional level task (Northup & Gully, 2001). An instructional level task (typically defined as completed at 70–90% accuracy) is assumed to represent an optimal classroom condition. However, a difficult task condition (completed with less than 60% accuracy) that results in high levels of disruptive behavior may suggest a need for curriculum modifications.

Peers

Children with ADHD are by definition easily distracted. Peers, especially, may easily distract these children and may often reinforce problem behaviors. Broussard and Northup (1997) conducted experimental functional analyses for 4 students with ADHD-related disruptive behaviors in a regular education classroom. A previous descriptive analysis indicated that teacher attention, peer attention, and escape from academic assignments were all likely to follow disruptive behavior. A subsequent experimental functional analysis indicated that the presence of peers and contingent attention from peers both influenced disruptive behavior for each student. Subsequent interventions that allowed peer interactions contingent on appropriate behavior were shown to reduce disruptive behavior to zero and to increase on-task behavior for all students.

Negative Reinforcement

Negative reinforcement contingencies, such as escape from academic tasks, may also take a variety of forms for a child with ADHD in a typical classroom. It may be uncommon for a classroom teacher to literally remove an academic demand following disruptive behavior, as is the typical procedure in a traditional functional analysis. However, students may escape academic demands in a typical classroom in a variety of other ways. Perhaps most commonly they may simply remain off-task. In some instances it may appear as if a teacher's presentations of academic tasks have completely extinguished, as when work incompletion is ignored, especially if the student is not otherwise disruptive. As a possible analogue to a teacher ignoring work incompletion, Magee and Ellis (2000) conducted a unique demand condition that consisted of the therapist leaving the room for 30 s contingent on the occurrence of target behaviors. This form of contingent escape maintained the target behavior of one of two students.

For more disruptive behaviors it is not uncommon for a teacher to use some form of timeout. Although intended as a mild punishment, the procedure may in fact function as negative reinforcement by removing the child from an otherwise aversive academic task. For this reason Northup et al. (1995, 1997) used brief, nonexclusionary timeouts as a test for negative reinforcement. The children were first seated, given a difficult academic task, and then given an instruction to sit quietly and complete the task. Contingent on any target behavior, the child's chair was turned away from the task and all people and activities for 30s. This condition served the dual purpose of also being a test of the effectiveness of timeout as a mild punishment. For some children disruptive behavior was reduced to zero or near zero in the timeout condition only, suggesting that this procedure did in fact appear to function as mild punishment.

Instructions

An additional variation concerns whether or not children are given prior instructions regarding the specific contingencies associated with different analog assessment conditions. Some studies using functional analyses with verbal children have provided no prior instructions (Broussard & Northup, 1995), whereas others have provided specific instructions regarding the contingencies associated with each condition (e.g., "If you talk, I will have to remind you to work quietly"; Northup et al., 1997). Children with well-developed verbal repertoires could be greatly influenced by a description of the subsequent contingencies (Shimoff, Matthews, & Catania, 1986). Northup, Kodak, Grow, Lee, and Coyne (2004) conducted a functional analysis using analog assessment conditions with a common contingency. The three conditions varied only by three different instructions describing the contingency. In one condition the contingency was described as "taking a break," in another condition it was described as "timeout," and no description of the contingency was provided in a third condition. A typically developing 5-year-old child with ADHD participated. Rates of inappropriate behavior varied substantially across the three conditions as a function of the prior instructions. This study suggested that instructions may be a procedural variable that could substantially influence functional analysis outcomes for some children.

FUNCTIONAL ANALYSIS BASED ON REINFORCER DIMENSIONS AND TEMPORAL DISCOUNTING

Barkley (1997, 1998) conceptualized ADHD as a problem of self-control or impaired behavioral inhibition. He posited that children with ADHD are deficient in the capacity for their behavior to be influenced by delayed consequences. Behavior that is more sensitive to immediate than to remote consequences suggests temporal discounting. Temporal discounting is said to occur when the value of a desired outcome diminishes as a function of delay to that outcome. For example, interest in taking pictures might be higher with a digital camera in which the outcome (a picture) can be seen immediately after the picture is taken than with a Polaroid camera in which the picture develops gradually. Temporal discounting can be measured in the context of choices between concurrently available response alternatives that are associated with different outcomes. For example, in behavior analytic research, impulsivity has been operationally defined as choices of response options that result in smaller, sooner reinforcers (SSRs) rather than larger, later reinforcers (LLRs). Self-control is the opposite (Ainslie, 1974; Logue, Pena-Correal, Rodriguez, & Kabela, 1986; Rachlin, 1974). An individual who consistently chooses a response alternative that produces a small reward now, rather than one that yields a larger reward later, would be said to demonstrate impulsivity. These objective and precise measures provide a functional account of impulsivity as an alternative to the topographically defined and often subjective measures typically used to diagnose ADHD.

An example of this approach is illustrated by Sonuga-Barke, Taylor, Sembi, and Smith (1992) who examined choices between SSR and LLR by children with and without ADHD. Differences between children with and without ADHD were revealed in a comparison of time constraints in which more points could be earned by choosing SSR and trial constraints in which more points could be earned by choosing LLR. Children with ADHD were less likely than those without ADHD to choose LLR when it was the best point-earning strategy. This suggests that children with ADHD were more likely to be influenced by reductions in the overall delay than by maximizing the reward amount.

Neef et al. (2005) conducted a conceptually similar study with 58 children with and without a diagnosis of ADHD. The researchers used brief, computer-based assessments involving choices of concurrently presented arithmetic problems associated with competing reinforcer dimensions to assess impulsivity. During each session, reinforcer immediacy [I], rate [R], quality [Q], and response effort [E] were placed in direct competition with one another. For example, in the Q vs I condition, students could choose between math problems that produced points exchangeable for a highly preferred item the next day or a less preferred item available immediately. In the I vs E condition, students could choose between difficult problems for immediate rewards and easy problems for delayed rewards. All possible pairs of dimensions were presented across the 6 assessment conditions (R vs Q, R vs I, R vs E, Q vs I, Q vs E, I vs E). Impulsive choices were defined as those controlled primarily by reinforcer immediacy, relative to the other dimensions. The choices of children with ADHD were most influenced by reinforcer immediacy and quality and least by rate and effort, suggesting impulsivity, whereas the choices of children in the non-ADHD group were most influenced by reinforcer quality. Neef, Bicard, and Endo (2001) reported similar findings with three students with ADHD.

These studies developed a methodology for objectively and precisely measuring a key construct associated with a diagnosis of ADHD and suggest that measures based on temporal discounting may offer a means of diagnosing children with ADHD. In addition, they suggest direction for intervention to promote self-control and a means of evaluating treatments according to the extent to which the treatments reduce children's sensitivity to delay.

FUNCTIONAL ASSESSMENT OF IMPULSIVITY

A functional analysis of impulsivity involves objective measures of the extent to which the value or effectiveness of a consequence is a function of its immediacy relative to other possible dimensions (i.e., quality, rate, magnitude, probability, or effort). Assessments are conducted by arranging choices in which the response options are between those that result in immediate consequences that are less favorable with respect to another dimension (e.g., immediate low-quality reinforcers) and delayed consequences that are more favorable with respect to the same alternative dimensions (e.g., delayed high-quality reinforcers). Table 5.1 illustrates

TABLE 5.1 The possible choices that are given in an assessment of the effects of Reinforcer Immediacy, Reinforcer Quality, Reinforcer Rate, Reinforcer Magnitude, and Response Effort on impulsive and nonimpulsive choosing in children with ADHD

Conditions Dimensions	R vs I Problem		I vs Q Problem		M vs I Problem		E vs I Problem	
	Set 1	Set 2	Set 1	Set 2	Set 1	Set 2	Set 1	Set 2
Reinforcer Immediacy (I)	*Delay*	*Imm.*	**Imm.**	**Delay**	*Delay*	*Imm.*	**Delay**	**Imm.**
Reinforcer Rate (R)	**High**	**Low**	Med.	Med.	**High**	**Low**	Med.	Med.
Reinforcer Quality (Q)	High	High	*Low*	*High*	Low	High	High	High
Reinforcer Magnitude (M)	Med.	Med.	Med.	Med.	*High*	*Low*	**Med.**	**Med.**
Response Effort (E)	Med.	Med.	Med.	Med.	Med.	Med.	*Low*	*High*

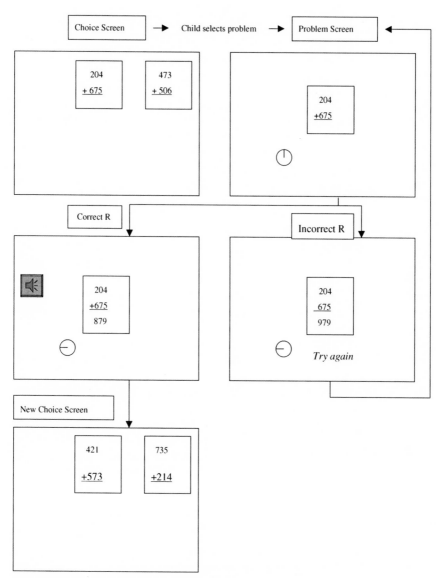

FIGURE 5.1   Illustration of a trial sequence used to assess impulsive versus nonimpulsive choosing during computer-based assessment of impulsivity.

possible choices, and Figure 5.1 shows an example of a choice sequence. Typically, assessments have been automated with computer programs (e.g., successively presented choices of mathematics problem alternatives associated with competing reinforcer dimensions as used in the series of studies by Neef and colleagues). The extent to which behavior is influenced by immediacy or another dimension can be determined by measuring the percentage of choices that are made to the respective alternatives, as well as examining response patterns. For example, if the student consistently chooses the response that produces immediate reinforcement even when that response results in less preferred reinforcers, fewer reinforcers, or requires more effort relative to the alternative, his or her responding would be characterized as highly impulsive. A student who demonstrates self-control, on the other hand, might favor the response option that produces a more highly preferred reinforcer even when it is delayed, which suggests that reinforcer quality is more potent than reinforcer immediacy. Alternatively, a student might demonstrate self-control by allocating his or her responding across the two alternatives in a way that yields the most reinforcement from each.

Inspection of response patterns can be informative as well. For example, a student might initially choose the response option that produces high-quality delayed reinforcers; however, once he or she has obtained a preferred reinforcer, the student might switch to the alternative that produces immediate but lower quality reinforcers. In this situation, the student's response pattern suggests self-control even though the majority of his or her later choices might have been allocated to the option that produced immediate reinforcement.

Several considerations are important for the assessment to yield valid findings. First, it is important to ensure that the student discriminates differences in each of the dimensions being compared (e.g., response options that result in immediate reinforcement from those that result in delayed reinforcement). If the student does not discriminate favorable and unfavorable levels of the dimension, choices will be influenced by arbitrary variables and the influence of the dimension of interest cannot be determined. Therefore, we recommend conducting a baseline in which the only difference between the two response options is with respect to the dimension of interest (e.g., choices between responses that result in a high rate versus a low rate of reinforcement, and subsequently choices between responses that result in immediate versus delayed reinforcement, and so on). Second, it is important to ensure both the integrity and discriminability of the dimensions being compared. For example, in order to ensure sufficient differentiation between high- and low-quality reinforcers, a preference assessment in which the student chooses the reinforcers he or she most wants to earn should be conducted immediately beforehand. Third, the values of the reinforcer dimensions used in the assessment should correspond to those in the natural environment. For example, the delay alternative should not involve waiting only a few seconds if the student must typically wait much longer to receive a reinforcer. Fourth, the response alternatives used in the assessment must be able to be performed by the student. If one of the response alternatives involves a behavior that would result in immediate reinforcement but which the student cannot perform successfully, for example, the immediacy of reinforcement will be moot because the reinforcer cannot be obtained. In that situation, responding would likely be biased to the other alternative because it is the only one that would result in any reinforcement at all.

Several variations of this assessment methodology are possible. For example, an adjusting delay procedure might be used in which the duration of the delay associated with the LLR response alternative is decreased until the LLR and the SSR are selected an equal percentage of the time (the "indifference point"). Indifference points could then be compared. Although this procedure has been used extensively in basic research, it has not yet been evaluated as a means of assessing children with ADHD. Another variation involves manipulating the delay to conditioned reinforcers (e.g., points) rather than the delay to the exchange of points for awards (terminal reinforcers). However, there is little evidence to date that the former is more influential than the latter. In addition, analog assessments under controlled conditions might be informed by descriptive assessments of reinforcer dimensions that occur in the natural environment, paralleling those of descriptive assessments of reinforcement contingencies that maintain behavior. Although additional research will undoubtedly yield further advances in functional assessments of impulsivity, the basic paradigm described here has potential advantages not only as a diagnostic tool but as a means of identifying influential dimensions of reinforcement that can be used to guide and enhance the effectiveness of interventions.

ASSESSMENT OF MULTIPLY CONTROLLED BEHAVIOR

Different behaviors may of course serve very different functions (Derby et al., 1994). However, a single behavior may also serve multiple functions. That is, a behavior may function to access one source of reinforcement in one context and another source of reinforcement in another context (Day, Horner, & O'Neill, 1994; Haring & Kennedy, 1990; Iwata et al., 1982/1994). The wide variety of behaviors that define ADHD may be particularly likely to serve different functions or serve different functions in different settings. For example, being out-of-seat may be maintained only by escape from tasks, and inappropriate vocalizations may be maintained only by peer attention. Alternatively, being out-of-seat may be maintained by both escape and peer attention.

In general, the display of multiply controlled behaviors is consistent with the previous discussion of choice and reinforcer dimensions. That is, altering the value of a reinforcer dimension (i.e., rate, effort, immediacy, quality) for one behavior may alter the probability of the occurrence of some other behavior (Neef, Mace, & Shade, 1993). For example, if a teacher ignores inappropriate vocalizations, a student may instead engage in an alternative behavior such as getting out of his or her seat to gain teacher attention. Similarly, if the teacher ignores inappropriate vocalizations, the student may attempt to gain peer attention by engaging in the same behavior.

Multiply controlled behaviors may greatly increase the complexity of functional analysis and treatment. The clear identification of all maintaining contingencies for multiply controlled behaviors may require extended assessments across a variety of contextual conditions. Multiply controlled behaviors may also require different interventions for different behaviors that can be rapidly alternated as needed.

INTERVENTIONS FOR ADHD

OVERVIEW OF TREATMENTS FOR ADHD

The most common treatment for ADHD is stimulant medication (Barkley, 1998; Purdie, Hattie, & Carroll, 2002), which has increased dramatically in recent years; approximately 2 million children are treated with these medications annually (Greenhill, Halperin, & Abikoff, 1999; Pincus et al., 1998; Purdie et al., 2002; Safer, Zito, & Fine, 1996). Some authorities have argued that stimulant medication should be the predominant treatment for ADHD (Barkley, 1997; The MTA Cooperative Group, 1999). For example, the Multimodal Treatment Study of Children with ADHD (MTA) concluded that medication management alone was more effective than psychosocial treatments used alone on ratings of core ADHD symptoms (inattention and hyperactivity); furthermore, combined medication management and intensive behavioral treatment added only modestly to the benefits of medication management. Combined medication management and psychosocial treatment were found to be superior to either treatment alone on measures of classroom behavior, ratings of social skills and parent-child relationships, peer sociometric ratings, and academic achievement.

Some authors have noted that even though improvements have been found in some aspects of behavior of 70–80% of children with ADHD who receive medication, such improvements are often inadequate, typically remaining one standard deviation above the norm (Pelham, Wheeler, & Chronis, 1998; Purdie et al., 2002). Moreover, researchers have pointed out that stimulant medication has not been found to produce improvements in academic performance or learning (see Purdie et al., 2002). Finally, most studies have used parent and teacher ratings or other indirect measures that are either subject to influence by, or are confounded with, other behavioral changes (e.g., sustained attention, level of activity, vocalizations).

A concern is that, for many children with ADHD, "educational solutions to their difficulties at school are either not contemplated or take second place to medication" (Purdie et al., 2002, p. 86). Some have argued strongly that pharmacological intervention can be an important first step, but in order to promote the educational success of students with ADHD, strategies must be developed that address their academic difficulties (DuPaul & Eckert, 1997; Purdie et al., 2002). Based on a review of the literature, Ervin, Bankert, and DuPaul (1996) concluded that interventions are most likely to be effective when they include behavioral contingencies that focus on specific training corresponding to the desired performance changes in the contexts and settings where problems occur. Unfortunately, very few investigations of classroom-based interventions for students with ADHD have been reported in the literature; in their meta-analysis of research on interventions for ADHD, Purdie et al. identified only 8 studies conducted in school settings from 1990 to 1998.

UNIVERSAL BEHAVIORAL INTERVENTIONS

Functional interventions for children with ADHD first assume the use of basic universal classroom management practices and the integrity of less intensive, more selected interventions (Witt, Vanderheyden, & Gilbertson, 2004). Universal interventions are those practices that are intended to effectively manage the classroom behavior of all students and thus provide an optimal environment for teaching and learning. A particular emphasis is placed on preventing problem behaviors and on the reinforcement of prosocial behavior. These basic practices include at the least classroom rules that are established, posted, actively taught, and—perhaps most importantly—consistently enforced. Other examples include providing effective instructions, an appropriate amount of structure, and a consistent response to misbehavior that does occur. An especially important consideration is the provision of quality instruction at an appropriate instructional level. If universal classroom management procedures are not in place, then the first strategy should be to train staff to implement these procedures (Witt et al., 2004).

Extensive research has also documented the effectiveness of a variety of more selected behavioral interventions for children with ADHD. DuPaul and Eckert (1997) conducted a meta-analysis which found that contingency management procedures that are conducted directly in the classroom result in significant positive changes in behavior for many children with ADHD. Contingency management procedures that have been demonstrated to be effective include differential reinforcement with teacher attention, token economies, home-school contingencies, group contingencies, and mild punishment procedures such as response-cost and timeout (Abramowitz & O'Leary, 1991). One or more of these basic procedures are often tried first as a supplement to universal classroom management procedures for children who experience only mild problems associated with ADHD. Only if a student continues to exhibit significant problem behaviors should an individual functional analysis be considered.

FUNCTIONAL CONTINGENCY-BASED INTERVENTIONS

As with functional analysis and assessment in general, functional approaches to treatment for children with ADHD do not differ conceptually from functional approaches to problem behaviors with other populations. That is, the goal of functional analysis is to better understand the contingencies that maintain problem behavior. This is the primary reason to conduct a functional analysis. The essential implication is that intervention, and especially reinforcement-based interventions, must match the results of the functional analysis on an individual basis.

Interventions that are selected without regard to behavioral function may fail for a number of reasons. First, the intervention may in fact be contraindicated and inadvertently reinforce the problem behavior through either positive or negative reinforcement. Second, the intervention may simply be functionally irrelevant to the target behavior. Third, the intervention may not provide functional reinforcement for an alternative appropriate behavior. The essence of functional intervention is that the source of reinforcement identified as maintaining a problem behavior be removed for the problem behavior (extinction) and that the same type of reinforcement be provided for an appropriate alternative behavior.

Once the problem behavior is placed on extinction, the same reinforcement for an alternative appropriate behavior may be provided in a variety of ways. Most commonly, a teacher, parent, or other caregiver takes responsibility for directly providing the reinforcement on a predetermined schedule. For example, a teacher may praise successful work completion by a child whose problem behavior was maintained by attention, or the teacher may provide brief breaks every 15 or 30 min for a child whose behavior was maintained by escape from academic tasks. We have often supplemented this procedure by allowing students to earn token coupons representing varying amounts of teacher or peer attention or escape contingent on an appropriate alternative behavior. The coupons may then be "cashed in" as in a traditional token economy (Broussard & Northup, 1997).

A limitation of such procedures is that they may be time-consuming and place excessive demands on some caregivers, especially for students with more frequent problem behaviors. However, many, if not most, students with a diagnosis of ADHD can become actively involved in their own treatment, including very young children. For example, strategies to teach students to effectively recruit their own reinforcement have been well documented (Alber & Heward, 2001). Ervin, DuPaul, Kern, and Friman (1998) used functional assessments to develop self-evaluation procedures for two adolescents that were both effective and highly acceptable to the students and their teachers. In addition, the inclusion of peers in functional-based interventions has been shown to help facilitate prosocial behavior (Flood, Wilder, Flood, & Masuda, 2002; Jones, Drew, & Weber, 2000).

TEACHING SELF-CONTROL

Once influential reinforcer dimensions have been identified in an assessment, the information can be used to guide interventions to promote self-control. For example, Neef et al. (2001) conducted an assessment with three students diagnosed with ADHD, which revealed that their choices were influenced principally by reinforcer immediacy, suggesting impulsivity. The assessment also showed that the second most influential dimension was reinforcer quality for two of the students and reinforcer rate for one of the students. Therefore, the alternative influential dimension was arranged to compete with reinforcer immediacy. Two of the students were allowed to choose between math problems associated with immediate low-quality reinforcers and delayed high-quality reinforcers; the other student was allowed to choose between math problems associated with immediate reinforcers delivered at a low rate and delayed reinforcers delivered at a high rate. In all cases, the delay for the high-quality (or high-rate) reinforcement alternative was minimal to encourage students to select that option. The delay to reinforcement was then progressively increased. This procedure resulted in students allocating the majority of their time to the delayed reinforcement response alternative, even when the delay to reinforcement was increased to 24 hr. In other words, the students demonstrated self-control.

In another study, Neef and Lutz (2001) used assessment results to develop classroom-based procedures for reducing disruptions. For example, assessment results for one student showed that immediacy of reinforcement was the most influential dimension. Therefore, an intervention involving immediate reinforcement for not exceeding a specified number of disruptions (i.e., differential reinforcement of low-rate behavior) was compared with delayed reinforcement. In all cases, a reversal design showed a substantial decrease in the rate of disruptions in the conditions involving favorable levels of the influential reinforcer dimension (e.g., immediate as opposed to delayed reinforcement).

CASE STUDY[1]

Chang was a 10-year-old student with ADHD who was in the fourth grade of an urban public elementary school. He was referred by his teacher

---

[1]This case study is based on one of the participants in a dissertation completed by Summer J. Ferreri.

and the school principal because of low academic achievement and disruptive classroom behavior. He was not taking any medication.

A computer-based assessment was conducted as described earlier. The assessment involved three conditions (one per each 5-min session) in which immediacy of reinforcement was placed in competition with rate of reinforcement, quality of reinforcement, and response effort, respectively. In I vs R sessions, Chang chose between identical math problems of medium difficulty, one of which resulted in a high rate of points for correct problem completion that were exchangeable for rewards the next day, and the other of which resulted in a lower rate of points that were exchangeable for rewards immediately. In I vs Q sessions, Chang chose between identical math problems, one of which resulted in points exchangeable for immediate low-quality reinforcers (from "Store B") and one which produced points exchangeable for delayed high-quality reinforcers (from "Store A"). High- and low-quality reinforcers were determined by a preference test, in which Chang had ranked 20 available items. The five items he selected as most desired were placed in Store A, and less preferred items were placed in Store B. In I vs E sessions, Chang chose between difficult acquisition-level problems that resulted in points immediately exchangeable for preferred rewards, and easy fluency-level problems that resulted in points exchangeable for preferred rewards after a 24-hr delay. Easy and difficult problems were determined by accuracy and rate of problem completion on a pretest. The assessment was preceded by four baseline sessions, in which Chang chose between math problems that produced immediate versus delayed delivery of reinforcers, a high rate versus low rate of points for problem completion, highly preferred versus less preferred reinforcers, and between math problems that were easy and difficult, respectively. Chang's choices of the problems that resulted in the favorable consequences during baseline verified that he discriminated the differences for each dimension.

During all conditions of the assessment, Chang always chose the math problems that resulted in immediate delivery of the reinforcer, even when those problems were more difficult, produced a lower rate of reinforcement, or lower quality reinforcers, respectively, relative to the alternative problem option. These results demonstrated a high degree of impulsivity.

Assessment results were used to design a classroom intervention that addressed Chang's low academic productivity and accuracy and his classroom disruptions during independent seatwork. Typically, during independent seatwork the teacher distributed a worksheet that contained items from workbooks for math, language arts, and test preparation. Following independent seatwork, the teacher collected the worksheets, wrote a check mark or a minus on the sheet depending on whether or not the student had made an attempt to write in answers, and reviewed the answers in class. On the rare occasions in which Chang attempted answers to any questions, his answers were always incorrect, and he was off-task most of the time (see Figures 5.2 and 5.3).

A modification was then made to the worksheet ("Control Intervention"). The front was duplicated on the back where the correct answers were written with Invisible Changeable Crayola Markers™; after completing a problem on the front, Chang could reveal the correct answer by coloring over the identical space on the back of the page. Given that Chang's assessment results indicated that his behavior was influenced principally by immediacy of reinforcement, this modification to the worksheet allowed determination of whether immediate reinforcement in the form of revealing the correct answer and his response matching that answer if correct alone would be sufficient to increase his productivity and accuracy.

Chang began to attempt more problems, some of which were correct, and there was a reduction in his off-task behavior. In an effort to further enhance performance, immediate conditioned reinforcers were added ("Experimental Intervention"). Specifically, one fourth of the answers on the back page were randomly designated with a star, which was also revealed by Chang coloring over the space. If Chang's answer to a worksheet problem matched the correct answer on the back of the page and it was designated with a star, he earned a point. Four points could be

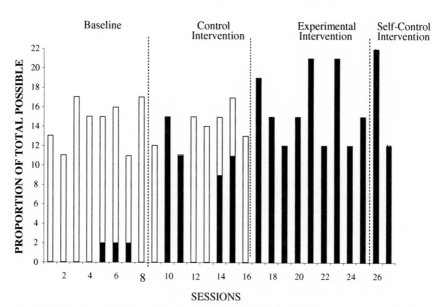

FIGURE 5.2 Proportion of total problems (top bar) attempted and correct (shaded portion of bar) during Baseline, Control, Experimental, and Self-Control interventions across conditions for Chang "D" indicates selection of the delayed reinforcer. See text for further description of conditions.

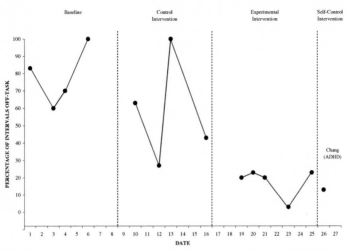

FIGURE 5.3   The percentage of intervals of off-task behavior during Baseline, Control, Experimental, and Self-Control interventions across conditions for Chang. See text for further description of conditions.

exchanged for a reward at the end of the period, and Chang recorded the number of points he had earned. This procedure allowed immediate reinforcement for correct problem completion without relying on the teacher to provide it; Chang himself performed the behaviors associated with delivery of the conditioned reinforcers. To help ensure that Chang attempted to answer the problems on his own, he was informed at the start of the period that revealing the correct answer before making an entry would result in point forfeiture. No violations were observed. Chang attempted all of the problems, and there was a dramatic increase in accuracy. Concomitantly, there was a substantial reduction in his off-task behavior.

A commitment component was subsequently added ("Self-Control Intervention") to promote the development of self-control (i.e., increase tolerance for reinforcement delay). Chang was given a card that listed 16 prizes. He ranked them by writing in a number next to the item. The card also contained a table with the days of the week with two options under each day: The choices were either "I will choose a prize from numbers 1–5 at the end of the day" (representing high-quality delayed reinforcement) and "I will choose a prize from numbers 6–10 when I finish my seatwork" (representing low-quality immediate reinforcement). Before beginning seatwork, Chang circled the option that would then be in effect for each subsequent day. On both days, Chang chose the delayed option (representing self-control). He attempted 100% of the problems, the majority of which were completed correctly, and his off-task behavior remained at low levels. Chang's teacher reported being highly satisfied and impressed with the dramatic changes in Chang's behavior, and she sought guidance in continuing the intervention during the subsequent school year.

REFERENCES

Abramowitz, A., & O'Leary, S. (1991). Behavioral interventions for the classroom: Implications for students with ADHD. *School Psychology Review, 20*, 220–234.

Ainslie, G. W. (1974). Impulse control in pigeons. *Journal of the Experimental Analysis of Behavior, 21*, 485–489.

Alber, S., & Heward, W. (2001). Teaching students to recruit positive reinforcement. *Journal of Behavioral Education, 10*, 177–204.

American Psychiatric Association. (2000). *Diagnostic and statistical manual of mental disorders* (4th ed.). Washington DC: Author.

Atkins, M. S., & Pelham, W. E. (1991). School-based assessment of attention deficit-hyperactivity disorder. *Journal of Learning Disabilities, 24*, 197–225.

Barkley, R. A. (1997). *ADHD and the nature of self control.* New York: Guilford Press.

Barkley, R. A. (1998). *Attention deficit hyperactivity disorder.* New York: Guilford Press.

Barkley, R. A., Fischer, M., Edelbrocke, C. S., & Smallish, L. (1990). The adolescent outcome of hyperactive children diagnosed by research criteria: I. An 8-year prospective follow-up study. *Journal of American Academy of Child and Adolescent Psychiatry, 29*, 546–557.

Boyajian, A. E., DuPaul, G. J., Handler, M., Eckert, T., & McGoey, K. (2001). The use of classroom-based brief functional analysis with preschoolers at-risk for attention deficit hyperactivity disorder. *School Psychology Review, 30*, 278–293.

Broussard, C., & Northup, J. (1995). An approach to functional assessment and analysis of disruptive behavior in regular education classrooms. *School Psychology Quarterly, 10*, 151–164.

Broussard, C., & Northup, J. (1997). The use of functional analysis to develop peer interventions for disruptive classroom behavior. *School Psychology Quarterly, 12*, 65–76.

Critchfield, T. S., & Kollins, S. H. (2001). Temporal discounting: Basic research and the analysis of socially important behavior. *Journal of Applied Behavior Analysis, 34*, 101–122.

Day, H. M., Horner, R. H., & O'Neill, R. E. (1994). Multiple functions of problem behaviors: Assessment and intervention. *Journal of Applied Behavior Analysis, 27*, 279–289.

Derby, K. M., Wacker, D. P., Peck, S., Sasso, G., DeRaad, A., Berg, W., et al. (1994). Functional analysis of separate topographies of aberrant behavior. *Journal of Applied Behavior Analysis, 27*, 267–278.

DuPaul, G. J., & Eckert, T. L. (1997). The effects of school-based interventions for attention-deficit hyperactivity disorder: A meta-analysis. *School Psychology Review, 26*, 5–27.

DuPaul, G. J., Eckert, T., & McGoey, K. (1997). Interventions for students with attention deficit hyperactivity disorder: One size does not fit all. *School Psychology Review, 26*, 369–381.

Ervin, R. A., Bankert, C. L., & DuPaul, G. J. (1996). Treatment of attention-deficit/hyperactivity disorder. In M. A. Reinecke & F. M. Dattilio (Eds.), *Cognitive therapy with children and adolescents: A casebook for clinical practice* (pp. 38–61.) New York: Guilford Press.

Ervin, R. A., DuPaul, G. J., Kern, L., & Friman, P. C. (1998). Classroom-based functional and adjunctive assessments: Proactive approaches to intervention selection for adolescents with attention deficit hyperactivity disorder. *Journal of Applied Behavior Analysis, 31,* 65–78.

Flood, W. A., Wilder, D. A., Flood, A. L., & Masuda, A. (2002). Peer-mediated reinforcement plus prompting as treatment for off-task behavior in children with attention deficit hyperactivity disorder. *Journal of Applied Behavior Analysis, 35,* 199–204.

Forness, S. R., Kavale, K. A., Sweeney, D. P., & Crenshaw, T. M. (1999). The future of research and practice in behavioral disorders: Psychopharmacology and its school implications. *Behavioral Disorders, 24,* 305–318.

Greenhill, L. L., Halperin, J. M., & Abikoff, H. (1999). Stimulant medications. *Journal of the American Academy of Child and Adolescent Psychiatry, 38,* 503–512.

Gulley, V., & Northup, J. (1997). Comprehensive school-based assessment of the effects of methylphenidate. *Journal of Applied Behavior Analysis, 30,* 627–638.

Gunter, P. L., Jack, S. L., Shores, R. E., Carrell, D. E., & Flowers, J. (1993). Lag sequential analysis as a tool for functional analysis of student disruptive behavior in classrooms. *Journal of Emotional and Behavioral Disorders, 1,* 138–148.

Haring, T. G., & Kennedy, C. H. (1990). Contextual control of problem behavior in students with severe disabilities. *Journal of Applied Behavior Analysis, 23,* 235–243.

Iwata, B. A., Dorsey, M. F., Slifer, K. J., Bauman, K. E., & Richman, G. S. (1994). Toward a functional analysis of self-injury. *Journal of Applied Behavior Analysis, 27,* 197–209. (Reprinted from *Analysis and Intervention in Developmental Disabilities, 2,* 3–20, 1982).

Jones, K. M., Drew, H. A., & Weber, N. L. (2000). Noncontingent peer attention as a treatment for disruptive classroom behavior. *Journal of Applied Behavior Analysis, 33,* 343–346.

Kodak, T., Grow, L., & Northup, J. (2004). A functional analysis of elopement for a young child with attention deficit hyperactivity disorder. *Journal of Applied Behavior Analysis, 37,* 229–232.

Kollins, S. H., Ehrhardt, K., & Poling, A. (2000). Clinical drug assessment. In A. Poling & T. Byrne (Eds.), *Introduction to behavioral pharmacology* (pp. 191–218). Reno, NV: Context Press.

Logue, A. W., Pena-Correal, T. E., Rodriguez, M. L., & Kabela, E. (1986). Self-control in adult humans: Variation in positive reinforcer amount and delay. *Journal of the Experimental Analysis of Behavior, 46,* 159–173.

Magee, J., & Ellis, S. (2000). Modifications to basic functional analysis procedures to school settings: A selective review. *Behavioral Interventions, 19,* 205–228.

The MTA Cooperative Group. (1999). A 14-month randomized clinical trial of treatment strategies for attention-deficit/hyperactivity disorder. *Archives of General Psychiatry, 56,* 1073–1086.

National Institutes of Health. (1998). *Diagnosis and treatment of attention-deficit/hyperactivity disorder.* Washington, DC: US Government Printing Office.

Neef, N. A., Bicard, D. F., & Endo, S. (2001). Assessment of impulsivity and the development of self-control by students with attention deficit hyperactivity disorder. *Journal of Applied Behavior Analysis, 34,* 37–60.

Neef, N. A., & Lutz, M. (2001). Assessment of variables affecting choice and application to classroom interventions. *School Psychology Quarterly, 16,* 239–252.

Neef, N. A., Mace, F. C., & Shade, D. (1993). Impulsivity in students with serious emotional disturbance: The interactive effects of reinforcer rate, delay, and quality. *Journal of Applied Behavior Analysis, 26,* 37–52.

Neef, N. A., Marckel, J., Ferreri, S. J., Bicard, D. F., Endo, S., Aman, M. G., et al. (2005). Behavioral assessment of impulsivity: A comparison of children with and without attention deficit hyperactivity disorder. *Journal of Applied Behavior Analysis, 38,* 23–37.

Northup, J., Broussard, C., Jones, K., George, T., Vollmer, T., & Herring, M. (1995). The differential effects of teacher and peer attention for children with attention deficit hyperactivity disorder. *Journal of Applied Behavior Analysis, 28*, 225–227.

Northup, J., Fusilier, I., Swanson, V., Huete, J., Bruce, T., Freeland, J., et al. (1999). Further analysis of the separate and interactive effects of methylphenidate and common classroom contingencies. *Journal of Applied Behavior Analysis, 32*, 35–50.

Northup, J., & Gulley, V. (2001). Some contributions of functional analysis to the assessment and treatment of behaviors associated with attention deficit hyperactivity disorder. *School Psychology Review, 30*, 227–239.

Northup, J., Gulley, V., Edwards, S., & Fountain, L. (2001). The effects of methylphenidate in the classroom: What dosage, for which children, for what problems? *School Psychology Quarterly, 16*, 303–323.

Northup, J., Jones, K., Broussard, C., DiGiovanni, G., Herring, M., Fusilier, I., et al. (1997). A preliminary analysis of interactive effects between common classroom contingencies and methylphenidate. *Journal of Applied Behavior Analysis, 30*, 121–125.

Northup, J., Kodak, T., Grow, L., Lee, J., & Coyne, A. (2004). Instructional influences on analogue functional analysis outcome. *Journal of Applied Behavior Analysis, 38*, 509–512.

Pastor, P. N., & Reuben, C. A. (2002). Attention deficit disorder and learning disability: United States 1997–1998. *Vital Health Statistics, 10*, 1–12.

Pelham, W. E., Jr., Wheeler, T., & Chronis, A. (1998). Empirically supported psychosocial treatments for attention deficit hyperactivity disorder. *Journal of Clinical Child Psychology, 27*, 190–205.

Pincus, H. A., Tanielian, T., Marcus, S., Olfson, M., Zarin, D., Tompson, J., et al. (1998). Prescribing trends in psychotropic medications. *Journal of the American Medical Association, 279*, 526–531.

Purdie, N., Hattie, J., & Carroll, A. (2002). A review of the research on interventions for attention deficit hyperactivity disorder: What works best? *Review of Educational Research, 72*, 61–99.

Rachlin, H. (1974). Self-control. *Behaviorism, 2*, 94–107.

Rowland, A. S., Umbach, D. M., Catoe, K. E., Long, S., Raginer, D., Naftel, A. J., et al. (2001). Studying the epidemiology of attention-deficit hyperactivity disorder: Screening method and pilot results. *Canadian Journal of Psychiatry, 46*, 931–940.

Rowland A. S., Umbach, D. M., Stallone L., et al. (2002). Prevalence of medication treatment for attention deficit-hyperactivity disorder among elementary school children in Johnston County, North Carolina. *American Journal of Public Health, 92*, 231–234.

Safer, D. J., Zito, J. M., & Fine, E. M. (1996). Increased methylphenidate usage for ADD in the 1990s. *Pediatrics, 98*, 1084–1088.

Scahill, L., & Schwab-Stone, M. (2000). Epidemiology of ADHD in school-age children. *Child and Adolescent Psychiatric Clinics of North America, 9*, 541–555.

Shimoff, E., Matthews, B. A., & Catania, E. (1986). Human operant performance sensitivity and pseudosensitivity to contingencies. *Journal of the Experimental Analysis of Behavior, 46*, 149–157.

Shinn, M. (1995). Best practices in curriculum based measurement and its use in a problem solving model. In A. Thomas & J. Grimes (Eds.), *Best practices in school psychology-III* (pp. 547–567). Washington, D.C.: National Association of School Psychologists.

Sonuga-Barke, E. J. S., Taylor, E., Sembi, S., & Smith, J. (1992). Hyperactivity and delay aversion I: The effects of delay on choice. *Journal of Child Psychology and Psychiatry, 33*, 387–398.

Stoner, G., Carey, S. P., Ikeda, M. J., & Shinn, M. R. (1994). The utility of curriculum-based measurement for evaluating the effects of methylphenidate on academic performance. *Journal of Applied Behavior Analysis, 27*, 101–113.

Umbreit, J. (1995). Functional assessment and intervention in a regular classroom setting for the disruptive behavior of a student with attention deficit hyperactivity disorder. *Behavioral Disorders, 20*, 267–278.

Weiss, G. H., & Hechtman, L. T. (1993). *Hyperactive children grown up: ADHD in children, adolescents, and adults.* New York: Guilford.

Witt, J., VanDerHeyden, A., & Gilbertson, D. (2004). Troubleshooting behavioral interventions: A systematic process for finding and eliminating problems. *School Psychology Review, 33*, 363–383.

Wolraich, M. L., Hannah, J. N., Baumgaertal, A., & Feurer, I. D. (1998). Examination of DSM-IV criteria for attention-deficit/hyperactivity disorder in a county-wide sample. *Journal of Developmental and Behavioral Pediatrics, 19*, 162–168.

6

CHAOS, COINCIDENCE, AND CONTINGENCY IN THE BEHAVIOR DISORDERS OF CHILDHOOD AND ADOLESCENCE

ROBERT G. WAHLER

The University of Tennessee

FUNCTIONAL ANALYTIC MODEL

In 1977, B. F. Skinner published a brief paper in which he argued: "Coincidence is the heart of operant conditioning. A response is strengthened by certain kinds of consequences, but not necessarily because they are actually produced by it" (p. 4). He then went on to describe his work on superstition in pigeons along with anecdotes about human vulnerability to this phenomenon, particularly when social contingencies are involved. He closed his intriguing paper with these sentences: "The genetic endowment responsible for our behavioral processes cannot fully protect us from the whims of chance, and the statistical and scientific measures we devise to bring our behavior under the more effective control of nature are not adequate for the extraordinarily complex sample space in which we live. Science has not ignored some underlying order; it has not yet devised ways of protecting us against spurious evidence of order" (p. 6).

Skinner's paper suggests that people must study their environmental experiences in order to discern the differences between consequences that are truly produced by their behavior, compared to those consequences that

Functional Analysis in Clinical Treatment

are merely coincidental. If we do not discriminate carefully over sufficient time periods, we are apt to be vulnerable to superstitious or adventitious learning experiences that will add to problems in managing our day-to-day lives. Skinner's own laboratory work with pigeons demonstrated the power of superstition and as he argued in his 1977 paper, "Many myths appear to represent this function" (p. 5).

Parents help their young children to assess the lawfulness of environmental experiences in two ways. Early in children's lives, parents orchestrate playful turn-taking experiences in which infants and toddlers can clearly detect the impact of their responses on those of the parents, and they can also see caregivers' social support of their various responses (Raver, 1996). Later, as the children develop language in the preschool years, the turn-taking involves conversations about recent experiences, and the focus shifts to teaching the youngster to clearly articulate sequential events in these happenings (Welch-Ross, 1997). When the children are helped to develop coherent narrative accounts of who, what, where, when, and how, they gain further competence in detecting contingency patterns in day-to-day life.

However, life is at best only periodically predictable, and parents are not always able or willing to orchestrate these crucial teaching experiences for their children (Holden & Miller, 1999). In fact, some children grow up in home environments in which inconsistent parenting is the mode of life (Dumas et al., 2005). In these families, Skinner's 1977 outline of the tendency to find order when it does not exist appears to be a blueprint in understanding the learning experiences of chronically troubled children, particularly those who are developing progressively more serious forms of oppositional behavior. These children represent hallmark examples of Skinner's contention that discriminating and using social contingencies to promote personal growth are indeed awesome tasks. The research evidence to be reviewed in this chapter will show that chronically oppositional children live in home environments that are even more chaotic than Skinner imagined; the children contribute to this chaos through their difficult temperaments, and they are likely to have memory and attentional deficits blocking their searches for any sort of environmental order. Given these handicaps, it is easy to imagine the children being influenced by adventitious social consequences, sure to keep them still further from acquiring the relationship-building skills needed to find an adaptive place in their families and larger communities of peers. If these children cannot detect the true contingencies within their family settings, it seems likely that they will eventually learn to use their temperament-driven coercive behavior to force environmental order. Should this happen, their oppositional behavior could become a social control tactic, negatively reinforced by the enhanced predictability of parental and sibling behavior. Though these brief episodes of predictable consequences are composed of harsh verbalizations and even physical abuse, they also constitute temporary escapes from what seems to be a randomly arranged world (Wahler & Dumas, 1989). Most of these children appear to recognize their abilities to generate this sort of impact on family members, and with this recognition, they are likely to develop verbal myths that might maintain their oppositional tactics. These myths are actual child narratives about personal power that are readily offered by the youth as self-stated guidelines for their behavior. Since the guideline is effective in predicting desired consequences in the short term, these disruptive children and youth see little reason to believe their parents, teachers, and counselors, all of whom urge the youngsters to learn to get along with people.

Today's family intervention strategies for children with disruptive behavior disorders focus on teaching the children and youth to look more closely at their home environments, newly structured through parent training procedures designed to help the parents to create appropriate order in their family interactions. Given these changes, molecular prosocial contingencies are highlighted, thus making it feasible for the children and youth to begin this new learning process. Of course, the parents have also been taught appropriate limit setting or discipline tactics designed to reduce oppositional behavior and to maintain consistency in family interactions (Cavell & Strand, 2002).

Today's parent trainers are also aware of how fragile these newly learned child and youth self-management skills are, due to the fact that social contingencies will be harder to find after the formal parent training ends, and due to the fact that the youngsters are well practiced in imposing contingencies in most any loosely structured social group. Because of this fragility, parent trainers are beginning to adopt additional strategies designed to guide parents in helping their children and youth to more clearly understand the roots of myth-related oppositionality and its overstated power. Through trainer-guided dialogue between parent and child, it is hoped that both individuals will better tolerate the ambiguity of immediate life experiences and that they will also discover that order does indeed exist, at least in the hindsight of dialogue about these earlier experiences. This form of dialogue aimed at restructuring of children and youth myths, while still in its clinical infancy, has an empirical base in studies of how typical children begin to question their personal beliefs. A review of these latter findings suggests that parents can help their normal children to expand these beliefs to include the views of other people, and the expansion appears to be associated with the children's self-reports and nonverbal demonstrations of empathy along with a willingness to follow parental guidance. These latter speculations on broadening the parent training enterprise will conclude this chapter.

DIAGNOSIS AND RELATED CHARACTERISTICS

While most parents expect toddlers to be disobedient, inattentive, and impulsive, these problematic characteristics of young children are lessened through appropriate parenting and the youngsters' gradual gains in age-related maturity. Some, however, intensify and maintain these ways of engaging family members and peers to such an extent that their parents and teachers ask for guidance from mental health professionals. When this happens, the professionals are apt to diagnose Oppositional Defiant Disorder (ODD) and, depending on the child's attentional performance, a comorbid diagnosis of Attention Deficit Hyperactivity Disorder (ADHD). This latter diagnosis includes impulsivity, but for our purposes the attentional problems are more significant, given the references to impulsivity already present in ODD (e.g., often loses temper, is touchy or easily annoyed by others; see American Psychiatric Association, 2000). The attentional problems described in ADHD add a significant hardship for children with ODD because their abilities to shift attention is impaired, making it difficult for them to use the full scope of environmental information in choosing what to do next (see Barkley, 1997). As expected, the combination of these disorders in children forecasts the more severe form of conduct disorder in their later years (Lahey, McBurnett, & Loeber, 2000).

Temperament ratings of infants and young children have proven to be modest but consistent predictors of their later tendencies to act in disruptive ways with their parents, siblings, peers, and teachers (Caspi, Henry, Moffitt, & Silva, 1995; Rothbart & Bates, 1998). The rating categories "low behavioral inhibition" and "novelty seeking" have been shown to covary in expected directions with the children's willingness to cooperate with their parents. Thus, Kochanska (1993) found inverse correlations between low behavioral inhibition and young children's sensitivity to their mothers' socialization tactics, and Frick, Lilienfeld, Ellis, Loney, and Silverthorn (1999) found connections between older children's thrill seeking and their externalizing problems.

Temperament is widely considered to be the physiological and emotional root of children's developing personality traits (Caspi, 1998; Rothbart & Bates, 1998). As expected, the findings from longitudinal studies covering infancy and early childhood to adolescence and adulthood have revealed covariations between time 1 temperament ratings and time 2 ratings of personality traits, including externalizing problems (Capsi, 1998; Schwartz, Snidman, & Kagan, 1999). Of course, while temperament is clearly a self-driven influence on children's life course well-being, these child effects have also been shown to be moderated as well as mediated by parenting practices. Thus, Dodge (2002) found that the predictive power of these "difficult" temperament ratings was significantly weakened when measures of harsh and inconsistent parenting were added to the respective regression models, suggesting that temperament wields its deleterious effects by provoking poor parenting practices. Another look at this interaction of child temperament and parenting practices is shown in correlational studies aimed at revealing the extent to which temperament effects are moderated by parenting. In these investigations, Bates, Pettit, Dodge, and Ridge (1998) and Stoolmiller (2001) found that the expected linkage between children's temperament and the disruptive behavior depended on poor parenting practices.

These results lend a note of encouragement to parents who struggle with the challenges presented by youngsters with difficult temperaments. The studies suggest that appropriate parenting of a toddler rated as having "low behavioral inhibition" can moderate the expressions of this child characteristic, as long as the parents can stay the course. Parents will tell you that this is easier said than done because the daily bombardment of the youngster's impulsivity and inattention creates a form of combat fatigue in which the best laid parenting plan is apt to crumble. Nevertheless, many parents find the resolve to stay on track through the support of a spouse, friend, or extended family member, or simply because of their own stamina, knowledge, and spirit. Others, who may also experience a lack of social support, economic instability, interpersonal conflict with adults, and harmful personal habits, such as substance abuse, may become depressed, angry, and are apt to lose heart in their parenting tasks (Wahler & Dumas, 1989).

When parents get off track in their parenting tasks, they cease their work as objective observers and are then likely to respond inconsistently to what their children say and do (Dix, 1993; Patterson, 1997). If this happens, a series of cascading events is apt to follow, leading to a worsening of the parent-child relationship, the child's enhanced skill in coercion and oppositionality, and the gradual formation of narrative myths that justify this skill. Thus, when the home environment becomes disorganized through socioeconomic stressors, parents' personal problems, and children's difficult temperament, it is hard for both adults and children to find order in that environment (Dumas et al., 2005).

Some of the parent-child interaction patterns formed under chaotic home conditions clearly seem to be acquired through the dyad's joint production of reinforcement contingencies. Patterson's classic book *Coercive Family Process* (1982) outlined data sets and interpretations showing escalation of mutually aversive parent and child interactions in which both parties are negatively reinforced for their coercive tactics. The child's behavior is reinforced when a parent discontinues a futile push for compliance, and the parent is reinforced by the youngster's cessation of complaints about the parent. This adversarial process begins when parents fail to monitor their children's behavior, putting these caregivers at a loss to detect and reinforce child prosocial responses, as well as in spotting and stopping the children's initiations of coercion (Snyder & Stoolmiller, 2002).

Under the previously described chaotic home conditions, children and parents are also likely to develop habits and beliefs that seem due more to coincidence than to contingency. Thus, parents who are poor observers and who demonstrate harsh and inconsistent discipline habits will freely offer narratives comprising hostile beliefs about their children, and these adults show a linkage between their stated beliefs and their poor discipline tactics (Dix, 1993; Miller & Prinz, 2003; Nix et al., 1999; Snyder, Cramer, Afrank, & Patterson, 2005). While the causal and directional properties of this linkage are not yet clear, Snyder et al. (2005) offered the logical speculation that these parent attributions operate as verbal rules, guiding discipline practices that are not based on the children's actions and words. As long as the children continue to show instances of disruptive behavior, "coincidence suffices" and the parents' poor practices will be maintained (Skinner, 1977, p. 4). Not surprisingly, the children are prone to develop their own hostile beliefs that covary with their coercive actions and temper outbursts (Dodge, 1986, 1993; Dodge, Bates, & Pettit, 1990). Similar to their parents, these disruptive children and youth appear to be governed by the hostile beliefs as much as by reinforcement contingencies. In fact, the term *hostile aggression* is often used to differentiate rule-governed conduct problems from those that represent instrumental or contingency-based aggression.

As reviewed thus far, the research literature points to young children's disruptive behavior disorders as coconstructed by the children and their parents. The pathological scenario pictures a combination of ODD and ADHD as driven by child temperament and poor parenting. Once underway, this worst-case developmental process is likely to inadvertently shape inattentive, argumentative, and noncompliant behavior into the more hurtful actions comprising conduct disorder (APA, 2000). Inattention can now be more clearly defined in neuropsychological measures of immediate and spatial memory impairments (Raine et al., 2005), and oppositionality is shaped into physical assault, theft, and running away from home (Burke, Loeber, & Birmaher, 2002). These developmental transitions in disruptive behavior, while typically occurring in adolescence, seem to be emerging earlier in the lives of troubled children. One national survey of police arrests of very young juveniles aged less than 13 years from 1988 to 1997 revealed a 17% drop in property crimes, but a 45% increase in violent crimes. Simple assaults increased 79%, weapons law violations increased 76%, drug abuse violations increased 165%, and curfew violations increased 121% (Snyder, Espiritu, Huuzinga, Loeber, & Petechuk, 2003).

These "early starters" (Lahey & Loeber, 1994) are also those preadolescents who are likely to become "life course persistent" antisocial individuals (Moffitt, 1993). As expected, they show all of the markers of persistence described earlier in this chapter (i.e., oppositionality, inattention, and impulsivity), and their home-based social interactions with parents are harsh and inconsistent. In addition, a portion of this chronically disruptive group display what Frick and Ellis (1999) described as "callous-unemotional traits." Compared to the larger proportion of early starters, these preadolescents have been shown to be higher in thrill-seeking motivation and lower in concern for the welfare of other people. Like the rest of their conduct-disordered group mates, they are inattentive, impulsive, and oppositional, but their lack of emotional reactivity, including fearlessness, sets them apart on a dimension of dangerousness (Blair, 1999; Caputo, Frick, & Brodsky, 1999).

FUNCTIONAL ASSESSMENT AND ANALYSIS

While all disruptive children and adolescents create problems for themselves and for their families and communities, the early starters are noteworthy because of their chronicity. Including those with callous-unemotional traits, early starters represent a target group manifesting the full range of behavioral and cognitive markers of DSM disruptive disorders. Were we to more fully understand these markers through functional analyses, new and more effective clinical strategies should emerge. As seen in the previous section of this chapter, these markers include the development of coercive behavior patterns known to be governed by negative and positive reinforcement contingencies in families and peer environments (Dishion, McCord, & Poulen, 1999; Stoolmiller & Snyder, 2004). Once these patterns are established, the governance of these patterns appears to broaden into rules that are little influenced by reinforcement contingencies (Crick & Dodge, 1996; Dodge, 1993; Hayes & Ju, 1995). This process appears to involve the children's memories of past experiences (Burks, Laird, Dodge, Pettit, & Bates, 1999) and is therefore subject to the youngsters' memory abilities and degree of order in the actual experiences. Since we know that early starters are likely to have memory impairments and their family experiences are likely to have been chaotic, it seems reasonable to assume that their recall of past happenings will be more ambiguous than those formulated by less troubled peers.

CHILDREN'S MEMORIES OF THE PAST AND THEIR CONSTRUCTION OF MYTHS

Recalling Skinner's 1977 speculation on "the force of coincidence," his earlier laboratory studies of adventitious reinforcement are relevant in our efforts to fully understand the behavior of chronically disruptive children and adolescents. If coincidence is the heart of operant conditioning and if this form of adventitious learning leads to the development of myths (Skinner, 1977, p. 5), one would wonder if myth formation could be a maintenance factor for disruptive behavior—particularly the verbal behavior of disruptive children and youth.

According to Webster (1975), a myth is "a traditional story of unknown authorship, ostensibly with a historical basis, but serving usually to explain some phenomenon of nature, the origin of man, or the customs, institutions, religious rites, etc., of a people." Viewed in this way, myths are created by individuals and by groups of individuals to make sense out of their life experiences. Notice in the definition that these stories ostensibly have a historical basis, meaning that myths are not necessarily valid pictures of what happened. Even those individuals with superb memory capacities will embellish, omit, and add to the developing stories that will eventually be claimed as true (see the validation studies surveyed by Henry, Moffitt, Caspi, Langley, & Silva, 1994).

During conversations about the past, parents have opportunities to help their young children to formulate personal memories into *coherent* accounts of what happened (Hudson, Fivush, & Kuebli, 1992). Coherence refers to the structural organization of a narrative or the degree to which the story makes sense and yields a detailed picture of the happenings (Fivush, 1991; Hudson, 1990; Nelson, 1993). The teaching process usually gets underway during the children's later preschool years (3 to 5 years), or whenever they begin to use language to represent themselves in the past as well as in the present (Nelson, 1993).

Being a parent "editor" for a fledgling narrator requires almost the same order of skill and patience as that needed to teach the youngster appropriate social skills (Fivush & Fromhoff, 1988; Hudson, 1990). By adopting an "elaborative" style when engaging their youngsters in conversations about the past, parents prompt the youngsters' recall of personal experiences, including details of these happenings (e.g., who, where, when, why, and how), all of which are then organized in story format (i.e., beginning, middle, and end). Not surprisingly, children who participate in this teaching process learn to recount their personal experiences through narratives that are more clearly described and more appropriately organized than is true of children whose parents do not offer this guidance (Reese, Haden, & Fivush, 1993).

The fruits of becoming a competent narrator in the preschool years are evident when these children are asked to make conclusions about their memories. Generally speaking, those whose memories are better organized and more detailed are also more apt to realize that their conclusions are not necessarily the same as other people's conclusions about the same happening (Welch-Ross, 1997). In other words, some preschoolers can understand that they and others may sometimes experience false beliefs (Astington, 1993). This ability and willingness to question the validity of one's narrative accounts of the past lead a youngster to wonder about the how and why of any remembered happening (Fivush, 1994; Miller, 1994). With the continual help of a parent editor, who encourages the linkage of recent memories with similar experiences in the more distant past, the youngster becomes progressively more sophisticated in causal reasoning (Perner & Ruffman, 1995; Welch-Ross, 1995). As a result, the child's ability to sort out happenings based on coincidence versus contingency should improve.

It seems evident that effective parenting involves bouts of systematic child-parent turn-taking in both action (Raver, 1996) and in conversation about past actions (Welch-Ross, 1997). We know from the research literature that children who seem to be developing disruptive behavior disorders are not regular participants in this turn-taking teaching enterprise. Due to the home chaos caused by parents and by their children, the youngsters will not acquire competence in building social skills or in building coherent autobiographical narratives. Instead, the children enhance their abilities to coerce others, and their personal stories are unquestioned facts of life. While there is very little empirical evidence linking children's coercive actions and their narratives, Burks et al. (1999) found correlational evidence of a link between children's "hostilely-oriented knowledge structures," their perceptions of hostility in the actions of others in the here and now, and parent/teacher ratings of the children's aggression. The Burks group's definition of knowledge structures as based on the children's memories of past experiences offers a reasonable fit with our use of the narrative concept.

FUNCTIONAL ANALYTIC-BASED INTERVENTIONS

The coercive actions and ambiguous narratives of preadolescents on the road to conduct disorder might well be supported through social experiences based on both coincidence and contingency. If the latter mechanism were solely responsible for these behavior disorders, therapeutic interventions would clearly fit today's parent training procedures based on contingency management. However, given that coincidence is also a contributor, we should consider broadening these procedures to include training parents in how to help their children to study personal memories of their various social experiences. Following guidelines generated by the research just reviewed (Fivush, 1991; Welch-Ross, 1997), parents would prompt their children to review the past in narrative format focused on the children's views of what, when, where, who, why, and how. Since these children have already constructed their stories, a parent's role as "editor" is to encourage "narrative *restructuring*" by improving story coherence (Wahler

& Castlebury, 2002). Thus, the parent's job is not to directly change story content but to help the author improve these tellings through enhancing detail and organization of the happenings. In other words, the goal of narrative restructuring is to promote structural change of the child's ambiguous memories in the hope that this youngster will begin to question his or her conclusions about these happenings (Welch-Ross, 1997). Should this doubt occur, the power of these conclusions as setting events for disruptive behavior ought to be weakened. As of yet, there are no published studies examining this intervention hypothesis, and thus, the following steps are solely based on our clinical experiences.

The sequence of interventions in our parent training strategy first involves contingency management composed of appropriate discipline and positive reinforcement (Patterson & Chamberlain, 1994; Webster-Stratton, 1998). Then, when the child acquires the cooperative behavior necessary for conversations about the past, the parent training can be expanded to include the yet to be tested narrative restructuring. This second step is usually more difficult for parents because of what they hear from their children, and because of our clinical protocol limiting parent responses to the narratives. Most parents want to correct their children's recall of happenings, and they especially want to alter the children's conclusions based on the youngsters' views of how and why things happened. As one might expect (Burks et al., 1999), the typical conclusions by disruptive disordered children and youth contain hostile content, blaming other people, and self-justification in being oppositional.

Fortunately, our clinical use of this expanded parent training model suggests that clinicians gain credibility in parents' eyes after the standard contingency management procedures work. Given parents' success in reducing child oppositionality and coercion and the generation of cooperation, parents are more willing to accept another parenting protocol that seems counterintuitive to most of them. Why, some parents ask, should I accept more lies about what happened along with this baloney about not being responsible? The only satisfying answer to this question is often found in three points: (1) You cannot alter your child's memories and beliefs by imposing those of your own; (2) you have already taken charge of your child's life, and he/she will continue to be a willing participant as long as you continue to practice appropriate contingency management; and (3) your child can and should gain a clear and credible picture of his/her past, and this new picture might stabilize his/her willingness to reconsider personal beliefs about why and how things happened.

Narrative restructuring with disruptive behavior disordered children will soon be studied through experimental analyses of its enhanced coherence of the youngsters' autobiographical narratives. While a number of correlational studies with normal and disruptive children reveal the expected covariation between the children's narrative coherence and prosocial behavior (Goin & Wahler, 2002; Oppenheim, Nuz, Warren, & Emde, 1997), the functional properties of this connection have been pursued largely through clinical practice, as in our research group in Tennessee.

CASE STUDY

To illustrate our clinical practice of parent training in contingency management and narrative restructuring, consider this referral to our clinic. Carlos was a 9-year-old Hispanic male, of average intelligence, brought to our clinic by his widowed 26-year-old mother who lived with her only child in subsidized housing. Carlos met criteria as an "early starter" conduct disordered child due to bullying peers, being cruel to animals, destroying property, being truant from school, and breaking into a neighbor's house. His mother petitioned the juvenile court for help, and the court referred mother and child to our clinic.

In a baseline diagnostic phase of 3 weeks, 1-hr mother-child home interactions were videotaped twice per week, and the dyad came to our clinic once per week for 30-min videotaped sessions in which mother prompted Carlos to talk about any of his recent and distant experiences at home, school, or community. The six home observations were coded by observers using our Standard Observation Codes-Revised (SOC-R; Cerezo, 1988), and the three narrative transcripts were coded for coherence of Carlos's narrative accounts, and mother's prompts were coded as elaborative, repetitive, or directive by raters using the Castlebury and Wahler (1997) guidelines.

A summary of these baseline home sessions showed Carlos to be coercive and offering little in the way of prosocial activity. In her turn, mother was inconsistent in her use of positive, neutral, and negative social attention. In the baseline clinic sessions, Carlos' narrative scores reflected low coherence, and his mother seldom asked him to elaborate his tellings. Thus, our parent training intervention was aimed at helping Carlos' mother to reduce her harsh, negative attention and to use timeout along with consistent use of her neutral and positive attention following the boy's prosocial responses. Following this intervention, we planned to teach mother to increase the coherence of Carlos' narratives by prompting his elaboration of ambiguous and confusing story segments.

Parent training occurred at home in weekly sessions following the baseline and began with mother's contingency management of Carlos' coercive demands, interruptions, and verbal insults, a pattern that was clearly evident across the baseline sessions. The training began in our clinic through mother's reading of a pamphlet on discipline. This was followed by two sessions in which mother and clinician reviewed baseline videotapes to identify Carlos' coercive responses, his prosocial responses, and his mother's reactions, as well as the clarity of her own instructions. Then, mother was instructed and given feedback in the use of timeout for Carlos' coercive responses and praise for the boy's prosocial actions. Mother also made it clear to Carlos through her stated rules that he would no longer be allowed to roam the neighborhood, and his whereabouts at school would be monitored by teachers. The contingency management system was then implemented at home with the clinician making twice weekly visits to provide feedback to mother and to Carlos. Following each session, mother and clinician reviewed segments of the videotape to point out mother's mistakes and to praise her appropriate use of social attention and timeout. Observers coded all of these videotaped sessions. The 30-min weekly video-taped clinic sessions in which Carlos and his mother had conversations about the past continued, and the videotape transcripts were rated for Carlos coherence and mother's role as "editor."

Mother's parent training performance as measured by SOC-R became appropriate and consistent within 3 weeks as indicated by observer coding of her contingent social attention and timeout. Compared to her baseline performance, her harsh responses declined from 20% of the SOC-R time intervals to 3%. Her appropriate and consistent use of social attention increased from 60% to 93%. As expected, Carlos' coercive responses declined from 18% to 2% and his prosocial responses, including compli-ance, increased from 11% to 22%. Following another 3 weeks of similar home performance by Carlos and his mother, the home-based parent train-ing stopped, home videotaping and coding continued, and the narrative restructuring phase began in our clinic.

A look at the ratings of Carlos's narratives and his mother's ways of prompting and responding to his recall of recent and earlier experiences showed the following baseline pattern over the six clinic sessions. Carlos did tell stories about his personal experiences with his mother, grandpar-ents, deceased father, peers, and teachers. Coherence ratings were com-posed of *clarity* (having a central point, detail in the happenings, all happenings relevant to the central point) and *credibility* (time and/or place, feelings and/or thoughts, ownership of how and why). Mother was simply told to prompt Carlos to tell about his experiences, and her contributions were rated as directives, such as corrections and telling him what hap-pened, prompting him to elaborate his tellings, and repetitive prompts, such as "Uh huh" and "I see."

Carlos' coherence ratings were low (clarity mean was 2.0 out of 5 points, and credibility was 1.2 out of 5 points). Mother's ways of initiating and supporting Carlos' stories were repetitive (mean = 76%), directive (mean = 21%), and elaborative (mean = 3%). Most of his story central points were about himself as a macho male (e.g., "I kick butt"), a risk taker (e.g., "I'm not afraid to go anywhere or do anything"), and rejection (e.g., "Nobody likes me, so what?"). Often his tone of voice was hostile, and he often interrupted his mother's prompts.

The narrative restructuring protocol started with mother's reading of a pamphlet describing how parents help their children to improve the coherence (i.e., clarity and credibility) of their personal stories. Mother was then shown a listing of Carlos' low coherence scores for the stories he told through her prompting in the clinic baseline sessions. She was also shown a corresponding list of her scores as "editor" when Carlos related his experiences; she was not surprised to see that her primary mode of editing involved repetitive and directive responses (97% of all her utterances). She was surprised when told by her parent trainer that she must markedly increase her prompts for Carlos' elaboration of his tellings (baseline mean = 3%). Amidst mother's protests that she would endorse his macho image and ruin the gains in his home-based prosocial behavior, the parent trainer assured mother that she could still handle his home behavior through her newly developed contingency management skills.

Carlos' mother's fears were confirmed when Carlos enthusiastically elaborated his stories following her prompts for clarity and credibility. As he gradually improved his story coherence scores, he did indeed test her authority at home, but with her trainer's support, she proved her ability to quell his episodes of coercion and oppositionality. Not surprisingly, her calm and consistent limit setting was met by his insults (e.g., "You two-faced bitch. You know who I am!"). Despite his pronouncements that he would ignore her, he proved to be a willing storyteller during the mother-child clinic sessions, and his narrative coherence scores steadily increased. Over the next 3 months, mother actually became intrigued by Carlos' views of the past, and her worries about his misconceptions were replaced by interest in his well-thought-out memories and glimmers of problem ownership along with emerging self-doubts about his competence.

Carlos periodically tested his mother's authority in numerous disruptive episodes over the next year of our maintenance phase (monthly clinic sessions, two home observations, and mother's monthly telephone reports). There were no further reports of his more serious conduct offenses such as cruelty to animals, breaking and entering, and bullying peers. By all accounts, the mother-child relationship was fairly positive. We were particularly impressed by Carlos' mother's authoritative parenting, Carlos' more cooperative behavior at home and at school, and continued revisions of his well-organized and detailed memories.

CONCLUSIONS

Coincidence is unavoidable in one's experiences with other people. When it occurs during social interactions, any one or all of the participants could acquire habits based purely on this "spurious evidence of order" (Skinner, 1977, p. 6). We know that coincidence covaries with the usual unpredictability of moment-to-moment social exchanges, and so it stands to reason that this phenomenon will be more prevalent in chaotic environments. We also know that seriously troubled disruptive children and youth tend to grow up in such home and community environments and, thus, it is likely that their problem behaviors will to some extent be acquired and maintained through coincidental as opposed to instrumental consequences.

When these children and youth are asked to talk about why they behave as they do, most will justify their actions through ambiguous narratives that are readily dismissed by adult listeners as refusals to accept ownership. While this conclusion by parents, teachers and counselors is valid, it misses the likely heart of the refusals—the ambiguity of the narratives. Well-constructed narratives by children are usually due to cooperative efforts by parents and youth working together in their respective roles as editor and narrator. Of course, oppositionality is a hallmark feature of disruptive children, meaning that they are as unlikely to construct clear and credible life stories as they are to acquire prosocial repertoires of behavior.

Most of what is concluded in this chapter is based on correlational evidence and the author's speculations on developmental trends and intervention strategies. As such, these writings are ideas and hypotheses, presented for the consideration of researchers and clinicians. Functional analysis always begins with baseline patterns of covarying behavior and stimuli, along with theoretical notions about the functional significance of these correlations. B. F. Skinner presented some of these notions in his 1977 paper, leading this author to lay out the framework of this chapter.

REFERENCES

American Psychiatric Association (2000). *Diagnostic and statistical manual of mental disorders* (text revision). Washington, DC: Author.

Astington, J. W. (1993). *The child's discovery of mind.* Cambridge, MA: Harvard University Press.

Barkley, R. A. (1997). Behavioral inhibition, sustained attention, and executive functions: Constructing a unifying theory of ADHD. *Psychological Bulletin, 121,* 65–94.

Bates, J. E., Pettit, G. S., Dodge, K. A., & Ridge, B. (1998). Interaction of temperamental resistance to control and restrictive parenting in the development of externalizing behavior. *Developmental Psychology, 34,* 982–995.

Blair, R. J. R. (1999). Responsiveness to distress cues in the child with psychopathic tendencies. *Personality and Individual Differences, 27,* 135–145.

Burke, J. D., Loeber, R., & Birmaher, B. (2002). Oppositional defiant disorder and conduct disorder: A review of the past 10 years, part II. *Journal of the American Academy of Child and Adolescent Psychiatry, 41,* 1275–1293.

Burks, V. S., Laird, R. D., Dodge, K. A., Pettit, G. S., & Bates, J. E. (1999). Knowledge structures, social information processing, and children's aggressive behavior. *Social Development, 8*, 220–236.

Caputo, A. A., Frick, P. J., & Brodsky, S. L. (1999). Family violence and juvenile sex offending: Potential mediating roles of psychopathic traits and negative attitudes toward women. *Criminal Justice and Behavior, 26*, 338–356.

Caspi, A. (1998). Personality development across the life courses. In W. Damon (Series Ed.) & N. Eisenberg (Vol. Ed.), *Handbook of child psychology: Vol. 3. Social, emotional, and personality development* (5th ed., pp. 311–388). New York: Wiley.

Caspi, A., Henry, B., Moffitt, T. E., & Silva, P. A. (1995). Temperamental origins of child and adolescent behavior problems: From age 3 to age 15. *Child Development, 66*, 55–68.

Castlebury, F. D., & Wahler, R. G. (1997). *Guidelines in coding the personal narratives of children, parents, and teachers.* Unpublished manuscript, University of Tennessee, Knoxville.

Cavell, T. A., & Strand, P. S. (2002). Parent-based interventions for aggressive, antisocial children: Adapting to a bilateral lens. In L. Kuczynski (Ed.), *Handbook of dynamics in parent-child relations* (pp. 395–419). Thousand Oaks, CA: Sage.

Cerezo, M. A. (1988). Standardized observation codes. In M. Hersen & A. S. Bellack (Eds.), *Dictionary of behavioral assessment techniques* (pp. 442–445). New York: Pergamon Press.

Crick, N. R., & Dodge, K. A. (1996). Social information processing mechanisms in reactive and proactive aggression. *Child Development, 67*, 993–1002.

Dishion, T. J., McCord, J., & Poulen, F. (1999). When interventions harm: Peer groups and problem behavior. *American Psychologist, 54*, 755–764.

Dix, T. H. (1993). Attributing dispositions to children: An interactional analysis of attribution in socialization. *Personality and Social Psychology Bulletin, 19*, 633–643.

Dodge, K. A. (1986). A social information processing model of social competence in children. In M. Perlmutter (Ed.), *Minnesota symposium on child psychology* (Vol. 8, pp. 77–126). Hillsdale, NJ: Erlbaum.

Dodge, K. A. (1993). Social-cognitive mechanisms in the development of conduct disorder and depression. *Annual Review of Psychology, 44*, 559–584.

Dodge, K. A. (2002). Mediation, moderation, and mechanisms in how parenting affects children's aggressive behavior. In J. G. Borkowski, S. L. Ramew, & M. Bristol-Power (Eds.), *Parenting and the child's world: Influence on academic, intellectual and social development* (pp. 215–229). Mahwah, NJ: Erlbaum.

Dodge, K. A., Bates, J. E., & Pettit, G. S. (1990). Mechanisms in the cycle of violence. *Science, 250*, 1678–1683.

Dumas, J. E., Nissley, J., Nordstrom, A., Smith, E. P., Prinz, R. J., & Levine, D. W. (2005). Home chaos: Sociodemographic, parenting, interactional, and child correlates. *Clinical Child and Adolescent Psychology, 34*, 93–104.

Fivush, R. (1991). The social construction of personal narratives. *Merrill-Palmer Quarterly, 37*, 59–82.

Fivush, R., & Fromhoff, F. A. (1988). Style and structure in mother-child conversations about the past. *Discourse Processes, 11*, 337–355.

Fivush, R. C. (1994). Constructing narrative, emotion and self in parent-child conversation about the past. In U. Neisser & R. Fivush (Eds.), *The remembering self: Construction and accuracy in the self narrative* (pp. 136–157). New York: Cambridge University Press.

Frick, P. J., & Ellis, M. L. (1999). Callous-unemotional traits and subtypes of conduct disorder. *Clinical Child and Family Psychology Review, 2*, 149–168.

Frick, P. J., Lilienfeld, S. O., Ellis, M. L., Loney, B. R., & Silverthorn, P. C. (1999). The association between anxiety and psychopathy dimensions in children. *Journal of Abnormal Child Psychology, 27,* 381–390.

Goin, R., & Wahler, R. G. (2001). Children's personal narratives and their mothers' responsiveness as covariates of the children's compliance. *Journal of Child and Family Studies, 10,* 439–447.

Hayes, S. C., & Ju, W. (1995). The applied implications of rule-governed behavior. In W. O'Donohue & L. Krasner (Eds.), *Theories of behavior therapy: Exploring behavior change* (pp. 374–391). Washington, DC: American Psychological Association.

Henry, B., Moffett, T. E., Caspi, A., Langley, J., & Silva, P. A. (1994). On the "remembrances of things past": A longitudinal evaluation of the retrospective method. *Psychological Assessment, 6,* 92–101.

Holden, G. W., & Miller, P. C. (1999). Enduring and different: A meta analysis of the similarity in parents' child rearing. *Psychological Bulletin, 125,* 223–254.

Hudson, J. A. (1990). The emergence of autobiographic memory in mother-child conversation. In R. Fivush & J. A. Hudson (Eds.), *Knowing and remembering in young children* (pp. 166–196). Cambridge, England: University Press.

Hudson, J. A., Fivush, R., & Kuebli, J. (1992). Scripts and episodes: The development of event memory. *Applied Cognitive Psychology, 6,* 483–505.

Kochanska, G. (1993). Toward a synthesis of parental socialization and child temperament in early development of conscience. *Child Development, 64,* 325–347.

Lahey, B. B., & Loeber, R. (1994). Framework for a developmental model of oppositional defiant disorder and conduct disorder. In D. K. Routh (Ed.), *Disruptive behavior disorders in childhood* (pp. 139–180). New York: Plenum.

Lahey, B. B., McBurnett, K., & Loeber, R. (2000). Are attention-deficit/hyperactivity disorder and oppositional defiant disorder developmental precursors to conduct disorder? In A. Sameroff, M. Lewis, & S. M. Miller (Eds.), *Handbook of developmental psychopathology* (2nd ed., pp. 431–446). New York: Plenum Press.

Miller, G. E., & Prinz, R. J. (2003). Engagement of families in treatment for childhood conduct problems. *Behavior Therapy, 34,* 517–534.

Miller, P. J. (1994). Narrative practices: Their role in socialization and self-construction. In U. Neisser & R. Fivush (Eds.), *The remembering self: Construction and accuracy in self-narratives* (pp. 158–179). NY: Cambridge University Press.

Moffitt, T. E. (1993). Adolescence-limited and life-course-persistent antisocial behavior: A developmental taxonomy. *Psychological Review, 100,* 674–701.

Nelson, K. (1993). The psychological and social origins of autobiographical memory. *Psychological Science, 1,* 1–8.

Nix, R. L., Pinderhughes, E. E., Dodge, K. A., Bates, J. E., Pettit, G. S., & McFadyen-Ketchum, S. A. (1999). The relation between mothers' hostile attribution tendencies and children's externalizing behavior problems: The mediating role of mothers' harsh discipline practices. *Child Development, 70,* 896–909.

Oppenheim, D., Nuz, A., Warren, S., & Emde, R. N. (1997). Emotion regulation in mother-child narrative co-construction: Associations with children's narratives and adaptation. *Developmental Psychology, 33,* 284–294.

Patterson, G. R. (1982). *Coercive family process* (Vol. 3). Eugene, OR: Castalia.

Patterson, G. R. (1997). Performance models for parenting: A social interactional perspective. In J. Grusec & L. Kuczynski (Eds.), *Parenting and the socialization of values: A handbook of contemporary theory* (pp. 193–235). New York: Wiley.

Patterson, G. R., & Chamberlain, P. (1994). A functional analysis of resistance during parent training therapy. *Clinical Psychology: Science and Practice, 1,* 53–70.

Perner, J., & Ruffman, T. (1995). Episodic memory and autonoetic consciousness: Developmental evidence and a theory of childhood amnesia. *Journal of Experimental Child Psychology, 59,* 516–548.

Raine, A., Moffitt, T. E., Caspi, A., Loeber, R., Stouthamer-Loeber, M., & Lynam, D. (2005). Neuropsychological impairments in conduct disordered youth. *Journal of Abnormal Psychology, 114*, 38–49.

Raver, C. C. (1996). Relations between social contingency in mother-child interaction and two year-olds' social competence. *Developmental Psychology, 32*, 850–859.

Reese, E., Haden, C. A., & Fivush, R. (1993). Mother-child conversations about the past: Relationships of style and memory over time. *Cognitive Development, 8*, 403–430.

Rothbart, M. K., & Bates, J. (1998). Temperament. In W. Damon (Series Ed.) & N. Eisenberg (Vol. Ed.), *Handbook of child psychology: Social, emotional, and personality development* (5th ed., pp. 105–176). New York: Wiley.

Schwartz, C. E., Snidman, N., & Kagan, J. (1999). Adolescent social anxiety as an outcome of inhibited temperament in childhood. *Journal of the American Academy of Child and Adolescent Psychiatry, 38*, 1008–1015.

Skinner, B. F. (1977). The force of coincidence. In B. Etzel, J. LeBlanc, & D. Baer (Eds.), *New developments in behavioral research: Theory, method, and application* (pp. 3–6). Hillsdale, NJ: Lawrence Erlbaum.

Snyder, H. N., Espiritu, R. C., Huuzinga, D., Loeber, R., & Petechuk, D. (2003). Risk and protective factors of child delinquency. *Child Delinquency: Bulletin Series, March*, 1–8.

Snyder, J., & Stoolmiller, M. (2002). Reinforcement and coercion mechanisms in the development of antisocial behavior: The family. In J. B. Reid, G. R. Patterson, & J. Snyder (Eds.), *Antisocial behavior in children and adolescents: A developmental analysis and model for intervention* (pp. 65–100). Washington, DC: American Psychological Association.

Snyder, J. S., Cramer, A., Afrank, J., & Patterson, G. R. (2005). The contributions of ineffective discipline and parental hostile attributions of child misbehavior to the development of conduct problems at home and school. *Developmental Psychology, 41*, 30–41.

Stoolmiller, M. (2001). Interaction of child manageability problems and parent-child discipline tactics in predicting future growth in externalizing behavior for boys. *Developmental Psychology, 37*, 814–825.

Stoolmiller, M., & Snyder, J. (2004). A multilevel analysis of parental discipline and child antisocial behavior. *Behavior Therapy, 35*, 365–402.

Wahler, R. G., & Castlebury, F. D. (2002). Personal narratives as maps of the ecosystem. *Clinical Psychology Review, 22*, 297–314.

Wahler, R. G., & Dumas, J. E. (1989). Attentional problems in dysfunctional mother-child interactions: An inter-behavioral model. *Psychological Bulletin, 105*, 116–130.

Webster, N. (1975). *Webster's new twentieth century dictionary of the English language*. London: Collins World Publishing Co.

Webster-Stratton, C. (1998). Preventing conduct problems in Head Start children: Strengthening parenting competencies. *Journal of Consulting and Clinical Psychology, 66*, 715–730.

Welch-Ross, M. K. (1995). An integrative model of the development of autobiographical memory. *Developmental Review, 15*, 338–365.

Welch-Ross, M. K. (1997). Mother-child participation in conversation about the past: Relationships to preschoolers' theory of mind. *Developmental Psychology, 33*, 618–629.

7

FUNCTION-BASED ASSESSMENT AND TREATMENT OF PEDIATRIC FEEDING DISORDERS

CATHLEEN C. PIAZZA

Munroe Meyer Institute
University of Nebraska Medical Center

AND

LAURA R. ADDISON

Marcus Institute

DIAGNOSIS AND RELATED CHARACTERISTICS

A feeding disorder is identified when a child is unable or refuses to eat or drink a sufficient quantity or variety of food to maintain proper nutrition (Babbitt et al., 1994). The greatest period of risk from poor nutrition occurs in early infancy, a time of rapid physical growth and brain development. Without sufficient nutrition, the child may experience the negative side effects of malnutrition and/or dehydration. In addition, insufficient calories or nutrients place the child at long-term risk for behavioral and developmental problems (Scrimshaw, 1998).

Approximately 25–35% of typically developing children and approximately 33–80% of children with developmental delays exhibit feeding problems (Gouge & Ekvall, 1975; Palmer & Horn, 1978). However, these prevalence estimates should be viewed with caution due to the heterogeneity of feeding disorders. That is, a feeding disorder may be characterized by a variety of different topographical presentations such as total refusal to eat, dependence on supplemental feedings such as a gastrostomy tube (G-tube), inappropriate mealtime behavior (e.g., crying, batting at the spoon), failure to thrive (FTT), and selectivity by type and texture.

According to the *Diagnostic and Statistical Manual of Mental Disorders, 4th Edition (DSM-IV-TR)*, Feeding Disorder of Infancy and Childhood (307.59) is characterized by a "persistent failure to eat adequately, as reflected in significant failure to gain weight or significant weight loss over at least 1 month" (American Psychiatric Association, 1994, p. 99). Additional diagnostic criteria for a feeding disorder specify that no medical condition severe enough to account for the feeding disturbance exists, and the feeding disturbance is not better accounted for by another mental disorder or by lack of available food. The onset of a feeding disorder according to *DSM-IV-TR* is prior to age 6 years. In contrast, the *International Classification of Diseases, 9th Revision, Clinical Modification (ICD-9-CM)* uses the term *Feeding Difficulties and Mismanagement* (783.3), which excludes problems in the newborn and problems of a nonorganic nature (Practice Management Information Corporation, 2005).

Attempts to dichotomize feeding disorders into medical versus environmental etiologies have not been supported by the literature. That is, a number of studies have shown that the etiology of feeding disorders is complex and multifactorial (Burklow, Phelps, Schultz, McConnell, & Rudolph, 1998; Rommel, DeMeyer, Feenstra, & Veereman-Wauters, 2003). For example, Rommel et al. (2003) evaluated 700 children referred for the assessment and treatment of severe feeding difficulties. Feeding disorders were characterized as medical (86%), oral-motor (61%), and/or behavioral (18%). Combined causes of feeding problem occurred in over 60% of children.

The high prevalence of medical conditions and oral-motor dysfunction in children with feeding disorders suggests that biological factors play an important role in the etiology of feeding problems. Children with chronic medical problems such as food allergies, malabsorption, gastroesophageal reflux disease (GERD), delayed gastric emptying, metabolic anomalies, or congenital defects of the gastrointestinal tract may associate eating with pain, general malaise, or nausea. For example, in children with GERD, eating may be associated with vomiting and the pain that occurs when excess acid is released into the stomach or esophagus. These children might then develop behavior problems to avoid eating. Many times, even after the painful medical condition is treated, the child may continue to refuse food because if the child never or rarely eats, he or she never learns that eating is no longer painful. Nausea is particularly important in the development of aversions to food (Schafe & Bernstein, 1997). When nausea is paired with eating, aversions to tastes may develop after only one or a limited number of trials, may generalize to many foods, and may be highly resistant to treatment.

The presence of chronic medical problems may also contribute to the etiology of feeding problems because infants with complex medical histories are often subjected to numerous invasive diagnostic tests and procedures that may involve manipulation of the face and mouth (e.g., laryngoscope). Therefore, the child may associate the presentation of objects to the face and mouth (e.g., a spoon) with these early negative experiences. Parents of hospitalized and medically fragile children often report oral aversions that affect feeding and other behaviors associated with the face and mouth (e.g., tooth brushing, face washing).

Oral-motor dysfunction may include problems such as difficulty swallowing, inability to lateralize food (move it from side to side), tongue thrust, and difficulty sucking that may affect the child's ability to eat. The high prevalence of prematurity among children with feeding disorders underscores the importance of early opportunities to feed in the development of appropriate oral-motor skills. Preexisting oral-motor dysfunction may be exacerbated when the child refuses to eat, which contributes further to the child's failure to develop appropriate oral-motor skills. That is, if the child does not have the opportunity to practice the skill of eating, the child does not develop the oral-motor skills to become a competent eater. Furthermore, refusal to eat may lead to failure to thrive. Ironically, undernourished children may lack the energy to become capable eaters (Troughton & Hill, 2001).

Caregiver responses to child behavior during meals also contribute to the maintenance of feeding problems. Piazza, Fisher et al. (2003) conducted descriptive assessments of child and caregiver behaviors during meals. The descriptive assessments suggested that caregivers responded to child inappropriate behaviors with one or more of the following consequences: (a) allowing escape from bites of food or the entire meal, (b) coaxing or reprimanding (e.g., "Eat your peas, they are good for you."), or (c) providing the child with a toy or preferred food.

The effects of these caregiver consequences on child behavior were then tested systematically using a modification of the analogue functional analysis (Iwata, Dorsey, Slifer, Bauman, & Richman, 1994/1982). During the analog functional analysis, bites of food were presented on a fixed-time (FT) 30-s schedule throughout four conditions; attention, escape, tangible, and control. During the attention condition, inappropriate behaviors (e.g., batting at the spoon, head turning) produced attention (e.g., brief verbal reprimands). In the escape condition, inappropriate behaviors produced a break from the spoon presentation. Access to a tangible item (e.g., preferred food) was provided for inappropriate behaviors during the tangible condition. Alternatively, no differential consequences were provided during the control condition. Results of the functional analyses were differentiated for 67% (10 out of 15) of the participants. Of the 10 children whose functional analyses were differentiated, 90% displayed behaviors sensitive to negative reinforcement. Multiple functions (e.g., inappropriate behaviors maintained by access to adult attention or tangible items) were identified for 80% of the children who showed differential responding during functional analyses, suggesting that many feeding problems are often maintained by negative reinforcement. In addition, a significant number of children with feeding disorders may also be sensitive to other sources of reinforcement (e.g., access to attention).

FUNCTIONAL ASSESSMENT AND ANALYSIS

A major question, which arises during the assessment of children with feeding problems, is when is a feeding disorder of sufficient severity to warrant treatment? Part of the difficulty in answering that question is that feeding disorders are a heterogeneous group of problems, and there is no one agreed-upon rule for initiating treatment. In fact, feeding difficulties are quite common among infants and toddlers, and many professionals will advise parents, "Don't worry, he'll grow out of it."

One method for determining the necessity of treatment is to evaluate the child's growth parameters. Most children's height and weight are obtained regularly during well child visits with their pediatrician. The child's pediatrician will plot these growth parameters on a "growth curve." These growth curves have been standardized such that one can compare the child's height and weight to the height and weight of other children of the same age and sex. Growth parameters conform to an approximate bell-shaped curve such that most children will grow along the 50th percentile; although some children's growth parameters will plot above the 50th percentile and some children will plot below the 50th percentile. The general expectation is that any given child will grow along his or her own curve.

A good rule of thumb is that treatment should be initiated when the child's growth either plateaus or decelerates (i.e., the child fails to grow along his or her curve). Although the criteria for diagnosing FTT vary among professionals, a diagnosis of FTT might be another indicator for initiating treatment. Generally, the application of this term indicates a significant failure to grow as expected based on the child's chronological age (i.e., below the 3rd or 5th percentile).

However, as indicated previously, some children may be small for their age but grow along their own curve. In this case, the child's physician may complete additional tests to determine if the child's small size is normal or pathological. Additional considerations for initiating treatment include (a) dehydration or malnutrition, particularly when emergency treatment is required; (b) the presence of a nasogastric (NG) tube with no increase in the percentage of calories obtained via oral feeding for 3 consecutive months; (c) the presence of a G-tube with no increase in the percentage of calories obtained via oral feeding for 6 consecutive months; (d) feeding behavior that is not age appropriate (e.g., a bottle-dependent 5-year-old child, a 3-year-old who does not self-feed); and (e) the child exhibits behavior at meal time that causes distress to the family.

Treatment decisions might be especially problematic when assessing a child who is a picky eater. How do we know when a child is receiving sufficient nutrients, particularly when the child is growing according to expectations? Several strategies might be employed when assessing the necessity of treatment for these children. First, physical examination may be useful for detecting dietary insufficiencies, malnutrition, and/or dehydration; thus, consultation with the child's pediatrician should be the starting point for decision making. Second, a referral to a nutritionist is important to assess the child's diet from a nutrient standpoint. An additional strategy is to evaluate the extent to which the child's eating habits differ significantly from that of the family (i.e., does the child require meals that are different from what is served by the family?) or to evaluate the extent to which the child's eating habits limit the child or family socially (e.g., does the family have to alter vacation plans to accommodate the child's diet?).

An interdisciplinary treatment that focuses on all of the components (i.e., biological, oral-motor, and psychological) contributing to feeding problems should be implemented because children have feeding problems for a variety of reasons. A thorough medical work-up is essential prior to initiating treatment for any child. Recall that Rommel et al. (2003) showed that 86% of children with feeding problems had an identifiable medical problem. Thus, initiation of treatment prior to resolution of medical problems is likely to be counter-therapeutic, particularly if treatment results in the ongoing pairing of oral intake with pain. In addition, some medical conditions require that the child follow a carefully regimented diet, and failure to do so may have dire consequences. For example, a child with a glycogen storage disorder requires careful monitoring of blood sugar levels and a diet that is restricted relative to simple sugars. Drops in blood sugar levels may result in seizures, coma, and even death.

Referral to a speech and/or occupational therapist also may be indicated to ensure that it is safe to feed the child orally. Some children aspirate solids or liquids; thus, for these children eating may be life threatening. In addition, a referral to a nutritionist is important to determine the guidelines for the quality and quantity of foods that should comprise the child's diet.

A structured interview and direct observation of a feeding should be conducted once the child has been cleared from a medical, oral-motor, and nutrition standpoint. The goal of the structured interview is to obtain information that can be used to develop hypotheses about why the feeding problem is occuring. The information from the interview then can be used to develop a data-based assessment of the problem.

Following the structured interview, we generally conduct a functional analysis (Piazza, Fisher et al., 2003). The functional analysis begins with several observations of the caregiver feeding the child as he or she would at home. During these observations, we collect data on operationally defined parent and child behaviors. Upon completion of the caregiver-fed meals, we then conduct an analogue functional analysis using the information obtained from the caregiver-fed meals. The conditions we conduct during the functional analysis are based on the consequences used by the caregivers during the caregiver-fed meals. So, for example, if the parent provides attention and escape following child inappropriate behavior, then we would conduct attention and escape conditions during our functional analysis.

Recently, we have been conducting our functional analyses using a pairwise design in which each test condition (i.e., attention, escape, and tangible) is compared to the control condition. Each session consists of 5 bites, which are presented on an average of once every 30s. If the child does not take the bite and does not engage in inappropriate behavior, the bite is removed after 30s and a new bite is presented. Across all conditions, acceptance (defined as the child opening his/her mouth and the entire bite being deposited within 5s of the initial bite presentation) and mouth clean (defined as no food or drink larger than the size of a pea is visible in the child's mouth within 30s of initial acceptance) result in brief praise ("Good job"). We generally have 4 foods available (fruit, starch, vegetable, protein), and we rotate through the foods in a random order. In the control condition, preferred toys based on the results of a preference assessment (Fisher et al., 1992) and adult attention are available continuously. In this condition, inappropriate behavior results in no differential consequence. In the attention condition, inappropriate behavior results in 20s of attention and the bite is removed at the end of the 20-s interval. In the escape condition, inappropriate behavior results in 20s of escape (i.e., the spoon is removed). Lastly, in the tangible condition, inappropriate behavior results in 20s of access to a tangible item, and the bite is removed at the end of the 20-s interval. We conduct a tangible condition only if we observe the caregiver delivering a tangible item following inappropriate behavior during the caregiver-fed meals. We use visual inspection to determine the conditions under which inappropriate behavior is greatest, which then informs treatment.

Additional analyses of motivating operations may provide useful information about specific stimuli, which alter the efficacy of the reinforcers identified during the functional analysis (Michael, 1982; Smith & Iwata, 1997). One method for evaluating the effects of motivating operations is to alter the antecedent conditions while maintaining constant reinforcement conditions (Smith & Iwata, 1997). A variety of food-related stimuli may be appropriate for evaluation such as feeding utensils (e.g., cup vs. spoon), seating (e.g., highchair vs. regular chair), and feeder (e.g., mom vs. dad), to name a few. Observations during the caregiver-fed meals may be helpful in the identification of stimuli that may be appropriate for this type of evaluation.

For example, evaluations of food type or texture (Munk & Repp, 1994) may aid in the identification of specific types of foods or textures that are associated with differential levels of acceptance or inappropriate behavior (Patel, Piazza, Santana, & Volkert, 2002). Patel et al. (2002) assessed the effects of food type and texture on levels of expulsion for one child. The initial assessment of food type showed that levels of expulsion were higher in the presence of meat relative to other foods (i.e., fruit, starch, vegetable) at the same texture, suggesting that type of food (meat) was responsible for the differential levels of expulsion. However, when texture of meats, but not other foods, was decreased, levels of expulsion also decreased. Taken together, these results suggested that both type and texture of food influenced levels of expulsion. In this case, when different types and textures of foods were presented and consequent conditions remained constant, the child's motivation to expel was altered by the presentation of different types and textures of food.

FUNCTION-BASED INTERVENTIONS

One consideration when developing treatments is that children with food refusal often have a different history or experience with food than typically eating children. That is, even though many typically eating children will have bouts of feeding problems from time to time, their problems resolve without intervention. By contrast, the feeding problems of children with food refusal often persist and worsen over time (Lindberg, Bohlin, & Hagekull, 1991). Therefore, the techniques or recommendations that are used for typically eating children may not apply to children with food refusal.

Treatment of feeding problems should be goal-driven and outcome-oriented. Feeding behaviors (e.g., amounts consumed, acceptance of bites of food) should be measured objectively, and treatment decisions should be data-based. Outcomes should be assessed regularly throughout the duration of treatment. A primary focus of treatment should be to meet the child's nutritional needs in calories, nutrients, or both. A second major goal should be to improve the child's feeding behavior, which may include decreasing inappropriate behavior or increasing age-appropriate skills such as self-feeding skills or advancing texture. Finally, a third goal should be to train caregivers to implement the treatment procedures in the home and in other environments.

A number of studies have demonstrated that treatments based on functional analyses are more effective than treatments that are not preceded by a functional analysis (Iwata, Zarcone, Vollmer, & Smith, 1994). Surprisingly, the functional analysis method has been applied infrequently to the assessment and treatment of pediatric feeding disorders. Piazza, Fisher et al. (2003) evaluated the extent to which recommended treatments matched the results of differentiated functional analyses for 10 children. A function-based treatment was developed for 8 out of the 10 participants whose functional analysis results were differentiated. At least one of the functional reinforcers (escape, attention, tangible) was used in a reinforcement-based treatment (i.e., either differential positive, differential negative, or noncontingent reinforcement) for 100% of the participants. An escape extinction procedure (either nonremoval of the spoon or physical guidance) was used with 6 out of the 9 (67%) participants for whom escape was identified as a reinforcer. Inappropriate behavior was ignored for all participants who demonstrated sensitivity to attention as reinforcement. However, the unique contribution of the functional analysis results could not be isolated because the treatments consisted of multiple components, and the effects of each component were not isolated. This is a common problem in the feeding literature in that most studies have implemented treatments as packages (e.g., differential reinforcement plus escape extinction), without the efficacy of the individual components of the package being evaluated.

The simplest approach to treatment of a feeding problem such as food refusal or food selectivity would be to use reinforcement to increase the acceptance of food. Several studies have suggested that reinforcement-based procedures alone may be effective for increasing consumption (Riordan, Iwata, Finney, Wohl, & Stanley 1984; Riordan, Iwata, Wohl, & Finney, 1980). For example, Riordan and colleagues (1980, 1984) suggested that positive reinforcement alone was responsible for increases in acceptance for 3 participants. However, these data are difficult to interpret because refusal behaviors produced escape during baseline but were ignored during the positive reinforcement treatment, and physical guidance was required to increase acceptance for one participant.

By contrast, studies by Hoch, Babbitt, Coe, Krell, and Hackbert (1994); Ahearn, Kerwin, Eicher, Shantz, and Swearingin (1996); Patel, Piazza, Martinez, Volkert, and Santana (2002); Piazza, Patel, Gulotta, Sevin, and Layer (2003); and Reed et al. (2004) showed that acceptance did not increase initially with positive reinforcement alone. In all of these studies, a putative escape extinction procedure was necessary to increase consumption initially. In Hoch et al. (1994), the putative escape extinction procedure consisted of presenting the bite to the child's lips and holding the bite at the child's lips until the bite could be deposited into the child's mouth (non-removal of the spoon). Ahearn et al. (1996) compared nonremoval of the spoon (NRS) with another putative escape extinction procedure, physical guidance (PG). During PG, if the child did not accept the bite within 5 s of presentation, then gentle pressure was placed at the mandibular joint to open the child's mouth. The feeder then deposited the bite into the child's mouth. Both procedures were equally effective in increasing acceptance and decreasing inappropriate behavior. In addition, PG may have been associated with slightly fewer negative side effects relative to NRS.

Patel, Piazza, Martinez et al. (2002) compared the effectiveness of differential positive reinforcement for acceptance (the first behavior in the chain of eating) relative to differential positive reinforcement for mouth clean (the terminal behavior in the chain of eating) for increasing consumption. Neither differential reinforcement procedures increased consumption. When differential positive reinforcement was combined with escape extinction (EE), both acceptance and mouth clean increased, independent of whether reinforcement was provided for acceptance or mouth clean. Taken together, Hoch et al. (1994); Ahearn et al. (1996); and Patel, Piazza, Martinez et al. (2002) suggested that EE might be important for increasing consumption. However, the relative contribution of extinction was not clear because EE was never implemented in the absence of reinforcement.

By contrast, Cooper et al. (1995) used a component analysis procedure to evaluate the components of a treatment package responsible for maintenance of eating. When the putative escape extinction procedure (NRS) was removed for 3 participants, acceptance decreased, suggesting that NRS was necessary for the maintenance of acceptance. However, it was not clear if NRS was necessary to produce initial increases in acceptance because NRS was implemented initially in conjunction with multiple components (e.g., differential reinforcement). Cooper et al. (1995) also evaluated the effectiveness of differential and noncontingent reinforcement with toys and social attention with 2 participants; however, the effects of the reinforcement components were less clear because the phases were not replicated.

To clarify the contribution of negative and positive reinforcement in treatment of feeding problems, Piazza, Patel et al. (2003) examined the individual contribution of positive reinforcement and putative escape extinction (NRS or PG) procedures as treatment for the feeding problems of 4 children. Acceptance increased only when putative escape extinction was used in treatment, independent of the presence or absence of differential positive reinforcement. Similarly, Reed et al. (2004) showed that noncontingent reinforcement in the absence of putative escape extinction was not associated with reductions in inappropriate behavior or increases in acceptance. Of note, however, was that the differential reinforcement of alternative behaviors (DRA) and noncontingent reinforcement (NCR) components appeared to contribute to treatment for some children when combined with the putative escape extinction procedures by reducing extinction bursts, crying, and other inappropriate behavior. These results suggested that putative-negative-reinforcement-based procedures such as NRS and PG play a central role in the treatment of feeding problems and that positive reinforcement when combined with putative escape extinction may be beneficial for some children.

We referred to NRS and PG as putative escape extinction procedures for several reasons. First, none of the studies evaluating NRS or PG used a functional analysis to identify the reinforcer(s) maintaining problem behavior prior to initiation of treatment. Second, procedures such as NRS and PG include putative attention and tangible extinction as well because no other reinforcement (e.g., attention, tangible items) was delivered following inappropriate behavior. Thus, even though behavioral change occurred for all participants in all of the studies that used NRS and/or PG, the function of these putative escape extinction procedures is unclear.

A number of studies have suggested that putative escape extinction procedures are effective; however, these procedures should be used cautiously in therapy for several reasons. First, as indicated previously, a child's safety for oral feeding should be determined prior to treatment initiation. Second, consultation with a speech and/or occupational therapist may be helpful in determining the most appropriate texture, feeding utensil, and volume of solid or liquid presentation for the child. For example, higher textures are often associated with higher response effort for the child. That is, once the bite is deposited, the child then has to chew and swallow the higher textured bites. In some cases, aspiration risk may increase if the child is distressed during the meal and fails to chew the bite adequately. Moreover, higher response effort may be associated with lower levels of consumption (Kerwin, Ahearn, Eicher, & Burd, 1995). Third, proper positioning is also critical in feeding, so consultation with a physical therapist is essential for those children who require special seating arrangements to maintain proper positioning. Lastly, consideration should be given to conducting initial extinction sessions in a therapeutic environment to observe the child's response to the procedure prior to having a caregiver implement the procedure without professional supervision.

Caregivers should be informed fully about the effects and potential side effects of extinction-based treatments such as extinction bursts and increases in emotional behavior (Lerman & Iwata, 1995; Lerman, Iwata, & Wallace, 1999). Most children with feeding problems are often under age 8 years; therefore, we typically limit the length of sessions to between 30 to 60 min, depending on the age of the child and parental input. Theoretically, terminating a session prior to the child taking the bite may be counter-therapeutic. However, our experience is that parents do not tolerate excessively long sessions (e.g., greater than 1 hr) with their toddlers.

In addition, our preliminary pilot data suggest that the probability of bite acceptance does not increase after 1 hr (i.e., if the child has not accepted the bite within 1 hr, it is unlikely he or she will accept the bite at all). However, these data have not been subjected to scientific scrutiny, so that conclusion should be accepted with caution.

Despite these cautionary notes, the literature suggests that putative escape extinction procedures such as NRS and PG are effective as treatments for feeding problems. One possible reason for the success (and perhaps necessity in some cases) of escape extinction is that like any avoidance behavior, repeated non-reinforced experiences with the aversive stimulus minimize the opportunity for extinction (Spector, Smith, & Hollander, 1981, 1983). That is, a child who refuses food never learns that eating is not aversive. In fact, animal studies have shown that under some circumstances, extinction of food avoidance occurs only when the aversive taste is infused into the oral cavity (Spector et al., 1981, 1983). Similarly, for some children, exposure to food via escape extinction may be necessary to increase acceptance.

Given the importance of negative reinforcement relative to feeding problems, it is surprising that differential negative reinforcement (DNR) has not been used more frequently in treatment. A notable exception is a study by Kelley, Piazza, Fisher, and Oberdorff (2003) who compared the effects of positive reinforcement (access to a preferred food), negative reinforcement (presentation of a nonpreferred food), and a combination of positive and negative reinforcement to increase consumption of liquids from a cup. Consumption increased in all conditions; thus, the individual contribution of negative reinforcement could not be determined. The few additional studies that have been conducted on the effects of negative reinforcement as treatment of food refusal have lacked adequate experimental control (Kahng, Boscoe, & Byrne, 2003; Kitfield & Masalsky, 2000). Thus, the effects of negative reinforcement as treatment for food refusal should be investigated further.

A number of other procedures have shown promise either alone or as adjuncts to putative escape extinction in the treatment of feeding problems. Piazza et al. (2002) showed that the simultaneous presentation of preferred and nonpreferred food was more effective than sequential presentation of preferred and nonpreferred food (Kern & Marder, 1996) in the treatment of food selectivity. In the simultaneous presentation condition, a bite of nonpreferred food (e.g., green bean) was presented at the same time (e.g., on top of) a bite of the preferred food (e.g., chip). In the sequential condition, the preferred food was presented following consumption of the nonpreferred food.

Mueller, Piazza, Patel, Kelley, and Pruett (2004) and Patel, Piazza, Kelly, Ochsner, and Santana (2001) evaluated variations of the simultaneous presentation procedure by combining simultaneous presentation with fading and NRS. Mueller et al. (2004) identified two children who would consume some foods (labeled preferred) but not others (labeled nonpreferred) with NRS. During treatment, the preferred and nonpreferred foods were blended together in specific ratios (e.g., 90% preferred food, 10% nonpreferred food). Then the amount of nonpreferred food was increased gradually by 10% until the participants were consuming the nonpreferred foods in the absence of the preferred foods.

The study by Patel et al. (2001) was similar except that treatment involved first blending water and Carnation Instant Breakfast® (CIB), and then adding milk to the water and CIB. Fading in conjunction with NRS or PG also has been used to increase texture (Shore, Babbitt, Williams, Coe, & Snyder, 1998), to increase variety (Freeman & Piazza, 1998), and to treat liquid refusal (Hagopian, Farrell, & Amari, 1996). Shore et al. (1998) combined higher with lower textures and gradually increased texture in 25% increments. Freeman and Piazza (1998) gradually increased the amount (in 5% increments) and type (fruit, protein, starch, vegetable) of foods consumed. Similarly, Hagopian et al. (1996) added backward chaining to fading plus reinforcement to treat the liquid refusal of one child.

Several other procedures have received limited attention in the literature as treatments for feeding problems and warrant additional study. Kerwin et al. (1995) showed that levels of acceptance varied as a function of spoon volume (e.g., higher levels of acceptance were associated with lower spoon volumes). Dawson et al. (2003) showed that a high-probability procedure did not contribute to the treatment of one child's feeding problem either alone or in conjunction with EE. Finally, preliminary data by Kahng, Tarbox, and Wilke (2001) suggested that response cost (removal of tangible items following problem behavior and/or refusal) might be useful as treatment.

Up to this point, the focus of the chapter has been on procedures that target increased acceptance of food and decreased inappropriate behavior. However, these behavior changes may be necessary but not sufficient to treat a feeding problem because eating is not a single response. Rather, consumption consists of a chain of behaviors that may include not only accepting, but also chewing, swallowing, and retaining food or drink. Sevin, Gulotta, Sierp, Rosica, and Miller (2002) demonstrated that problems in consumption may occur at any point along the chain. They increased the acceptance of one child using NRS. The increases in acceptance following the implementation of NRS were accompanied by increases in expulsion. Treatment of expulsion (re-presenting expelled food) was associated with the emergence of packing. The authors treated packing with a redistribution procedure (placing the food back on the tongue).

Depositing expelled food back in a child's mouth (re-presentation) appears to be a promising intervention for expulsion based on the limited published data (Coe et al., 1997; Sevin et al., 2002). Patel, Piazza, Santana, et al. (2002) showed that an assessment of type and texture (presentation of different types of foods at different texture; Munk & Repp, 1994) could be used to prescribe treatment for expulsion. The assessment of type and texture suggested that expulsion occurred more often with meats compared to other foods (fruits, vegetables, starches). Decreasing the texture of meats, but not the other foods, resulted in decreased expulsion.

Packing (holding or pocketing food in the mouth) is another behavior problem that has been described, but infrequently evaluated, in the feeding literature. Patel, Piazza, Layer, Coleman, and Swartzwelder (2005) treated the packing of three children who were diagnosed with inadequate weight gain. Each child was presented with a variety of different textures of food and the texture associated with the lowest levels of packing was presented as treatment. In cases where lowering texture is not a treatment option (e.g., caregivers want the child to maintain an age-appropriate texture), food redistribution may be a viable treatment option for packing (Gulotta, Piazza, Patel, & Layer, 2005; Sevin et al., 2002). The redistribution procedure involves gathering the packed food on a spoon or an oral gum stimulator (Nuk®) and placing it back on the child's tongue according to a fixed-time schedule.

Swallow elicitation is another procedure that has been used to facilitate swallowing of food. Hoch et al. (1994) and Lamm and Greer (1988) showed that a posterior placement of food on the tongue either using a Nuk® brush or a finger resulted in swallow elicitation, which could result in a faster swallow. However, this procedure should be used with caution. The swallow is a reflex, which once triggered, will occur independent of whether the airway is closed or not.

PARENT TRAINING

The majority of published studies on the treatment of feeding disorders have used trained individuals as therapists. However, a few studies have examined the extent to which parents could be trained to implement treatment (Anderson & McMillan, 2001; Mueller et al., 2003; Werle, Murphy, & Budd, 1993), which is important because, presumably, most children consume the majority of their meals in the home. Thus, the extent to which caregivers implement treatment procedures accurately and consistently will affect the long-term success of treatment. Werle et al. (1993) used verbal discussion, handouts, role play, behavioral rehearsal during mealtimes, verbal feedback after meals, and periodic videotape review of previous training sessions to train caregivers. Similarly, Anderson and McMillan (2001) used verbal and written instructions, modeling, videotape review, and performance feedback during and after in-home outpatient feeding services. Both Werle et al. (1993) and Anderson and McMillan (2001) used multicomponent strategies for training; therefore, it was unclear which specific strategies (e.g., modeling, feedback) were necessary for successful training.

Mueller and colleagues (2003) evaluated the effects of four different multicomponent training packages to increase the treatment integrity of parents implementing pediatric feeding protocols. The training package used in Study 1 consisted of written protocols (baseline), verbal instructions, therapist modeling, and rehearsal training, which resulted in increases in treatment integrity for all participants. Therefore, Study 2 evaluated the efficacy of components of the parent-training packages used in Study 1: (a) written protocols, verbal instructions, and modeling; (b) written protocols, verbal instructions, and rehearsal; and (c) written protocols and verbal instructions. All three training packages produced high treatment integrity. Mueller and colleagues (2003) concluded that successful training should consist of multiple reiterations of the procedure and that the form of the reiteration (e.g., modeling vs. rehearsal) was not as important as the fact that the procedures are reiterated.

OUTCOME DATA

Few studies have been published on either the short- or long-term effects of programs for the treatment of pediatric feeding disorders. Piazza and Carroll-Hernandez (2004) presented a preliminary analysis of the outcome measures for a pediatric feeding disorders program, which showed that over 87% of the goals for treatment were met by the time of discharge. Analysis of follow-up data indicated that the majority of patients (87%) continue to be followed post-discharge from the day treatment program and 85% of those patients continued to make progress toward age-typical feeding. Similarly, Byars et al. (2003) showed that a behaviorally based, intensive interdisciplinary feeding program was successful in increasing intake and decreasing G-tube feedings for nine patients. Irwin, Clawson, Monasterio, Williams, and Meade (2003) showed that children with cerebral palsy and feeding problems improved in the number of bites accepted, weight, and height following intensive interdisciplinary treatment combining behavioral strategies and oral-motor techniques.

SUMMARY OF TREATMENT RECOMMENDATIONS

Children have feeding problems for a variety of reasons; therefore, assessment and treatment should focus on every component (i.e., biological, oral-motor, and psychological) that contributes to feeding problems. Currently, treatment strategies with the most scientific support are based on applied behavior analysis (Benoit, Wang, & Zlotkin, 2000; Kerwin, 1999). Procedures based on putative escape extinction such as nonremoval of the spoon (NRS) and physical guidance (PG) have the most empirical support in the literature as effective treatments for food refusal and selectivity. More recent literature also suggests that procedures based on positive reinforcement alone (e.g., DRA, NCR) may not be consistently effective as treatment, although positive reinforcement may enhance the effectiveness of NRS and PG. In addition, the literature on the use of fading-based procedures is growing and shows good promise. Data on the treatment of expulsion and packing is limited; however, re-presentation for expulsion, texture manipulations for both expulsion and packing, and redistribution for packing warrant further study. Finally, although the literature on parent training is limited, studies suggest that parents can be trained to implement pediatric feeding protocols with some combination of written protocols, verbal instructions, therapist modeling, and rehearsal training. Preliminary outcome data for intensive, interdisciplinary programs, which use behaviorally based treatments, suggest that successful outcomes can be produced for the majority of patients treated.

DIRECTIONS FOR FUTURE RESEARCH

The opportunities for additional investigations in the area of assessment and treatment of pediatric feeding disorders is vast given the dearth of the literature in this area. Studies are needed in the area to further describe and define the problem and prevalence, to understand its etiology, to describe its course, to describe effective assessment and treatment methods, and to provide long-term follow-up data on these methods.

Specifically, in the areas of behavioral assessment and treatment, no studies that we could find have used a functional analysis to prescribe treatment for feeding disorders. Studies in this area would provide greater specificity in terms of the identification of the function of inappropriate behavior and the functional properties of treatment. That is, a pretreatment functional analysis will provide information about the relative importance of including individual components such as escape extinction, attention extinction, and/or positive and negative reinforcement in treatment. Future studies should examine function-based treatments for treating other problem behaviors associated with pediatric feeding disorders such as expulsions, packing, gagging, and vomiting.

Application of the functional analysis methodology may also be helpful in better understanding why some children develop feeding disorders while others do not. It may be the case that children who develop feeding disorders have some particular sensitivity to escape as reinforcement and/or have differential exposure to escape during eating. In addition, other variables (e.g., attention, access to preferred foods or activities) may also function as reinforcement for food refusal (Piazza, Fisher et al., 2003). Furthermore, future research should examine the risk factors for the development of feeding problems. Children identified to be at risk for feeding problems could then be followed for an extended period of time to examine their reinforcement histories. This research might be useful in determining if differential sensitivity or exposure to escape as reinforcement influences the development of feeding problems. Future research should also examine how biological (e.g., GERD) and behavioral variables interact in both the etiology and treatment of feeding problems.

CASE STUDY

BACKGROUND INFORMATION

Peyton was a 3-year-old girl admitted to an intensive program for the assessment and treatment of poor oral intake and food selectivity by type and texture. Peyton was born prematurely and had a diagnosis of developmental delays. She engaged in inappropriate mealtime behavior such as batting at the spoon, head turning, and negative vocalizations.

Prior to her participation in the program, Peyton underwent a thorough interdisciplinary evaluation to assess any underlying physical causes for her feeding problems. The medical team conducted diagnostic studies and assessed the safety of oral feeding by ensuring that she had the appropriate skills to swallow and prevent aspiration of ingested foods. Furthermore, the nutritionist evaluated Peyton's nutritional and metabolic needs to promote weight gain for her age and size.

FUNCTIONAL ASSESSMENT

Upon Peyton's admission to the program, a structured interview was conducted with Peyton's caregivers to develop a hypothesis as to why the behavior problems were occurring. Following this interview, several observations were conducted of Peyton's mother feeding Peyton as she typically did at home. During these observations, Peyton was seated in a highchair in front of her mother who fed her chicken and rice at a pureed texture with a spoon. Results of these observations indicated insufficient oral intake, a low rate of acceptance, and high rates of inappropriate mealtime behavior (e.g., head turns and bats). Following these observations, goals were set for Peyton and an analogue functional analysis was conducted.

FUNCTIONAL ANALYSIS

Based on the observations of the caregiver-fed meals, the analogue functional analysis consisted of three conditions (i.e., attention, escape, and control as described earlier in the chapter). Peyton's functional analysis was conducted using a pair-wise experimental design in which each test condition (i.e., attention and escape) was compared with the control condition.

The results of Peyton's functional analysis indicated elevated levels of inappropriate mealtime behavior (e.g., batting at the spoon and head turning) during the escape condition relative to the control condition, suggesting that Peyton engaged in inappropriate mealtime behavior to gain access to escape from the bite presentation (see Figure 7.1). However, Peyton engaged in undifferentiated levels of inappropriate mealtime behavior in both the attention and control conditions, suggesting that attention did not function as a reinforcer for her inappropriate mealtime behavior. Furthermore, Peyton had zero levels of acceptance (defined as the child opening his/her mouth and the entire bite being deposited in the child's

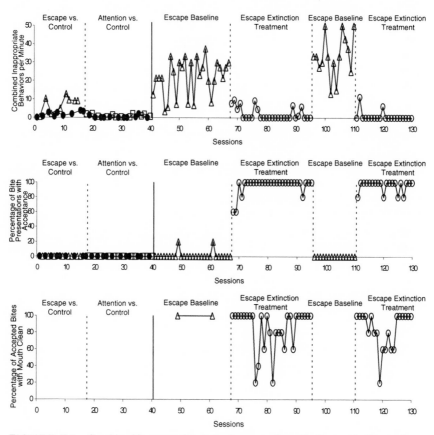

FIGURE 7.1 Combined inappropriate behaviors per minute for Peyton during the functional analysis and treatment conditions (top panel). Percentage of bite presentations followed by acceptance for Peyton during the functional analysis and treatment conditions (middle panel). Percentage of accepted bites followed by mouth cleans for Peyton during the functional analysis and treatment conditions (bottom panel).

mouth within 5s of the initial presentation) in all of the conditions with the exception of one acceptance in a control condition. Visual inspection of Peyton's functional analysis indicated that she engaged in the highest rates of inappropriate behaviors when these behaviors resulted in escape from the bite presentation.

FUNCTION-BASED INTERVENTION

Since Peyton's inappropriate behavior was negatively reinforced by escape from the spoon and eating, a treatment based on escape extinction (i.e., nonremoval of the spoon) was implemented. Treatment resulted in an increase in acceptance and mouth cleans (defined as no food or drink larger than the size of a pea being visible in the child's mouth within 30s of initial acceptance) above 80% and a decrease in inappropriate mealtime behavior to below 1 inappropriate behavior per minute. Figure 7.1 displays both Peyton's inappropriate mealtime behavior (i.e., top panel) as well as her acceptance (i.e., middle panel) and mouth cleans (i.e., bottom panel) during her functional analysis and treatment conditions.

Prior to discharge from the intensive program, Peyton's parents and teachers were trained to accurately implement the feeding protocol designed for Peyton. Treatment sessions were conducted in the session room, as well as at home and at Peyton's school to program for generalization. To assist in the transition from the intensive program, the team nutritionist developed a meal plan for Peyton. Peyton successfully met all of her goals prior to being discharged from the program.

Peyton's progress is continuing to be monitored via a monthly outpatient program. During these visits, any caregiver concerns are addressed, and Peyton's caregivers are observed feeding Peyton to ensure treatment integrity. Peyton's parents report that they are extremely happy with her progress. She no longer engages in problem behaviors during mealtime, and the length of mealtimes has decreased dramatically.

REFERENCES

Ahearn, W. H., Kerwin, M. E., Eicher, P. S., Shantz, J., & Swearingin, W. (1996). An alternating treatments comparison of two intensive interventions for food refusal. *Journal of Applied Behavior Analysis, 29,* 321–332.

American Psychiatric Association. (1994). *Diagnostic and statistical manual of mental disorders* (4th ed.). Washington, DC: Author.

Anderson, C. M., & McMillan, K. (2001). Parental use of escape extinction and differential reinforcement to treat food selectivity. *Journal of Applied Behavior Analysis, 34,* 511–515.

Babbitt, R. L., Hoch, T. A., Coe, D. A., Cataldo, M. F., Kelly, K. J., Stackhouse, C., et al. (1994). Behavioral assessment and treatment of pediatric feeding disorders. *Developmental and Behavioral Pediatrics, 15,* 278–291.

Benoit, D., Wang, E. L., & Zlotkin, S. H. (2000). Discontinuation of enterostomy tube feeding by behavioral treatment in early childhood: A randomized controlled trial. *Journal of Pediatrics, 137*, 498–503.

Burklow, K. A., Phelps, A. N., Schultz, J. R., McConnell, K., & Rudolph, C. (1998). Classifying complex pediatric feeding disorders. *Journal of Pediatric Gastroenterology and Nutrition, 27*, 143–147.

Byars, K. C., Burklow, K. A., Ferguson, K., O'Flaherty, T., Santoro, K. A., & Kaul, A. (2003). A multicomponent behavioral program for oral aversion in children dependent on gastrostomy feedings. *Journal of Pediatric Gastroenterology and Nutrition, 37*, 473–480.

Coe, D. A., Babbit, R. L., Williams, K. E., Hajimihalis, C., Snyder, A. M., Ballard, C., et al. (1997). Use of extinction and reinforcement to increase food consumption and reduce expulsion. *Journal of Applied Behavior Analysis, 30*, 581–583.

Cooper, L. J., Wacker, D. P., McComas, J. J., Brown, K., Peck, S. M., Richman, et al. (1995). Use of component analyses to identify active variables in treatment packages for children with feeding disorders. *Journal of Applied Behavior Analysis, 28*, 139–153.

Dawson, J. E., Piazza, C. C., Sevin, B. M., Gulotta, C. S., Lerman, D., & Kelley, M. L. (2003). Use of high-probability instructional sequence and escape extinction in a child with food refusal. *Journal of Applied Behavior Analysis, 36*, 105–108.

Fisher, W. W., Piazza, C. C., Bowman, L. G., Hagopian, L. P., Owens, J. C., & Slevin, I. (1992). A comparison of two approaches for identifying reinforcers for persons with severe and profound disabilities. *Journal of Applied Behavior Analysis, 25*, 491–498.

Freeman, K. A., & Piazza, C. C. (1998). Combining stimulus fading, reinforcement, and extinction to treat food refusal. *Journal of Applied Behavior Analysis, 31*, 691–694.

Gouge, A. L., & Ekvall, S. W. (1975). Diets of handicapped children: Physical, psychological and socioeconomic correlations. *American Journal of Mental Deficiency, 80*, 149–157.

Gulotta, C. S., Piazza, C. C., Patel, M. R., & Layer, S. A. (2005). Using food redistribution to reduce packing in children with severe food refusal. *Journal of Applied Behavior Analysis, 38*, 39–50.

Hagopian, L. P., Farrell, D. A., & Amari, A. (1996). Treating total liquid refusal with backward chaining and fading. *Journal of Applied Behavior Analysis, 29*, 321–332.

Hoch, T. A., Babbitt, R. L., Coe, D. A., Krell, D. M., & Hackbert, L. (1994). Contingency contacting: Combining positive reinforcement and escape extinction procedures to treat persistent food refusal. *Behavior Modification, 18*, 106–128.

Irwin, M. C., Clawson, E. P., Monasterio, E., Williams, T., & Meade, M. (2003). Outcomes of a day feeding program for children with cerebral palsy. *Archives of Physical Medicine and Rehabilitation, 84*, A2.

Iwata, B. A., Dorsey, M. F., Slifer, K. J., Bauman, K. E., & Richman, G. S. (1994). Toward a functional analysis of self-injury. *Journal of Applied Behavior Analysis, 27*, 197–209. (Reprinted from *Analysis and Intervention in Developmental Disabilities, 2*, 3–20, 1982).

Iwata, B. A., Zarcone, J. B., Vollmer, T. R., & Smith, R. G. (1994). Assessment and treatment of self-injurious behavior. In E. Schopler & G. B. Mesibov (Eds.), *Behavioral issues in autism* (pp. 131–159). New York: Plenum Press.

Kahng, S., Boscoe, J. H., & Byrne, S. (2003). The use of an escape contingency and a token economy to increase food acceptance. *Journal of Applied Behavior Analysis, 36*, 349–353.

Kahng, S., Tarbox, J., & Wilke, A. E. (2001). Use of a multicomponent treatment for food refusal. *Journal of Applied Behavior Analysis, 34*, 93–96.

Kelley, M. E., Piazza, C. C., Fisher, W. W., & Oberdorff, A. J. (2003). Acquisition of cup drinking using previously refused foods as positive and negative reinforcement. *Journal of Applied Behavior Analysis, 36*, 89–93.

Kern, L., & Marder, T. J. (1996). A comparison of simultaneous and delayed reinforcement as treatments for food selectivity. *Journal of Applied Behavior Analysis, 29,* 243–246.

Kerwin, M. E. (1999). Empirically supported treatments in pediatric psychology: Severe feeding problems. *Journal of Pediatric Psychology, 24,* 193–214.

Kerwin, M. E., Ahearn, W. H., Eicher, P. S., & Burd, D. M. (1995). The costs of eating: A behavioral economic analysis of food refusal. *Journal of Applied Behavior Analysis, 28,* 245–260.

Kitfield, E. B., & Masalsky, C. J. (2000). Negative reinforcement-based treatment to increase food intake. *Behavior Modification, 24,* 600–608.

Lamm, N., & Greer, R. D. (1988). Induction and maintenance of swallowing responses in infants with dysphagia. *Journal of Applied Behavior Analysis, 21,* 143–156.

Lerman, D. C., & Iwata, B. A. (1995). Prevalence of the extinction burst and its attenuation during treatment. *Journal of Applied Behavior Analysis, 28,* 93–94.

Lerman, D. C., Iwata, B. A., & Wallace, M. D. (1999). Side effects of extinction: Prevalence of bursting and aggression during the treatment of self-injurious behavior. *Journal of Applied Behavior Analysis, 32,* 1–8.

Lindberg, L. G., Bohlin, G., & Hagekull, B. (1991). Early feeding problems in a normal population. *International Journal of Eating Disorders, 10,* 395–405.

Michael, J. (1982). Distinguishing between discriminative and motivational functions of stimuli. *Journal of the Experimental Analysis of Behavior, 37,* 149–155.

Mueller, M. M., Piazza, C. C., Moore, J. W., Kelley, M. E., Bethke, S. A., Pruett, A. E., et al. (2003). Training parents to implement pediatric feeding protocols. *Journal of Applied Behavior Analysis, 36,* 545–562.

Mueller, M. M., Piazza, C. C., Patel, M. R., Kelley, M. E., & Pruett, A. E. (2004). Increasing variety of foods consumed by blending nonpreferred foods into preferred foods. *Journal of Applied Behavior Analysis, 37,* 159–170.

Munk, D. D. & Repp, A. C. (1994). Behavioral assessment of feeding problems of individuals with severe disabilities. *Journal of Applied Behavior Analysis, 27,* 241–250.

Palmer, S., & Horn, S. (1978). Feeding problems in children. In S. Palmer & S. Ekvall (Eds)., *Pediatric nutrition in developmental disorders* (pp. 107–129). Springfield, Il: Charles C. Thomas.

Patel, M. R., Piazza, C. C., Kelly, M. L., Ochsner, C. A., & Santana, C. M. (2001). Using a fading procedure to increase fluid consumption in a child with feeding problems. *Journal of Applied Behavior Analysis, 34,* 357–360.

Patel, M. R., Piazza, C. C., Layer, S. A., Coleman, R., & Swartzwelder, D. M. (2005). A systematic evaluation of food textures to decrease packing and increase oral intake in children with pediatric feeding disorders. *Journal of Applied Behavior Analysis, 38,* 89–100.

Patel, M. R., Piazza, C. C., Martinez, C. J., Volkert, V. M., & Santana, C. M. (2002). An evaluation of two differential reinforcement procedures with escape extinction to treat food refusal. *Journal of Applied Behavior Analysis, 35,* 363–374.

Patel, M. R., Piazza, C. C., Santana, C. M., & Volkert, V. M. (2002). An evaluation of food type and texture in the treatment of a feeding problem. *Journal of Applied Behavior Analysis, 35,* 363–374.

Piazza, C. C., & Carroll–Hernandez, T. A. (2004). Assessment and treatment of pediatric feeding disorders. *Encyclopedia on early childhood development.* Montreal, Quebec: Centre of Excellence for Early Childhood Development; 1–7. Retrieved August 31, 2005, from http://www.excellence-earlychildhood.ca/documents/Piazza-Carroll-HernandezANGxp.pdf

Piazza, C. C., Fisher, W. W., Brown, K. A., Shore, B. A., Patel, M. R., Katz, R. M., et al. (2003). Functional analysis of inappropriate mealtime behaviors. *Journal of Applied Behavior Analysis, 36,* 187–204.

Piazza, C. C., Patel, M. R., Gulotta, C. S., Sevin, B. M., & Layer, S. A. (2003). On the relative contributions of positive reinforcement and escape extinction in the treatment of food refusal. *Journal of Applied Behavior Analysis, 36*, 309–324.

Piazza, C. C., Patel, M. R., Santana, C. M., Goh, H., Delia, M. D., & Lancaster, B. M. (2002). An evaluation of simultaneous and sequential presentation of preferred and nonpreferred food to treat food selectivity. *Journal of Applied Behavior Analysis, 35,* 259–270.

Practice Management Information Corporation. (2005). *ICD-9-CM International Classification of Diseases, 9th Rev: Clinical Modification, Vols. 1, 2, & 3.* Los Angeles, CA: Author.

Reed, G. K., Piazza, C. C., Patel, M. R., Layer, S. A., H. H., Bachmeyer, M. H., Bethke, S. D., et al. (2004). On the relative contributions of noncontingent reinforcement and escape extinction in the treatment of food refusal. *Journal of Applied Behavior Analysis, 37*, 27–41.

Riordan, M. M., Iwata, B. A., Finney, J. W., Wohl, M. K., & Stanley, A. E. (1984). Behavioral assessment and treatment of chronic food refusal in handicapped children. *Journal of Applied Behavior Analysis, 17,* 327–341.

Riordan, M. M., Iwata, B. A., Wohl, M. K., & Finney, J. W. (1980). Behavioral treatment of food refusal and selectivity in developmentally disabled children. *Applied Research in Mental Retardation, 1,* 95–112.

Rommel, N., DeMeyer, A. M., Feenstra, L., & Veereman-Wauters, G. (2003). The complexity of feeding problems in 700 infants and young children presenting to a tertiary care institution. *Journal of Pediatric Gastroenterology and Nutrition, 37,* 75–84.

Schafe, G. E., & Bernstein, I. L. (1997). Development of the enhanced neural response to NaCl in Fischer 344 rats. *Physiology & Behaviour, 61,* 775–778.

Scrimshaw, N. S. (1998). Malnutrition, brain development, learning, and behavior. *Nutrition Research, 18,* 351–379.

Sevin, B. M., Gulotta, C. S., Sierp, B. J., Rosica, L. A., & Miller, L. J. (2002). Analysis of response covariation among multiple topographies of food refusal. *Journal of Applied Behavior Analysis, 35,* 65–68.

Shore, B. A., Babbitt, R. L., Williams, K. E., Coe, D. A., & Snyder, A. (1998). Use of texture fading in the treatment of food selectivity. *Journal of Applied Behavior Analysis, 31,* 621–633.

Smith, R. G., & Iwata, B. A. (1997). Antecedent influences on behavior disorders. *Journal of Applied Behavior Analysis, 30,* 343–375.

Spector, A. C., Smith, J. C., & Hollander, G. R. (1981). A comparison of dependent measures used to quantify radiation-induced taste aversion. *Physiology and Behavior, 27,* 887–901.

Spector, A. C., Smith, J. C., & Hollander, G. R. (1983). The effect of post-conditioning CS experience on recovery from radiation-induced taste aversion. *Physiology and Behavior, 30,* 647–649.

Troughton, K. E., & Hill, A. E. (2001). Relation between objectively measured feeding competence and nutrition in children with cerebral palsy. *Developmental Medicine and Child Neurology, 43,* 187–190.

Werle, M. A., Murphy, T. B., & Budd, K. S. (1993). Treating chronic food refusal in young children: Home-based parent training. *Journal of Applied Behavior Analysis, 26,* 421–433.

8

TIC DISORDERS AND TRICHOTILLOMANIA

RAYMOND G. MILTENBERGER

North Dakota State University

DOUGLAS W. WOODS
MICHAEL B. HIMLE

University of Wisconsin-Milwaukee

DIAGNOSIS AND RELATED CHARACTERISTICS

Tic disorders are characterized by sudden, repetitive, nonrhythmic motor and vocal tics. The three primary categories of tic disorders are Transient Tic Disorder, Chronic Tic Disorder, and Tourette's Disorder (TD). Transient Tic Disorder is diagnosed if one or more motor and/or vocal tics have been present for at least 4 weeks but less than 12 months. Chronic Tic Disorder is diagnosed if motor or vocal tics (but not both) have been present for at least 12 months. If both motor and vocal tics are present for 12 months, then a diagnosis of TD is given (American Psychiatric Association, 1994).

Tics are commonly categorized as either simple or complex. Simple tics involve a single muscle group and include actions such as eye-blinking; facial movements; head, arm, and leg jerks; and sounds such as grunting, sniffing, coughing, or single words. Complex tics involve coordinated actions involving multiple muscle groups and have a purposeful appearance. Examples include touching, tapping, straightening objects, and multiple-word verbalizations, including echolalia and palilalia (repetition of the same words over and over). Over time, an individual's tics may vary in body location and/or topography and are likely to wax and wane in frequency and intensity (Leckman, King, & Cohen, 1999).

Functional Analysis in
Clinical Treatment

In addition to the tics, many individuals report the presence of unpleasant somatic sensations, often referred to as "premonitory urges" or "sensory tics," that immediately precede tics (Leckman, Walker, & Cohen, 1993). Many individuals who experience premonitory urges report that the aversive sensation is alleviated by the performance of the tic and that their tics are at least partially volitional responses performed to alleviate the premonitory urge (Kwak, Dat Vuong, & Jankovic, 2003).

Trichotillomania (TTM), listed as an impulse control disorder in the *DSM-IV* (APA, 1994), involves the recurrent pulling of one's hair with noticeable hair loss or the potential for hair loss if left untreated. To receive a diagnosis of TTM, the individual also must report an increased sense of tension, often described as an "urge," prior to pulling and pleasure, gratification, or relief while pulling (see Chapter 19, "Personality Disorders," for functional approaches to other impulse control disorders).

Common sites of hair pulling include but are not limited to the scalp, eyebrows, eyelashes, and pubic regions (Christenson, Mackenzie, & Mitchell, 1991). Most individuals pull their hair with their thumb and fingers, but some use an instrument such as tweezers. After pulling, individuals often manipulate the hair in some way (e.g., by looking at it, twirling it, touching it to their lips, putting it in their mouth, or eating it).

While topographically distinct, there are functional similarities between tics and hair pulling, as well as other habit disorders such as chronic skin picking, thumb and finger sucking, and fingernail biting (Woods & Miltenberger, 1996). Although they are not discussed here, it is useful to know that the conceptualization of these behaviors is consistent with the functional model described for tics and hair pulling and that these behaviors can be assessed and treated using many of the techniques described in this chapter.

FUNCTIONAL ANALYTIC MODEL

Problem behaviors may be strengthened if they produce positive reinforcers or if they prevent or terminate aversive stimuli. In addition, Iwata, Vollmer, Zarcone, and Rodgers (1993) suggested that problem behaviors may be reinforced socially, through the actions of others, or automatically, through changes in the external or internal environment caused directly by the behavior. As observable behaviors with the potential to contact social contingencies, tics and hair pulling could be maintained by any of the four types of reinforcement: social positive reinforcement (e.g., attention), automatic positive reinforcement (e.g., sensory stimulation), social negative reinforcement (e.g., escape from tasks or aversive activities), or automatic negative reinforcement (e.g., relief from aversive experiences such as pain, unpleasant sensations, or negative emotions). However, most of the evidence suggests that tics are maintained primarily by automatic negative reinforcement (Carr, Sidener, Sidener, & Cummings, 2005; Evers & van de Wetering, 1994; Yates, 1958) and that hair pulling is maintained primarily by automatic negative reinforcement (Miltenberger, 2005) or automatic positive reinforcement (Miltenberger, Long, & Rapp, 1998; Rapp, Miltenberger, Galensky, Ellingson, & Long, 1999).

TICS

Evidence for automatic negative reinforcement of tics comes primarily from self-report data in which individuals describe the occurrence of unpleasant sensations such as sensory tics or premonitory urges that occur prior to the occurrence of the tic with momentary relief from the sensations upon the occurrence of the tic (Findley, 2001). These sensations have been reported as both physical and mental (Miguel et al., 2000). Physical sensations are described as tension, pressure, tickling, itching, or other bodily sensations in the skin, bones, muscles, and joints (Bliss, 1980; Bullen & Hemsley, 1983; Evers & van de Wetering, 1994; Findley, 2001). Mental sensations have been described as generalized uncomfortable feelings or urges (Miguel et al., 2000). These physical or mental sensations are experienced as unpleasant and appear to function as establishing operations (EOs) that make the occurrence of the tic more likely because the tic functions to produce brief escape from the sensations.

It is also possible that the occurrence of tics may enter into social contingencies. If the occurrence of a tic produces attention from significant individuals, for example, in the form of concern, annoyance, scolding, or other reactions, then the tic may continue to occur as a function of social positive reinforcement (Watson & Sterling, 1998). Likewise, if the occurrence of the tic results in escape from, or avoidance of, aversive activities or interactions, the tic may continue to occur as a function of social negative reinforcement. In such cases, the tic may have been maintained by automatic reinforcement initially, but over time, became at least partially maintained by social contingencies involving attention or escape as well.

HAIR PULLING

Evidence for an automatic negative reinforcement function of hair pulling comes primarily from adults who pull their hair and who report the presence of unpleasant emotions or affective experiences, such as tension, anxiety, anger, sadness, frustration, prior to pulling with momentary relief from the unpleasant emotions while hair pulling (Christenson & Mansueto, 1999; Christenson, Risvedt, & Mackenzie, 1993; Miltenberger, Rapp, & Long, 2001). Consistent with the conceptualization of automatic negative reinforcement for tics, the occurrence of unpleasant emotions functions as an EO that makes hair pulling more likely because hair pulling provides immediate, but short-lived, alleviation or reduction in the unpleasant emotions.

Evidence for an automatic positive reinforcement function of hair pulling comes from two sources: the self-reported experiences of adults who pull their own hair (Stanley, Borden, Mouton, & Breckenridge, 1995) and the functional analysis of hair pulling in children and adults with mental retardation (Rapp et al., 1999). Stanley et al. reported that hair pulling was most likely to occur for some individuals when they were bored (during periods of low ambient stimulation or when involved in sedentary activities) and that hair pulling provided stimulation. Functional analysis research likewise has shown that hair pulling exhibited by children and adults with mental retardation is more likely to occur when they are alone and not engaged in activities (Miltenberger et al., 1998; Rapp et al., 1999; Rapp, Miltenberger, & Long, 1998). Furthermore, the observation that hair pulling was followed reliably by periods of hair manipulation suggested that the sensory stimulation produced by manipulating pulled hair reinforced hair pulling (Miltenberger et al., 1998; Rapp et al., 1999).

FUNCTIONAL ASSESSMENT AND ANALYSIS

Functional assessment procedures are designed to identify the immediate antecedents and consequences of the problem behavior. Researchers have developed three approaches to conducting a functional assessment: indirect assessments, consisting of interviews, questionnaires, or rating scales; direct observation of the antecedents and consequences as the target behavior occurs; and functional analysis, in which antecedents and consequences are manipulated to demonstrate a functional relationship with the target behavior (Iwata et al., 1993; Lennox & Miltenberger, 1989). Although functional assessment methods were originally developed for use with problem behaviors exhibited by children and individuals with mental retardation (Iwata, Dorsey, Slifer, Bauman, & Richman, 1982; Sturmey, Carlson, Crisp, & Newton, 1988), functional assessment research also has been conducted with tics and hair pulling.

TICS

Most research investigating factors influencing the occurrence of tics has involved indirect assessment in which individuals with tic disorders are asked to describe the events they experience before the occurrence of the tic and how those events change as the tic occurs (Kwak et al., 2003; Leckman et al., 1993). Because the events in question most often are covert, sensory experiences, indirect assessment typically is used for assessing possible antecedents and consequences of tics. As indicated previously, when indirect assessments are used, subjects often report unpleasant sensations as antecedents with momentary relief from the unpleasant sensations as the consequence for tics (Kwak et al., 2003. Indirect assessments, such as self-report checklists, also have been used to identify external or contextual stimuli that are associated with fluctuations in tic frequency (O'Connor, Briesbois, Brault, Robillard, & Loiselle, 2003; Silva, Munoz, Barickman, & Friedhoff, 1995). This strategy can be useful for exploring a wide variety of social stimuli that may be functionally related to an individual's tics. However, we recommend that, whenever possible, practitioners confirm functional relationships using direct observation and/or functional analysis procedures.

Some researchers have employed functional analysis methodologies to determine whether social reinforcement is involved in the maintenance of tics. Carr, Taylor, Wallander, and Reiss (1996) and Scotti, Schulman, and Hojnacki (1994) used functional analysis to investigate the influence of attention and escape from tasks as possible social reinforcers for tics exhibited by individuals with mental retardation. They found that tics were most frequent in attention and escape conditions, but also in an alone condition, suggesting that tics were probably not under the specific control of social reinforcement. Instead, this undifferentiated pattern suggested control by automatic reinforcement. In a more recent investigation, Watson and Sterling (1998) conducted a functional analysis of the vocal tic of a 4-year-old and showed that the tic was most likely to occur in the condition involving contingent parental attention. The subsequent effectiveness of an intervention consisting of attention extinction and DRO using attention delivered for the absence of the tic further supported this conclusion. Other researchers have demonstrated antecedent influences on tics (e.g., Malatesta, 1990; Woods, Watson, Wolfe, Twohig, & Friman, 2001) or control of tics through social contingencies (e.g., Woods & Himle, 2004), but no other functional analysis research has been reported with tics. Malatesta (1990) measured a child's motor tic when the father was present and when the father was absent and found a higher rate of tics when the father was present. However, Malatesta offered no explanation for this finding. Woods et al. (2001) showed that vocal tics of two boys occurred at a higher rate during conditions involving tic-related talk relative to conditions in which there was no tic talk. The authors did not offer an explanation for why tics may have been higher in the tic talk condition. Woods and Himle (2004) showed that the tics exhibited by four children with TD occurred at a lower rate during a DRO condition in which they received tokens for the absence of tics. Woods and Himle (2004) and Meidinger et al. (2005) showed that the delivery of verbal instructions to suppress tics was far less effective in reducing the rate of tics for individuals with TD. The results of Woods and Himle (2004) do not demonstrate the function of tics for these children, but do show that the tics can be influenced by reinforcement contingencies.

TRICHOTILLOMANIA

Similar to the research on tics, most functional assessment research with hair pulling involves indirect assessment in which individuals with TTM report the antecedents and consequences of hair pulling. While direct functional assessment methods may be desired, direct observation of antecedents and consequences of pulling is problematic for two reasons. First, hair pulling is often a behavior that is not performed in the presence of others, possibly because of a history of social punishment. Second, the functional antecedents and consequences to pulling are often private and not directly available for observation by others.

Indirect functional assessments have identified being alone, sedentary activities, and a variety of private events, such as tension, anxiety, or other unpleasant emotions, as the primary antecedents to hair pulling (Christenson et al., 1991; Christenson & Mansueto, 1999; Stanley et al., 1995). Tactile stimulation to the fingers or scalp and/or manipulation of the pulled hair (Miltenberger et al., 1998; Rapp et al., 1999) and escape from an aversive private state, such as tension reduction or anxiety reduction or stimulation (Miltenberger, 2005), have been identified as the most common consequences to pulling.

Unfortunately, few studies have utilized functional analysis methods with hair pulling. In one investigation with a young child and an adult with mental retardation, Miltenberger et al. (1998) conducted a functional analysis and showed that hair pulling was more probable when the individuals were alone. When attention or escape was made contingent on the behavior, little or no hair pulling occurred. In addition, both individuals manipulated the hair each time they pulled it, suggesting that tactile stimulation arising from hair manipulation was reinforcing hair pulling. Rapp et al. (1999) conducted a functional analysis of the hair pulling exhibited by an adolescent with mental retardation. Similar to Miltenberger et al., Rapp et al. showed that hair pulling occurred almost exclusively during the alone condition and that hair manipulation always followed hair pulling (see Figure 8.1). To further demonstrate that hair pulling was reinforced by tactile stimulation, Rapp et al. masked the stimulation by having the subject wear a latex glove. Hair pulling ceased. In addition, when already pulled hairs were made available to the subject, she manipulated those hairs and did not engage in further hair pulling, further demonstrating that tactile stimulation was reinforcing (see Figure 8.2).

FUNCTIONAL ANALYTIC INTERVENTIONS

Several behavioral strategies have been shown to be effective for reducing tics and hair pulling (Woods & Miltenberger, 1995). Although some of

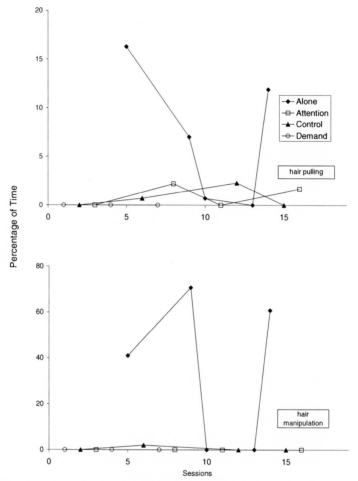

FIGURE 8.1   The percentage of time the participant engaged in hair pulling (top panel) and hair manipulation (bottom panel) across alone, attention, control, and demand conditions. Social consequences were provided for hair pulling only.

the interventions are based on functional assessment results, others are not. In the following sections, we review five primary treatment approaches for tics and hair pulling: extinction, antecedent control, operant techniques, habit reversal training, and exposure-based treatments.

EXTINCTION

Although tics and hair pulling often are maintained by automatic reinforcement, social contingencies also can influence their expression in some

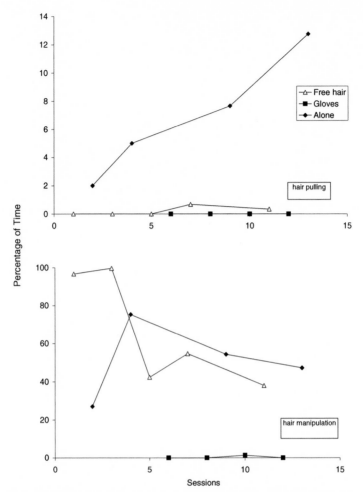

FIGURE 8.2   The percentage of time the participant engaged in hair pulling (top panel) and hair manipulation (bottom panel) across alone, free hair, and glove conditions.

individuals. As such, the manipulation of the automatic or social reinforcement shown to be maintaining the habit behavior constitutes a functional intervention. For example, when Watson and Sterling (1998) discovered that an individual's tics were worsened by contingent attention, they eliminated the attention for the tic and provided attention for the absence of the tic, thereby decreasing the frequency of the tic. Similarly, when Rapp et al. (1999) demonstrated that tactile stimulation from hair play was the reinforcer for hair pulling, they decreased hair pulling with sensory extinction. In this case, the use of a latex glove masked the stimulation arising from hair manipulation.

Unfortunately, the use of extinction has limited utility for problem behaviors maintained by automatic negative reinforcement. The use of extinction is limited because the reinforcing consequence—relief from unpleasant private events—is covert, and cannot be withheld contingent on the behavior (Miltenberger, 2005).

ANTECEDENT CONTROL TECHNIQUES

Another functional intervention strategy that has been used to treat tics and hair pulling is antecedent control. Antecedent control involves the systematic manipulation of antecedents that are functionally related to the habit (see Silva et al. [1995] and O'Connor et al. [2003] for a reviews of antecedent stimuli associated with tics and Christensen et al. [1991] for a review of antecedent stimuli associated with hair pulling). There are two primary antecedent control techniques. The first is to eliminate any antecedent stimuli that may function as discriminative stimuli for the habit (Miltenberger, 2005). For example, individuals who engage in hair pulling only in their bathroom and with grooming instruments may benefit from removal of those instruments from the bathroom and/or the residence altogether. If a detailed functional assessment identifies that an individual's tics are much more frequent and bothersome during periods of boredom, then the scheduling of activities may be recommended in order to reduce boredom. Similarly, if it is found that an individual is more likely to pull hair when alone in the morning, then it may be possible to have a family member accompany that person throughout the high-risk time, thereby eliminating being alone as an antecedent.

A second antecedent control procedure is to decrease or eliminate the EO for the behavior. In the case of hair pulling in which negative emotional experiences function as an EO, eliminating the EO would require procedures to alleviate the negative emotions (Miltenberger, 2005). These procedures could include relaxation procedures, anger management procedures, pleasant events scheduling, treatment for depression (see Chapter 15, Mood Disorders by Hopko, Hopko and Lejuez), or other procedures. It is not clear whether the sensory experiences that function as an EO for tics can be altered through behavioral means.

Clearly, there is great variability in the number and type of antecedents that might exacerbate a person's tics or hair pulling. Consistent with a functional-analytic approach, the use of stimulus control techniques should be tailored to each individual and the specific technique selected based on thorough functional assessment. For a more comprehensive explanation of stimulus control techniques and recommendations for specific techniques that can be used to treat common patterns in hair pulling see Penzel (2003) and Keuthen, Stein, and Christenson (2001). Unfortunately, no such resource is available for tics, but practitioners with a working knowledge of stimulus control techniques can easily adapt procedures such as those outlined by Penzel (2003) for use with tics.

OPERANT TECHNIQUES

Several treatments employing reinforcement and/or punishment have been successful with habit disorders. Although the treatments described here were not based on the results of prior functional assessments, they have been effective nonetheless (see Elliott & Fuqua, 2001, and Adesso & Norberg, 2001, for reviews). Lahey, McNees, and McNees (1973) successfully treated a child's obscene vocal tic with exclusionary timeout. Varni, Boyd, and Cataldo (1978) successfully employed self-monitoring, differential reinforcement for low-rate behavior (DRL), and exclusionary timeout to treat a child's multiple tics. Similarly, Gray (1979) used reinforcement for non-pulling and punishment procedures to treat a young child with severe hair pulling. Likewise, Wagaman, Miltenberger, and Williams (1995) showed that reinforcement for the absence of the behavior (DRO) greatly decreased a child's vocal tic. In a more recent study, Rapp et al. (1998) used an alarm-sounding device to decrease pulling in an individual with treatment-resistant hair pulling. Each time the individual's wrist came within close proximity of the head, the alarm sounded, and when the wrist moved away from the head, the alarm stopped. The authors proposed that the alarm served to punish the hair pulling response. However, it is possible that the alarm increased awareness of hair pulling (prompted the individual to identify or tact each occurrence) or that the termination of the alarm negatively reinforced the competing response of moving the hand away from the head. Ristvedt and Christenson (1996) instructed an individual with trichotillomania to apply sensitizing cream to her scalp each day in order to increase pain contingent upon hair pulling and found that pulling was reduced significantly at 4-month follow-up. However, it is unclear if punishment or awareness enhancement (identifying each occurrence of the behavior) contributed to the effectiveness of the procedure. Finally, Rapp et al. (2000) used response blocking/brief restraint and DRO (attention for absence of hair pulling) to decrease the hair pulling exhibited by an adolescent with mental retardation.

Although there is some evidence that reinforcement and punishment procedures can be effective, several caveats should be noted. First, the research evaluating these procedures has been conducted almost exclusively with children and adults with disabilities. Thus, their effectiveness with typically developed adults with habit disorders is unknown. Second, it should be noted that the long-term effectiveness of punishment procedures and differential reinforcement may be limited if the contingencies maintaining tics or hair pulling are not identified and manipulated (Miltenberger, 2005). In such cases, the habit may persist despite the delivery of programmed contingencies or may return when the contingencies are removed.

HABIT REVERSAL TRAINING

Habit reversal training (HRT) is a multicomponent treatment procedure that is based on the rationale that habits can be decreased if an individual identifies each occurrence of relevant discriminative stimuli, such as premonitory urges or early components of the habit response chain, and performs a physically competing response to prevent or interrupt the habit. Azrin and Nunn (1973) contended that repeatedly engaging in the competing behavior not only prevents the habit, but strengthens the muscles antagonistic to the tic, thereby allowing the competing behavior to replace the maladaptive habit if practiced sufficiently. In addition to strengthening antagonistic muscles, Miltenberger, Fuqua, and Woods (1998) proposed that the competing response may replace the habit behavior because it is differentially reinforced and/or because the habit is self-punished by contingent performance of an effortful competing response. According to the differential reinforcement hypothesis, the competing response is reinforced by socially mediated contingencies (e.g., praise) or the competing response serves the same function (i.e., produces the same reinforcing consequences) as the habit.

Azrin and Nunn's (1973) original HRT package included awareness training, competing response training, and a variety of motivation and generalization techniques, including inconvenience reviews, social support (i.e., prompts and social reinforcement for use of the competing response), and public practice. Component analyses conducted in the past 20 years have identified the combination of awareness training, competing response training, and social support as sufficient for HRT to be effective. The abbreviated package has been referred to as simplified habit reversal training (SHRT). SHRT has been shown to be equally as effective as the original package and easier to administer, as it can often be conducted in only two to eight, 1-hr sessions. Woods and Miltenberger (2001) provided step-by-step manuals outlining SHRT for use with tics, hair pulling, and oral-digital habits, such as fingernail biting and skin picking. Numerous research studies support the effectiveness of SHRT for tics (Azrin & Nunn, 1973; Azrin & Peterson, 1990: Miltenberger, Fuqua, & McKinley, 1985; Wilhelm et al., 2003; Woods, Twohig, Flessner, & Roloff, 2003) and hair pulling (Azrin, Nunn, & Frantz, 1980; Miltenberger & Fuqua, 1985; Mouton & Stanley, 1996; Rapp, Miltenberger, Long, Elliott, & Lumley, 1998a). Carr and Chong (2005) recently reviewed 20 studies that used HRT to treat a total of over 100 individuals with tics and found the procedure to be "probably efficacious" according to Task Force on Promotion and Dissemination of Psychological Procedures (1995) guidelines.

Studies evaluating the mechanism(s) by which HRT is effective have been sparse. Miltenberger and Fuqua (1985) demonstrated that performing the competing response contingent upon the habit was more effective than noncontingent performance of the competing response. This finding is consistent with the punishment analysis provided by Miltenberger et al. (1998) and suggests that HRT does not simply strengthen atrophied muscles antagonistic to the habit, as proposed by Azrin and Nunn (1973). In other studies, Sharenow, Fuqua, and Miltenberger (1989) and Woods et al. (1999) compared the effectiveness competing responses that were and were not physically incompatible with the target habit and found the two procedures to be equally effective. The finding that the competing response does not have to physically compete with the habit is difficult to explain by Azrin and Nunn's model. In addition, this finding suggests that the competing response does not simply replace the tic as a functionally equivalent behavior, one of the possibilities proposed by Miltenberger et al. (1998). Clearly, more research is needed to determine the mechanism by which HRT works.

EXPOSURE-BASED TREATMENTS

Another treatment that has stemmed from behavioral models is cue-exposure and response prevention (ERP; Verdellen, Keijsers, Cath, & Hoogduin, 2004). During ERP, which has been successful in the treatment of OCD, the person is exposed to anxiety-producing stimuli and is then prevented from neutralizing or escaping the anxiety until the person habituates to the anxiety (Grayson, Foa, & Steketee, 1982). The rationale for the use of ERP with tics and hair pulling is based on the automatic negative reinforcement model of these disorders. ERP involves creating a graded hierarchy of stimuli that elicit the premonitory urge and then introducing these stimuli in a systematic manner while simultaneously preventing the performance of the habit behavior. Theoretically, if an individual experiences the urge and is prevented, or prevents him/herself, from performing the habit, he/she will habituate to the premonitory urge and the habit will no longer be functional.

ERP has only recently been applied to tics. Hoogduin, Verdellen, and Cath (1997) delivered ERP to four individuals with TD and found that three of them reported a reduction in the urge to tic during the session. In addition, tics were significantly reduced for these three individuals. Verdellen et al. (2004) extended this work by randomly assigning 43 TD individuals to either ERP or HRT. Both treatments were equally effective. In another study, Wetterneck and Woods (in press) used ERP to treat complex tics in an adolescent male with multiple motor and vocal tics. They found that ERP not only reduced the tic that was targeted for intervention, but

TABLE 8.1 Treatments for Tics and Hair Pulling

- When possible, use extinction and implement stimulus control procedures.
- Utilize habit reversal for tics and hair pulling exhibited by adults.
- Consider the use of operant procedures and stimulus control procedures with children or individuals with disabilities.
- ERP is a promising procedure in need of further study.

---

also generalized to other complex tics. Unfortunately, research has yet to evaluate whether ERP is effective in the treatment of hair pulling. Clearly, research on the use of ERP for tics and hair pulling is preliminary, and conclusions about the procedure's effectiveness cannot yet be drawn. In addition, further research is needed to clarify the mechanism by which such procedures are effective (see Table 8.1).

CASE STUDY

DIAGNOSIS AND DESCRIPTION

What follows is a case description of Brad, a 13-year-old typically developing male who was referred for evaluation and treatment recommendations related to a possible tic disorder. According to his mother, Brad's tics began when he was seven years old. His first tics were simple sniffing and forceful eyeblinking, which his parents attributed to allergies. He was treated with antihistamines with little success. At age nine years, Brad's sniffing and eyeblinking decreased considerably during the summer but did not completely remit. About every two weeks, Brad experienced a bout of tics during which the tics increased to bothersome levels. Each of these episodes lasted approximately two days and usually corresponded to a stressful or exciting event, such as a friend's birthday party, relatives visiting the house, or a vacation. Brad's tics worsened considerably when he resumed school in the fall of the following year. In addition to sniffing and eyeblinking, he began exhibiting a loud coughing tic and several head and facial movements, including jerking his head forward, wrinkling his nose, and pursing his lips. Brad reported that children at school began to tease him, which made his tics more frequent and more noticeable and forceful. Between the ages of 10 and 13 years, Brad's tics continued a course of waxing and waning in intensity and frequency and changed in complexity and topography. Between the ages of seven and 13 years, Brad continuously exhibited multiple motor and phonic tics which qualified him for a *DSM-IV* diagnosis of Tourette's Disorder.

FUNCTIONAL ASSESSMENT

During the initial interview, Brad reported that his tics were less frequent in the summer when he was active and were the most frequent during the first few weeks of school each semester and when attending social gatherings. He also reported that talking about the tics "made him have to do the tic." Brad's mother reported that tics were especially frequent immediately after school. At the time of the evaluation, Brad's tics included both motor and vocal tics. His vocal tics included a loud cough, throat clearing, and a forceful "Ha" vocalization. His motor tics included lip pursing, arm extensions, jerking his head forward and back, and hitting himself in the chest with his fist. Brad reported that he could predict his tics because they were immediately preceded by an "energy feeling that just needed to get out," which the tic alleviated momentarily.

An indirect functional assessment revealed that Brad's tics were most likely maintained by automatic negative reinforcement by removal of the "energy" feeling contingent upon the tic. In addition, several social antecedent and consequence stimuli were hypothesized to be functionally related to the tics. Using a semistructured interview, Brad and the examiner identified four antecedents associated with increased tic frequency: social gatherings, talking about tics, after-school activities, and stress/excitement. It was also discovered that social gatherings that were stressful or exciting, or those during which Brad's tics were noticed and discussed, such as when meeting new people, were especially likely to be associated with worsening of tics. Brad's tics also worsened significantly immediately after school. It was discovered that Brad usually worked on his homework after school, but lately his mother had allowed him an hour to "unwind" and "get his tics out" by going to his room where Brad usually watched television or played videogames. This routine had been causing problems because often Brad's mother argued with Brad in order to get him to begin his homework, even after the free time had expired, and although his tic decreased during free time, they again worsened when asked to begin his homework. This sequence of events made it difficult for Brad to get his homework finished in an appropriate amount of time. It was hypothesized that Brad's tics were being reinforced, at least in part, by escape from his homework.

FUNCTIONAL ANALYSIS

In order to further evaluate possible social functions of Brad's tics, a brief functional analysis was conducted in which Brad was observed for five, 5-min conditions, each repeated five times. The conditions were as follows: alone, talking about tics, non-tic-related conversation, escape from homework, and attention for tics. It was hypothesized that Brad's tics would be evident across all conditions, supporting the automatic negative reinforcement view, but would be highest during the tic-related conversation and escape conditions. Brad displayed tics in all conditions, including the alone condition, providing direct support for the automatic negative reinforcement conceptualization. In addition, Brad's tics were noticeably worse when talking about tics, suggesting that tic-related talk was functionally related to the tics; however, the exact function remained unknown. The other three conditions did not differ from each other. Interestingly, the escape-from-homework condition did not differ from the rest of the conditions. A follow-up functional analysis compared the rate of tics during alone, escape from homework to no activity, and escape from homework to a desired activity conducted in Brad's home. This second functional analysis showed that Brad's tics were highest when allowed to escape to a desired activity (such as a videogame). This finding supported the view that Brad's tics were reinforced by access to preferred activities during homework time after school.

INTERVENTION

The intervention included three components: Habit reversal training, stimulus control, and extinction/differential reinforcement. First, habit reversal training was introduced. Using awareness training techniques, Brad was taught to identify each occurrence of his premonitory sensations and other pre-tic signals, such as initial movements involved in arm-raising (see Woods, 2001, for a detailed treatment manual). A competing response was introduced for each tic. Brad was instructed to engage in the competing response for 3 min each time he experienced the "warning sign" for the tic and/or immediately following each occurrence of the tic. Competing responses for each of Brad's tics are provided in Table 8.2.

In addition to Brad's awareness training and competing response training, Brad's mother was recruited to provide social support. She was taught to praise Brad for engaging in the competing response and prompt him to

TABLE 8.2   Brad's Tics and the Competing Responses That Were Taught for Each

Tic	Competing Response
• Arm extension and hitting self • Lip pursing • Head Jerking • Vocal tics	• Hold at 90-degree angle, bent at elbow, and press elbow against side of body. • Curl lips into mouth and gently clench front teeth onto lips. • Press chin to chest and tense neck muscles. • Close mouth, clench teeth together, and regulate breathing through nose.

use the competing response when she observed a tic. In addition, Brad was instructed to practice the competing response during structured periods each day and was given tangible reinforcers for compliance.

After Brad had mastered habit reversal training, an intervention based on stimulus control was introduced to reduce or eliminate antecedents that were associated with a worsening of tics. For Brad, antecedent stimuli included social gatherings, stress/excitement, and talking about tics, and especially situations that included all three: stressful social gatherings that involved talking about his tics. In order to reduce stress/excitement, Brad was taught relaxation techniques. In order to minimize talking about tics, Brad was taught skills to allow him to inform others about his tics such that questions and discussion from others would be minimal. In addition, Brad was gradually introduced to more stressful social situations while practicing the skills he learned in habit reversal training.

The final component of treatment was extinction and differential reinforcement. The functional analysis suggested that Brad's tics were being reinforced by access to preferred activities during scheduled homework time. In order to reverse this contingency, Brad was allowed 30 min of "unwinding" time immediately after school, but rather than free access to preferred activities, he was instructed to practice his relaxation training and habit reversal skills. Access to his videogames was no longer contingent on the occurrence of tics but instead was provided contingent upon treatment compliance and completion of his homework.

REFERENCES

Addesso, V. J., & Norberg, M. N. (2001). Behavioral interventions for oral-digital habits. In D. W. Woods & R. G. Miltenberger (Eds.), *Tic disorders, trichotillomania, and other repetitive behavior disorders: Behavioral approaches to analysis and treatment* (pp. 223–240). Boston: Kluwer.

American Psychiatric Association. (1994). *Diagnostic and statistical manual of mental disorders* (4th ed.). Washington, D.C.: Author.

Azrin, N. H., & Nunn, R. G. (1973). Habit reversal: A method of eliminating nervous habits and tics. *Behaviour Research and Therapy, 11*, 619–628.

Azrin, N. H., Nunn, R. G., & Frantz, S. E. (1980). Treatment for hairpulling (trichotillomania): A comparative study of habit reversal and negative practice training. *Journal of Behavior Therapy and Experimental Psychiatry, 11*, 13–20.

Azrin, N. H., & Peterson, A. L. (1990). Treatment of Tourette syndrome by habit reversal: A wait-list control group comparison. *Behavior Therapy, 21*, 305–318.

Bliss, J. (1980). Sensory experiences in Gilles de la Tourette syndrome. *Archives of General Psychiatry, 37*, 1343–1347.

Bullen, J. G., & Hemsley, D. R. (1983). Sensory experiences as a trigger in Gilles de la Tourette's syndrome. *Journal of Behavior Therapy and Experimental Psychiatry, 14*, 197–201.

Carr, J. E., & Chong, I. M. (2005). Habit reversal treatment of tic disorders: A methodological critique of the literature. *Behavior Modification, 29*, 858–875.

Carr, J. E., Sidener, T. M., Sidener, D. W., & Cummings, A. R. (2005). Functional analysis and habit-reversal treatment of tics. *Behavioral Interventions, 20,* 185–202.

Carr, J. E., Taylor, C. C., Wallander, J. J., & Reiss, M. L. (1996). A functional analytic approach to the diagnosis of transient tic disorder. *Journal of Behavior Therapy and Experimental Psychiatry, 27,* 291–297.

Christenson, G. A., Mackenzie, T. B., & Mitchell, J. E. (1991). Characteristics of 60 adult hair pullers. *American Journal of Psychiatry, 148,* 365.

Christenson, G. A., & Mansueto, C. S. (1999). Trichotillomania: Descriptive characteristics and phenomenology. In D. J. Stein, G. A. Christenson, & E. Hollander (Eds.), *Trichotillomania* (pp. 1–41). Washington, D.C.: American Psychiatric Association Press.

Christenson, G. A., Ristvedt, S. L., & Mackenzie, T. B. (1993). Identification of trichotillomania cue profiles. *Behaviour Research and Therapy, 31,* 315–320.

Elliot, A. J., & Fuqua, R. W. (2001). Behavioral interventions for trichotillomania. In D. W. Woods & R. G. Miltenberger (Eds.), *Tic disorders, trichotillomania, and other repetitive behavior disorders: Behavioral approaches to analysis and treatment* (pp. 151–170). Boston: Kluwer.

Evers, R. A. F., & van de Wetering, B. J. M. (1994). A treatment model for motor tics based on specific tension-reduction technique. *Journal of Behavior Therapy and Experimental Psychiatry, 25,* 255–260.

Findley, D. B. (2001). Characteristics of tic disorders. In D. W. Woods & R. G. Miltenberger (Eds.), *Tic disorders, trichotillomania, and other repetitive behavior disorders: Behavioral approaches to analysis and treatment* (pp. 53–72). Boston: Kluwer.

Gray, J. J. (1979). Positive reinforcement and punishment in the treatment of childhood trichotillomania. *Journal of Behavior Therapy and Experimental Psychiatry, 10,* 125–129.

Grayson, J. B., Foa, E. B., & Steketee, G. (1982). Habituation during exposure treatment: Distraction vs. attention focusing. *Behaviour Research and Therapy, 20,* 323–328.

Hoogduin, K., Verdellen, C., & Cath, D. (1997). Exposure and response prevention in the treatment of Gilles de la Tourette's syndrome: Four case studies. *Clinical Psychology and Psychotherapy, 4,* 125–137.

Iwata, B. A., Dorsey, M., Slifer, K., Bauman, K., & Richman, G. (1982). Toward a functional analysis of self-injury. *Analysis and Intervention in Developmental Disabilities, 2,* 3–20. (Reprinted from *Analysis and Intervention in Developmental Disabilities, 2,* 3–20, 1982).

Iwata, B. A., Vollmer, T. R., Zarcone, J. R., & Rogers, T. A. (1993). Treatment classification and selection based on behavioral function. In R. Van Houten & S. Alexrod (Eds.), *Behavior Analysis and Treatment* (pp. 101–125). New York: Plenum Press.

Keuthen, N. J., Stein, D. J., & Christenson, G. A. (2001). *Help for hairpullers: Understanding and coping with trichotillomania.* Oakland, CA: New Harbinger.

Kwak, C., Dat Vuong, K. D., & Jankovic, J. (2003). Premonitory sensory phenomenon in Tourette's syndrome. *Movement Disorders, 18,* 1530–1533.

Lahey, B. B., McNees, M. P., & McNees, M. C. (1973). Control of an obscene "verbal tic" through timeout in an elementary classroom. *Journal of Applied Behavior Analysis, 6,* 101–104.

Leckman, J. F., King, R. A., & Cohen, D. J. (1999). Tics and tic disorders. In J. F. Leckman & D. J. Cohen (Eds.), *Tourette's Syndrome—Tics, obsessions, compulsions: Developmental psychopathology and clinical care* (pp. 23–42). New York: John Wiley & Sons.

Leckman, J. F., Walker, D. E., & Cohen, D. J. (1993). Premonitory urges in Tourette's syndrome. *American Journal of Psychiatry, 150,* 98–102.

Lennox, D. B., & Miltenberger, R. G. (1989). Conducting a functional assessment of problem behavior in applied settings. *Journal of the Association for Persons with Severe Handicaps, 14,* 304–311.

Malatesta, V. J. (1990). Behavioral case formulation: An experimental assessment study of transient tic disorder. *Journal of Psychopathology and Behavioral Assessment, 12*, 219–232.

Meidinger, A. L., Miltenberger, R. G., Himle, M. B., Omvig, M., Trainor, C., & Crosby, R. (2005). An investigation of tic suppression and the rebound effect in Tourette Syndrome. *Behavior Modification, 29*, 716–745.

Miguel, E. C., do Rosario-Campos, M. C., Prado, H. S., do Valle, R., Rauch, S. L., Coffey, B. J., et al. (2000). Sensory phenomena in obsessive-compulsive disorder and Tourette's disorder. *Journal of Clinical Psychiatry, 61*, 150–156.

Miltenberger, R. G. (2005). The role of automatic negative reinforcement in clinical problems. *International Journal of Behavioral and Consultation Therapy, 1*, 1–11.

Miltenberger, R. G., & Fuqua, R. W. (1985). A comparison of contingent vs. non-contingent competing response practice in the treatment of nervous habits. *Journal of Behavior Therapy and Experimental Psychiatry, 16*, 195–200.

Miltenberger, R. G., Fuqua, R. W., & McKinley, T. (1985). Habit reversal with muscle tics: Replication and component analysis. *Behavior Therapy, 16*, 39–50.

Miltenberger, R. G., Fuqua, R. W., & Woods, D. W. (1998). Applying behavior analysis to clinical problems: Review and analysis of habit reversal. *Journal of Applied Behavior Analysis, 31*, 447–469.

Miltenberger, R. G., Long, E. S., & Rapp, J. T. (1998). Evaluating the function of hair pulling: A preliminary investigation. *Behavior Therapy, 29*, 211–219.

Miltenberger, R. G., Rapp, J. R., & Long, E. S. (2001). Characteristics of trichotillomania. In D. Woods & R. Miltenberger (Eds.), *Tic disorders, trichotillomania, and repetitive behavior disorders: Behavioral approaches to analysis and treatment* (pp. 133–150). Boston: Kluwer.

Mouton, S. G., & Stanley, M. A. (1996). Habit reversal training for trichotillomania: A group approach. *Cognitive and Behavioral Practice, 3*, 159–182.

O'Connor, K. P., Brisebois, H., Brault, M., Robillard, S., & Loiselle, J. (2003). Behavioral activity associated with onset in chronic tic disorder and habit disorder. *Behaviour Research and Therapy, 41*, 241–249.

Penzel, F. (2003). *The hair pulling problem: A complete guide to trichotillomania.* New York: Oxford University Press.

Rapp, J. T., Miltenberger, R., Galensky, T., Ellingson, S., & Long, E. (1999). A functional analysis of hair pulling. *Journal of Applied Behavior Analysis, 32*, 329–337.

Rapp, J. T., Miltenberger, R. G., Galensky, T. L., Ellingson, S. A., Stricker, J., & Garlinghouse, M. (2000). Treatment of hair pulling and hair manipulation maintained by digital tactile stimulation. *Behavior Therapy, 31*, 381–393.

Rapp, J. T., Miltenberger, R. G., & Long, E. S. (1998). Augmenting simplified habit reversal with an awareness enhancement device. *Journal of Applied Behavior Analysis, 31*, 665–668.

Rapp, J. T., Miltenberger, R., Long, E., Elliott, A., & Lumley, V. (1998). Simplified habit reversal for chronic hair pulling in three adolescents: A clinical replication with direct observation. *Journal of Applied Behavior Analysis, 31*, 299–302.

Ristvedt, S. L., & Christenson, G. A. (1996). The use of pharmacologic pain sensitization in the treatment of repetitive hair pulling. *Behaviour Research and Therapy, 34*, 647–648.

Scotti, J. R., Schulman, D. E., & Hojnacki, R. M. (1994). Functional analysis and unsuccessful treatment of Tourette's syndrome in a man with mental retardation. *Behavior Therapy, 25*, 721–738.

Sharenow, E. L., Fuqua, R. W., & Miltenberger, R. G. (1989). The treatment of muscle tics with dissimilar competing response practice. *Journal of Applied Behavior Analysis, 22*, 35–42.

Silva, R. R., Munoz, D. M., Barickman, J., & Friedhoff, A. J. (1995). Environmental factors and related fluctuation of symptoms in children and adolescents with Tourette's disorder. *Journal of Child Psychology & Psychiatry, 36*, 305–312.

Stanley, M. A., Borden, J. W., Mouton, S. G., & Breckenridge, J. K. (1995). Nonclinical hairpulling: Affective correlates and comparison with clinical samples. *Behaviour Research and Therapy, 33*, 179–186.

Sturmey, P., Carlson, A., Crisp, A. G., & Newton, J. T. (1988). A functional analysis of multiple aberrant responses: A refinement and extension of Iwata et al.'s (1982) methodology. *Journal of Mental Deficiency Research, 32*, 31–46.

Task Force on Promotion and Dissemination of Psychological Procedures. (1995). Training in and dissemination of empirically validated psychological treatments. *Clinical Psychologist, 48*, 3–23.

Varni, J. W., Boyd, E. F., & Cataldo, M. F. (1978). Self-monitoring, external reinforcement, and timeout procedures in the control of high rate tic behaviors in a hyperactive child. *Journal of Behavior Therapy and Experimental Psychiatry, 9*, 353–358.

Verdellen, C. W. J., Keijsers, G. P. J., Cath, D. C., & Hoogduin, C. A. L. (2004). Exposure with response prevention versus habit reversal in Tourette's syndrome: A controlled study. *Behaviour Research and Therapy, 42*, 501–511.

Wagaman, J., Miltenberger, R., & Williams, D. (1995). Treatment of a vocal tic by differential reinforcement. *Journal of Behavior Therapy and Experimental Psychiatry, 26*, 35–39.

Watson, T. S., & Sterling, H. E. (1998). Brief functional analysis and treatment of a vocal tic. *Journal of Applied Behavior Analysis, 31*, 471–474.

Wetterneck, C. T., & Woods, D. W. (In press). An evaluation of exposure and response prevention for treating repetitive behaviors associated with Tourette's syndrome. *Journal of Applied Behavior Analysis.*

Wilhelm, S., Deckersbach, T., Coffey, B. J., Bohne, A., Peterson, A. L., & Baer, L. (2003). Habit reversal versus supportive psychotherapy for Tourette's disorder: A randomized controlled trial. *American Journal of Psychiatry, 160*, 1175–1177.

Woods, D. W. (2001). Habit reversal treatment manual for tic disorders. In D. W. Woods & R. G. Miltenberger (Eds.), *Tic disorders, trichotillomania, and other repetitive behavior disorders: Behavioral approaches to analysis and treatment* (pp. 97–132). Boston: Kluwer.

Woods, D. W., & Himle, M. B. (2004). Creating tic suppression: Comparing the effects of verbal instruction to differential reinforcement. *Journal of Applied Behavior Analysis, 37*, 417–420.

Woods, D. W., & Miltenberger, R. G. (1995). Habit reversal: A review of applications and variations. *Journal of Behavior Therapy and Experimental Psychiatry, 26*, 123–131.

Woods, D. W., & Miltenberger, R. G. (1996). Are persons with nervous habits nervous? A preliminary examination of habit function in a nonreferred population. *Journal of Applied Behavior Analysis, 29*, 259–261.

Woods, D. W., & Miltenberger, R. G. (2001). *Tic disorders, trichotillomania, and other repetitive behavior disorders: Behavioral approaches to analysis and treatment.* Boston: Kluwer.

Woods, D. W., Murray, L. K., Fuqua, R. W., Seif, T. A., Boyer, L. J., & Siah, A. (1999). Comparing the effectiveness of similar and dissimilar competing responses in evaluating the habit reversal treatment for oral-digital habits in children. *Journal of Behavior Therapy and Experimental Psychiatry, 30*, 289–300.

Woods, D. W., Twohig, M. P., Flessner, C., & Roloff, T. (2003). Treatment of vocal tics in children with Tourette syndrome: Investigating the efficacy of habit reversal. *Journal of Applied Behavior Analysis, 36*, 109–112.

Woods, D. W., Watson, T. S., Wolfe, E., Twohig, M. P., & Friman, P. C. (2001). Analyzing the influence of tic-related talk on vocal and motor tics in children with Tourette's syndrome. *Journal of Applied Behavior Analysis, 34*, 353–356.

Yates, A. J. (1958). The application of learning theory to the treatment of tics. *Journal of Abnormal and Social Psychology, 56*, 175–182.

9

ENCOPRESIS AND ENURESIS

W. LARRY WILLIAMS
University of Nevada, Reno

MARIANNE JACKSON
University of Nevada, Reno

AND

PATRICK C. FRIMAN
Girls and Boys Town

The endpoint of the alimentary process involves elimination of urine and feces. Two of the most common presenting complaints in primary medical care for children involve disordered elimination, specifically enuresis (urine) and encopresis (feces). Prevalence estimates range as high as 2% of 5-year-old children for encopresis and 25% of 6-year-old children for enuresis (Friman & Jones, 1998). These disorders usually occur independently but can co-occur. There is a broad range of medical conditions that can cause encopresis and enuresis, but these causes are rare. They are real, however, and need to be ruled out prior to going forward with behavioral assessment and treatment. The vast majority of cases are functional and their comprehensive assessment readily yields the identity of functionally relevant variables that can either be modified through behavioral intervention or manipulated to bring about modifications in behavior related to elimination.

A range of definitions for the two disorders exists, and a unifying theme involves the inappropriate deposit of waste in terms of location, timing, or frequency. For example, the *Diagnostic and Statistical Manual (fourth edition) (DSM-IV)* of the American Psychiatric Association ([APA], 1994) defines encopresis as (a) repeated passage of feces into inappropriate places, (b) at least once a month for at least 3 months, (c) by a child of at least 4 years of age (mental age of 4 years if developmentally delayed), and

(d) "the fecal incontinence cannot be exclusively due to the physiological effects of a substance (e.g., laxatives) or a general medical condition except through a mechanism involving constipation" (p. 106). The *DSM-IV* further classifies encopresis in terms of whether it is associated with constipation or not and whether afflicted children have previously been fully continent for an extended period, in which case the term *secondary* is used, or if continence has never been achieved, in which case the term *primary* is used.

The *DSM-IV* diagnostic conditions for enuresis include repeated voiding of urine into clothing or bed at least twice a week for at least 3 months. If the frequency is smaller than that but the voiding is a cause of significant distress or impairment to social, academic, or occupational functioning, it satisfies diagnostic criteria. The child must be at least 5 years of age or exhibit that level of developmental ability if developmental delays are present. The condition cannot be directly due to the physiological effects of a substance (e.g., diuretics) or a general medical condition. As with encopresis, the *DSM-IV* further classifies enuresis into primary and secondary cases. Additionally, the *DSM-IV* subdivides enuresis into three subtypes: nocturnal, diurnal, and combined nocturnal and diurnal. Enuresis has a well-established genetic basis. Approximately 75% of affected children have a first-degree biological relative who has had the disorder, and it is more prevalent in monozygotic than dizygotic twins.

A BIOBEHAVIORAL PERSPECTIVE ON ENCOPRESIS AND ENURESIS

The elimination disorders are complex from a behavior analysis standpoint because physiology plays such a prominent role in their etiology and course. Diagnostic assessment yields information on behavioral and physiological variables, and effective treatment typically takes both into account. For example, the bladder in enuretic children is often overresponsive to filling. Treatment therefore often involves reducing this hyperresponsivity through exercises which attempt to stretch the bladder. As another example, the majority of cases of encopresis involve constipation. Effective treatment almost always involves the ingestion of substances that soften stools, such as increased fiber in the diet and stool-softening medications (Friman, 2004; Friman & Jones, 1998; Houts, 1991; Levine, 1982; Mellon & Houts, 1995; Mellon & McGrath, 2000).

The increased understanding of the interplay between physiological processes in the onset and course of incontinence has resulted in a virtual revolution in professional and lay interpretations of the relevant conditions, and the contemporary view is now biobehavioral. For example, as Friman (2004) noted in a discussion of functional encopresis (FE)

... functional encopresis has been misunderstood, misinterpreted, and mistreated for centuries. During the last half of the twentieth century, however, and particularly towards its end, a fuller, bio-behavioral understanding of FE's causal conditions was obtained and an empirically supported approach to its treatment established. The bio-behavioral understanding and approach to FE is dramatically different than the psychogenic understanding and approach of history. The bio-behavioral approach addresses the physiology of defecation primarily and addresses the psychology of the child as a set of variables that are not causal but can be critical to active participation in treatment." (p. 51)

Similar comments about contemporary understanding of enuresis are also increasingly available (e.g., Friman & Jones, 1998; Houts, 1991; Mellon & McGrath, 2000). Consistent with the contemporary biobehavioral approach to incontinence, the discussions of encopresis and enuresis that follow will begin with a brief description of physiological factors and then proceed to psychological and behavioral assessments, biobehavioral treatment, and end with a description of the role of functional assessment and analysis.

ENCOPRESIS

MEDICAL ASSESSMENTS AND TREATMENT

Multiple physiological factors are associated with encopresis, the most important of which are colonic motility, constipation, and fecal retention. Multiple dietary behavioral variables contribute to these factors. The most important of these are (1) insufficient roughage or bulk in the diet; (2) irregular diet; (3) insufficient oral intake of fluids; (4) medications that may have a side effect of constipation; (5) unstructured, inconsistent, and/or punitive approaches to toilet training; and (6) toileting avoidance by the child. Table 9.1 provides a summary of the best practice features for addressing encopresis.

Because the physiological variables are so central to the condition, the initial goal of assessment should involve thorough medical examination. Among the many goals of this examination should be the determination of the extent of stool retention. Chronic retention can lead to fecal impaction, which results in enlargement of the colon. Colon enlargement results in decreased motility of the bowel system and, occasionally, in involuntary passage of large stools and frequent soiling due to seepage of soft fecal matter. The seepage is often referred to as paradoxical diarrhea because the children retain large masses of stool and thus are functionally constipated, but their colon allows passage of soft stool around the mass, which results in diarrhea. Parents, unaware of these processes, can improvidently administer medication for diarrhea to their already constipated children, thus slowing their bowel motility even more (Friman, 2004; Friman & Jones, 1998).

TABLE 9.1   A Sample Biobehavioral Treatment Plan

---

1. Refer to appropriately trained physician for evaluation.
2. Demystify bowel movements and problems and eliminate all punishment.
3. Completely evacuate bowel. Procedures are prescribed and overseen by physician.
4. Establish regular toileting schedule. Ensure that child's feet are on a flat surface during toileting.
5. Establish monitoring and motivational system.
6. Require child participation in clean up.
7. Teaching appropriate wiping and flushing.
8. Implement dietary changes that include regularity of meals and increases in fluid and fiber intake.
9. Utilize facilitative medication. What, when, and how much to be established by physician.
10. Establish method for fading facilitative medication.

From Friman (2004). A bio-behavioral, bowel and toilet training treatment for functional encopresis. In W. Odonohue, J. Fisher, & S. C. Hayes (Eds.), *Cognitive behavior therapy: Applying empirically supported techniques in your practice* (p. 53). New York: John Wiley.

Another very important reason for the medical examination involves ruling out medical causes for soiling. Rare anatomic and neurologic problems can lead to fecal retention and soiling. Anatomic problems include a variety of malformations and locations of the anus which are detectable on physical exam and require medical management (Hatch, 1988). Hirschsprung's disease or congenital aganglionosis is a disorder in which the nerves that control the muscles in the wall of part or all of the colon are absent, causing severe constipation (Christophersen & Mortweet, 2001; Levine, 1982). Its incidence is approximately 1 in 25,000, and it usually causes severe symptoms in infancy (Levine, 1975). Thus, the clinical presentation itself should prevent the astute clinician from mistaking one for the other. The possible exception is ultra short segment Hirschsprung's disease, which has a more subtle clinical picture. However, the existence of this condition is controversial and, even if it does exist, proper collaboration between physician and behavioral psychologist should ensure timely diagnosis.

PSYCHOLOGICAL ASSESSMENTS FOR ENCOPRESIS

As indicated earlier, historically, incontinence and especially encopresis were first viewed as a problem in the character or personality of the individual, and thus it was common to treat the condition with social disapproval or even more extreme forms of punishment (Henoch, 1889). Initially, assessment and clinical interpretations stemmed primarily from a psychodynamic perspective, and psychopathological interpretations were common.

However, as we have noted, the primary causal variable for soiling is fecal retention. And in most cases, retention is not caused by characterological or psychopathological problems, and encopresis is not associated with significant increases in other psychological problems (Friman, Mathews, Finney, & Christophersen, 1988; Gabel, Hegedus, Wald, Chandra, & Chaponis, 1986). However, encopresis can be associated with some behavioral problems, especially oppositional behavior, and thus this class of behavior should be assessed (Landman & Rappaport, 1985).

BEHAVIORAL TREATMENT FOR ENCOPRESIS

There have been a wide number and variety of strictly behaviorally oriented treatments for encopresis which are not explicitly based on functional assessment or analysis, such as those using simple instructions and positive reinforcement for appropriate voiding (Ashkenazi, 1975; Ayllon, Simon, & Wildman, 1975), contracting and self-management (Plachetta, 1976), using negative reinforcement or overcorrection (Crowley & Armstrong, 1977; Rolider & Van Houten, 1985) or exclusionary timeout (O'Brien, Ross, & Christophersen, 1986). Encopresis has been treated in adults with intellectual disabilities (Smith, 1994), in school situations (Dixon & Saudargas, 1980; George, Coleman, & Williams, 1977), and in outpatient therapy contexts (Boon & Singh, 1991; Friman & Jones, 1998; Gelber & Meyer, 1964; Neale, 1963; Ritterband et al., 2003).

The current biobehavioral treatments for encopresis focus on immediate medical assessment and treatment of the condition. This typically will involve dietary, fluid, and activity changes, which are often referred to as cathartic treatments; practice at sitting on the toilet several times a day; and reinforcement for maintaining a voiding schedule and especially for successful defecation in the toilet. Although it is clear from the literature that parental reaction to the condition can be a crucial maintaining variable, there are few reports we are aware of specifically describing simple socially mediated encopresis (Conger, 1970). However, it should be no surprise that having the parents involved in a positive program for appropriate defecation together with medically reduced discomfort should result in regular continence. Indeed, these are the results of several large-scale studies incorporating these general procedures (Stark, Owens-Stively, Spirito, Lewis, & Guevremont, 1990; Stark et al., 1997).

FUNCTIONAL BEHAVIORAL ANALYSIS

A behavioral analysis of the processes involved in encopresis addresses the role of physiological activity in the normal functioning of the colon, rectum, and associated smooth muscles and sphincters. The typical child is taught to recognize the need to void and is taught to accomplish this in

increasingly independent and appropriate ways. If anatomical or physiological problems develop, these may result in inadvertent Pavlovian
association of movement of feces to the rectum or the process of voiding
with discomfort or pain. It is also possible that early experience of voiding
can be associated with extreme emotional distress, embarrassment, or
other social unpleasantness. In retentive encopresis, this may negatively
reinforce the voluntary (operant) avoidance of the normal elimination
process, which can then in turn lead to a decrease in the colon's motility,
constipation, and possible eventual impacted bowel. From a functional
analytic viewpoint, then, the establishment of the encopretic condition
could be due to escape or avoidance of so-called automatic or sensory
consequences or avoidance of prior mediated social aversives associated
with voiding.

If medical causes are less relevant, encopresis may have socially mediated causes. To reveal possible socially mediated causes of encopresis,
assessment from a functional analytic viewpoint should gather information
on the toileting history of the child and current frequency of soiling. It is
pertinent in this process to provide support to the child and parents in a
matter of fact approach emphasizing the problem as a medical condition
that many children have. Assessment of any and all behaviorally related
information, such as antecedents and consequences of inappropriate
voiding, may reveal environmental events that may have had influence on
toilet refusal. Therefore, the current reactions to the problem by the
parents, other adults, or peers are central to such an analysis especially if
their attention and general upset are indeed a maintaining variable for the
problem. It should also be determined if tangible consequences are associated with maintaining the problem, or if issues with incontinence function
to allow the child to escape certain contexts or people.

Similarly, current environmental antecedent and consequence conditions need to be identified for maintaining a regular schedule for defecation
while the bowel is stabilized in retentive encopresis, or sooner in nonretentive encopresis. Are there clear prompts for regular toileting attempts?
What is the current response cost for attempting regular voiding? What
consequences could be introduced and systematically faded for monitoring
one's self and maintaining regular voiding, appropriate dieting changes, or
consumption of medications or stool softeners?

A CASE STUDY OF THE TREATMENT
OF ENCOPRESIS

Steege and Harper (1989) reported a case study of John, an 11-year-old
boy, who had a history of noncompliance with a series of previously well-
designed outpatient-based treatment efforts and who was exhibiting an
average of 2.5 soiling accidents per day. Extensive and repeated medical

examinations, including rectal manometry, did not reveal organic pathology. The authors implemented various treatments including (a) fleet enemas; (b) suppositories; (c) milk of magnesia; (d) positive reinforcement for the nonoccurrence of soiling accidents; (e) positive reinforcement for the occurrence of appropriate bowel movements; (f) a high fiber diet; and (g) positive practice within the home environment over the next 6 months. John's parents kept accurate records of his appropriate bowel movements and soiling accidents. However, his parents reported that he soiled 1.5 times/week while averaging two appropriate bowel movements/day. At that time, his parents reported that he was becoming very resistive by refusing enemas, having tantrum behaviors, and crying to the earlier treatment procedures, particularly the use of cathartics. In addition, he was hiding his soiled underwear, thus making detection of soiling accidents difficult.

The maintenance of accidents after extensive medical treatment and John's clear resistance to these procedures, such as having tantrums, crying, and hiding of soiled underwear, were probably escape from the treatment. Whereas there were possible medical reasons for soiling, the role of socially mediating variables was unclear. Perhaps the social reinforcement for appropriate elimination was not sufficient to overcome the apparent inconvenience of soiling, or perhaps soiling functioned to maintain parental attention in general.

In order to address theses issues, the authors used a treatment package that eliminated enema use and enabled monitoring of soiled or clean underwear. The method incorporated (1) positive reinforcement procedures, including points earned daily and cashed in for items previously agreed upon as desirable by John, (2) including differential reinforcement of appropriate behavior (successful voids), (3) differential reinforcement of other behavior (clean underwear at checks), (4) self-evaluation, positive practice using a shower and washing underwear after accidents, (5) milk of magnesia, and (6) a high fiber diet. Numbered underwear was used to increase the frequency of appropriate bowel movements and to eliminate hiding of soiled underwear.

Following John's treatment in a hospital inpatient setting, the therapist reviewed the effectiveness of the procedures with John and his parents. Next, the specific treatment procedures were reviewed and explained to them. The acceptability of the treatment was demonstrated by (1) parental statements regarding the applicability of the treatment package to the home environment; (2) child and parental understanding of the components of the treatment packages, which was evaluated initially by asking the parents to verbally rehearse the treatment procedures; and (3) child and parental compliance with the treatment regimen during a 3-day home visit phase wherein the treatment was implemented in the home environment. John returned to the hospital for 5 further days of treatment and then returned home. Note that the return to hospital before commencing outpatient treatment at home may have acted as an establishing operation for behaviors of both John and his parents that led to program success in order to avoid the relative inconvenience of hospitalization. The mean weekly frequency of the target behaviors was measured for 20 weeks following discharge, with additional follow-up contacts conducted at 6 and 12 months post-discharge.

After an increase in bowel movement frequency and one instance of soiling on the third day at home, John was able to have on average three successful bowel movements per day with no soiling, and this performance was maintained for a year (see Figure 9.1).

ENURESIS

MEDICAL ASSESSMENT AND TREATMENT

Acquisition of urinary continence is a complex physiological process (Muellner, 1951; Vincent, 1974). Normal continence is attained through appropriate voluntary elimination via sphincter release upon the lowering of the bladder neck when it is full and preventing micturition by contraction of pelvic floor muscles, which raises the top of the bladder. Continence

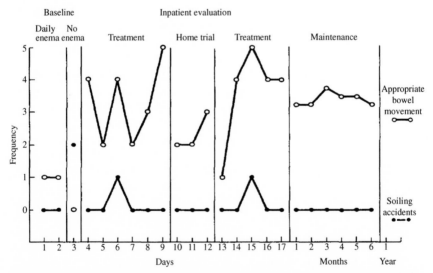

FIGURE 9.1   Frequency of appropriate bowel movements and soiling accidents during baseline, hospital assessment and treatment, and subsequent home treatment and maintenance. (Steege & Harper. [1989]. Enhancing the management of secondary encopresis by assessing acceptability of treatment: A case study. *Journal of Behavior Therapy and Experimental Psychiatry, 20*, 333–341.) Reprinted with permission.

involves an appropriate bladder capacity, and the development of stimulus control of a full bladder over prevention of micturition until an appropriate situation for urination is present. This involves becoming aware of the need to urinate to avoid the emergency condition of "urgency" where micturition is eminent, as well as the inhibition of urination while awake or sleeping. Incontinence can result from physical anomalies, neurological anomalies, and in their absence the lack of training oneself to recognize full bladder stimulation and act upon it.

There are numerous well-known potential physiopathologic causes of enuresis, including urinary tract infection, urinary tract anomaly, bladder instability, occult spina bifida, epilepsy, diabetes mellitus, and sleep apnea. Most of these causes can be ruled out by complete history, physical exam, and urinalysis. When unanswered questions remain, other more elaborate laboratory examinations, such as voiding cystourethrogram or polysomnographic evaluation are available (Friman & Jones, 1998; Gross & Dornbusch, 1983).

Pharmacological treatments have typically involved tricyclic antidepressants such as imipramine, which reduces premature contractions of the bladder following partial filling and thereby increases functional bladder capacity (Stephenson, 1979). Imipramine, in doses between 25 and 75 mg given at bedtime, produces initial reductions in wetting in a majority of children, often within the first week of treatment (Blackwell & Currah, 1973). However, any increase in continence appears only while the child is on the drug, and both short- and long-term studies show enuresis usually recurs when tricyclic therapeutic agents are withdrawn. The permanent elimination of enuresis produced with imipramine is reported to be 25%, ranging from 5% to 40% (Blackwell & Currah, 1973; Houts, Berman, & Abramson, 1994). More recently, desmopressin (DDAVP), an intranasally administered vasopressin analogue, has been shown to reduce nocturnal enuretic episodes in children. However, once it is removed, only 25% of children remain continent (Norgaard, Pedersen, & Djurhuus, 1985). It is unknown if its effectiveness is due to its renal effects of increasing urine concentration and therefore lessening nocturnal bladder volume or some other process. It is important to understand that medications do not teach continence skills, but may be used for short-term continence success as part of a comprehensive training intervention, for example, to aid in achieving continence for an overnight camping experience or sleepover.

BEHAVIORAL TREATMENTS FOR ENURESIS

There has been a proliferation of behaviorally oriented studies concerning nocturnal enuresis and to a lesser extent diurnal enuresis which are not explicitly based on functional assessment or analysis. Children and adolescents have been reported to have been successfully treated with different versions of the original Mowrer (1938) wet alarm pad (Finley, Besserman, Bennet, Clapp, & Finley, 1973; Finley, Wansley, & Blenkarn, 1977; Friman & Vollmer, 1995; Taylor & Turner 1975), the operant-oriented "dry-bed" methods (Azrin & Foxx, 1974; Azrin, Sneed, & Foxx, 1974; Barman, Katz, O'Brien, & Beauchamp, 1981; Bollard & Nettlebeck, 1981; Bollard & Woodroffe, 1977), bladder training (Averink, Melein, & Duker, 2005; Harris & Purohit, 1977; Kimmel & Kimmel, 1970; Paschalis, Kimmel, & Kimmel, 1972), progressive awakenings (Singh, Phillips,& Fischer 1976), token systems (Popler, 1976), and bibliotherapy (Van Londen, Van-Londen-Barentsen, Van Son, & Mulder, 1995). Enuresis treatments have also been widely reported for differing populations such as seniors (Adkins & Mathews, 1997; Schnelle et al., 1983; Spangler, Risley, & Bilyew, 1984), persons with intellectual disabilities (Azrin, Bugle, & O'Brien, 1971; Azrin & Foxx, 1971; Mahoney, Van Wagenen, & Meyerson, 1971; Phibbs & Wells, 1982), children with autism (LeBlanc, Carr, Crossett, Bennett, & Detweiler, 2005), and persons with acquired brain injury (Papworth, 1989.) The current biobehavioral treatment approach (Friman & Jones, 1998; Houts, 1991) is the result of development over the years of more and more information regarding the underlying mechanisms of incontinence simultaneously with the empirical demonstrations of environmental influences via Pavlovian and operant processes. The interested reader is referred to several cogent reviews that best describe these developments (see Bollard & Nettlebeck, 1981; Fielding, 1982; Finley et al., 1973; Lovibond, 1963; Morgan, 1978; Mountjoy, Ruben, & Bradford, 1984).

PSYCHOLOGICAL ASSESSMENT OF ENURESIS

Enuresis has been recognized for centuries (Mountjoy et al., 1984), but only recently have biobehavioral analyses become the predominant assessment and treatment. Mowrer (1938) is most often cited as the beginning of the behavioral analysis of enuresis with his introduction of a wet alarm bed pad device for treating nocturnal enuresis, although Mountjoy et al. (1984) reported on a U.S. patent office award for a wet alarm as early as 1905. Treatment involved having the enuretic child sleep on a pad which could detect moisture and set off an alarm sound awakening the child so that urination could be stopped and completed in the toilet. Mowrer and Mowrer (1938) proposed a Pavlovian conditioning explanation as the underlying behavioral mechanism in which an unconditioned stimulus (UCS), a bell or buzzer sound, was paired with a currently neutral stimulus (NS), the physiological stimulation of the full bladder, such that the full bladder eventually came to cause awakening as a conditioned stimulus (CS) and thereby preventing bed wetting (CR). Martin and Kubly's (1955) explanation was that "As time goes on it is also hoped that the somewhat more temporarily removed response of sphincter control will become associated with bladder tension, thus allowing the child to sleep through the night dry" (1963, p. 17).

However, Place (1954) challenged the Pavlovian explanation on the grounds that conditioning should not remain intact after removal of the alarm; i.e., respondent extinction should occur. That is, a formerly neutral stimulus (NS), such as the feeling of a full bladder, when presented over several conditioning trials just before an unconditioned stimulus (UCS), such as a bell or pad alarm, may acquire the automatic control of that UCS over its associated unconditioned response (awakening and awareness of full bladder and pelvic floor muscle tension). Thus, after several appropriate pairings, the alarm may indeed come to elicit the now conditioned response (CR) of muscle contraction. However, as in all respondent preparations, the CS will eventually lose its ability to elicit the CR unless it is occasionally paired with the UCS. Therefore, once the alarm is discontinued, there is no reason for the awareness and appropriate muscle contraction to continue.

Lovibond (1963) provided an explanation and demonstration of the extant wet alarm procedures in terms of Konorski's (1948) type II, variety III conditioned reflex, which was essentially a description of a behavior acquired because it allowed escape from an aversive event (negative reinforcement or the strengthening of a behavior because its consequences reduce, eliminate, or prevent an aversive situation). In this conceptualization, the relaxation of the bladder sphincter causes the onset of the alarm, which is an aversive stimulus that is escaped from and eventually avoided, through contraction of the sphincter. Lovibond demonstrated the superiority of this explanation by citing Crosby (1950) who compared (a) the Mowrer apparatus; (b) a similar device that delivered shock; (c) his own twin alarm pad which, upon urination, provided an obnoxious horn for 1 s; and (d) the regular awakening alarm buzzer delayed by 1 min, which was also associated with requiring the child or an attendant to reset it. The twin signal device was demonstrated to be more effective than both the Mowrer and Crosby devices in arresting wetting. Indeed, it did appear that the CR was sphincter contraction.

By the end of the 1960s, there were three possible explanations for how bell and pad procedures worked: classical conditioning of sphincter contraction, operant punishment of sphincter relaxation, and operant escape of the alarm through sphincter contraction (Peterson, Wright, & Hanlon, 1969). Common to all procedures, however, was the approximately 35% relapse rate following treatment. That is, respondent extinction seemed to play a role in the non-maintenance of bladder control once the aversive bell or horn was eventually no longer present. Soon thereafter, researchers found that intermittent reinforcement by having the alarm sound for only some occurrences of urinating, rather than all occurrences, resulted in greatly reduced relapse (Finley et al., 1973).

With a marked deviation from the classical conditioning emphasis, Azrin et al. (1974) reported on their "dry-bed training" treatment for enuresis. In two experiments they demonstrated that (a) a second alarm in the parents' room to awaken them was more effective than just an alarm in the enuretic child's room, and that (b) an alarm only in the parents' room was sufficient to stop wetting in a few days. Thus, because there was no UCS, no Pavlovian conditioning component was present. There was only an operant escape/avoidance arrangement, where the child's wetting receives social disapproval and the child must clean up and practice lying on the bed and counting to 50, before going to the bathroom. One night of intensive training involved practicing getting up and attempting to urinate in the toilet, ingestion of extra fluids before sleep, hourly awakenings to go to the toilet, having the child inhibit urination for 1 hr if possible, and returning to bed and verifying that it is dry. The dry bed method reduced accidents sharply and kept them at zero rates compared to the alarm-only method, which did not show significant reductions over the 2–3 weeks of the study. The Azrin et al. (1974) study complimented two now classical publications in which overcorrection procedures, increased fluid intake, and intensive initial practice sessions were successful in toilet training adults with intellectual disabilities (Azrin & Foxx, 1971) and as a toilet training method for parents of typical children (Azrin & Foxx, 1974).

FUNCTIONAL BEHAVIORAL ANALYSIS

The theoretical explanation of successful enuresis treatment shifted from a Pavlovian model that sought to rectify aberrant or underdeveloped physiological reflexes controlling the urinary process through use of alarms to an operant model encompassing diverse environmental events that provided aversive consequences to be avoided via negative reinforcement and other consequences that strengthened appropriate voiding via positive reinforcement. Aversive consequences were social disapproval and overlearning or positive practice overcorrection (required mass repeating of a correct behavior for each occurrence of the undesirable behavior). Positive consequences were social and tangible consequences for following a schedule, voiding in the toilet, and maintaining a dry bed. These treatments have evolved with physiological-relevant interventions such as increasing bladder capacity and resistance to urgency cues, training discrimination of subtle cues that signal eminent urination, and awakening procedures to form the current biobehavioral approach.

An important aspect of this development is seen in Fielding's (1980, 1982) work, which focused attention on the overall developmental process of continence as it occurs in most children with respect to the crucial role of the environment in that development. Fielding observed differences in the effectiveness of the wet bed alarm treatment alone, and together with retention control training (RCT; Kimmel & Kimmel, 1970) for children with nocturnal enuresis as opposed to both diurnal and nocturnal enuresis. Fielding (1980) found that alarm training alone was effective for nocturnal-alone enuretics, but that alarm plus RCT was better. Alarm plus RCT was also better for day and night enuresis, but was not as effective as with nocturnal-alone enuretic children. Children who wet day and night responded more slowly and had higher relapse rates. Bladder capacity was not associated with success, and thus Fielding questioned if diurnal and nocturnal enuresis were products of different processes. Fielding (1982) followed up on these findings and reported on diurnal and nocturnal enuretic children versus normal controls with respect to their associated posturing with respect to bladder volume, "urgency," and accidents. Enuretic children appeared to void sooner; showed less overall urgency behaviors, such as crossing the legs and squeezing them together, holding one's genital area, etc.; and often had accidents within minutes of toileting. Fielding postulated a developmental process in children in which awareness of the need to void and inhibition of micturition before strong urgency behaviors occur develops slowly over years and in coordination with parental prompts to void after relevant "urgency" behaviors are observed by parents.

For children with enuresis, self-awareness, inhibition and finally adequate emptying of the bladder are not under appropriate stimulus control of either the distended bladder, nor their associated urgency behavioral posturing. Enuresis then could result from a lack of appropriate discrimination of any of several crucial points in the micturition process of bladder distention, accompanying postural movements and parents not discriminating these postures during normal daily child rearing. Fielding's studies imply that response chains may exist in some children, in which voiding is reliably preceded by various motor behaviors suggesting that interventions earlier in the response chain may be an effective treatment option. For example, an adult may request the child go to the bathroom and void early, or the child might be taught to accurately discriminate his/her own postural behavior and learn to void when exhibiting motor behaviors that reliably predict urination. Although we are not aware of any subsequent research, it is clear that recent procedures for dealing with diurnal enuresis would affect this postulated development of self-awareness in the child with enuresis through reinforcement of regular toileting scheduling based on baseline frequency of voiding, monitoring of dry pants, and therefore heightened observation by parents and reinforcement for appropriate voiding.

Specific behavior analytic examples are available of functional environmental variables related to urinary continence. For example, in terms of operant consequences, Friman and Volmer (1995) reported on the reduction of urinary accidents in a 15-year-old girl by use of a wet pad alarm.

They described the effects of their procedure as negative reinforcement. Hansen (1979) reported on successful use of a twin alarm bed pad using a fading procedure for stopping night wetting in two children over 200 days and with no accidents at a 1-year follow-up. This intervention involved escaping the alarm and subsequent avoidance of it.

With respect to antecedent events, Hagopian, Fisher, Piazza, and Wierzbicki (1993) reported on successful reduction in urinary accidents in a 9-year-old boy with profound mental retardation. Their intervention involved frequent sitting on the toilet and pouring 5 oz. of warm water over the genitals to prompt urination. In a recent demonstration of stimulus control of context and environment over urination, Tarbox, Williams, and Friman (2004) reported a case study involving continence of a man diagnosed with severe mental retardation. They demonstrated through a reversal design that diapers exert stimulus control over accidents. There is no consequence for discriminating the need to urinate, nor for inhibiting urination, while wearing a diaper, and diaper wearing increases uninhibited urination. Therefore, to attain continence, diapers should be completely removed. This long-known but undemonstrated issue has relevance for keeping children in diapers and similar functioning clothing longer than necessary, as well as for maintaining continence in adults as they age or with disabilities, for the convenience of caregivers.

CASE STUDY OF ENURESIS

Lassen and Fluet (1979) combined conditioning, a modification of overlearning, and a token economy based on naturally occurring reinforcers to treat successfully a 10-year-old girl, Sally, who had a lifetime history of nocturnal enuresis. Previous conditioning alone, psychiatric treatment, and medication (Tofranil) had not been successful. The treatment for enuresis began after treatment of a token system for a variety of behaviors that Sally's parents had been unsuccessful at getting her to complete, such as chores and homework. Baseline data showed that she wet the bed 12 out of 14 nights. A structured point system for earning rewards that were relevant for Sally was demonstrated to increase chore completion and homework assignments. Prior complaining and admonitions on the part of the parents for Sally not completing these tasks had no effect on these behaviors. This would indicate that inadvertent parental attention or control through disruption of family life may have maintained Sally's nonperformance of these behaviors. The physical consequences (accidents) for not engaging in these behaviors were apparently not significant enough to counteract their noncompletion through avoidance. Parental concern and attention also may have been directly maintaining incontinence.

A bell and pad conditioning procedure was instituted. Sally was instructed that when the bell was triggered, she was to turn off the alarm, go to the bathroom, wash her face with cold water, and void. She was then to wake her mother, who was to observe her change her sheets and reset the alarm. Sally received points for following the procedure and for each dry night. When this procedure did not result in significant dry nights, the therapists discovered that Sally's mother was actually helping her change the sheets upon awakening. This was stopped and Sally stayed dry for 15 of 19 nights and then 17 nights in a row. At this point, Young and Morgan's (1972) bladder training procedure to decrease relapse was added in which Sally drank 16 oz. of fluid within an hour before bedtime. Sally then wet 5 out of 7 nights. Due to this relapse, Sally was instructed to drink only 6 oz. of fluid during the hour before bedtime. She was dry for 10 consecutive nights with this amount and remained dry for the next 10 nights after her fluid intake was increased to 10 oz. and finally 12 oz. Sally remained dry throughout this overlearning procedure. The fluid intake, points, and bell and pad were discontinued at this time, with Sally having been dry 30 consecutive nights. Three-, 6- and 12-month follow-up contacts with the parents indicated no relapse.

CONCLUSION

Enuresis and encopresis have significant physiological dimensions that exert a functional influence on the development of continence skills. Additionally, a broad range of environmental events affect these physiological dimensions as well as the development of the continence skills. Therefore, we have endorsed the biobehavioral approach to both problems. It is simply not possible to identify a single causal variable for the typical cases of encopresis and enuresis. Both are multiply determined. However, a functional perspective on the multiple determinative sources appears optimal to us because it leads so directly to interventions that are functionally relevant to the causes and expressions of incontinence. Presently, there is little research on the role functional assessment and analysis can play in the assessment of incontinence except in rare cases typically involving extreme developmental disability and a constellation of behavior problems, of which incontinence typically is only one member. That functional assessment would be relevant for cases of encopresis that do not include constipation or stool retention seems self-evident. At present, there is no literature describing the cause of these cases, and professionals working them up typically resort to hypothetical constructs that are popular in their orientation to psychology, such as aberrant family dynamics, psychopathology, or post-traumatic stress disorder. From a scientific perspective, however, at present very little can be said about cause. Research into function could fill this gap.

That a functional perspective would be useful for secondary cases of enuresis and encopresis also seems self-evident. Some children acquire continence skills and then appear to lose those skills. Physiological variables have not been implicated, at least not persuasively, and thus information from functional assessments and analyses could yield variables with a determinative role and thus lead to more effective treatments.

Even in the routine cases of encopresis or enuresis, a functional perspective is highly relevant and valuable and, as we have indicated, underemployed at present. For example, resistance, passive and active, plays a significant role in encopresis and diurnal enuresis. For some children, cessation of an ongoing activity is much more costly than for others. For example, a child with low social status on the playground who has been invited to play a game pays a much higher social cost for leaving the game to use the toilet than does a high status child who is regularly invited to play. Incontinent children are, other things being equal, lower in social status than continent children. Therefore, social variables may play a role. As another example, children who are the "architect and victim" (Patterson, 1982) of coercive family processes may resist toileting as a more general pattern of resistance to authority. Research into such resistance has not been explored in association with incontinence, at least not to our knowledge.

Even nocturnal enuresis is associated with a range of functional variables that have yet to be explored. For example, we are aware of no research on the seasonal expression of enuresis. Yet, other things being equal, rising to use the bathroom on a cold winter's night is much more aversive than doing so on a warm summer one. As another example, allowing children to wear pull-ups or diapers to bed reduces the aversive properties of accidents and may reduce motivation to achieve continence skills thereby (Tarbox et al., 2004). There are many other examples. More generally, there is a dearth of research into social and automatically generated consequences for incontinence, and pursuing that research more vigorously could not only enhance understanding and treatment of it, but also underscore the value of a behavior analytic perspective on child biobehavioral problems.

REFERENCES

Adkins, V. K., & Mathews, R. M. (1997). Prompted voiding to reduce incontinence in community-dwelling older adults. *Journal of Applied Behavior Analysis, 30*, 153–156.

American Psychiatric Association. (1994). *Diagnostic and statistical manual of mental disorders* (4th ed.). Washington, D.C.: Author.

Ashkenazi, Z. (1975). The treatment of encopresis using a discriminative stimulus and positive reinforcement. *Journal of Behavior Therapy and Experimental Psychiatry, 6*, 155–157.

Averink, M., Melein, L., & Duker, P. C. (2005). Establishing diurnal bladder control with the response restriction method: Extended study on its effectiveness. *Research in Developmental Disabilities, 26*, 143–151.

Ayllon, T., Simon, S. J., & Wildman, R. W. (1975). Instructions and reinforcement in the elimination of encopresis: A case study. *Journal of Behavior Therapy and Experimental Psychiatry, 6*, 235–238.

Azrin, N. H., Bugle, C., & O'Brien, F. (1971). Behavioral engineering: Two apparatuses for toilet training retarded children. *Journal of Applied Behavior Analysis, 4*, 249–253.

Azrin, N. H., & Foxx, R. M. (1971). A rapid method of toilet training the institutionalized retarded. *Journal of Applied Behavior Analysis, 4*, 89–99.

Azrin, N. H., & Foxx, R. M. (1974). *Toilet training in less than a day.* New York: Simon and Schuster.

Azrin, N. H., Sneed, T. J., & Foxx, R. M. (1974). Dry bed training: Rapid elimination of childhood enuresis. *Behaviour Research and Therapy, 12*, 147–156.

Barman, B., Katz, R., O'Brien, F., & Beauchamp, K. (1981). Treating irregular enuresis in developmentally disabled persons: A study in the use of overcorrection. *Behavior Modification, 5*, 336–346.

Blackwell, B., & Currah, J. (1973). The psychopharmacology of nocturnal enuresis. In I. Kolvin, R. C. MacKeith, & S. R. Meadow (Eds.), *Bladder control and enuresis* (pp. 231–257). Philadelphia: Lippincott.

Bollard, J., & Nettlebeck, T. (1981). A comparison of dry-bed training and standard urine-alarm conditioning treatment of childhood bedwetting. *Behaviour Research and Therapy, 19*, 215–226.

Bollard R. J., & Woodroffe, O. (1977). The effect of parent-administered dry-bed training on nocturnal enuresis in children. *Behaviour Research and Therapy, 15*, 159–165.

Boon, F. L., & Singh, N. N. (1991). A model for the treatment of encopresis. *Behavior Modification, 15*, 355–371.

Christophersen, E. R., & Mortweet, S. L. (2001). *Treatments that work with children: Empirically supported strategies for managing childhood problems.* Washington, D.C.: APA Books.

Conger, J. (1970). Treatment of encopresis by the management of social consequences. *Behavior Therapy, 1*, 386–390.

Crosby, N. D. (1950). Essential enuresis: Successful treatment based on physiological concepts. *Medical Journal of Australia, 7*, 533–543.

Crowley, C. P., & Armstrong, P. M. (1977). Positive practice, overcorrection and behavior rehearsal in the treatment of 3 cases of encopresis. *Journal of Behavior Therapy and Experimental Psychiatry, 8*, 411–416.

Dixon, J. W., & Saudargas, R. A. (1980). Toilet training, cueing, praise, and self-cleaning in the treatment of classroom encopresis: A case study. *Journal of School Psychology, 18*, 135–140.

Fielding, D. (1980). The response of day and night wetting children and children who wet only at night to retention control training and the enuresis alarm. *Behaviour Research and Therapy, 18*, 305–317.

Fielding, D. (1982). Analysis of the behaviour of day- and night-wetting children: Towards a model of micturition control. *Behaviour Research and Therapy, 20*, 49–60.

Finley, W. W., Besserman, R. L., Bennett, L. F., Clapp, R. F., & Finley P. M. (1973). The effect of continuous, intermittent, and "placebo" reinforcement on the effectiveness of the conditioning treatment for enuresis nocturnal. *Behaviour Research and Therapy, 2*, 289–297.

Finley, W. W., Wansley R. A., & Blenkarn, M. M.(1977). Conditioning treatment of enuresis using a 70% intermittent reinforcement schedule. *Behaviour Research and Therapy, 15*, 419–427.

Friman, P. C. (2004). A bio-behavioral, bowel and toilet training treatment for functional encopresis. In W. Odonohue, J. Fisher, & S. C. Hayes (Eds.), *Cognitive behavior therapy:*

Applying empirically supported techniques in your practice (pp. 51–58). New York: John Wiley.

Friman, P. C., & Jones, K. M. (1998). Elimination disorders in children. In S. Watson & F. Gresham (Eds.), *Handbook of child behavior therapy* (pp. 239–260). New York: Plenum.

Friman, P. C., Mathews, J. R., Finney, J. W., & Christophersen, E. R. (1988). Do children with encopresis have clinically significant behavior problems? *Pediatrics, 82*, 407–409.

Friman, P. C., & Vollmer, D. (1995). Successful use of the nocturnal urine alarm for diurnal enuresis. *Journal of Applied Behavior Analysis, 28*, 89–90.

Gabel, S., Hegedus, A. M., Wald, A., Chandra, R., & Chaponis, D. (1986). Prevalence of behavior problems and mental health utilization among encopretic children. *Journal of Developmental and Behavioral Pediatrics, 7*, 293–297.

Gelber, H., & Meyer, V. (1964). Behaviour therapy and encopresis: The complexities involved in treatment. *Behaviour Research and Therapy, 2*, 227–231.

George, T. W., Coleman, J. J., & Williams, P. S. (1977). The systematic use of positive and negative consequences in managing classroom encopresis. *Journal of School Psychology, 15*, 250–254.

Gross, R. T., & Dornbusch, S. M. (1983). Enuresis. In M. D. Levine, W. B. Carey, A. C. Crocker, & R. T. Gross (Eds.), *Developmental-behavioral pediatrics* (pp. 575–586). Philadelphia, PA: Saunders.

Hagopian, L. P., Fisher, W., Piazza, C. C., & Wierzbicki, J. J. (1993). A water-prompting procedure for the treatment of urinary incontinence. *Journal of Applied Behavior Analysis, 26*, 473–474.

Hansen, G. D. (1979). Enuresis control through fading, escape, and avoidance training. *Journal of Applied Behavior Analysis, 12*, 303–307.

Harris L. H., & Purohit, A. P. (1977). Bladder training and enuresis: A controlled trial. *Behaviour Research and Therapy, 15*, 485–490.

Hatch, T. F. (1988). Encopresis and constipation in children. *Pediatric Clinics of North America, 35*, 257–281.

Henoch, E. H. (1889). *Lectures on children's diseases* (*Vol. 2*; J. Thompson, Trans.). London: New Syndenham Society.

Houts, A. C. (1991). Nocturnal enuresis as a bio-behavioral problem. *Behavior Therapy, 22*, 133–151.

Houts, A. C., Berman, J. S., & Abramson, H. (1994). Effectiveness of psychological and pharmacological treatments for nocturnal enuresis. *Journal of Consulting and Clinical Psychology, 62*, 737–745.

Kimmel, H. D., & Kimmel, E. C. (1970). An instrumental conditioning method for the treatment of enuresis. *Journal of Behavior Therapy and Experimental Psychiatry, 1*, 121–123.

Konorski, J. (1948). *Conditioned reflexes and neuron organization*. Cambridge: Cambridge University Press.

Landman, G. B., & Rappaport, L. (1985). Pediatric management of severe treatment-resistant encopresis. *Development and Behavioral Pediatrics, 6*, 349–351.

Lassen, M., & Fluet, N. (1979). Multifaceted behavioral treatment for nocturnal enuresis. *Behavior Therapy & Experimental Psychiatry, 10*, 155–156.

LeBlanc, L. A., Carr, J. E., Crossett, S. E., Bennett, C. M., & Detweiler D. D. (2005). Intensive outpatient behavioral treatment of primary urinary incontinence of children with autism. *Focus on Autism and Other Developmental Disabilities, 20*, 98–105.

Levine, M. D. (1975). Children with encopresis: A descriptive analysis. *Pediatrics, 56*, 407–409.

Levine, M. D. (1982). FE: Its potentiation, evaluation, and alleviation. *Pediatric Clinics of North America, 29*, 315–330.

Lovibond, S. H. (1963). The mechanism of conditioning treatment of enuresis. *Behaviour Research and Therapy, 1*, 17–21.

Mahoney, K., Van Wagenen, R. K., & Meyerson, L. (1971). Toilet training of normal and retarded children. *Journal of Applied Behavior Analysis, 4*, 173–181.

Martin, B., & Kubly, D. (1955) Results of treatment of enuresis by a conditioned response method. *Journal of Consulting Psychology, 19*, 71–73.

Mellon, M. W., & Houts, A. C. (1995). Elimination disorders. In R. T. Ammerman and M. Hersen (Eds.), *Handbook of child behavior therapy in the psychiatric setting* (pp. 341–366). New York: John Wiley.

Mellon, M. W., & McGrath, M. L. (2000). Empirically supported treatments in pediatric psychology: Nocturnal enuresis. *Journal of Pediatric Psychology, 25*, 193–214.

Morgan, R. T. T. (1978). Relapse and therapeutic response in the conditioning treatment of enuresis: A review of recent findings on intermittent reinforcement, over learning and stimulus intensity. *Behaviour Research and Therapy, 16*, 273–279.

Mountjoy, P., Ruben, D. H., & Bradford, T. (1984). Recent technological advances in the treatment of enuresis: Theory and commercial devices. *Behavior Modification, 8*, 291–315.

Mowrer, O. H. (1938). Apparatuses for the study and treatment of enuresis. *American Journal of Psychology, 51*, 163–165.

Mowrer, O. H., & Mowrer, W. M. (1938). Enuresis: A method for its study and treatment. *American Journal of Orthopsychiatry, 18*, 436–459.

Muellner, S. R. (1951). The physiology of micturition. *The Journal of Urology, 65*, 805–813.

Neale, D. H. (1963). Behaviour therapy and encopresis in children. *Behaviour Research and Therapy, 1*, 139–149.

Norgaard, J. P., Pedersen, E. B., & Djurhuus, J. C. (1985). Diurnal antidiuretic hormone levels in enuretics. *Journal of Urology, 134*, 1029–1031.

O'Brien, S., Ross, L. V., & Christophersen, E. R. (1986). Primary encopresis: Evaluation and treatment. *Journal of Applied Behavior Analysis, 19*, 137–145.

Papworth, M. A. (1989). The behavioral treatment of nocturnal enuresis in a severely brain-damaged client. *Behaviour Therapy & Experimental Psychiatry, 30*, 365–368.

Paschalis, A., Kimmel H. D., & Kimmel, E. (1972). Further study of diurnal instrumental conditioning in the treatment of enuresis nocturnal. *Behavior Therapy & Experimental Psychiatry, 3*, 253–256.

Patterson, G. R. (1982). *Coercive family processes.* Eugene, OR: Castalia.

Peterson, R. A., Wright, R. L. D., & Hanlon, C. C. (1969). The effects of extending the CS-UCS interval on the effectiveness of the conditioning treatment for nocturnal enuresis. *Behaviour Research and Therapy, 7*, 351–357.

Phibbs, J., & Wells, M. (1982). The treatment of nocturnal enuresis in institutionalized retarded adults. *Behavior Therapy & Experimental Psychiatry, 13*, 245–249.

Place, U. T. (1954, May). *Conditioning and the treatment of enuresis: A theoretical discussion.* Read to the South Australian Group, British Psychological Society, London.

Plachetta, K. E. (1976). Encopresis: A case study utilizing contracting, scheduling and self-charting. *Journal of Behavior Therapy and Experimental Psychiatry, 7*, 195–196.

Popler, K. (1976). Token reinforcement in the treatment of nocturnal enuresis: A case study and six month follow-up. *Behavior Therapy & Experimental Psychiatry, 7*, 83–84.

Ritterband, L. M., Cox, D. J., Walker, L. S., Kovatchev, B., McKnight, L., Patel, K., et al. (2003). An internet intervention as adjunctive therapy for pediatric encopresis. *Journal of Consulting and Clinical Psychology, 71*, 910–917.

Rolider, A., & Van Houten, R. (1985). Treatment of constipation-caused encopresis by a negative reinforcement procedure. *Journal of Behavior Therapy and Experimental Psychiatry, 16*, 67–70.

Schnelle, J. F., Traughber, B., Morgan, D. B., Embry, J. E., Binion, A. F., & Coleman, A. (1983). Management of geriatric incontinence in nursing homes. *Journal of Applied Behavior Analysis, 16*, 235–241.

Singh, R., Phillips, D., & Fischer, S. C. (1976). The treatment of enuresis by progressively earlier waking. *Behavior Therapy & Experimental Psychiatry, 7*, 277–278.

Smith, L. J. (1994). A behavioral approach to the treatment of nonretentive nocturnal encopresis in an adult with severe learning disabilities. *Journal of Behavior Therapy and Experimental Psychiatry, 25*, 81–86.

Spangler, P. F., Risley, T. R., & Bilyew, D. D. (1984). The management of dehydration and incontinence in nonambulatory geriatric patients. *Journal of Applied Behavior Analysis, 17*, 397–401.

Stark, L., Owens-Stively, J., Spirito, A., Lewis, A., & Guevremont, D. (1990). Group behavioral treatment of retentive encopresis. *Journal of Pediatric Psychology, 15*, 659–671.

Stark, L. J., Opipari, L. C., Donaldson, D. L., Danovsky, M. R., Rasile, D. A., & DelSanto, A. F. (1997). Evaluation of a standard protocol for retentive encopresis: A replication. *Journal of Pediatric Psychology, 22*, 619–633.

Steege, M. W., & Harper, D. C. (1989). Enhancing the management of secondary encopresis by assessing acceptability of treatment: A case study. *Journal of Behavior Therapy and Experimental Psychiatry, 20*, 333–341.

Stephenson, J. D. (1979). Physiological and pharmacological basis for the chemotherapy of enuresis. *Psychological Medicine, 9*, 249–263.

Tarbox, R., Williams, W. L., & Friman, P. C. (2004). Extended diaper wearing: Effects on continence in and out of the diaper. *Journal of Applied Behavior Analysis, 37*, 97–100.

Taylor, P. D., & Turner, R. K. (1975). A clinical trial of continuous, intermittent and overlearning "bell and pad" treatments for nocturnal enuresis. *Behaviour Research and Therapy, 13*, 281–293.

Van Londen, A., Van Londen-Barentsen, M. L., Van Son, M. J. M., & Mulder, G. (1995). Relapse rate and subsequent parental reaction after successful treatment of children suffering from nocturnal enuresis: A $2^1/_2$ year follow-up of bibliotherapy. *Behaviour Research and Therapy, 33*, 309–311.

Vincent, S. A. (1974). Mechanical, electrical and other aspects of enuresis. In J. H. Johnston & W. Goodwin (Eds.), *Reviews in pediatric urology* (pp. 280–313). New York: Elsevier.

Young, G. C., & Morgan, R. T. (1972). Overlearning in the conditioning treatment of enuresis. *Behaviour Research and Therapy, 10*, 147–151.

APPENDIX 9.1

ONLINE RESOURCES FOR ENURESIS AND ENCOPRESIS

There are a variety of web sources with information regarding products and information in general

http://www.AllegroMedical.com (This site has products such as bed alarms, etc.)

http://www.bedwettingstore.com (This site has products such as alarms and pads for the bed, etc

http://www.paediatrics warehouse.com

http://www.bed-wetting-prevention.com (This site has some information on what to use and how to use it, etc.)

http://www.keea.org.nz (This site provides organization giving facts, resources, etc.)

http://www.soilingsolutions.com (This site identifies treatments and resources for encopresis.)

10

STEREOTYPIC MOVEMENT DISORDER

CRAIG H. KENNEDY

Vanderbilt University

Stereotypic movements are one of the most common forms of behavior deemed "problematic," particularly among people with developmental disabilities. However, not all stereotypic movements are associated with negative learning and quality of life outcomes and, thus, do not always require intervention. These behaviors often occur for a variety of reasons that researchers are only beginning to understand. Therefore, interventions for stereotypy, when appropriate, take a variety of forms related to the function(s) of the behavior. From a clinical standpoint, all of these observations suggest stereotypic movements are a challenging behavior to treat appropriately and effectively.

In this chapter, I will review the epidemiology of stereotypy, noting the prevalence, demographics, and forms associated with these behaviors. This will then lead to a discussion of how stereotypy is conceptualized and the emerging functional perspective that is being used. A review of functional assessment approaches used to identify the causes of stereotypy will then be presented. This will be followed by a review of function-based interventions derived from functional assessments. Finally, the chapter will conclude with a case study illustrating some of the ideas presented.

EPIDEMIOLOGY OF STEREOTYPY

Stereotypic movements are repetitive behaviors varying little in topography that do not appear to serve a meaningful purpose (Sprague & Newell, 1996). Examples can include body rocking, hand waiving, finger flicking, head bobbing, and hair twirling. Although each of these behaviors is distinctly different in form, such behaviors are deemed stereotypic if they occur repeatedly in a restricted pattern. In addition, nonfunctional rituals/ routines (e.g., taking three steps forward and two steps back before turning left) and persistent preoccupation with stimuli (e.g., spinning the wheels of toy cars) are increasingly being included as stereotypies (South, Ozonoff, & McMahon, 2005). Although a precise structural definition of stereotypy is difficult to determine, leaving some ambiguity regarding topographical descriptions, experienced clinicians can readily identify these behaviors (Bodfish, Parker, Lewis, Sprague, & Newell, 2001).

The emergence of stereotypic movements is part of normal child development. However, these behaviors typically subside by 18 months of age (Thelen, 1996). Examples of stereotypies emerging during typical development include body rocking, hair twirling, head banging, and tooth grinding. If such behavior persists beyond 18 months of age, they tend to be labeled as stereotypic movements. For individuals with developmental disabilities, prevalence estimates range from 34% to 82% (Berkson, Tupa, & Sherman, 2001). Some disabilities have a higher prevalence rate of stereotypy than others. For example, stereotyped movements occur so frequently in autism spectrum disorders (ASD) that they are part of the diagnostic criteria for the condition (Bodfish, Symons, Parker, & Lewis, 2000).

Several decades of research have clearly indicated an inverse relation between level of intellectual functioning and the occurrence of stereotypy. In general, a lower estimated IQ is correlated with an increased likelihood of stereotypic movements (Berkson & Davenport, 1962). In addition to a higher prevalence rate among people with more severe developmental disabilities, once stereotypies emerge, they are more likely to persist throughout the life span of the person (Ballaban-Gil, Rapin, Tuchman, & Shinnar, 1996; Berkson, 2002). Another correlation with functioning level is the form or complexity of the stereotyped behaviors. For individuals with higher functioning levels who engage in stereotypies, topographies tend to be more symbolic (e.g., a fascination with dinosaurs), leading some researchers to refer to these behaviors as "circumscribed interests" or "obsessions" (South et al., 2005). Whether stereotyped behaviors interfere with typical development or are sequelae of arrested development is currently not well understood (Thelen, 1996).

FUNCTIONAL CONCEPTUALIZATION
OF STEREOTYPY

Stereotypy has been presumed to serve self-stimulatory or perceptual consequence functions by a range of researchers and clinicians (see Lovaas, Newsom, & Hickman, 1987). This assumption is so firmly entrenched in practice that many individuals assume stereotypic movements serve to produce physical sensations that are rewarding without any assessment evidence. From a behavior analytic standpoint, such an assumption is based on stereotypy serving to produce *nonsocial positive reinforcement* (see Kennedy, 2002). By "nonsocial" what is meant is that the behavior of another individual is not required for reinforcement to occur. Therefore, the individual who engages in stereotypy is capable of producing the stimulation herself by emitting a specific response. By "positive reinforcement" what is meant is that the behavior produces a stimulus that increases the probability of the response. Examples of nonsocial positive reinforcement might include putting pressure on your eye to produce visual stimulation or clicking your tongue to produce sound (e.g., Kennedy & Souza, 1995).

A second type of nonsocial contingency that can maintain stereotypy is *nonsocial negative reinforcement.* As with nonsocial positive reinforcement, nonsocial negative reinforcement is based on a contingency in which the person himself can alter the status of a stimulus with his own behavior, regardless of the behavior of others. In terms of reinforcement, negative reinforcement occurs when a response reduces, alters, or eliminates a stimulus and the behavior increases in probability. Unlike positive reinforcement, negative reinforcement does not produce a stimulus, but instead escapes or avoids stimulation. Examples might include covering your ears to reduce the decibel level of a loud sound or tapping your head to reduce the pain from otitis media (e.g., O'Reilly, Lacey, & Lancioni, 2000; Tang, Koppekin, Caruso, & Kennedy, 2002).

Together, nonsocial positive and negative reinforcement comprise the traditional functional conceptualization of stereotypic movements (Lovaas et al., 1987). However, as functional assessment methods have emerged and matured, the functional properties of stereotypy have required an expansion. In addition to nonsocial reinforcement maintaining stereotypy, there is ample evidence that social reinforcers can also be a source of stimulation for stereotypy. In general, "social" reinforcement requires the behavior of another person for reinforcement to occur. One type of social reinforcement is *social positive reinforcement*, which entails the presentation of a stimulus by another person that functions to increase the probability of responding. Examples of this might include your engaging in finger flicking to gain the attention of another person or body rocking to be allowed access to a computer (e.g., Kennedy, Meyer, Knowles, & Shukla, 2000).

As with nonsocial reinforcement, the second general type of social reinforcer is negative reinforcement. In *social negative reinforcement*, responding increases in probability because it results in another person reducing, altering, or removing some type of stimulation. In such instances, stereotypy occurs because it results in another person changing the status of a noxious stimulus. Examples of this process might include hair twirling, resulting in a teacher terminating instructional demands, or humming loudly, occasioning another person to turn down the radio (e.g., Tang, Patterson, & Kennedy, 2003).

It has been theoretically postulated that stereotypy evolves from serving a perceptual consequence function to serving to obtain or avoid socially mediated reinforcers (Guess & Carr, 1991; Kennedy, 2002). From this perspective, stereotypical movements emerge as a process of typical development (Thelen, 1996) but persist because of developmental delays. Because stereotypies can occur so frequently, particularly for people with severe developmental disabilities (Berkson, 2002), the behaviors can come into contact with a range of adventitious social reinforcement contingencies. The result is that the variables maintaining responding can transfer from nonsocial reinforcement to social reinforcement and eventually include individual topographies of stereotypy that can serve a range of reinforcer functions. Indeed, this adventitious reinforcement process appears to also be a point of genesis for self-injurious behaviors from stereotypical movements (Kennedy, 2002; Richman & Lindauer, 2005). For example, instances have been documented in which a stereotypy such as hand waiving topographically evolved into head hitting, resulting in abrasions and contusions to the individual's forehead (Richman & Lindauer, 2005).

FUNCTIONAL ASSESSMENT OF STEREOTYPY

The conceptualization of stereotypy just presented poses several challenges to clinicians interested in functionally assessing these responses. A first issue to consider is whether stereotypical movements require intervention. There is evidence that stereotypy can interfere with skill acquisition and use of adaptive behaviors, as well as serve to stigmatize the individual (Symons, Sperry, Dropik, & Bodfish, 2005). However, many instances of stereotypical movements are benign and may not warrant the time and cost of assessment and intervention. Typically, it is the role of a comprehensive team of individuals, including care providers, professionals, and the person engaging in stereotypy, who need to decide whether the stereotypical movements require intervention (Horn, Thompson, & Nelson, 2004).

If a decision is made to reduce stereotypy, then a *functional behavioral assessment* (FBA) of these behaviors is required. Several techniques, developed by a range of researchers, can be used in an FBA (Iwata, Roscoe, Zarcone, & Richman, 2002; Miltenberger, 2003; O'Neill, Horner, Albin, Storey, & Sprague, 1996). The techniques vary in terms of their precision and effort, with more precise techniques requiring greater effort. In this chapter, FBA techniques will be classified into three general categories: (a) record reviews/interviews, (b) descriptive assessments, and (c) experimental analyses. The overall goal of an FBA is to identify what behaviors are of concern and when and why they occur.

Record reviews/interviews require the least amount of effort but also provide the least precise information. However, they are a necessary element in the FBA process and provide the basis for much of the activities used to conduct descriptive assessments and experimental analyses. Generally, record reviews entail reading previous evaluation reports provided by a range of professionals (e.g., speech language pathologists, primary care providers, behavior analysts), previous and current data relating to stereotypy, previous and current interventions, support plans, and other information as appropriate. This information is used to gather a detailed history of the behaviors, any interventions that may have been used, and any pertinent diagnostic information.

Interviews involve semistructured interactions with care providers and others who are knowledgeable of the person and behaviors. A range of interview instruments is available from a variety of sources (see Alberto & Troutman, 2002; Miltenberger, 2003; O'Neill et al., 1996). Typically, requested information from the informant focuses on what behaviors are of concern and why, the topographies of the behaviors, contexts in which the behaviors do and do not occur, conditions that may increase or decrease the behaviors, and what functional properties the behaviors may serve. As a result of interviews and record reviews, the clinician should have an understanding of what specific behaviors are of concern, the conditions under which they occur, and the functional reasons for their occurrence. This information can then be used to conduct descriptive assessments and/or experimental analyses.

Descriptive assessments involve the direct observation of stereotypy in the environments in which the person who emits them lives, works, and recreates. The goal of descriptive assessments is to develop correlational information about antecedent and consequent events that may be related to the stereotypical movements. Often these events can be identified from record reviews/interviews, but sometimes critical events are not identified until the behaviors are observed in the environments in which they occur. A range of data collection protocols has been developed to conduct descriptive assessments, but all share a common focus on collecting direct observation data on the antecedents and consequences relating to the stereotypy (see Bijou, Peterson, & Ault, 1968; O'Neill et al., 1996; Thompson, Felce, & Symons, 1999; Touchette, MacDonald, & Langer, 1985).

In relation to stereotypy, descriptive assessments should identify what levels and types of environmental stimulation are occurring when stereotypy is observed. For example, does stereotypy occur only when low levels of environmental stimulation are present, or does stereotypy occur only when high levels of environmental stimulation are present? The former pattern might suggest that stereotypy occurs to produce sensory stimulation in an otherwise understimulating context (i.e., nonsocial positive reinforcement) or gain access to increased levels of stimulation (i.e., social positive reinforcement). The latter pattern might suggest that stereotypy occurs to reduce excessive stimulation (i.e., nonsocial negative reinforcement) or escape/avoid overly stimulating contexts (i.e., social negative reinforcement). By directly observing what occurs before and after stereotypical movements, the clinician can make further refinements regarding hypotheses of the functional properties of stereotypy. Because functional interventions are based on the functional consequences of stereotypy, identifying plausible sources of reinforcement is critical.

In instances in which descriptive assessments in conjunction with record reviews/interviews do not yield clear hypotheses regarding sources of reinforcement, then experimental analyses of the stereotypy can be used. Experimental analyses, as the name implies, are small-scale experiments using some type of single-case design (Kennedy, 2005). Typically, conditions are explicitly arranged allowing the testing of specific reinforcement contingencies in relation to behavior (Iwata, Dorsey, Slifer, Bauman, & Richman, 1982/1994; Wacker, Berg, Harding, & Cooper-Brown, 2004). That is, this FBA technique exposes stereotypy to experimentally arranged environments to test hypotheses regarding antecedents and consequences (Luiselli, 2006). For example, social attention might be made contingent upon stereotypy, or demands may be withdrawn contingent upon stereotypy. By conducting experimental analyses, a very precise set of hypotheses can be tested in relation to the occurrence of stereotypical movements to gain additional information about why the behaviors are occurring (Kennedy et al., 2000).

Often, experimental analyses are conducted following record reviews/ interviews and descriptive assessments. The information from the earlier FBA techniques can then be used to increase the contextual validity of the experimental analysis conditions. In particular, experimental analyses can be useful in two instances to identify the operant functions of stereotypy. First, if an undifferentiated pattern of stereotypy is identified during record reviews/interviews and descriptive assessments, experimental analyses can be used to identify specific reinforcement contingencies maintaining behavior (Kennedy, 2000; McCord & Neef, 2005; Vollmer, Marcus, Ringdahl, & Roane, 1995). For example, stereotypy may occur so frequently in naturalistic conditions that discrete antecedents and consequences cannot be clearly identified. Through the isolation of specific stimulus events and the arrangement of a contingent relation with stereotypy, discrete operant reinforcers can be tested and identified. A second pattern that warrants an experimental analysis is instances in which stereotypy may be multiply determined. By "multiply determined," what is meant is that an individual behavior (e.g., hand waiving) occurs for two or more reinforcement contingencies. For example, hand waiving might occur to both produce nonsocial positive reinforcement during low levels of environmental stimulation and to escape from academic instruction.

As a collection, the FBA techniques just reviewed are designed to identify environmental events having functional effects on the occurrence of stereotypical movements. Clinicians typically use record reviews/interviews in conjunction with descriptive assessments. If additional clarity or refinement is needed in order to understand why stereotypy is occurring, then experimental analyses may also be conducted. However, the overall goal is to develop clear hypotheses regarding why stereotypical movements occur so that functional interventions can then be derived from this assessment information.

FUNCTIONAL INTERVENTION FOR STEREOTYPY

Early in the development of the field of behavior analysis, interventions were not based on assessment data (Kazdin, 1978). Instead, clinicians selected from a range of evidence-based interventions that they perceived likely to reduce problem behaviors. This nonfunctional approach resulted in frequent success, but many individuals did not respond to intervention. These intervention failures were not well understood and forced researchers to consider the functional properties of behavior in order to gain a more complete understanding of treatment successes and failures (Carr, 1977; Iwata et al., 1982/1994; Touchette et al., 1985). This functional orientation led to the development of FBA approaches and spawned the current focus on functional assessments leading to function-oriented interventions (e.g., Luiselli, 2006; O'Neill et al., 1996; Scotti & Meyer, 1999; Sigafoos, Arthur, & O'Reilly, 2003).

Functional interventions, by definition, are based on the results of the FBA process. Therefore, functional interventions are determined by whether the stereotypy is maintained by (a) nonsocial positive reinforcement, (b) nonsocial negative reinforcement, (c) social positive reinforcement, and/or (d) social negative reinforcement (see Kennedy, 2002). The specific type of intervention(s) depends on which type of reinforcer(s) is involved. Typically, an intervention is focused on each reinforcer function maintaining stereotypy. For example, if stereotypy is maintained by nonsocial positive reinforcement and social negative reinforcement, interventions will need to address each of these functions.

For instances of stereotypy maintained by nonsocial positive reinforcement, interventions need to consider the basis for these behaviors occurring. When nonsocial positive reinforcement is identified as a maintaining condition for a response, the inference has been made that some type of sensory consequence produced by the stereotypy is producing positive reinforcement (Kennedy, 1994). To date, this has been a predominant source of reinforcement associated with stereotypical movements, although no large-scale epidemiological data are currently available. Behaviors maintained by nonsocial positive reinforcement also pose the challenge of being freely available to an individual and can be produced any time the stereotypy is emitted.

The most effective interventions for nonsocial positively reinforced stereotypy consider the behavior within a *concurrent operants* framework (Herrnstein, 1970; Rapp, Vollmer, St. Peter, Dozier, & Cotnoir, 2004). Concurrent operants are an empirically derived means of understanding choice behavior. Variables such as reinforcer frequency, reinforcer magnitude, reinforcer delay, and response effort for two, or more, available choices influence whether an individual chooses one option (e.g., stereotypy) or another (e.g., communication to access a preferred activity). Thus, the individual has a range of responses she can emit, each with reinforcers associated with its occurrence. Therefore, stereotypy serving a nonsocial positive reinforcement function can be conceptualized as self-delivery of a positive reinforcer in relation to other available reinforcers in the environment. If additional sources of positive reinforcement are available for other behaviors, then the individual may allocate her responding to those options rather than stereotypy. For example, you might choose to emit stereotypy in an environment that is otherwise not stimulating to produce auditory stimulation (e.g., humming as a stereotypical response), but you might opt not to engage in stereotypy if you can request access to music as an alternative form of auditory stimulation.

Following a concurrent operants perspective, the occurrence of stereotypy for nonsocial positive reinforcement can be viewed as an indication that the person is in an understimulating environment and opting to provide her own stimulation via stereotypical movements. Two general strategies are suggested for reducing stereotypy serving this function. First, *environmental enrichment* (e.g., noncontingent positive reinforcement) can be used (Horner, 1980; Vollmer, 1999). Using this technique, care providers give supplemental stimulation in a person's environment (e.g., music, toys), with the stimuli often derived from a preference assessment to increase the probability the events will function as positive reinforcers (Ahearn, Clark, DeBar, & Florentino, 2005). Second, *living skills instruction* can be provided (Farlow & Snell, 2000; Halle, Chadsey, Lee, & Renzaglia, 2004). With this strategy, the individual is taught new skills, typically using task analysis and systematic prompting procedures, giving the person the ability to contact new sources of stimulation (e.g., playing the piano, drawing a picture) that might function as positive reinforcers. Such a technique can be used to teach an individual to replace the stimulation she receives from stereotypy with more appropriate play or leisure activities.

For stereotypy maintained by nonsocial negative reinforcement, the focus of intervention is on the noxious stimulus associated with the stereotypical movements. When this behavioral function is implicated in the

occurrence of stereotypy, the behaviors are occurring to reduce or eliminate some source of noxious stimulation. It is important to remember that
what functions as a noxious stimulus for one individual may not be noxious
for another person. For example, a lawnmower may be very aversive to one
person, but a neutral or even pleasing sound to another. In addition, some
disabilities (e.g., autism) have unusual sensory sensitivities associated with
them (Schreibman, 2005). In these instances, stereotypy can be used by
an individual to escape or avoid stimuli that others may not consider bothersome (e.g., touch, sound, taste, light, smell).

Interventions for stereotypy maintained by nonsocial negative reinforcement focus on strategies for changing noxious stimuli, thus obviating the
need for the individual to emit the stereotypical movements. Two general
procedures are typically used. The first focuses on *environmental modifications* to reduce, alter, or eliminate the noxious stimulus (Myles, 2005).
For example, if a person covers his ears when his mother mows the lawn,
then a muffler might be installed in the lawnmower (reduction), an electric
lawnmower may be used (alteration), or the lawn can be mowed when the
child is not at home (elimination). A second approach to functional intervention uses a variation on living skills instruction discussed previously for
nonsocial positive reinforcement. In the case of nonsocial negative reinforcement, living skills instruction focuses on teaching the person new
skills to reduce, alter, or eliminate the stimulus that is noxious to him. For
example, if a person finds bright light aversive, he can be taught to find and
put on sunglasses before going outside or how to turn down lighting in a
room that is too brightly lit.

Instances of social positive reinforcement involve a person using stereotypy to gain access to some type of stimulation that otherwise would not
be available. This necessarily involves stereotypical movements occurring
to change the behavior of another person and, hence, can be viewed as a
form of communication. For example, a person's care providers may not
allow him access to his iPod unless he engages in vigorous body rocking
and humming. However, when he engages in these behaviors, his care
providers give him the iPod to "quiet him down" and, unwittingly, have
taught him to use stereotypy to access a preferred item that is otherwise
unavailable.

The primary intervention for instances of stereotypy maintained by
social positive reinforcement is *functional communication training* (FCT;
Carr & Durand, 1985). Using FCT, the person is taught an alternative form
of behavior (e.g., signing, visual symbols, speaking) to functionally replace
the stereotypical response. That is, for instances when the individual might
use stereotypy to gain access to a positive reinforcer, she is taught an alternative form of communication producing the same type of positive reinforcer. For example, if a person uses body rocking to get the attention
of other people, she could be taught to use a picture symbol to recruit

attention so that the stereotypy no longer is required to gain access to the positive reinforcer (Charlop-Christy, Carpenter, Le, LeBlanc, & Kellet, 2002). By using FCT, a person is taught a more appropriate and efficient means of accessing preferred items or people, the need for engaging in stereotypy is eliminated, and the intervention focus is shifted to enhancing personal communication.

When stereotypy is emitted to avoid or escape stimuli by changing the behavior of other people, it is referred to as occurring for social negative reinforcement. In these instances, stereotypy is occurring as a form of communication to change the environment via another individual. For example, during instruction, a student might engage in arm waiving and bouncing in her seat, with the result being that the teacher terminates instruction and has the student sit down somewhere else to "calm down." If instruction is a noxious event for the student, then the teacher has negatively reinforced the occurrence of stereotypy by removing the child from the instructional situation contingent upon the stereotypical movements.

For stereotypy maintained by social negative reinforcement, interventions again focus on FCT as a means of teaching more preferable forms of communication. However, for instances of negative reinforcement, the focus of instruction is on communication leading to the reduction, alteration, or elimination of the noxious stimulus. For example, in the previous example of arm waiving and bouncing, the student could be taught to request a brief break or an opportunity to work on a more preferred task for a brief period of time (Kennedy et al., 2000). By doing this, the person is taught to use a more appropriate form of communication to alter events in his environment that are nonpreferred, resulting in increased communicative development and reductions in stereotypical movements.

When FBA'a show stereotypical movements to be multiply determined, the current recommended course of action is to tailor interventions for each specific reinforcer function maintaining stereotypy (Kennedy et al., 2000; Tang et al., 2003). For example, if head bobbing was determined to serve nonsocial positive reinforcement and social negative reinforcement functions, an intervention would need to be identified for each. In addition, if behaviors are determined to be situation specific, individual interventions may need to be targeted for the appropriate environment to ensure that the most appropriate intervention is being used in a particular setting.

CASE STUDY

To help integrate the information we have reviewed regarding the functional conceptualization, assessment, and intervention of stereotypy, I will finish the chapter with a case study. In this case, I will review the assessment and intervention procedures for a 10-year-old boy named James who had autism and moderate intellectual disabilities. James spoke in short sentences and could follow three-step verbal requests. His stereotypic behavior was waiving his left hand in a rhythmic and repeated fashion in front of his face. James was referred to the Vanderbilt Kennedy Center Behavior Analysis Clinic by his local school district at the request of his parents and teacher because of concerns that his stereotypy was interfering with his learning and performance.

Record reviews indicated this topography of stereotypy had been included in his Individualized Educational Plans for the entire time he had been in public schools (i.e., since 4 years of age). Pediatrician, speech language pathologist, clinical psychologist, and physical therapist evaluations all mentioned hand waiving as a predominant behavior that his parents and teachers were concerned with because of its frequency of occurrence. All written reports noted that the behavior served a sensory consequence function (i.e., "self-stimulation"). Interviews with James's parents and current teacher yielded information that was consistent with the record review. At this point, we opted to observe the child at home and in school using a descriptive assessment approach.

The descriptive assessment component of the FBA involved the use of an event-by-interval recording system used to note the occurrence of stereotypy, antecedent events, and changes in the environment following stereotypy (see O'Neill et al., 1996). James was observed at home for 3 hr each evening for 4 days and 2 hr each day at school for 4 days, yielding 20 hr of direct observation data. The results showed that the stereotypy was emitted during 65% of observation intervals, with no consistent antecedents or consequences correlating with the stereotypy. At this point, it was concluded that an undifferentiated pattern had emerged during the FBA, and we elected to incorporate an experimental analysis approach into the process.

With the consent of his parents and the agreement of his teacher, we engaged in the experimental analysis of James's stereotypy at his school in an empty classroom. Four conditions were tested: (a) Attention (social positive reinforcement), (b) Demand (social negative reinforcement), (c) No Attention (nonsocial positive reinforcement), and (d) Recreation (control used in case stereotypy is multiply determined). For specific procedural details, the reader is referred to Kennedy et al. (2000). We completed one set of conditions per day for 6 days. The results of the experimental analysis are shown in Figure 10.1. Throughout the analysis, stereotypy was low in the Recreation condition, but showed sensitivity to the reinforcement contingencies used in each of the other three conditions (i.e., Attention, Demand, and No Attention). We interpreted these data as indicating that James's stereotypy was multiply determined. Specifically,

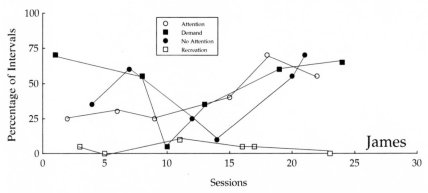

FIGURE 10.1   Occurrence of stereotypy across experimental analysis conditions (see legend). Data are arrayed as the percentage of intervals of stereotypy for James. [Source: Kennedy, C. H., Meyer, K. A., Knowles, T., & Shukla, S. (2000). Analyzing the multiple functions of stereotypical behavior for students with autism: Implications for assessment and treatment. *Journal of Applied Behavior Analysis, 33*, 559–571. Copyright 2000 by the Society for the Experimental Analysis of Behavior. Reproduced with permission.]

his hand waiving occurred to gain attention from adults (i.e., social positive reinforcement), escape from academic instruction (i.e., social negative reinforcement), and "self-stimulation" (i.e., nonsocial positive reinforcement).

We then tested each of the identified functions of behavior using FCT interventions tailored to each function of stereotypy in the same setting used for the experimental analysis. That is, we developed a communication intervention that allowed him to gain attention using an American Sign Language (ASL) symbol, briefly escape instruction using another ASL symbol, and gain access to a preferred activity during break using a third ASL symbol. We assessed the FCT interventions using a multiple baseline across concurrent operants design and 10-s partial-interval recording (Kennedy, 2005). The results are shown in Figure 10.2. James learned to use different signs for different reinforcer functions depending on the context (i.e., social skills training, academic instruction, or break time). This information confirmed that James's stereotypical behavior was multiply determined and required interventions matched to each of the three reinforcer functions we had identified during the FBA.

We then met as a team (i.e., behavior analyst, parents, special educator, general educator, paraprofessionals, speech language pathologist, and physical therapist) to discuss the results of the FBA. The team suggested that the intervention have three components to match the three behavioral functions of stereotypy and that FCT was an appropriate intervention approach. In addition, the team, following the advice of the speech language pathologist, agreed that ASL was the most desirable response form

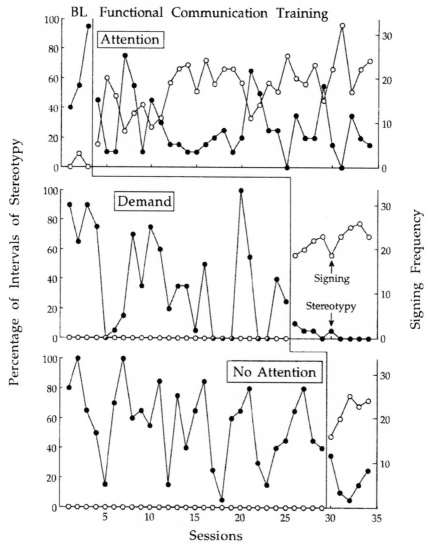

FIGURE 10.2 Occurrence of stereotypy for James across Attention, Demand, and No Attention conditions. Data are arrayed as the percentage of intervals of stereotypy on the left y axis and number of signs per session on the right y axis. [Source: Kennedy, C. H., Meyer, K. A., Knowles, T., & Shukla, S. (2000). Analyzing the multiple functions of stereotypical behavior for students with autism: Implications for assessment and treatment. *Journal of Applied Behavior Analysis, 33,* 559–571. Copyright 2000 by the Society for the Experimental Analysis of Behavior. Reproduced with permission.]

for communication training based on previous successes in using signing and parental preference for this approach to communication. Finally, it was decided that the intervention should be implemented simultaneously throughout school and also at home, in order to maximize opportunities for James to use ASL.

We then worked with James's team *in situ* to implement these FCT interventions consistently across home and school. We identified different contexts to be associated with different interventions: (a) Systematic instruction contexts were chosen for requesting assistance (social negative reinforcement); (b) social contexts, such as leisure activities and meals, were targeted for requesting attention (social positive reinforcement); and, (c) "down times" were selected for requesting preferred activities (nonsocial positive reinforcement). In addition, all support providers were taught that when James spontaneously emitted an ASL communication attempt regardless of context, it should be honored as a means of facilitating generalized use of the FCT (Stokes & Baer, 1977). The *in situ* technical assistance included a review of the behavior intervention plan (written in nontechnical language), modeling of intervention techniques (identifying appropriate contexts, prompting, consequences, and honoring ASL communications), and feedback on support provider performance (Clark, Cushing, & Kennedy, 2004). This was done twice a week for 2 weeks in school and home settings by which time all support providers were accurately implementing the intervention. James's speech language pathologist then worked with team members to elaborate on the FCT interventions to increase his communication across a range of settings. Although 6 months later James's stereotypy still occurred 5% of the time, mainly during "down time", it was greatly reduced and his symbolic communication had greatly increased.

CONCLUSION

Stereotypical movements occur for a variety reasons, but all are a function of some type of reinforcement produced directly by the response (nonsocial reinforcement) or by changing the behavior of others (social reinforcement). In this chapter, I have reviewed this functional conceptualization of stereotypy, which differs from earlier ideas about the nature of stereotypical movements. In addition, I described various approaches used in an FBA to identify specific functions that a person's stereotypy may serve. The development of interventions in a functional assessment of stereotypy is dependent on assessment results. I discussed various functional interventions that are recommended for specific reinforcer functions. The chapter was concluded by a case history illustrating the processes outlined in this chapter.

REFERENCES

Ahearn, W. H., Clark, K. M., DeBar, R., & Florentino, C. (2005). On the role of preference in response competition. *Journal of Applied Behavior Analysis, 38*, 247–250.

Alberto, P. A., & Troutman, A. C. (2002). *Applied behavior analysis for teachers* (6th ed.). Saddle River, NJ: Prentice Hall.

Ballaban-Gil, K., Rapin, I., Tuchman, R., & Shinnar, S. 1996). Longitudinal examination of the behavioral, language, and social changes in a population of adolescents and young adults with autistic disorder. *Pediatric Neurology, 15*, 217–223.

Berkson, G. (2002). Early development of stereotyped and self-injurious behaviors: II. Age trends. *American Journal on Mental Retardation, 107*, 468–477.

Berkson, G., & Davenport, R. K. (1962). Stereotyped movements of mental defectives. I. Initial survey. *American Journal on Mental Deficiency, 66*, 849–852.

Berkson, G., Tupa, M., & Sherman, L. (2001). Early development of stereotyped and self-injurious behaviors: I. Incidence. *American Journal on Mental Retardation, 106*, 539–547.

Bijou, S. W., Peterson, R. F., & Ault, M. H. (1968). A method to integrate descriptive and experimental field studies at the level of data and empirical concepts. *Journal of Applied Behavior Analysis, 1*, 175–191.

Bodfish, J. W., Parker, D. E., Lewis, M. H., Sprague, R. L., & Newell, K. M. (2001). Stereotypy and motor control: Differences in the postural stability dynamics of persons with stereotyped and dyskinetic movement disorders. *American Journal on Mental Retardation, 106*, 123–134.

Bodfish, J. W., Symons, F. J., Parker, D. E., & Lewis, M. H. (2000). Varieties of repetitive behavior in autism: Comparisons to mental retardation. *Journal of Autism and Developmental Disorders, 30*, 237–43.

Carr, E. G. (1977). The motivation of self-injurious behavior: A review of some hypotheses. *Psychological Bulletin, 84*, 800–816.

Carr, E. G., & Durand, V. M. (1985). Reducing behavior problems through functional communication training. *Journal of Applied Behavior Analysis, 18*, 111–126.

Charlop-Christy, M. H., Carpenter, M., Le, L., LeBlanc, L. A., & Kellet, K. (2002). Using the picture exchange communication system (PECS) with children with autism: Assessment of PECS acquisition, speech, social-communicative behavior, and problem behavior. *Journal of Applied Behavior Analysis, 35*, 213–231.

Clark, N. M., Cushing, L. S., & Kennedy, C. H. (2004). An intensive onsite technical assistance model to promote inclusive practices. *Research and Practice for People with Severe Disabilities (formerly JASH), 29*, 253–262.

Farlow, L. J., & Snell, M. E. (2000). Teaching basic self-care skills. In M. E. Snell & F. Brown (Eds.), *Instruction of students with severe disabilities* (4th ed., pp. 331–380). New York: Merrill.

Guess, D., & Carr, E. G. (1991). Emergence and maintenance of stereotypy and self-injury. *American Journal on Mental Retardation, 96*, 299–319.

Halle, J. W., Chadsey, J., Lee, S., & Renzaglia, A. (2004). Systematic instruction. In C. H. Kennedy & E. Horn (Eds.), *Inclusion of students with severe disabilities* (pp. 54–76). Boston: Allyn & Bacon.

Herrnstein, R. J. (1970). On the law of effect. *Journal of the Experimental Analysis of Behavior, 13*, 243–266.

Horn, E., Thompson, B., & Nelson, C. (2004). Collaborative teams. In C. H. Kennedy & E. V. Horn (Eds.), *Including students with severe disabilities* (pp. 17–30). Boston, MA: Allyn & Bacon.

Horner, R. D. (1980). The effects of an environmental enrichment program on the behavior of institutionalized profoundly retarded children. *Journal of Applied Behavior Analysis, 13*, 473–491.

Iwata, B. A., Dorsey, M. F., Slifer, K. J., Bauman, K. E., & Richman, G. S. (1994). Toward a functional analysis of self-injury. *Journal of Applied Behavior Analysis, 27*, 197–209. (Reprinted from *Analysis and Intervention in Developmental Disabilities, 2*, 3–20, 1982).

Iwata, B. A., Roscoe, E. M., Zarcone, J. R., & Richman, D. M. (2002). Environmental determinants of self-injurious behavior. In S. R. Schroeder, M. L. Oster-Granite, & T. Thompson (Eds.), *Self-injurious behavior* (pp. 93–104). Washington, D.C.: American Psychological Association.

Kazdin, A. E. (1978). *History of behavior modification*. Baltimore: University Park Press.

Kennedy, C. H. (1994). Automatic reinforcement: Oxymoron or hypothetical construct? *Journal of Behavioral Education, 4*, 387–396.

Kennedy, C. H. (2000). When reinforcers for problem behavior are not readily apparent: Extending functional assessments to biological reinforcers. *Journal of Positive Behavior Interventions, 2*, 195–202.

Kennedy, C. H. (2002). The evolution of stereotypy into self-injury. In S. Schroeder, M. L. Oster-Granite, & T. Thompson (Eds.), *Self-injurious behavior: Gene-brain-behavior relationships* (pp. 133–143). Washington, D.C.: American Psychological Association.

Kennedy, C. H. (2005). *Single-case designs for educational research*. Boston, MA: Allyn & Bacon.

Kennedy, C. H., Meyer, K. A., Knowles, T., & Shukla, S. (2000). Analyzing the multiple functions of stereotypical behavior for students with autism: Implications for assessment and treatment. *Journal of Applied Behavior Analysis, 33*, 559–571.

Kennedy, C. H., & Souza, G. (1995). Functional analysis and treatment of eye-poking. *Journal of Applied Behavior Analysis, 28*, 27–37.

Lovaas, O. I., Newsom, C., & Hickman, C. (1987). Self-stimulatory behavior and perceptual reinforcement. *Journal of Applied Behavior Analysis, 20*, 45–68.

Luiselli, J. K. (2006). *Antecedent intervention: Recent developments in community focused behavioral support* (2nd ed.). Baltimore: Paul H. Brookes.

McCord, B. E., & Neef, N. A. (2005). Leisure items as controls in the attention condition of functional analyses. *Journal of Applied Behavior Analysis, 38*, 417–426.

Miltenberger, R. G. (2003). *Behavior modification: Principles and procedures*. Independence, KY: Wadsworth Publishing.

Myles, B. S. (2005). *Children and youth with Asperger syndrome: Strategies for success in inclusive settings*. Thousand Oaks, CA: Corwin Press.

O'Neill, R. E., Horner, R. H., Albin, R. W., Storey, K., & Sprague, J. R. (1996). *Functional assessment and program development for problem behavior: A practical handbook*. Independence, KY: Wadsworth Publishing.

O'Reilly, M. F., Lacey, C., & Lancioni, G. E. (2000). Assessment of the influence of background noise on escape-maintained problem behavior and pain behavior in a child with Williams syndrome. *Journal of Applied Behavior Analysis, 33*, 511–514.

Rapp, J. T., Vollmer, T. R., St. Peter, C., Dozier, C. L., & Cotnoir, N. M. (2004). Analysis of response allocation in individuals with multiple forms of stereotyped behavior. *Journal of Applied Behavior Analysis, 37*, 481–501.

Richman, D. M., & Lindauer, S. E. (2005). Longitudinal assessment of stereotypic, proto-injurious, and self-injurious behavior exhibited by young children with developmental delays. *American Journal on Mental Retardation, 110*, 439–450.

Scotti, J. R., & Meyer, L. H. (1999). *Behavioral intervention principles, models, and practices*. Baltimore: Paul H. Brookes.

Schreibman, L. (2005). *The science and fiction of autism*. Cambridge, MA: Harvard University Press.

Sigafoos, J., Arthur, M., & O'Reilly, M. (2003). *Challenging behavior and developmental disability*. Baltimore: Paul H. Brookes.

South, M., Ozonoff, S., & McMahon, W. M. (2005). Repetitive behavior profiles in Asperger syndrome and high-functioning autism. *Journal of Autism and Developmental Disorders, 35,* 145–158.

Sprague, R. L., & Newell, K. M. (1996). *Stereotyped movements: Brain and behavior relationships.* Washington, D.C.: American Psychological Association.

Stokes, T. F., & Baer, D. M. (1977). An implicit technology of generalization. *Journal of Applied Behavior Analysis, 10,* 349–367.

Symons, F. J., Sperry, L. A., Dropik, P. L., & Bodfish, J. W. (2005). The early development of stereotypy and self-injury: A review of research methods. *Journal of Intellectual Disability Research, 49,* 144–158.

Tang, J.-C., Koppekin, A., Caruso, M., & Kennedy, C. H. (2002). Functional analysis of stereotypical ear-covering in a child with autism. *Journal of Applied Behavior Analysis, 35,* 95–98.

Tang, J.-C., Patterson, T. G., & Kennedy, C. H. (2003). Identifying specific sensory modalities maintaining the stereotypy of students with multiple profound disabilities. *Research in Developmental Disabilities, 24,* 433–451.

Thelen, E. (1996). Normal infant stereotypes: A dynamic systems approach. In R. L. Sprague & K. M. Newell (Eds.), *Stereotyped movements: Brain and behavior relationships* (pp. 139–165). Washington, D.C.: American Psychological Association.

Thompson, T., Felce, D., & Symons, F. J. (1999). *Behavioral observation: Technology and applications in developmental disabilities.* Baltimore: Paul H. Brookes.

Touchette, P. E., MacDonald, R. F., & Langer, S. N. (1985). A scatter plot for identifying stimulus control of problem behavior. *Journal of Applied Behavior Analysis, 18,* 343–351.

Vollmer, T. R. (1999). Noncontingent reinforcement: Some additional comments. *Journal of Applied Behavior Analysis, 32,* 239–240.

Vollmer, T. R., Marcus, B. A., Ringdahl, J. E., & Roane, H. S. (1995). Progressing from brief assessments to extended experimental analyses in the evaluation of aberrant behavior. *Journal of Applied Behavior Analysis, 28,* 561–576.

Wacker, D. P., Berg, W., Harding, J., & Cooper-Brown, L. (2004). Use of brief experimental analyses in outpatient clinic and home settings. *Journal of Behavioral Education, 13,* 213–226.

11

A CONTEXTUAL MODEL OF RESTRAINT-FREE CARE FOR PERSONS WITH DEMENTIA

JANE E. FISHER
CLAUDIA DROSSEL
CRAIG YURY

AND

STACEY CHERUP
University of Nevada, Reno

DIAGNOSIS AND RELATED CHARACTERISTICS

The term *dementia* is used to describe a collection of symptoms that are caused by pathological changes in neurological function. Degenerative forms of dementia have devastating consequences for elderly persons and their families. Over the course of a degenerative dementia, severe losses occur across the person's entire repertoire including severe impairment in memory and new learning, verbal abilities, and motor functioning. There is currently no effective treatment for stopping or reversing the impairment caused by degenerative forms of dementia. Affected persons eventually become bedridden, unable to swallow, and completely dependent on others for their care. Dementia has many causes; the most common are Alzheimer's disease, vascular dementia, Lewy body disease, and frontotemporal forms of dementia. Reversible conditions that can cause symptoms of dementia include normal pressure hydrocephalus, medication reactions,

Functional Analysis in
Clinical Treatment

infection, dehydration, vitamin deficiency and other forms of poor nutrition, and head injury (National Institute of Neurological Diseases and Stroke, 2005).

Persons close to someone with dementia typically perceive the declines in the repertoire as major changes in "personality." The person with dementia is described as behaving in uncharacteristic ways due to changes in social behavior, irresponsibility in the management of finances, or acting in uninhibited ways, such as engaging in sexual acts in public. In advanced dementia the development of difficult behaviors is common, with over 50% of patients developing behaviors that warrant intervention (Chen, Borson, & Scanlan, 2000; Lyketsos et al., 2002). Problem behaviors include physical and verbal aggression; disruptive vocalizations, such as screaming or calling out for hours, repeating the same question hundreds of times per day; paranoia; and wandering and becoming lost in formerly familiar settings.

Dementia is an age-associated condition. Estimates of the prevalence among persons over the age of 65 affected with dementia range from 10% to 20%, with the number of people with the disease doubling every 5 years beyond age 65 (National Institute of Neurological Diseases and Stoke, 2005) to a prevalence of dementia among persons over age 85 estimated to be as high as 47% (Evans et al., 1989).

TRENDS IN THE CARE OF PERSONS WITH DEMENTIA

Medical Models

Historically, the medical model has dominated the treatment of behavioral changes that occur in dementia. The majority of studies investigating treatments for dementia involve pharmacological approaches to improving the cognitive symptoms of dementia or reducing behavior problems. Currently, five medications are approved by the U.S. Food and Drug Administration (FDA) for the treatment of the cognitive symptoms of Alzheimer's disease. The drugs fall into the functional categories of "acetylcholinesterase inhibitors" for mild to moderate or "N-methyl-D-aspartate (NMDA) receptor antagonists" for moderate to severe dementia of the Alzheimer's type (Cummings, 2004).

The treatment of the behavior problems that develop in dementia has been dominated by the use of antipsychotic and atypical antipsychotic medications, despite the fact that the FDA has not approved the use of any psychotropic medication for the treatment of any challenging behaviors in persons with dementia (Profenno & Tariot, 2004). The use of psychotropic medications for the treatment of behavior problems in elderly persons with dementia has been problematic for several reasons. Surveys have consistently documented the overmedication of elderly persons and/or the "off-label" use of medications for symptoms for which there was no FDA approval (Briesacher, Limcangco, & Simoni-Wastila, 2005). It was for these reasons that reliance on powerful psychotropic medications gained the label *chemical restraint*. The U.S. Department of Health and Human Services (2005) defines chemical restraints as "any drug used for discipline or convenience and not required to treat medical symptoms." The use of pharmacological restraint-based interventions with elderly persons with dementia is particularly problematic because of the potential adverse effect of eliminating or suppressing behavior in an individual who is already losing behaviors due to the underlying degenerative dementia. The consequences of behavior suppression for severely impaired patients are particularly disturbing. A person who is severely impaired and who experiences pain or physically uncomfortable drug side effects may not able to alert caregivers to his/her discomfort. In addition, behavioral changes due to medication side effects can be misinterpreted as reflecting a worsening of disease symptoms or as psychiatric symptoms.

Physical restraints are also used to manage problem behaviors in elderly persons with dementia. The United States Department of Health and Human Services (2005) defines *physical restraints* as "any manual method or physical or mechanical device, material, or equipment attached to or adjacent to the resident's body that the individual cannot remove easily which restricts freedom of movement or normal access to one's body." In 1987, Congress passed the Omnibus Reconciliation Act of Nursing Home Reform (OBRA) based in part on the recognition that the management of behavior problems in elderly dementia patients within nursing homes was primarily restraint-based, involving either pharmacological or mechanical intervention. OBRA mandated that all nursing facilities participating in the Medicare and Medicaid programs implement a restraint-reduction program, including decreased reliance on antipsychotic medications and the implementation of alternative approaches to medication including behavioral programming. Still, the dominance of the restraint-based methods continues. Surveys indicate that psychotropic medication use with persons with dementia has *increased* significantly in recent years (Briesacher et al., 2005; Stoppe & Staedt, 1999). Twenty-eight percent of Medicare beneficiaries in nursing homes received antipsychotics in 2000–2001—the highest reported rate in nearly a decade (Briesacher et al., 2005). The challenging behaviors that develop over the course of degenerative dementia continue to result in patients receiving restrictive pharmacological and mechanical restraint. Powerful psychotropic medications, primarily atypical antipsychotics, have potentially serious side effects, such as sedation and lethargy. Mechanical restraints, which restrict freedom of movement, increase the risk of falls, muscle atrophy, urinary retention, and strangulation by improper application of restraints (Rubenstein, 1997). Although mechanical restraints are often intended to prevent falls, restraints can actually contribute to fall-related injuries and deaths (American Geriatrics Society, 1997; Rubenstein, Josephson, & Robbins, 1994). Examination of the behaviors of persons with dementia indicates that mechanical restraints may produce the side effects of punishment, including increased levels of agitation.

A growing body of literature indicates that pharmacological approaches are expensive, are largely ineffective in reducing these behaviors, and carry significant dangers (Schneider, Dagerman, & Insel, 2005; Sink, Holden, & Yaffee, 2005). In 2005 the FDA issued a public health advisory regarding new safety information concerning the "off-label" use of atypical anti-psychotic drugs including Abilify (aripiprazole), Zyprexa (olanzapine), Seroquel (quetiapine), Risperdal (risperidone), Clozaril (clozapine), and Geodon (ziprasidone) for the treatment of behavior problems in elderly persons with dementia. An analysis of 17 placebo-controlled studies of four drugs in this class found that the rate of death was about 1.6 to 1.7 times that of placebo (FDA, 2005). A recent study found a higher increased risk of death for conventional antipsychotics compared with atypical antipsy-chotics (Wang et al., 2005).

Developments from Observational Research

Observational research has found that the behavior of persons with dementia continues to be influenced by environmental stimuli in a manner indicating that behaviors labeled as "problems" serve important adaptive functions (Buchanan & Fisher, 2002; Burgio et al., 1994; Hussian & Davis, 1985). Findings of a functional relationship between behavior and contextual stimuli have important implications for the treatment of persons with dementia and the evaluation of treatment effects. Research on pharmacological approaches to the treatment of behavior problems in dementia has relied almost exclusively on symptom reduction as the primary or sole dependent measure, rather than maintenance or increases in adaptive behaviors. Reduction in behavior as the primary or sole dependent variable is problematic when applied to persons with dementia given the fact that the behavioral repertoire of persons with dementia is significantly diminished and will continue to be reduced due to the underlying degenerative disease. Further, given the finding of the functionality of behavior in persons with dementia, a more appropriate approach to the evaluation of treatment effects is to focus on the maintenance of *any* adaptive behavior, regardless of its topography, and the assessment of what the person *is* doing when not engaging in a problem behavior.

The devastating losses within behavioral repertoire that occur in dementia result in declines in access to sources of reinforcement. Whenever possible, the *prevention* of problem behaviors by the provision of alternative sources of reinforcement should be pursued. Prevention of problem behavior in dementia is based on the principles of making the behavior unnecessary by altering establishing and discriminative conditions maintaining the problem behavior *and* meeting the person's needs through access to alternative sources of reinforcement. Preserving persons' functional repertoires and increasing their access to reinforcement serve as the foundations of a functional analytic model of dementia.

A FUNCTIONAL ANALYTIC MODEL OF DEMENTIA

Behavior analytic strategies hold significant promise for sustaining the fragile and diminishing behavioral repertoires of persons with dementia. The contextual model of treatment stems from the view that behavior is context-bound—that is, that behavior is a production of the individual and its interaction with the environment that produces a behavior (Catania, 1996). The contextual model of treatment is based on the assumption that environments can be changed through having an understanding of an individual's learning history and manipulating the relevant features of the environment.

Behavior can occur in a context that is verbal or nonverbal. A verbal context is one in which a behavior is verbally instructed, which results in a rule-governed behavior. A nonverbal context is one in which an individual comes into contact with contingencies that shape behavior without explicit verbal instruction. These contingencies are not easily described by the person. Many behaviors are learned and initially maintained by rule governance but become so second nature that they become dually maintained by nonverbal contingencies—early in our lives we are both told and shown how to act, and the behaviors produced are maintained by these "rules." However, as we grow older, we behave according to these rules without explicitly stating them to ourselves—the behavior becomes under the control of the cues in the environment (Chase & Danforth, 1991).

Changes in behavior that occur in dementia are characterized by the decline in rule-governed behavior and an increase in contingency-maintained behavior (see the body of research on "declarative" compared with "procedural," or "explicit" compared with "implicit" memory deficits in dementia; Eldridge, Masterman, & Knowlton, 2002). As a result, behavior that was under the control of more complex stimuli, such as the ability to respond to disguised verbal requests over time, may decline, while behavior that is under control of simple physical stimuli, such as removing clothing when one is warm, may be maintained. The rate of decline of rule-governed behavior varies across persons and might be influenced by factors such as educational and social history.

As stimuli continue to lose their discriminative properties, and the individual's repertoire diminishes further, stimulus-response chains may decline to where environmental stimuli evoke only emotional responses. The person's responses may appear topographically bizarre or "diseased," but when viewed within the context of declining abilities and environmental demands, they are revealed to be adaptive (Fisher, in press; Fisher, Cardinal, Yury, & Buchanan, In press). To illustrate, consider a situation in which a nursing assistant in an extended care facility attempts to undress and bathe a resident with dementia. If the resident does not recognize the nursing assistant or cannot tact being in a healthcare facility and receiving a service, the resident is likely to become frightened. He or she might engage in escape-motivated behaviors, such as screaming or kicking. Screaming, kicking, and biting tend to be exhibited by persons with dementia during interactions involving physical contact and the presentation of task demands, such as participation in activities of daily living. When such behaviors are observed without consideration of the historical and current contexts, the person with dementia is likely to be labeled "aggressive." In contrast, a contextual analysis considers that the behaviors are being exhibited by someone with impaired verbal abilities who can no longer effectively verbalize a request for the nursing assistant to stop touching him/her, during an interaction in which his/her clothing is being removed by a stranger. When examined in this context, the "problem" behaviors appear as adaptive, escape-motivated responses. A noncognitively impaired person would be expected to respond similarly if a stranger entered his/her bedroom. For a person with dementia, "aggressive" behaviors may in fact be the most effective responses available for terminating an aversive stimulus (Fisher, in press).

When they are viewed within the context of an increasingly fragile behavioral repertoire, it is imperative that functional behaviors in persons with dementia not be eliminated or restrained. Bridges-Parlet, Knopman, and Thompson (1994) found that the majority of instances of physically aggressive behavior among dementia patients in extended care facilities were directed at staff during caregiving activities or during attempts by staff to redirect the patient. Physical aggression is more common among moderately to severely demented patients and most often occurs during daily care routines (Hoeffer, Rader, McKenzie, Lavelle, & Stewart, 1997). Thus, it appears as though aggressive behavior is often reinforced by termination of an aversive interaction or demand. Unfortunately, escape-motivated behaviors exhibited by persons with dementia that are perceived as "aggressive" are highly likely to result in the administration of a psychotropic medication.

There is no evidence that atypical or typical antipsychotic medication for physically resistive behavior in persons with dementia affects the motivation to escape from aversive situations. Indeed, antipsychotic medications may disrupt avoidance responding, while leaving intact escape-maintained responding (Smith, Li, Becker, & Kapur, 2004). Findings from the animal literature on the effects of substances that function as dopamine antagonists suggest that they do not alter motor responses, but alter the organism's ability to respond (Blackburn, Pfaust, & Phillips, 1992; Blackburn & Phillips, 1989; Nowend, Arizzi, Carlson, & Salamone, 2001). This indicates that the administration of neuroleptics does not change an organism's motivation for engaging in escape behavior or alter the properties of aversive stimuli. Therefore, the function of the behavior has not changed. What has changed is that the organism is no longer able to respond as it once did to the stimuli. The implication for persons with dementia is that the situation may still be aversive, but the psychotropic medication suppresses their ability to engage in escape behavior. For severely impaired patients, physical aggression may by the only means available to terminate an aversive stimulus.

Historical factors may also be important for understanding the function of behavior in persons with dementia. Consider the example of an elderly combat veteran with a history of post-traumatic stress symptoms. The veteran may have been able to manage trauma-related anxiety through avoidance of specific stimuli throughout his adult life. With the development of dementia, he may no longer be able to control his contact with the trauma-related stimuli (Zeiss & Dickerman, 1989). An individual with this history who is placed in a crowded nursing facility may scream or yell until he/she is moved into a quiet place away from fear-evoking presence of strangers. Staff within the facility will likely interpret the screaming behavior as a symptom of dementia. Within a contextual model, the "problem" behavior is interpreted as an adaptive escape-motivated response that effectively results in the removal of an aversive stimulus.

RISK OF EXCESS DISABILITY

Elderly persons with dementia inevitably become severely disabled, but they are also at risk for *excess* disability. Excess disability occurs when the frequency of adaptive behavior is diminished prematurely. This may be due to insufficient reinforcement or aversive environmental consequences for adaptive behavior, inadequate stimulus control to set the occasion for adaptive behavior, such as lack of effective prostheses, poor lighting, or a confusing environment, or as the result of chemical or mechanical restraint (Fisher, in press). Persons with dementia are at high risk for contacting aversive consequences from the social community. Consider an example in which the verbal behavior of a person with dementia is repeatedly met by corrective feedback by well-intentioned family and friends. The feedback may punish the verbal behavior. Excess disability may develop if the person subsequently withdraws and as a consequence experiences a reduction in verbal behavior earlier than can be accounted for by the disease process (Gentry & Fisher, 2005).

Within the field of the care of persons with dementia, there is increasing recognition of the need for the development of restraint-free interventions. Functional analytic-based interventions are emerging as promising alternatives for supporting the repertoires of elderly persons with dementia and enhancing their quality of life.

CAREGIVING BEHAVIOR AS CONTEXT

Persons with dementia inevitably become dependent on others for their care. This requires that treatment planning addresses caregiver emotional and instrumental coping skills. Family caregivers of persons with dementia are at very high risk of negative effects on their psychological and physical health (Coon, Thompson, Steffen, Sorocco, & Gallagher-Thompson, 2003; Pinquart & Sorenson, 2003; Schulz, O'Brien, Bookwala, & Fleissner, 1995). Therefore, enhancing caregiver functioning may directly benefit the person with dementia (see Coon, Gallagher-Thompson, & Thompson, 2003, for a review of empirically supported treatments for caregivers).

FUNCTIONAL ASSESSMENT AND ANALYSIS

Functional assessment is based on the following assumptions: (1) each person with dementia has a unique biological and learning history; (2) dementing illnesses affect people differently; (3) no two persons will experience the exact same disease course; (4) a person's premorbid behavioral repertoire, general health, and current social and physical environment are important factors in determining the trajectory of decline in his/her functioning; and (5) the function of behavior may change and new functional behaviors may emerge as the disease process progresses and new learning occurs. Given these assumptions, individualized treatment planning is critical in order to correctly identify the function of the behavior and design interventions that preserve the person's functional repertoire for as long as possible.

RULING OUT PHYSIOLOGICAL FACTORS

Dementia is an age-associated disorder, and affected persons also experience other age-related conditions. It is therefore essential to rule out potentially treatable physiological conditions that could potentially contribute to a problem behavior. Medical records should be reviewed to assess for comorbid medical conditions, past injuries and/or illnesses that may result in chronic pain, and medication information. The potential side effects of medication should be carefully considered. Covariation of disruptive behavior and medication changes should be evaluated to rule out negative medication reactions.

Sensory impairments are especially prevalent in older adults. Approximately one third of elderly persons experience hearing impairment, approximately 20% experience vision impairment (Desai, Pratt, Lentzer, & Robinson, 2001), and approximately one third have mobility limitations (Freedman & Martin, 1998). Sensory aids, including eyeglasses and hearing aids with proper prescription, should be pursued, worn consistently, and used competently (see Lindsley, 1964, and Skinner & Vaughn, 1983, for discussions on the design of prosthetic environments for older adults). Also, caregivers can offer valuable information on whether the patient can independently care for physical ailments.

FUNCTIONAL ASSESSMENT

Martin and Pear (1999) outlined five major causes of disruptive behaviors from a behavior analytical perspective. First, positive reinforcement, such as social attention or tangible items, can develop and maintain behavioral excesses. Indicators of social attention as the function of the disruptive behavior are that attention reliably follows the behavior (e.g., the patient looks or approaches the individual offering the attention, and the patient smiles just before engaging in the disruptive behavior). For example, a person with dementia may be very verbose or call out repeatedly in order to attain attention from a caregiver. Second, negative reinforcement in the form of escape or avoiding aversive stimuli might maintain socially inappropriate behavior. A strong indicator of this cause is that the individual engages in the behavior only when certain types of requests are made of the patient. An example of this type of behavior would be a patient who resists when bathed by an unfamiliar caregiver as described previously. The patient may be aggressive with the caregiver until the caregiver stops attempting to bathe the patient. Thus, the patient's disruptive behavior has been negatively reinforced by the termination of being bathed by an unfamiliar caregiver. Third, automatic reinforcement occurs when stimulation of the body produces enjoyable sensations or reduces aversive sensations. Many repetitive behaviors serve this function. An indicator of this cause is that the behavior occurs at a steady rate, although it has no apparent effect on the patient or the environment. For example, in the case of automatic positive reinforcement, a patient may continually rub his/her leg in a sanding fashion, rub his/her hand over a wall while walking down a corridor, or pick or scratch his/her skin, possibly to the point at which physical harm is done. Automatic negatively reinforced behaviors may involve the termination of pain or anxiety. Fourth, external sensory reinforcement occurs when stimuli from the nonsocial environment maintain the disruptive behavior. An indicator of this function is that a disruptive behavior continues even though it appears to have no social consequences. Patients may bang on objects to produce sounds or may switch lights on and off repeatedly. Negative reinforcement in the form of the termination of aversive physical stimuli (e.g., a loud noise) may also maintain problem behavior. Lastly, some disruptive behavior appears to be elicited by stimuli in the environment, as opposed to being controlled by consequences of the behavior. Indicators of this cause of behavior are that the behavior appears consistently in the same environment or in the presence of certain stimuli, it is never followed by an identifiable consequence, and it appears involuntary. An example of an elicited cause of behavior would be a patient who is afraid of hypodermic needles who panics or becomes very anxious when he/she sees a needle.

The primary goal of functional assessment is to empirically determine what function or purpose the problem behavior serves for the person with dementia. The identification of the function of the behavior directly informs the treatment. Functional assessment involves either a descriptive functional assessment or a functional analysis. A descriptive functional assessment indirectly assesses the relationship between contextual stimuli and the problem behavior. A functional analysis, in contrast, directly manipulates potential controlling stimuli.

A major advantage of descriptive assessment in the natural environment is the opportunity to observe a large number of potential controlling variables with enhanced generality when compared to indirect assessment (Sasso et al., 1992). A major limitation of descriptive assessment is that it does not control extraneous variables, making correct identification of controlling variables more difficult. It also may be difficult to identify the reinforcing consequences of behaviors maintained on intermittent schedules (Sulzer-Azaroff & Mayer, 1977). Resulting data from descriptive analysis are inferential and only suggestive of functional relationships (Mace & Lalli, 1991).

Identifying Reinforcers

Identifying stimuli that function as reinforcers is often part of a functional assessment. Observation of a patient when he/she has free choice, in the absence of demands or otherwise limiting conditions, is a simple approach for identifying potential reinforcers. It is important to gain an adequate sample of preferences; in other words, one must observe the patient in different settings, at different times of the day, and under other conditions to identify several potential reinforcers. Interviewing persons who interact regularly with a patient regarding preferred objects and activities is an efficient approach to identifying potential reinforcers, although the accuracy of these reports has been questioned.

Structured approaches to stimulus preference assessment have been widely documented to be effective with impaired populations (Fisher et al., 1992). Stimulus preference assessment involves observing and recording the patient's approach responses to stimuli and/or recording duration of engagement with stimuli.

FUNCTIONAL ANALYSIS

Behavior screening tools can serve as a starting point for baseline investigations, as these screening tools refine the descriptions of the behaviors of interest, their frequency, duration, intensity, timing, and level of disruptiveness. The Cohen-Mansfield Agitation Inventory (Cohen-Mansfield & Billig, 1986) is an example of a behavior screening tool designed for use with persons with dementia. After pinpointing the behavior of interest, caregivers then may complete a questionnaire that assesses the functional patterns among antecedents, behaviors, and consequences (Lerman & Iwata, 1993). However, like all retrospective summary reports, such *descriptive functional analyses* are easy, accessible, and inexpensive assessment tools whose accuracy depends on the caregiver's skill as a historian and reporter. The use of direct observation tools results in comparatively greater accuracy: Promptly upon each occurrence of the behavior, the caregiver completes a diary card that questions the behavior's antecedent and consequent circumstances. In contrast to measures that assess the behavior retrospectively, summaries of the data on these diary cards—recorded over time—might reveal the behavior's function even when the caregiver was initially unable to detect a specific pattern among antecedents, behavior, and consequences (see also, Fisher, Harsin, & Hadden, 2000). Should the results be difficult to interpret, then an *experimental functional analysis* (EFA) (Iwata, Dorsey, Slifer, Bauman, & Richman, 1982/1994) can be conducted. An EFA systematically alters the consequences of the behavior. It assesses a behavior's communicative function by limiting access to social situations, preferred items or events, and to avoidance of nonpreferred tasks, items, or events to the occurrence of the behavior. If the experimental contingencies between behavior and consequences result in an increase in the rate of behavior, the behavior is thought to serve an expressive communicative function. An EFA rules out possibly confounding antecedent circumstances by observing the behavior in no access/alone and free access/social conditions, without providing explicit consequences upon occurrence of the behavior. Consequently, through descriptive or experimental functional analyses, the context in which the behavior occurs becomes known. After a baseline period, during which antecedents, behavior, and consequences of interest are recorded to enable a later comparison with the intervention condition, the behavior's context is altered. For a dangerous or low rate behavior, it may be more practical to use a descriptive assessment strategy to generate hypotheses regarding controlling functions. For other behaviors a functional analytic approach

TABLE 11.1   Practitioner Recommendations: Assessment

---

1) Assess medical history to rule out sensory deficits and comorbid conditions (e.g., pain, drug interactions) that may be contributing or causing disruptive behaviors.
2) Consult with caregivers to determine whether sensory aids and/or other prescribed environmental modifications are in use.
3) Caregiver reports can be efficient sources of information regarding the topography of the problem behavior and for generating hypotheses regarding controlling variables. Caregivers may overlook their own behavior and how it can affect the behavior of the dementia patient.
4) Experimental functional analysis has the greatest prescriptive value for problem behaviors; EFA may not be practical for dangerous behaviors.

---

may be more effective. A study by Buchanan and Fisher (2002) illustrates the use of EFA to identify the controlling function of disruptive vocalizations in elderly persons with dementia. Table 11.1 summarizes guidelines for practitioners.

FUNCTIONAL ANALYTIC INTERVENTIONS

PREVENT EXCESS DISABILITY

The functional analytic approach to dementia assumes that the behaviors observed are *adaptive* given a person's historical and current biopsychosocial circumstances. Given such functionality of all behavior, this approach aims at redesigning the context, such that a further narrowing of a person's repertoire will not occur. To illustrate, consider the use of profanity by a person with dementia: An intervention that reduces the rate with which profanity occurs is deemed inappropriate if it also further decreases a person's expressive capabilities. The goal of a functional analytic approach is to utilize context-altering strategies without accelerating physical or mental decline and without increasing the chance that a person might lose access to potentially vital repertoires.

Before designing a functional analytic intervention and as a first step in the prevention of excess disability, caregivers must assess sensory and motor functioning and ascertain the use of the proper aids (e.g., hearing aids, glasses, dentures). Second, a physician must rule out physiological conditions (delirium, adverse medication effects, etc.) that could limit a person's engagement in activities or contribute to behavioral problems. Third, because excess disability might be related to the negative social consequences and the increasing social isolation that accompany progressive cognitive decline, the functional analytic approach explicitly focuses on enhancing behavioral engagement by altering the context of behavior. Contextual antecedents and consequences are not triggers of behavior:

They describe the situations which most likely precede and follow the behavior of interest. The relationship among antecedents, behavior, and consequences is probabilistic. Consequently, antecedents do not have to be present at each occurrence of the behavior, and consequences also may not follow each occurrence. Rather, there is a pattern of antecedents and consequences that is reliably related to the behavior in question. Functional analyses investigate such patterns.

Importantly, patterns of antecedents, behaviors, and consequences are idiosyncratic. They depend on a person's history and current circumstances, including psychological and biological predispositions and the extent of physiological impairment. While one individual might engage in disruptive vocalizations mostly in loud or noisy situations, another individual might vocalize mostly when alone. The former person's vocalization might be maintained by the termination of noise, while the latter person's behavior might be maintained by access to social interactions. Although the vocalizations might sound the same, their functional patterns differ. This idiosyncratic nature of functional patterns requires individualized interventions, designed with the aim of preventing excess disability by leaving intact or enhancing socially communicative repertoires. Such individualized interventions (1) investigate the antecedents and consequences that are reliably related to the behavior of interest during a baseline condition and (2) alter these antecedents or the consequences during the treatment condition. Monitoring the frequency with which the behavior occurs in the unchanged baseline and changed treatment context results in data that speak to the potential effectiveness of the particular intervention.

DESIGN OF INTERVENTIONS

Antecedent-Based Interventions

One aspect of the functional analytic approach to interventions focuses on preventing the antecedent conditions that occasion the behavior. Interventions in this category include all environmental prompts that enhance the salience of cues and thereby reduce confusion, anxiety, and the necessity for self-protective behavior. Memory aids fall into this category (Bourgeois, 1993). Other examples are the use of visual barriers to prevent exit-seeking (Feliciano, Vore, LeBlanc, & Baker, 2004); the inclusion of individualized soft music to prevent self-protective behavior during toileting activities (Thomas, Heitman, & Alexander, 1997), to generally reduce agitation (Gerdner, 2000), or to increase eating (Ragneskog, Bråne, Karlsson, & Kihlgren, 1996); prompted voiding schedules to reduce incontinence (e.g., Adkins & Mathews, 1997; Engberg, Sereika, McDowell, Weber, & Brodak, 2002); and caregivers' use of age-appropriate speech adjusted to the individual's remaining verbal ability to prevent agitation (Hart & Wells, 1997).

A prosthetic environment also might include appropriate prompting procedures to maximize independent engagement in activities of daily living, such as dressing, eating, and toileting. Effective prompts during the course of a person's care might change from instructions to simplified one- or two-word prompts, to increasing physical guidance. Physical guidance should be limited to the initiation of the task to provide the individual with dementia with an opportunity for task completion and accomplishment. Engelman, Altus, Mosier, and Mathews (2003), for example, showed that training staff in the use of graduated prompts promoted dementia patients' engagement in personal care routines (see also Engelman, Altus, & Mathews, 1999; Engelman, Mathews, & Altus, 2002). Recently, personal-ized and automated prompting devices aimed at restoring independence in activities of daily living are being developed (Mihailidis, Barbenel, & Fernie, 2004).

Consequent-Based Interventions

All consequent-based interventions should aim to increase the rate of active engagement with the world. They use reinforcement procedures, which increase the rate of behavior and thereby build alternative reper-toires. At no time does the functional analytic approach resort to punish-ment procedures to reduce the rate of behavior (for a review of the side effects of punishment, see Sidman, 1989; Van Houten, 1983).

Promoting access to preferred items, activities, or events is the center-piece of the functional analytic approach. Caregiver reports such as the *Pleasant Event Schedule* (Teri & Logsdon, 1991) and reports of the person with dementia provide information about a person's preferences. Even severely impaired individuals are able to indicate their preferences by pointing, choosing, or spending more time with one item than another (see also DeLeon & Iwata, 1996; Fisher et al., 1992). As preferences differ between individuals, individualized assessments are necessary. In addition, preferences are conditional. They may depend on time of day, physiological well-being, and preceding events; for example, after having been engaged in stimulating activities, a quiet location might be preferred. For this reason, preference assessments should be individualized and conducted frequently.

If a descriptive or an experimental functional analysis has shown that a particular challenging behavior is maintained by a preferred event, one functional analytic solution is to provide the preferred event frequently and at random times. Thus, the presentation of the preferred event ceases to depend on the occurrence of the behavior; thereby, the behavior loses its communicative function. This procedure is termed noncontingent rein-forcement. Buchanan and Fisher (2002) used this procedure to decrease the rate of disruptive vocalizations. At the same time, caregivers might pay special attention to socially appropriate requests and reinforce them with access to the requested events and activities. This procedure is termed differential reinforcement. Heard and Watson (1999), for example, systematically reinforced all other behavior but wandering, the problematic target behavior. The identification of reinforcers was based on a preliminary descriptive functional assessment of residents' behavior and included attention, sweet foods, or sensory stimulation. As a result of this differential reinforcement procedure, the frequency of wandering decreased by 65–80%. Note that noncontingent and differential reinforcement procedures never deprive the person of social interaction or access to preferred items, activities, or events. Instead, caregivers make a conscious effort to eliminate the necessity for the problematic behavior and to engage the person in activities other than the problematic one. From a functional perspective, this engagement is similar to evidence-based procedures that have been shown to reduce depression in people with dementia (Teri et al., 2003; Teri, Logsdon, Uomoto, & McCurry, 1997; Teri & Uomoto, 1991); see also Chapter 15, *Mood disorders* by Hopko, Hopko & Lejuez for a discussion of behavioral activation (Jacobson, Martell, & Dimidjian, 2001).

The presentation of preferred events can also be made contingent on correct verbal behavior and thereby increase its accuracy. Henry and Horne (2000) provided access to preferred tangible items dependent on correct verbal responses and observed an increase in accuracy in individuals with severe dementia. Gentry and Fisher (2005) demonstrated that a shift in conversational style from interrogative-corrective to paraphrasing the demented individual's responses to ensure speaker and listener coherence (or joined stimulus control) increased the time the persons with dementia spent in conversation. Thus, "being understood"—engaging in turn-taking while similar environmental events are the subject of the speaker's and the listener's verbal behavior—might continue to function as a reinforcer even when individuals are severely impaired. In summary, consequent-based interventions involving preferred events are the key to maintaining individuals' social engagement.

MONITORING

Consequent-based functional analytic strategies are long-term strategies. The effects of noncontingent and differential reinforcement procedures might not be immediately observable, but they are evidenced by a change in the behavior's frequency with time. These long-term strategies counteract everyday staff practices, which typically lead to the immediate cessation of particularly challenging behaviors but have the unintentional effect of increasing the frequency of those behaviors in the long run (Hastings & Remington, 1994; see also Schnelle, Ouslander, & Cruise [1997] for a discussion of short-term compliance with minimal care standards versus adherence to strategies for improved long-term outcome).

TABLE 11.2   Practitioner Recommendations: Treatment

---

1) Act to prevent excess disability.
2) Maintain behavior in the person's repertoire for as long as possible.
3) Maximize the person's access to reinforcement.
4) Tailor interventions to the individual by altering the antecedents or the consequences of the problem behavior and by offering response alternatives.
5) Focus on the *prevention* of problem behavior by making the behavior unnecessary, by eliminating the motivation for the behavior, and by meeting needs through other sources of reinforcement.
6) Monitor the effectiveness of the intervention over time and anticipate that modifications will be necessary as the disease progresses.

Thus, continued monitoring of an intervention's effectiveness is important because it ensures adherence to the protocol, because the procedure might have to be refined if the preferences and the functional repertoires of the person with dementia change, and because monitoring helps caregivers come into contact with the continued success and to forgo short-term solutions that might inadvertently maintain the problem behavior. Table 11.2 summarizes practitioner recommendations for functional analytic-based treatment.

CASE STUDY

SETTING

The clients described in this case study received services through the Nevada Caregiver Support Center (NCSC), a state-funded program to support elderly Nevadans with dementia so that they can live safely and comfortably in their homes for as long as possible. NCSC services are designed to prevent excess disability in persons with dementia and to provide empirically supported educational and emotional support to family caregivers to prevent the stress-related physical and psychological problems commonly experienced by family caregivers.

The NCSC behavioral health services are individualized, recognizing that the needs of elderly persons with dementia and their family caregivers vary greatly and that a family's needs will change as the disease progresses and new problems emerge. Services are provided by a team of *Caregiver Coaches* who are doctoral students in the Clinical Psychology Program at the University of Nevada, Reno. The NCSC provides services to persons with dementia and their family caregivers from early stage through late stage dementia. Recognizing that access to services is difficult for family caregivers by virtue of their need to provide 24-hr supervision to their family member with dementia, NCSC services are provided in several locations including client homes; the NCSC office in Reno; and at venues in counties throughout northern Nevada such as county senior centers, assisted-living facilities, and libraries, etc. In addition, respite services are provided during home visits and group coaching sessions for caregivers whose access to service is limited by the lack of available alternative supervision for their family member with dementia.

NCSC behavioral health and wellness services are designed to prevent excess disability in persons across the course of the disease. Persons in the early stages of dementia are at very high risk for depression, suicidal ideation, and anxiety. Through individual and/or family coaching, clients receive referrals to healthcare and social service agencies to access support for the continued use of their skills and to reduce or prevent problems, such as depression and anxiety. Clients receive coaching in behavioral activation strategies for reducing depression and continuing to maintain engagement in preferred activities throughout the disease process and guidance in the use of environmental supports, such as prosthetic memory aids, in order to help them maintain their cognitive abilities for as long as possible. NCSC provides three kinds of services for elderly persons who have progressed to moderate or severe impairment. NCSC coaches their family caregivers in order to reduce or prevent behavior problems through evidence-based strategies for behavior management. NCSC reduces the risk of depression by adapting principles of behavioral activation and implementing pleasant activities for their family member with dementia. NCSC also enhances home comfort and safety by providing instruction in home modification for persons with memory impairment (Warner, 2000).

Caregiver Coaching

The NCSC Caregiver Coaching program is designed to target sources of caregiver strain through a combination of education, emotional support, and skills training. Coaches provide caregivers with education in evidence-based strategies for enhancing and maintaining the competence of their family member with dementia and preventing or reducing the behavior problems. An individualized care plan for reducing behavior problems is developed in collaboration with the family. Coaching in the implementation of the strategies and support in problem solving as they implement the strategies are provided. A 24-hr, toll-free helpline is available so that caregivers can access coaching in real-time.

Many caregivers also need support to help them deal with the severe emotional strain of caregiving and its consequences, including depression, anger, frustration, and anxiety (Ory, Yee, Tennstedt, & Schulz, 2000; Schulz et al., 1995). By the time many caregivers seek help, they are too emotionally distressed to benefit from didactic skills training or to participate in the problem solving needed for effective patient advocacy. For these caregivers, NCSC provides additional coaching in the form of individual, group, and/or family coaching sessions that are designed to increase caregivers' ability to accept and regulate their intense emotions and to maintain or increase the quality of their life through behavioral activation. When caregivers learn to engage in healthier coping strategies, distress decreases, mood improves, and they are then better able to provide safe and effective care for their family member (Cucciare & Fisher, 2005; Schulz et al., 2003).

The NCSC also provides group caregiver coaching workshops to families in Reno and rural communities across northern Nevada. In these workshops, groups of caregivers receive direct instruction in strategies for dealing with common caregiving challenges. Examples of workshop topics include communicating with persons with memory impairments and coping with the stress of caregiving. Coaches are available after each workshop for individualized coaching support.

The following brief snapshot of an elderly couple's experience in coping with dementia is presented in order to illustrate how the progression of dementia changes patients' and their family caregivers' needs. Two cross-sections of a 4-year course of treatment are presented to reflect how a contextual model was applied across the course of the disease progression.

CLIENTS

Intake

Mary was a 75-year-old woman who contacted the NCSC approximately 4 years ago after attending a caregiving coaching workshop on effective communication with persons with memory disorders. Mary and her husband, Thomas, had been married for 57 years when they began receiving services at the NCSC. Thomas, 80, had been diagnosed with Alzheimer's disease approximately 2 years prior to Mary's contacting the NCSC. Mary requested guidance regarding recent changes in Thomas' behavior. Thomas had been attending a day program for elderly persons with memory problems but was terminated from the program when the staff had difficulty redirecting him when he repeatedly tried to leave the building. The staff of the day program was concerned for their safety and the safety of other residents, as Thomas was becoming increasingly agitated when not allowed to leave the building.

Two coaches conducted a home visit. One met with Mary while the other met with Thomas. NCSC's assessment strategy is designed to provide information relevant to areas that most frequently impact quality of life for persons with memory disorders and their family members. The coaches conducted an idiographic assessment in order to understand Mary's and Thomas' functioning and the contextual factors that may be impacting their experience.

Assessment of Thomas's functioning focused on his mental status including behavioral strengths and impairments; physical health and medication usage; mood, such as evidence of depression, anger, and anxiety; and challenging behaviors. The assessment of challenging behaviors included a description of the topographical characteristics including frequency, duration, and intensity and a descriptive functional assessment of the conditions during which the behavior(s) occur. The coaches also assessed Thomas' preferred activities, including opportunities for and actual engagement in the activities.

Assessment of Mary's functioning focused on her mood, emotional and instrumental coping skill repertoire, and physical health. The coaches administered the *Revised—Ways of Coping Checklist* (RWCCL; Vitaliano, Russo, Carr, Maiuro, & Becker, 1985), a 41-item scale that assesses emotional caregivers' coping strategies, such as problem solving, social support-seeking behaviors, and avoidance strategies. They also assessed instrumental caregiving skills using the *Caregiver Task Checklist* (Gallagher-Thompson et al., 2000), a 29-item checklist that assesses the number of activities of daily living such as bathing and finances for which the caregiver provides assistance to the person with dementia. In addition, the assessment measure was augmented to assess reasons why the caregiver finds the activity difficult or upsetting.

The coaches also assessed the couple's functioning. Recognizing that families cope with dementia within a personal historical context, the NCSC assessment process includes an assessment of the dyad's relationship including a history affection versus conflict, the couple's history of division of labor for managing household and social activities such as driving and finances.

Finally, access to needed community and social support resources was assessed. This was done in order to determine whether referrals to community agencies such as Meals on Wheels, home health services, and respite programs are appropriate.

Case Conceptualization

Interview, questionnaire, and direct observation data indicated that Mary and Thomas had experienced significant changes in their relationship as a result of Thomas' declining abilities. Thomas had been a successful business executive prior to his retirement. Mary had been a homemaker throughout their marriage. Mary took over all of the major responsibilities that Thomas had handled during their marriage, including managing their finances, home and car repair, driving, and yard work, after Thomas' declining abilities. Thomas had become resistant to Mary's effort to do "his" jobs. Mary had repeatedly attempted to explain to Thomas why she needed to complete the tasks. Several times per day their interactions resulted in Thomas becoming very angry and physically agitated if Mary did not submit to his requests. Mary's acquiescence at times had put the couple in danger, for example, when she allowed Thomas to drive.

Mary described Thomas as becoming increasingly distressed if he could not see her. He had begun shouting her name if she left the room or he could not find her in their home or yard. Mary's opportunity to complete household tasks or have any time alone had been reduced to brief periods of a few minutes' duration.

Thomas had a history of excelling in social and athletic activities. He had been an avid tennis player and runner for most of his adult life. The couple lived near a health club where Thomas had played tennis competitively for years. In addition, he was an avid gardener. Prior to his diagnosis, Mary recognized that something was seriously wrong when Thomas was picked up by the police after trimming the flowers off all the rose bushes in a nearby park.

Mary reported feeling exhausted and overwhelmed by Thomas's care needs and his increasing dependence on her. She reported feeling hopeless now that Thomas was no longer allowed to attend the day program, as it was her only source of respite. She had become increasingly socially isolated in the months prior to contacting the NCSC. She and Thomas had no family living in the area, and Mary felt that it would be unreasonable to ask friends or neighbors to supervise Thomas so that she could occasionally leave their home without him in order to run errands. Friends had suggested that she consider a nursing home for Thomas which she adamantly resisted due to a reported sense of guilt over "abandoning him."

Assessment of Mary's Functioning

Mary demonstrated several strengths directly relevant to her role as a caregiver. She was observed to have a highly skilled repertoire for completing the majority of the instrumental tasks associated with Thomas' care. Mary was also observed to be very empathic toward Thomas, as reflected in her concern for his frustration with his inability to perform tasks he had excelled at throughout his life. Regarding her emotional coping, direct observation indicated that she frequently underestimated her competence in completing tasks. In addition, Mary was observed to be consistently self-critical. She often used derogatory language to describe her own behavior, in spite of much evidence of her effectiveness and heroic efforts on Thomas's behalf. She reported feeling very guilty about her occasional anger and frustration over Thomas's confusion. Mary's emotional coping primarily involved negative coping strategies, such as avoidance and escape strategies, and ruminating, rather than positive coping, such as self-soothing strategies and problem-solving in response to changeable situations.

Mary's access to pleasant events was at near zero level at baseline. She had engaged in only minimal self-care activities for several months. She was severely sleep deprived, had not exercised, and had gained about 15 pounds in the months before contacting the NCSC. She could not recall the last time she had had fun.

Thomas' Functioning

Thomas also was observed to have many strengths. At intake, he was able to engage in conversations about familiar topics from his past. He took great pride in showing the coaches and others the objects in his home and in giving tours of his garden. He was observed by the coaches to be interpersonally charming. He was also physically very healthy.

Thomas's access to sources of reinforcement had been severely restricted in recent months. He had continued to walk to the health club, but found that none of his former tennis friends were available to play with him. In addition, he had also experienced negative consequences during interactions at the local park when he had engaged in behavior that in the past had functioned to produce positive social attention. Thomas's efforts to engage in the financial and household tasks he had completed for decades were now met with verbal reprimands. He also increasingly appeared anxious when not at home unless Mary was present.

COACHING PLAN

In collaboration with Mary and Thomas, the coaches proposed a plan that involved increasing both her and Thomas's access to reinforcement. Mary agreed to the coaching plan summarized in Table 11.3. Over the next 2 years Mary and Thomas experienced significant improvements in the quality of their life relative to baseline. Mary quickly learned positive behavioral management strategies and repeatedly experienced the natural consequences of her effective use of distractions, embedding procedures, and benign misinformation. She developed increasingly creative ways of solving many of the problems described at intake. Both Mary's and Thomas's moods improved as their contact with pleasant experiences increased and contact with aversive consequences decreased.

Over the next 2 years Thomas remained physically healthy but experienced continued deterioration in his memory, executive planning, and verbal abilities. Mary decided to seek an alternative residence for Thomas, as she increasingly realized she was no longer able to manage the demands of his physical care due to the combination of his increased confusion and continued physical strength. A caregiving coach worked closely with Mary and Thomas as he transitioned to an assisted living facility (ALF). Mary chose to visit the ALF every day and continued to provide much of Thomas's hands-on care.

A few weeks after Thomas was admitted to the ALF, Mary observed a marked change in his speech and his ability to visually focus on and track when she spoke to him. A review of his medical record revealed that he

TABLE 11.3 A Summary of the Initial Coaching Plan for Mary and Thomas

Treatment Target	Intervention
Increase Mary's understanding of the effects of Alzheimer's disease on behavior in order to improve her ability to understand and predict Thomas's behavior and enhance her ability to take his perspective	Provided education in symptoms of dementia and the effects of Alzheimer's disease on behavior
Increase Mary's use of positive emotional coping strategies	Provided direct instruction and homework assignments in positive strategies for coping with what was an unavoidably stressful situation including training in self-empathy skills, self-soothing strategies, and active problem solving.
Reduce conflicts associated with communication	Provided Mary with instruction in how to communicate with Thomas in ways that would not provoke resistance and anger including not arguing or trying to rationally explain her decisions but instead relying on "benign misinformation."
Reduce or prevent occurrences of challenging behavior in Thomas	Provided Mary with instruction in positive methods of behavior management including how to operationalize behaviors in observable terms and recognize and monitor the antecedents and consequences of behavior to inform their purpose. Provided coaching in strategies for addressing Thomas's specific behaviors including distraction, noncontingent reinforcement, and embedding procedures.
Increase Thomas's access to reinforcement	Provided Mary with instruction in behavioral activation strategies for Thomas including how to modify activities so that he would experience more pleasure and less frustration (e.g., arrange for him to play tennis with a paid instructor, provide a spray bottle in place of pruning shears on walks in the park, provide photo albums containing photos of Mary to reduce his distress in her absence).
Increase Mary's access to reinforcement	Provided coaching in problem solving regarding options for respite care for Thomas so that Mary could engage in pleasant and health-promoting activities.
Reduce Mary's caregiving burden by increased access to community resources	Provided referrals to agencies including an eldercare attorney to assist in pursuing legal guardianship of Thomas and home health services.

had been administered Ativan and Haldol following staff observations of him slamming doors, raising his voice, and being resistant during showering. No physical aggression was observed. A descriptive functional assessment was conducted through direct observation of Thomas's behavior in the ALF. The assessment revealed that the door slamming and raised voice were attention-maintained and that his resistance during showering was escape-motivated. Over the next 2 weeks NCSC coaches assisted Mary in her attempts to promote the use of restraint-free strategies in the care of Thomas. An in-service in restraint-free care was provided by NCSC coaches to the staff of the ALF. The in-service focused on the adaptive nature of challenging behavior in persons with dementia and a description of the data that had been collected through the functional assessment. In addition, NCSC coaches provided staff with recommendations to Thomas's behavior including the use of embedding procedures during hands-on care to reduce escape-motivated behaviors.

During interactions with ALF staff, the NCSC coaches observed evidence of Thomas's stigmatization, including references to his "diseased brain" and his being labeled "the big scary guy." After repeatedly finding that Thomas appeared lethargic and sedated, Mary informed the administration of the ALF that she would not consent to Thomas receiving psychotropic medications. Shortly after this, she decided to move Thomas to a skilled nursing facility. NCSC coaches assisted in the transition to the skilled nursing facility. The NCSC staff provided the new facility with a summary of Thomas's positive response to restraint-free methods in an effort to counteract the records from the ALF that included several potentially stigmatizing references to Thomas' challenging behavior. The transition to the skilled nursing facility went smoothly. On several occasions Mary modeled the use of restraint-free strategies that she has found effective in preventing distress in Thomas and promoting his experience of pleasure. In addition, at Mary's request the psychotropic medications were discontinued. No instances of aggressive or other problem behaviors were reported by the staff of the skilled nursing facility.

REFERENCES

Adkins, V. K., & Mathews, R. M. (1997). Prompted voiding to reduce incontinence in community-dwelling older adults. *Journal of Applied Behavior Analysis, 30,* 153–156.

American Geriatrics Society. (2001). *Guidelines for restraint use.* Retrieved November 1, 2005, from http://www.americangeriatrics.org/products/position papers/restraint

Blackburn, J. R., Pfaust, J. G., & Phillips, A. G. (1992). Dopamine functions in appetitive and defensive behaviors. *Progress in Neurobiology, 39,* 247–279.

Blackburn, J. R., & Phillips, A. G. (1989). Blockade of acquisition of one-way conditional avoidance responding by haloperidol and metoclopramide but not by thioridazine or clozapine: Implications for screening new antipsychotic drugs. *Psychopharmacology, 98,* 453–459.

Bourgeois, M. S. (1993). Effects of memory aids on the dyadic conversations of individuals with dementia. *Journal of Applied Behavior Analysis, 26,* 77–87.

Bridges-Parlet, S., Knopman, D., & Thompson, T. (1994). A descriptive study of physically aggressive behavior in dementia by direct observation. *Journal of the American Geriatrics Society, 42,* 192–197.

Briesacher, B. A., Limcangco, M. R., & Simoni-Wastila, L. (2005). The quality of antipsychotic drug prescribing in nursing homes. *Archives of Internal Medicine, 165,* 1280–1285.

Buchanan, J. A., & Fisher, J. E. (2002). Functional assessment and noncontingent reinforcement in the treatment of disruptive vocalization in elderly dementia patients. *Journal of Applied Behavior Analysis, 35,* 68–72.

Burgio, L. D., Scilly, K., Hardin, J. M., Jankosky, J., Bonino, P., Slater, et al. (1994). Studying disruptive vocalization and contextual factors in the nursing home using computer-assisted real-time observation. *Journal of Gerontology: Psychological Sciences, 49,* 230–239.

Catania, A. C. (1996). On the origins of behavior structure. In T. R. Zentall & P. M. Smeets (Eds.), *Stimulus class formation in humans and animals* (pp. 3–12). Amsterdam, The Netherlands: Elsevier Science.

Chase, P. N., & Danforth, J. S. (1991). The role of rules in concept learning. In L. J. Hayes & P. N. Chase (Eds.), *Dialogues on verbal behavior* (pp. 205–222). Reno, NV: Context Press.

Chen, J. C., Borson, S., & Scanlan, J. M. (2000). Stage-specific prevalence of behavioral symptoms in Alzheimer's disease in a multi-ethnic community sample. *American Journal of Geriatric Psychiatry, 8,* 123–133.

Cohen-Mansfield J., & Billig, N. (1986). Agitated behavior in the elderly: A conceptual review. *Journal of the American Geriatrics Society, 34,* 711–721.

Coon, D. W., Thompson, L., Steffen, A., Sorocco, K., & Gallagher-Thompson, D. (2003). Anger and depression management: Psychoeducational skills training interventions for women caregivers of a relative with dementia. *The Gerontologist, 43,* 678–689.

Cucciare, M. A., & Fisher, J. E. (2005, November). *A prescriptive assessment approach to matching caregiver needs to services.* Paper presented at the annual meeting of the Gerontological Society of America, Orlando, FL.

Cummings, J. L. (2004). Alzheimer's disease. *New England Journal of Medicine, 351,* 56–67.

DeLeon, I. G., & Iwata, B. A. (1996). Evaluation of a multiple-stimulus presentation format for assessing reinforcer preferences. *Journal of Applied Behavior Analysis, 29,* 519–533.

Desai, M., Pratt, L. A., Lentzer, H., & Robinson, K. N. (2001). *Trends in vision and hearing among older Americans.* Hyattsville, MD: National Center for Health Statistics.

Eldridge, L. L., Masterman, D., & Knowlton, B. J. (2002). Intact implicit habit learning in Alzheimer's disease. *Behavioral Neuroscience, 116,* 722–726.

Engberg, S., Sereika, S., McDowell, B. J., Weber, E., & Brodak, I. (2002). Effectiveness of prompted voiding in treating urinary incontinence in cognitively impaired homebound older adults. *Journal of Wound, Ostomy & Continence Nursing, 29,* 252–265.

Engelman, K. K., Altus, D. E., & Mathews, R. M. (1999). Increasing engagement in daily activities by older adults with dementia. *Journal of Applied Behavior Analysis, 32,* 107–110.

Engelman, K. K., Altus, D. E., Mosier, M. C., & Mathews, R. M. (2003). Brief training to promote the use of less intrusive prompts by nursing assistants in a dementia care unit. *Journal of Applied Behavior Analysis, 36,* 129–132.

Engelman, K. K., Mathews, R. M., & Altus, D. E. (2002). Restoring dressing independence in persons with Alzheimer's disease: A pilot study. *American Journal of Alzheimer's Disease & Other Dementias, 17,* 37–43.

Evans, D. A., Funkenstein, H. H., Albert, M. S., Scherr, P. A., Cook, N. R., Chown, M. J., et al. (1989). Prevalence of Alzheimer's disease in a community population of older persons: Higher than previously reported. *Journal of the American Medical Association, 262*, 2551–2556.

Feliciano, L., Vore, J., LeBlanc, L. A., & Baker, J. C. (2004). Decreasing entry into a restricted area using a visual barrier. *Journal of Applied Behavior Analysis, 37*, 107–110.

Fisher, J. E. (In press). Behavioral interventions for promoting adaptive behavior in elderly persons with dementia. *Clinical Gerontologist.*

Fisher, J. E., Cardinal, C., Yury, C., & Buchanan, J. A. (In press). Dementia. In J. E. Fisher & O'Donohue W. T. (Eds.), *Practice guidelines for evidence based psychotherapy.* New York: Springer.

Fisher, J. E., Harsin, C. M., & Hadden, J. E. (2000). Behavioral interventions for dementia patients in long term care facilities. In V. Molinari (Ed.), *Professional psychology in long term care: A comprehensive guide* (pp. 179–200). New York: Hatherleigh Press.

Fisher, W., Piazza, C. C., Bowman, L. G., Hagopian, L. P., Owens, J. C., & Slevin, I. (1992). A comparison of two approaches for identifying reinforcers for persons with severe and profound disabilities. *Journal of Applied Behavior Analysis, 25*, 491–498.

Food and Drug Administration. (2005). *Deaths with antipsychotics in elderly patients with behavioral disturbances.* Retrieved April 12, 2005, from http://www.fda.gov/cder/drug/advisory/antipsychotics.htm

Freedman, V. A., & Martin, L. G. (1998). Understanding trends in functional limitations among older Americans. *American Journal of Public Health, 88*, 1457–1462.

Gallagher-Thompson, D., Lovett, S., Rose, J., McKibbin, C., Coon, D., & Futterman, A. (2000). Impact of psychoeducational interventions on distressed family caregivers. *Journal of Clinical Geropsychology, 6*, 91–110.

Gentry, R. A. & Fisher, J. E. (2005, November). *Facilitating conversation in persons with Alzheimer's disease.* Paper presented at the annual meeting of the Gerontological Society of America, Orlando, FL.

Gerdner, L. A. (2000). Effects of individualized vs. classical "relaxation" music on the frequency of agitation in elderly persons with Alzheimer's disease and related disorders. *International Psychogeriatrics, 12*, 49–65.

Hart, B. D., & Wells, D. (1997). The effects of language used by caregivers on agitation in residents with dementia. *Clinical Nurse Specialist, 11*, 20–23.

Hastings, R. P., & Remington, B. (1994). Rules of engagement: Toward an analysis of staff responses to challenging behavior. *Research in Developmental Disabilities, 15*, 279–298.

Heard, K., & Watson, T. S. (1999). Reducing wandering by persons with dementia using differential reinforcement. *Journal of Applied Behavior Analysis, 32*, 381–384.

Henry, L. M., & Horne, P. J. (2000). Partial remediation of speaker and listener behaviors in people with severe dementia. *Journal of Applied Behavior Analysis, 33*, 631–634.

Hoeffer, B., Rader, J., McKenzie, D., Lavelle, M., & Stewart, B. (1997). Reducing aggressive behavior during bathing cognitively impaired nursing home residents. *Journal of Gerontological Nursing, 23*, 16–23.

Hussian, R. A., & Davis, R. L. (1985). *Responsive care: Behavioral interventions with elderly people.* Champaign, IL: Research Press.

Iwata, B. A., Dorsey, M. F., Slifer, K. J., Bauman, K. E., & Richman, G. S. (1994). Toward a functional analysis of self-injury. *Journal of Applied Behavior Analysis, 27*, 197–209. (Reprinted from *Analysis and Intervention in Developmental Disabilities, 2*, 3–20, 1982).

Jacobson, N. S., Martell, C. R., & Dimidjian, S. (2001). Behavioral activation treatment for depression: Returning to contextual roots. *Clinical Psychology: Science and Practice, 8*, 255–270.

Lerman, D. C., & Iwata, B. A. (1993). Descriptive and experimental analyses of variables maintaining self-injurious behavior. *Journal of Applied Behavior Analysis, 26,* 293–319.

Lindsley, O. (1964). Geriatric behavioral prosthetics. In R. Kastenbaum (Ed.), *New thoughts on old age* (pp. 41–60). New York: Springer.

Lyketsos, C. G., Lopez, O., Jones, B., Fitzpatrick, A. L., Breitner, J., & DeKosky, S. (2002). Prevalence of neuropsychiatric symptoms in dementia and mild cognitive impairment: Results from the cardiovascular health study. *Journal of the American Medical Association, 288,* 1475–1483.

Mace, F. C., & Lalli, J. S. (1991). Linking descriptive and experimental analyses in the treatment of bizarre speech. *Journal of Applied Behavior Analysis, 24,* 553–562.

Martin, G., & Pear, J. (1999). *Behavior modification: What it is and how to do it (6ᵗʰ edition).* Upper Saddle River, NJ: Simon & Schuster.

Mihailidis, A., Barbenel, J. C., & Fernie, G. (2004). The efficacy of an intelligent cognitive orthosis to facilitate handwashing by persons with moderate to severe dementia. *Neuropsychological Rehabilitation, 14,* 135–171.

National Institute of Neurological Diseases and Stroke. (2005). *The dementias: Hope through research.* Retrieved November 1, 2005, from http://www.ninds.nih.gov/disorders/alzheimersdisease/detail

Nowend, K. L., Arizzi, M., Carlson, B. B., & Salamone, J. D. (2001). D1 or D2 antagonism in nucleus accumbens core or dorsomedial shell suppresses lever pressing for food but leads to compensatory increases in chow consumption. *Pharmacology, Biochemistry and Behavior, 69,* 373–382.

Pinquart, M., & Sorenson, S. (2003). Differences between caregivers and noncaregivers in psychological health and physical health: A meta-analysis. *Psychology and Aging, 18,* 250–267.

Profenno, L. A., & Tariot, P. N. (2004). Pharmacologic management of agitation in Alzheimer's disease. *Dementia and Geriatric Cognitive Disorders, 17,* 65–77.

Ory, M. G., Yee, J. L., Tennstedt, S. L., & Schulz, R. (2000). The extent and impact of dementia care: Unique challenges experienced by family caregivers. In R. Schulz (Ed.), *Handbook on dementia caregiving: Evidence-based interventions for family caregivers* (pp. 1–32). New York: Springer.

Ragneskog, H., Bråne, G., Karlsson, I., & Kihlgren, M. (1996). Influence of dinner music on food intake and symptoms common in dementia. *Scandinavian Journal of Caring Science, 10,* 11–17.

Rubenstein L. Z. (1997). Preventing falls in the nursing home. *Journal of the American Medical Association, 278,* 595–596.

Rubenstein, L. Z., Josephson, K. R., & Robbins, A. S. (1994). Falls in the nursing home. *Annals of Internal Medicine, 121,* 442–451.

Sasso, G. M., Reimers, T. M., Cooper, L. J., Wacker, D., Berg, W., Steege, M., et al. (1992). Use of descriptive and experimental analyses to identify the functional properties of aberrant behavior in school settings. *Journal of Applied Behavior Analysis, 25,* 809–821.

Schneider, L. S., Dagerman, K. S., & Insel, P. (2005). Risk of death with atypical antipsychotic drug treatment for dementia: Meta-analysis of randomized placebo-controlled trials. *Journal of the American Medical Association, 294,* 1934–1943.

Schnelle, J. F., Ouslander, J. G., & Cruise, P. A. (1997). Policy without technology: A barrier to changing nursing home care. *The Gerontologist, 37,* 527–532.

Schulz, R., Burgio, L., Burns, R., Eisdorfer, C., Gallagher-Thompson, D., Gitlin, L. N., et al. (2003). Resources for Enhancing Alzheimer's Caregiver Health (REACH): Overview, site-specific outcomes, and future directions. *The Gerontologist, 43,* 514–520.

Schulz, R., O'Brien, A. T., Bookwala, J., & Fleissner, K. (1995). Psychiatric and physical morbidity effects of dementia caregiving: Prevalence, correlates, and causes. *The Gerontologist, 35*, 771–791.

Sidman, M. (1989). *Coercion and its fallout.* Boston, MA: Authors Cooperative, Inc.

Sink, K. M., Holden, K. F., & Yaffe, K. (2005). Pharmacological treatment of neuropsychiatric symptoms of dementia: A review of the evidence. *Journal of the American Medical Association, 293*, 596–608.

Skinner, B. F., & Vaughn, M. E. (1983). *Enjoy old age: A practical guide.* New York: Norton.

Smith, A., Li, M., Becker, S., & Kapur, S. (2004). A model of antipsychotic action in conditioned avoidance: A computational approach. *Neuropsychopharmacology, 29*, 1040–1049.

Stoppe, G., & Staedt, J. (1999). Psychopharmacology of behavioral disorders in patients with dementia. *Gerontology & Geriatrics, 32*, 153–158.

Sulzer-Azaroff, B., & Mayer, G. R. (1977). *Applying behavior-analysis procedures with children and youth.* New York: Holt, Rinehart & Winston.

Teri, L., Gibbons, L. E., McCurry, S. M., Logsdon, R. G., Buchner, D. M., Barlow, W. E., et al. (2003). Exercise plus behavioral management in patients with Alzheimer Disease: A randomized controlled trial. *Journal of the American Medical Association, 290*, 2015–2022.

Teri, L., & Logsdon, R. G. (1991). Identifying pleasant activities for Alzheimer's disease patients: The pleasant events schedule—AD. *The Gerontologist, 31*, 124–127.

Teri, L., Logsdon, R. G., Uomoto, J. L., & McCurry, S. M. (1997). Behavioral treatment for depression in dementia patients: A controlled clinical trial. *Journals of Gerontology Series B: Psychological Sciences and Social Sciences, 52*, 159–166.

Teri, L., & Uomoto, J. L. (1991). Reducing excess disability in dementia patients: Training caregivers to manage patient depression. *Clinical Gerontologist, 10*, 49–63.

Thomas, D., Heitman, R., & Alexander, T. (1997). Effects of music on bathing cooperation for residents with dementia. *Journal of Music Therapy, 34*, 246–259.

United States Department of Health and Human Services. (2005). *Glossary: Definition of restraints.* Retrieved from http://www.medicare.gov/Glossary/ShowTerm.asp? Language=English&term=restraints

Van Houten, R. (1983). Punishment: From the animal laboratory to the applied setting. In S. Axelrod & J. Apsche (Eds.), *The effects of punishment on behavior* (pp. 13–42). New York: Academic Press.

Vitaliano, P. P., Russo, J., Carr, J. E., Maiuro, R. D., & Becker, J. (1985). The ways of coping checklist: Revision and psychometric properties. *Multivariate Behavioral Research, 20*, 3–26.

Wang, P. S., Schneeweiss, S., Avorn, J., Fischer, M. A., Mogun, H., Solomon, D. H., et al. (2005). Risk of death in elderly users of conventional vs. atypical antipsychotic medication. *New England Journal of Medicine, 353*, 2335–2341.

Warner, M. L. (2000). *The complete guide to Alzheimer's-proofing your home* (revised ed). West Lafayette, IN: Purdue University Press.

Zeiss, R. A., & Dickerman, H. R. (1989). PTSD 40 years later: Incidence and person-situation correlates in former POWs. *Journal of Clinical Psychology, 45*, 80–87.

12

BRAIN INJURY

MARK R. DIXON

AND

HOLLY L. BIHLER
Southern Illinois University

OVERVIEW OF BRAIN INJURY

Every 23 s one additional person receives a brain injury in the United States (Brain Injury Association of America, 2006, Traumatic Brain Injury). This estimate is derived from population-based studies in the United States which suggest that the incidence of brain injury ranges between 180 and 250 per 100,000 population per year (Bruns & Hauser, 2003). The terms *traumatic* and *acquired* brain injury are often used synonymously and may have similar effects on functioning and behavior. However, they are caused by different events. An acquired brain injury (ABI)

> [i]s an injury to the brain, which is not hereditary, congenital, degenerative, or induced by birth trauma. An acquired brain injury is an injury to the brain that has occurred after birth. An acquired brain injury commonly results in a change in neuronal activity, which effects the physical integrity, the metabolic activity, or the functional ability of the cell. An acquired brain injury may result in mild, moderate, or severe impairments in one or more areas, including cognition; speech-language communication; memory; attention and concentration; reasoning; abstract thinking; physical functions; psychosocial behavior; and information processing. (Brain Injury Association of America, 2005, Acquired Brain Damage Injury.)

A traumatic brain injury (TBI)

> [i]s an insult to the brain, not of a degenerative or congenital nature but caused by an external physical force that may produce a diminished or altered state of consciousness, which results in an impairment of cognitive abilities or physical functioning. It can also result in the disturbance of behavioral or emotional functioning. These impairments may be either temporary or permanent and cause

partial or total functional disability or psychosocial maladjustment. (Brain Injury
Association of America, 2005, Traumatic Brain Injury)

The primary forms of acquired brain injury result from a total lack of
oxygen (anoxic brain injury) and a reduction in the required amount of
oxygen (hypoxic brain injury). There are numerous causes of anoxic and
hypoxic brain injuries, including exposure to toxins, illegal drug use, carbon
monoxide poisoning, meningitis, certain diseases, and obstruction of the
airways that occurs as a result of choking, strangulation, near-drowning,
head or neck trauma, heart attack, aneurysm, stroke, or seizures. The
primary forms of traumatic brain injury occur when the brain moves within
the skull or the skull is broken and injures the brain. Both of these situations can occur as a result of a direct blow to the head. An injury to the
head can be the result of an automobile accident, physical violence, falling,
playing sports, or being shot with a firearm. In addition, to direct physical
contact, a traumatic injury can also occur as a result of a quick acceleration
and deceleration of the head. When this occurs, the brain moves from
where it was originally positioned and slams into one side of the skull and
then ricochets off and slams into the opposite side of the skull. As a result
of this rapid movement, nerve fibers are pulled apart, causing damage to
the brain tissue. This rapid acceleration and deceleration of the head can
occur during automobile accidents and during episodes of physical
violence.

Approximately 50% of traumatic injuries are the result of an accident
involving a motor vehicle, bicycles, or pedestrians. For individuals under
the age of 75 years, the majority of traumatic injuries are caused by these
types of accidents. However, falls cause the majority of traumatic injuries
among individuals 75 years and older. An estimated 20% of traumatic
injuries are due to violence, such as physical altercations, firearm assaults,
child abuse, and shaken baby syndrome. Less than 5% of traumatic injuries
are caused by sports injuries. Alcohol consumption also appears to be
involved in approximately half of traumatic injury incidents (National
Institute of Neurological Disorders and Stroke, 2005). Specific groups have
been identified as being at high risk for traumatic injuries. Certain age
groups including the very young, adolescents and young adults, and the
elderly are at increased risk. Other groups at risk include males and individuals with lower socioeconomic status. In fact, males are twice as likely
as females to suffer from a TBI than females (Mayo Clinic, 2004). Individuals with an existing brain injury are also much more likely to acquire
an additional brain injury compared to individuals with no preexisting
brain injury.

When a clinician operationally defines an individual's traumatic injury,
it is also common to classify whether the injury is the result of an open or
closed head injury. The terms *open* and *closed head injury* refer to the

condition of the skull at the time of the injury. When the skull is fractured or forced out of place, the individual has sustained an open head injury. Open head injuries allow the brain room to swell, preventing excess pressure on the brain tissues. However, it is also possible that fractured parts of the skull could contact and damage the brain. In addition, an open head injury can leave the brain vulnerable to further mechanical injury and infection.

The distinction between acquired and traumatic is based solely on cause of injury, and not the person's neurological, behavioral, or psychological repertoire postinjury. As a result, a functional analysis of brain injury necessitates an investigation of that person's repertoire as it exists postinjury, rather than emphasis being placed on the form of the injury. Furthermore, in recent years, the terms *acquired* and *traumatic* are used interchangeably by some (New York Times Company, 2006), while others in the area of brain injury are classifying all brain injuries as *acquired,* with *traumatic* being a type of an acquired injury (Wikimedia, 2006). While the debate continues, a functional analysis of brain injury is long overdue.

CONCEPTUALIZING THE FUNCTIONAL ANALYSIS

When attempting to gain a greater understanding of the functional relations involved in the repertoire of the person suffering from a brain injury, the clinician must realize that the brain injury itself is not assessed for functional relations. Rather, it is the behavior which results from the brain injury and its relationship to the environment that is functionally assessed. Because brain injury results in many varying forms of behavior, which may be maladaptive, in excess, or deficit, it is critical to conduct individual functional analyses.

Unlike some psychological disorders, brain injury is now largely a nonmodifiable state of the person: No amount of therapy or treatment can remove it from that person's body. Thus, it may be helpful to conceptualize the brain injury as a setting event which significantly alters the typical or at least the prior response-consequence relationships that previously existed for that person preinjury (Dixon, 2002). Regardless of cause, brain injury commonly results in changes at three levels of analysis: neurological, physical, and psychological. These three levels of analysis interact in various ways and produce various behavioral outcomes necessitating attention and modification to facilitate the rehabilitation process. For example, at the neurological level, the brain may be altered to modify an individual's reactions to various stimuli, such as noise and tactile stimulation. Changes in the ability to learn new skills may be seen, and postinjury epilepsy is very common (Mazzini et al., 2003). At the physical level, the individual's body may be impacted in ways that minimize or limit motor responses, produce tremors, complicate vision, eliminate or disrupt speech, and hinder language recognition. At the psychological level, individuals with brain injury are often diagnosed with personality disorders (Hibbard, Uysal, Kepler, Bogdany, & Silver, 1998), post-traumatic stress disorder, substance abuse, major depression, anxiety (Hibbard et al., 2000), and trauma-related conditions (National Institute of Neurological Disorders and Stroke, 2006), and behavior disorders (Max & Dunisch, 1997). Thus, the clinician working with an individual with brain injury will find this to be a very heterogeneous disorder which calls for expertise in a wide range of areas. However, when the primary goal of treatment remains to identify the functional relations of the behavior of interest, a clinician involved with treating an individual with brain injury will find consistency in the conceptual logic behind conducting functional analyses.

FUNCTIONAL ASSESSMENT AND ANALYSIS

Functional assessment and analysis following brain injury should be approached the same way a clinician would attempt to understand function of any other psychological disorder. Efforts should be made to minimize hypothesizing, speculation, and construction of inferences of the causes of behavior. Instead, the clinician should strive for objective definitions of the targeted behavior and look to external, modifiable events in the environment with large effects on the behaviors of interest which may be responsible for their emission. Although client-specific history, injury location, and the interactions between neurological, physical, and psychological deficits may play a role in the manifestation of a challenging behavior for the person with brain injury, these variables are not modifiable. Therefore, the clinician should look to the external environment as much as possible when identifying function. Too much emphasis on speculative internal mechanisms that may underlie the behavior will not be useful when attempting to produce behavior change as readily as if external agents are of central focus.

When the clinician seeks to identify environmental stimuli that may be functionally related to the behavior of interest, that clinician should examine to what degree the following functional relationships exist. Questions should be asked about the possible contingencies maintaining the behaviors of interest. These may include positive and negative social and nonsocial functions. First, is the behavior positively reinforced by access to attention, other social reinforcement, or tangible items? For example, does the individual become aggressive with staff during evenings when the lack of programmed activities or therapy sessions associated with attention-deprivation are scheduled in the postinjury treatment facility? If this appears to be the case, perhaps the behavior is maintained by the social reinforcement of the client with therapy staff, and not necessarily a static by-product of the injury itself. Does the behavior occur to gain access to other items or places? In other words, does the behavior seem to have a tangible component maintaining it? In the case of the person with brain injury, the clinician may discover that the behavior occurs in response to staff providing the person with snacks, providing a favorite picture, or playing a "calming" musical CD. Second, is the behavior maintained by negative reinforcement? Does the behavior function to escape or remove the client from demands placed on him/her? If this appears correct, then the clinician may learn that the client with brain injury is aggressive toward staff only when placed in physical therapy contexts that require increasing physical exertion and following complex instructions and makes the therapy staff stop such demands. The termination of therapy thereby inadvertently negatively reinforcing aggressive behavior. Third, is the behavior maintained by nonsocial consequences? Does the behavior have an automatic or sensory function? Here, a clinician may discover that the individual with brain injury is physiologically aroused, screaming and pounding his/her fists on the walls. The individual may have lost various forms of sensory perception, resulting in less intense forms of stimulation no longer being stimuli. These functions are often considered independently, and perhaps only one may maintain the targeted behavior of interest. However, it is possible that more than one function could maintain the behavior. In such a case, the clinician will need to identify the various functional relations and design treatment strategies to address each of these functions independently.

The sometimes infrequent emission of the targeted behaviors exhibited by persons with brain injury results in very difficult means by which the clinician can attempt to understand functional relations. Persons with brain injury, as opposed to other types of psychological disorders, may emit very dangerous or risky behavior at low rates (i.e., threatening suicide, making unwanted sexual advances, or walking out in traffic). Such behaviors may be very destructive, even with just one instance. Therefore, the clinician needs to quickly and accurately gain an understanding of the antecedents, contextual stimuli, and consequences which surround the problematic behavior. Direct observation of the behavior may or may not be possible; thus, the clinician often needs to use less objective measures in searching for function.

CRITICAL PREASSESSMENT INFORMATION

Any condition a clinician may treat requires the gathering of background information. In addition to the demographic nature of the individual and the history and severity of the behavior of interest, a great deal of preassessment information should be gathered. First, a detailed understanding of the accident which caused the brain injury should be summarized. Defining the individual's injury simply as a traumatic injury which destroyed portions of the frontal lobe of the brain is not enough. Instead, the same injury could be defined as being caused by a 9 mm gunshot wound to the head which took place during an attempted rape of that person by an unknown Asian male who broke into her house. While the outcome of the injury in both instances is the same, damage to the frontal lobe, the events surrounding the injury may shed additional light on behavioral and psychological conditions affecting this individual (e.g., a fear of Asian men, not wanting to be left alone at night).

Second, a full physical evaluation should be conducted which can indicate the current physical limitations of the individual, with emphasis on those caused by the injury. This evaluation should include sensory functioning, cognitive functioning, and motor abilities. Physical and occupational therapy tasks should be included to gain a comprehensive assessment of the changes in behavior following brain damage. The clinician will need to coordinate with a large number of specialists to achieve the necessary information. However, the time spent initially on such an evaluation will most likely be saved later when attempting to identify the functional variables surrounding the behavior of interest.

Third, an extensive neurological exam should be conducted to gain an understanding of internal physiological changes that have taken place within the person. Such changes cannot be seen by even the most skillful clinician, and rather than deduce hypotheses regarding the cause of behavior is due to neurological deficits, it would be best to understand what such deficits are and how those specific deficits impact observed functioning. The neurological exam should gather information such as length of coma, memory abilities, analytical abilities, verbal abilities, and spatial perceptual abilities. Furthermore, the neurological exam is most complete when coupled with brain imaging procedures, which can directly examine the various cranial nerves individually.

Fourth, the preassessment should include a listing of the client's current medications so that the clinician can understand how these medications affect behavior. Most clinicians understand that the interaction between medications and behavior is important. This is also true when clinicians treat individuals with brain injury. In fact, the types and amounts of medications consumed by persons with brain injury appear to be more diverse (Ashley & Krych, 1995, p. 69) and perhaps in even higher doses than other clinical populations. For example, an individual with a brain injury may take a painkiller for the injury from his motorcycle crash, a sleep inducer because individuals with brain injuries often suffer from insomnia, an antidepressant due to a mood disorder, and an antiseizure medication for newly developed epilepsy. None of these medications are primarily pre-scribed for the behavior problem of interest. However, all these medica-tions may impact the individual's behavior in various ways. Thus, a good deal of education on the interactions of drugs and behavior will be critical for the clinician to possess.

INDIRECT ASSESSMENT

Questionnaires

Many questionnaires are designed to help assist clinicians in better understanding the nature of behavior for their specific clients. These ques-tionnaires examine outcomes following injury (Jennett & Bond, 1975), the severity of disability (Rappaport, Hall, Hopkins, Belleza, & Cope, 1982), and functional independence (Granger et al., 1986). Persons with brain injury may also be assessed with additional questionnaires that were designed for various other clinical populations or to assess various person-ality traits or states. For example, persons with brain injury have been observed to show greater signs of depression, as measured by the Beck's Depression Inventory, initially following injury, which may improve over time (Rowland, Lam, & Leahy, 2005). Additionally, depression as mea-sured by the *Minnesota Multiphasic Personality Inventory-II* appears at a higher proportion for the brain injury population compared to non-brain-injured persons and may be in part due to somatic difficulties (Bush, Novack, Schneider, & Madan, 2004). Comorbidity of substance abuse within the brain injury population (i.e., drugs or alcohol abuse) is very common (Ashman, Schwartz, Cantor, Hibbard, & Gordon, 2004; Corri-gan, Bogner, Lamb-Hart, Heinemann, & Moore, 2005) and clinicians this as much as possible. Impulsivity, often considered a primary characteristic of brain injury, may also be assessed using hypothetical choice tasks found in various delay discounting procedures (Dixon et al., 2005).

If the clinician's primary objective is to assess function of the behavior, most questionnaires will need to be supplemented with additional assess-ments, as they may do more to establish severity or history, and less to shed light on behavioral function. Several questionnaires to identify behavioral functions have been developed for people with mental retardation. Several have also been used with people with brain injury. The Motivation Assess-ment Scale (MAS; Durand & Crimmins, 1988) is a 16-item behavior rating sale scored on a 7-point scale, anchored by the options "Never" and "Always." The MAS identifies function in four categories of behavior: attention, tangible, escape, and sensory. The MAS includes questions such as "Would the behavior occur continually, over and over, if the person was left alone for long periods of time, or does the behavior occur following a request to perform a difficult task?" Each question is designed to address a potential behavior function. In the preceding examples, the first type of question would, if answered as "never," perhaps rule out a potential sensory function that the behavior might serve, and the latter type of question would, if answered similarly, potentially rule out an escaping from demand function of the behavior. The therapist, the clienthim/herself, or by other people who are well acquainted with the behavior in question may complete the MAS. The MAS is one of the most frequently studied and used indirect functional assessments, despite its relatively poor psychometric properties (Bihm, Kienlen, Ness, & Poindexter, 1991; Sigafoos, Kerr, & Roberts, 1994; Sturmey, 1994).

The Questions About Behaviorial Function (QABF; Matson & Vollmer, 1995) is also a questionnaire initially designed for people with mental retardation; it evaluates the maintaining effects of attention, escape, tangible, physical discomfort, and nonsocial reinforcement. Questions and response options are similar to that of the MAS, but the QABF items are scored on a 5-point scale. The QABF has successfully identified the function of self-injurious behavior, aggression, and stereotypic movements in 84% of subjects in a study of nearly 400 people with mental retardation (Matson, Bamburg, Cherry, & Paclawaskyj, 1999). Investigations of the QABF's psychometric properties have revealed that it is both a reliable and valid tool for functional assessment (Matson & Vollmer, 1995; Paclawskyj, Matson, Rush, Smalls, & Vollmer, 2000).

Interviews

A number of structured or guided interview methods exist and can be helpful in gaining a greater functional understanding of the behavior problem of interest. Behavioral interviews focus on understanding the antecedents, establishing operations, setting events, and history of the problem, as well as the consequences which the behavior may serve. To perform this type of interview, the clinician must guide questions to the client along the lines of objective definitions and environmental events (Miltenberger & Fuqua, 1985). Questions which commonly occur in traditional clinical interviews can be rephrased to elicit functional information. For example, instead of asking the client *"How do you feel about losing your ability to use your left hand?"* the clinician should seek to discover functional relationships via questioning such as *"What seems to happen immediately after you have failed to pick up an item you used to be able to do so and become aggressive toward your caregivers? Do they then tend to just give you items to calm you down?"* Questions may also assess the magnitude of the disorder such as *"How long has your aggression toward others who try to assist you occurred?"* *"Has it increased in frequency?"* and so on. Again, the focus of these questions during the interview should be along the lines of objective measures by which to evaluate function.

Regardless of the cause or history surrounding the brain injury, the interview should take a general format of a brief introduction, assessment, and a brief closing. During the introduction, the clinician should recapitulate the problem behavior(s) which brought the client into treatment. The clinician should summarize all of the relevant information about the client that he/she currently has and ask the client to provide missing pieces to that initial information. Detailed questioning about functional relations should be left until initial rapport has been developed between parties, and reassurance should be given that the clinician is there to help. During the assessment part of the interview, the clinician should seek out means by which the client can describe the antecedents and consequences of the client's behavior of concern. At this time the therapist should explore the problem behavior's severity, intensity, history, and triggers. The therapist should also carefully explore the conditions under which the problem behavior does not occur, such as the places, people, and events that result in the problem not occurring. Such information will be useful when attempting to discover methods and strategies that can be used for treatment. During the closing of the behavioral interview, the clinician should summarize the information gathered during the interview, including initial analyses of the functional relationships that exist for the problem behavior. The clinician should stress the importance of seeing the problem as not an unmodifiable, internal flaw of personality of the person. Rather, it should be seen as an example of how situations in that person's life can arise which lend him/herself to the problem behavior occurring. The clinician should be empathic with the patient and inform him/her that there are others with similar disorders and, while difficult, behavior change can occur, and this will be the focus of subsequent therapy sessions.

A number of guided behavioral interviews have been developed for the functional assessment of behavior more generally and, while not specific for assessment in brain injury, hold a great deal of promise for understanding the functional relations. They include the *Functional Analysis Interview Form* (FAIF; O'Neil, Horner, Albin, Storey, & Sprague, 1990) and the *Functional Analysis Checklist* (FAC; Van Houten & Rolider, 1991). The FAIF is a structured interview designed to identify possible function and other relevant variables associated with the behavior. The FAIF provides an extensive amount of information and can aid the clinician to a great degree. The FAC is a 15-item interview that focuses on the physical environment, adjunctive behaviors, transitions, escape from demands, and positive reinforcement. These two behavioral interviews might be used together with brain injury-specific assessments.

DIRECT ASSESSMENT

Direct Observation

Direct observation provides a number of advantages over questionnaires and interviews. First, direct observation allows the clinician to objectively assess the behavior, the events which preceded the emission of the behavior, and the consequences which follow it. Questionnaires may not be specific enough to target specific incidents of problem behavior emission, and interviews may fail to uncover critical variables that the client did not report. Second, the indirect assessments require the client to remember what occurred and when it occurred, and hypothesize causes for why it occurred. Such issues are prone to distortion and bias. Direct observation removes such artifacts and allows the clinician to independently assess the critical features of the behavioral episode.

Direct observation may take one of a variety of forms. The clinician may physically travel with the client to observe the performance of the behavior in question. The client may be videotaped at high-risk times, so that behavioral functions can be deduced. Self-observation of the behavior may occur whereby the client him/herself records the behavior of interest and the variables which surround the episode of emission.

Potential limitations to direct observation include client and staff reactivity and practical problems in observing low-rate behaviors. Before ruling out direct observations, clinicians should attempt to identify the specific features of the behavior of concern for their individual client. It may be the case that noncompliance is frequent enough that a therapist may be able to capture multiple emissions on a video camera in the therapy room or that sexual advances are made to one particular staff person, even though at low rates (Sturmey, 1996). Thus, it may be difficult but not impossible to assess such behaviors in their natural environment via direct observation. Direct observation alone cannot yield causality regarding the functional relations for a given behavior. The clinician can only make hypotheses about what antecedents and consequences appear to be sustaining the behavior (Mace, 1994; Sprague & Horner, 1995). For example, upon witnessing a client fondle the breasts of a staff member over the course of a period of time, the clinician notes that the client appeared to smile excitedly not upon completing the act, but when the staff person reacted with a substantial verbal reprimand. That clinician can deduce that perhaps social attention is the function of the breast grasping and not some sexual arousal. Yet, without systematic manipulation of environmental variables—the essence of an experimental analysis—only unconfirmed hypotheses can be made about behavioral function. Given the severity of some forms of behavior exhibited by persons with brain injury, direct observation may be the closest approximation to identifying behavioral function that a clinician can attain. Such an approximation can be very useful when treating an individual but should be taken only as a tentative function until behavior change is clearly displayed by the client.

The antecedent-behavior-consequence assessment is an easy-to-administer direct observation method in which each episode of problem behavior, the environmental event(s) that precedes the behavior, and the environmental event(s) that follows the behavior are recorded and subsequently analyzed (Bijou, Peterson, & Ault, 1968). This type of assessment should be conducted before moving onto more time-consuming assessments. However, in order for this type of assessment to yield accurate results, it is important for individuals to consistently record instances of the behavior along with both the antecedents and consequences. An inconsistent or inaccurate assessment may result in incorrect identification of function and, subsequently, an ineffective treatment.

An additional method of direct observation that may hold promise for the assessment of function in the brain injury population is a scatterplot analysis (Touchette, MacDonald, & Langer, 1985). A scatterplot analysis consists of a grid with the time of day displayed along the y-axis and days along the x-axis. Analysis of scatterplot data identifies specific times or activities when behavior problems do and do not occur. For example, if an individual in a residential facility is primarily aggressive prior to meals, simply providing regular snacks or altering the meal schedule may significantly reduce the frequency of aggressive behavior.

Experimental Analyses

In an experimental functional analysis, stimulus conditions are systematically manipulated in order to assess the functional relationship between environmental events and the target behaviors (Iwata, Dorsey, Slifer, Bauman, & Richman, 1982/1994). The utility of the experimental manipulation of antecedents and consequences to reveal function has been well documented with several populations (Hanley, Iwata, & McCord, 2003; Furtz et al., 2003), including individuals with brain injury (Dixon et al., 2004; Fyffe, Khang, Fittro, & Russel, 2004).

Contingency-Based Analyses and Treatments

Fyffe et al. (2004) performed an experimental functional analysis on the inappropriate sexual behavior of a 9-year-old boy diagnosed with a TBI. The target behavior was defined as touching or attempting to touch others in the area of the groin, buttocks, or breasts. During the demand condition, compliance resulted in praise, and inappropriate sexual behavior resulted in a 30-s break from demands. During the social attention condition, sexual behavior resulted in a brief reprimand. In the toy play condition the participant was given access to preferred items and received noncontingent attention every 30s. The child emitted highest rates of sexual behavior when that behavior was followed by verbal reprimands delivered by the experiment; thus, the target behavior appeared to be positively reinforced by attention. Therefore, a treatment consisting of teaching an alternate response reinforced by attention and attention extinction was used. Functional communication training consisted of teaching the child to deliver a small card, resulting in a brief social interaction.

Attention extinction consisted of ignoring the sexual behavior. This intervention resulted in a 94% reduction in inappropriate sexual behavior over a brief period of time.

Dixon et al. (2004) conducted an experimental analysis of 4 adults with brain injury who regularly engaged in inappropriate verbal behavior including aggressive, suicidal, sexually inappropriate, and profane verbalizations. The individual experimental analyses revealed that the behavior of 2 of the 4 clients was maintained by social attention. The remaining 2 clients' behavior was maintained by escape from demands placed on them by the experimenter.

Two treatments were constructed which consisted of providing differential reinforcement of alternative behaviors which served the same functions as the target behaviors. Thus, for the two clients whose behavior was maintained by attention, social attention was provided contingent on the absence of the target behavior. For the two clients whose behavior was maintained by social attention, treatment consisted of a 2–3-s verbal statement from the experimenter contingent on each appropriate verbal utterance of the participant. For the two clients whose behavior was negatively reinforced maintained by escape from demands, escape was delivered contingent upon appropriate verbalizations and no consequences for inappropriate verbalizations. Treatment conditions for the two clients whose behavior was maintained by escape from demands varied. At the beginning of the session, one client was informed, "If you need to take a break, please ask appropriately." During the sessions, each appropriate verbal utterance resulted in a 30-s termination of prompts to engage in physical therapy exercises. Inappropriate verbal utterances were placed on extinction. As for the other client, every 15 s the experimenter asked questions about difficulties he was having in rehabilitation. Appropriate verbal utterances resulted in a 30-s termination of questioning and a statement that was relevant to the statement made by the participant. While the function of the problematic behaviors differed across these individual clients, the intervention resulted in decreases in inappropriate utterances for all four of the participants. In some instances appropriate verbalizations also increased.

Antecedent-Based Assessments and Treatments

Antecedent-based treatments may also be selected when, for practical and ethical reasons, the increased frequency and/or intensity often associated with extinction procedures may cause harm to the participant or others. Extinction procedures are often implicit in many treatments, such as the differential reinforcement treatments described earlier, and some antecedent-based interventions are relatively easy to implement.

Pace, Ivanic, and Jefferson (1994) demonstrated the utility of antecedent manipulation as a treatment for behavior maintained by escape. The participant, a man with a brain injury, displayed chronic use of obscene comments. During the experimental analysis, the participant was exposed to demand, social disapproval, and conversation conditions. During the demand condition, the experimenter presented simple requests every 15 s. Compliance resulted in praise, while engagement in the problem behavior resulted in a statement implying escape from that demand (e.g., "Okay, you don't have to do it"). During the conversation, or control condition, obscenity was ignored, leisure materials were accessible, and the experimenter initiated noncontingent social conversation every 15 s.

The participant's behavior was maintained by escape from demands. The treatment consisted of a stimulus fading procedure in which demands were initially removed, followed by gradual fading in of demands during 15-min sessions. During this demand fading condition, the experimenter engaged the participant in continual noncontingent social conversation, occasionally interrupting only to present demands. Instead of demands being provided every 15 s (as in baseline), only three demands were presented in the first session. Demands were gradually increased over each session until reaching the assessment frequency of demand presentation. Upon introduction of the demand fading, obscenity immediately decreased to zero. As the number of demands was gradually increased, the obscenity remained at near-zero levels. Furthermore, obscenity immediately increased during reversal conditions in which demands were rapidly increased to baseline rates and noncontingent attention was absent.

Multi-assessment Approaches

Another method that has led to effective functional interventions involves combining many different methods of assessment. Gardner, Bird, Maguire, Carreiro, and Abenaim (2003) employed several methods to assess the function of problem behaviors of two adolescent males with ABI. The authors used descriptive assessments instead of experimental analyses because the challenging behavior of the clients put themselves and others around them at significant risk.

The initial assessment began with direct observation on the clients; interviews with staff members, parents, and participants; and an examination of existing records. The functional assessment continued with formal structured interviews using the FAI with the family members. Observation tools, including ABC data, scatterplot diagrams, setting event checklists, and the Detailed Behavior Report (Groden, 1989), were also used.

The information gathered from these descriptive and indirect methods allowed for the formation of hypotheses regarding the function of each client's behaviors. These hypotheses led to the design of structured and controlled experimental conditions to experimentally test the hypothesized functions. The combination of assessment methods resulted in the identification of both immediate and distant potential evoking antecedents and maintaining consequences. For one client the observed problem behaviors included aggression, property destruction, pica behavior, fecal smearing, and the insertion of objects in his penis and rectum. Descriptive assessments identified reading activities, academic requests, new staff members, medication reductions, program changes, and family conflicts as potential immediate and distant antecedent variables contributing to the target behaviors. Analogue testing identified multiple functions of the client's problem behaviors, including attention from staff, escape from demands, and expression of boredom. The other client displayed frequent episodes of aggression and property destruction. Results of the descriptive assessments identified several immediate and distant variables that contributed to the target behaviors including family conflicts, domestic tasks, writing activities, academic requests, and future changes in the life span. Analogue testing suggested that this client's behaviors served as a way to escape demands, escape from staff or peers, or express disappointment about canceled visits with his parents.

Because the specific antecedents identified during descriptive assessments often led to increased agitation and frustration, which in turn often led to the targeted behaviors, the intervention consisted of both the management of antecedents and functional communication training specifically tailored to each of the participants based on the functional assessments. Both contingent (DRA and DRO) and noncontingent reinforcement procedures were also used because one of the initial goals of the intervention was to create an extremely rewarding environment for the participants. Extinction procedures were also implemented in order to ensure that problem behaviors were no longer resulting in the hypothesized desired consequences. For both participants, implementation of the treatment package initially resulted in a sudden spike in frequency of problem behaviors, followed shortly by a sharp reduction in the targeted problem behaviors. Over time, both participants' problem behaviors were at or near zero and remained at these low levels for the remainder of the intervention.

CASE STUDY

At the age of 18 years, Mike took a job as an auto mechanic immediately out of high school. This job provided Mike with a good amount of money and insurance for him and his new wife. Two months after starting his new job, Mike was in an accident. The hydraulic lift gave way, resulting in his head being crushed. After initial medical treatment, Mike was transferred to an acute rehabilitation facility, where he received residential care by skilled nursing staff, physical and occupational therapy, and vocational rehabilitation.

Upon entry into this facility, Mike did not speak much to other residents or his staff, had difficulty sleeping at night, woke up screaming, and eventually fell back asleep. Mike was not interested in physical therapy and often could be seen hiding his left hand behind his back. When asked to show his hand, Mike did so reluctantly and had extreme difficulty extending that hand to an open grasp position.

Beyond his physical body damage, Mike's head injury impacted several areas of the brain regulating speech production and recognition, motor movement of his legs, and memory skills. Mike also tended to often forget where he was and kept looking for his toolbox so he could fix cars in the parking lot of the facility. When told by staff that there was no toolbox, and he was not to attempt to fix any cars, Mike became aggressive and often attacked staff physically. In the most recent months, Mike had also become very sexually inappropriate with the female staff, in his verbal comments toward them as well as regularly grasping at their breasts and genitals.

Upon Mike's referral to his new behavioral clinician, the staff noted to Mike that if his behavior did not improve substantially, he would be removed to a psychiatric hospital with locked facilities and no specialized brain injury rehabilitation. Having a fair deal of experience with brain injury, Mike's clinician immediately ordered an extensive physical examination, the physical therapy and occupational therapy session records, a list of Mike's medications, and a neurological examination consisting of MRI and PET scans along with various cognitive tests. The clinician also gathered all staff incident reports of Mike's aggressive and sexual behavior, which the clinician hoped would aid in identifying relationships between behavior and the environmental events.

Before bringing Mike into the therapy session, the clinician met with the caregivers responsible for the majority of Mike's care. During this meeting, many insights were provided regarding Mike's problem behaviors. The following is a partial transcript of that interview:

Clinician: Thank you for rearranging your schedules to meet regarding Mike today. I know that he has been somewhat difficult to manage over the past few months, and it looks like he may be considered for an alternate placement if his problem behaviors cannot be brought under control. Before we begin talking about Mike, I want to define and provide examples of the target behavior of his that everyone appears most concerned with. The first behavior is his aggression towards others. This behavior may take the form of vocal outbursts towards others, yelling, screaming, and swearing. His aggression may also be in the form of physical violence such as hitting, slapping, or grabbing. Does this seem correct?

Staff members: Yes, this sounds correct. We agree.

Clinician: The second behavior we are concerned about is his sexual behavior that is rather inappropriate. This behavior takes the form of either sexual comments to female staff, along with the grasping of various body parts. Is this correct?

Staff members: Yes, this is correct.

Clinician: Okay, great. Now that we are in agreement about the two types of behavior that Mike is having difficulties with, I would like you to fill out a couple of brief forms regarding Mike's behaviors. As you will see, these forms have you rate your opinions about when the behavior occurs and under what conditions. For example, you might see a question that asks, "Does the behavior occur when no one else is around?" If you strongly agree with this statement, you would rate it a "5," and if you strongly disagree with this statement, you would rate it a "1." After we all get a chance to fill these forms out, I would like to take a 30-min break, assess the information, and return to further discuss Mike's behavior.

Staff completed the MAS and the QABF. It was important for the clinician to first meet with the staff prior to their filling out these questionnaires about the possible function of the behaviors because he wanted to make sure everyone was defining Mike's behaviors the same, as well as to make sure that each staff member knew to fill out the forms correctly, and only one form per problem behavior. Upon analysis of the data, it appeared to the clinician that Mike seemed to engage in aggressive behavior when the staff had made requests of a physical nature, such as using his injured hand, and his physical therapist rated aggression as always happening on one of the assessments more so than other staff members. In contrast, sexual inappropriate behavior tended to be emitted with staff members not really differentiating a possible functional relationship. Some staff rated the behavior occurring under demanding conditions, whereas others rated it as occurring frequently when staff provided Mike with a great deal of attention. Upon the conclusion of his initial analysis, the clinician determined that further interviewing of the staff was needed.

Clinician: From analysis of the data you provided me, along with the incident reports that were on file, I tend to think that Mike's aggression is often a response to situations which he would like to remove himself from. For example, as you can see in this latest incident report, Mike attacked Joan when she asked him to put on his shoes and tie them. Now remember, putting on shoes does not seem to be a big deal, but for Mike, his left hand was severely injured in his accident. Plus the results of his neurological exam suggest that he suffered damage to the cranial nerve responsible for fine motor movement. Together these disabilities may make the task of putting on his shoes more difficult than many of us might think.

In addition, we can see here from a summary of the questionnaires you completed, that all of you staff see the problem behavior of aggression occurring more frequently when you are asking Mike to do things. This is a great start to getting a handle on his problem behavior. What we do seem to have a problem with, however, is his sexual behavior. These data do not seem as clear. Thus, before we begin identifying a treatment for Mike, I want to interview Mike myself. Additionally in the meantime, I would like the staff that usually are the victims of his sexual advances to collect some observational data. Specially, I would like you to track the time, place, and the events surrounding the sexual behavior, what happens before, and what happens afterwards.

A week went by before the clinician was able to schedule an interview with Mike. This time period did allow for staff to record five instances of his sexual behavior and the environmental events which surrounded that behavior. Figure 12.1 displays the resulting data.

The clinician met with Mike and the following dialogue transpired.

Clinician: Mike, I have been examining your files, medical reports, and have met with your staff. You seem to be causing a lot of problems around here lately.

Mike: Yes, I guess you could say that. I really don't like it here. I really don't like it anywhere anymore. You suck too. I want to leave this place.

Clinician: That would be fine, Mike, but where would you go? Right now, the staff want to place you in a hospital where you will not get to go into the community anymore. Is that what you want?

Mike: No, I would rather go home. But, my wife won't look at me anymore. She has not even come here since I was hurt. She cannot stand the sight of me. I also got divorce papers in the mail my house staff said. I really can't read them anymore though. My brain is playing tricks on me.

Clinician: Well, that is certainly too bad. I am sorry to hear this, Mike. However, if you are going to get along here, you are going to need to try and work harder on obeying the rules and being nicer to staff.

The interview with Mike carried on for another hour. During that time, the clinician discovered that Mike was really depressed about his postinjury condition, that he had little hope of returning to work, and that his

Date and Time	What happened right before the behavior occurred?	What did the behavior look like?	What was your response to Mike when the behavior happened?
Friday 11/14 9pm	I asked Mike to do the dishes.	Mike pulled down his pants and touched my breast.	I told Mike not do to this, and that he better put his pants back on or he could not go to McDonald's tomorrow.
Saturday 11/15 noon	Mike was sitting at the table.	Mike rubbed his body against my butt.	I told Mike to stop it, and immediately walked away from him.
Monday 11/17 1pm	Mike was watching TV and a commercial came on.	Mike asked me to perform oral sex on him.	I ignored him and walked away.
Monday 11/17 1:15pm	I walked away from Mike (see above).	Mike was naked and chased me through thehouse.	I called my manager and we both restrained him for 10 min.
Tuesday 11/18 5pm	Mike was asked to set the table for dinner.	Mike picked up a pillow, started kissing it, and said he wished it was me.	I told Mike to leave me alone and not do that anymore.

FIGURE 12.1 Directions: Each time Mike displays inappropriate sexual behavior towards a staff member, please note the following information. Please complete the form immediately after his behavior takes place.

heart was broken by his wife's divorce filing. Upon the conclusion of that interview, the data obtained by the staff and the questionnaires completed, the clinician determined that it may be possible that Mike's aggressive behavior occurred in response to failures at physical therapy because he felt that he would never again work on cars. Additionally, the clinician thought that Mike's sexual behavior may be in response to his wife's wishes to leave him. Instead of it being the case that he engages in the behavior to upset the staff, the clinician concluded it may be possible that Mike engages in this behavior because it makes him feel like a female may again want to be with him in a sexual way, the way his wife was preinjury.

As a result of these conclusions, the clinician called for a final meeting to deliver a treatment protocol to the staff. Specifically, staff was to allow Mike small breaks during physical therapy tasks upon his requesting to need one. They were also to inform Mike that each task completed would make his hopes of getting back to work more possible. Mike also collected data himself on a daily basis for each task he completed independently. Staff was also told that when Mike made sexual comments for advances toward him, they were to respond as minimally as possible to terminate the situation. No big production of attention was to occur, but rather attention by the targeted female staff was to be delivered frequently to Mike when no sexual behavior occurred for periods of 6 hr. At those times, female staff was to tell Mike that he was doing really good and acting like a man. There was also a social skills program that was to start for Mike which included modules that taught appropriate and inappropriate social interactions with members of the opposite sex. Data were also collected following this meeting to assess the possible impact of treatment on the problem behavior. Figure 12.2 displays the initial baseline frequencies of Mike's emission of sexual comments, along with the frequencies observed during the implementation of the treatment plan.

In conclusion, while an experimental analysis was not conducted with Mike, the clinician utilized a number of different assessments in which possible functions of the behavior could be identified. Following the implementation of the multicomponent treatment plan, Mike's problematic behaviors were reduced dramatically, allowing him to remain in the residence. Mike also returned to work in a limited capacity.

This case study illustrates the complexity of problems that may occur for a person with brain injury. The interaction of neurological, physical, and psychological changes in the person all participated in observed behavior which necessitated a functional analysis. Through careful assessment and treatment recommendations, behavior of the person with brain injury can be altered in productive ways, leading to greater success in the rehabilitation process.

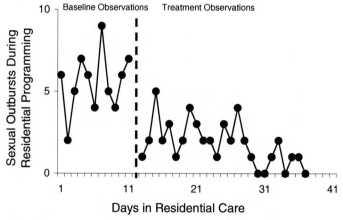

FIGURE 12.2 Frequency of sexual outbursts observed during pretreatment baseline and treatment sessions.

REFERENCES

Ashley, M. J., & Krych, D. K. (1995). *Traumatic brain injury rehabilitation.* New York: CRC Press.

Ashman, T., Schwartz, M., Cantor, J., Hibbard, M., & Gordon, W. (2004). Screening for substance abuse in individuals with traumatic brain injury. *Brain Injury, 18*, 191–202.

Bihm, E. M., Kienlen, T. L., Ness, M. E., & Poindexter, A. (1991). Factor structure of the Motivation Assessment Scale for persons with mental retardation. *Psychological Reports, 68*, 1235–1238.

Bijou, S. W., Peterson, R. F., & Ault, M. H. (1968). A method to integrate descriptive and field studies at the level of data and empirical concepts. *Journal of Applied Behavior Analysis, 1*, 175–191.

Brain Injury Association of America (n.d.). *Acquired brain injury.* Retrieved October 12, 2005, from http://www.biausa.org/Pages/types_of_brain_injury.html

Brain Injury Association of America (n.d.). *Traumatic brain injury.* Retrieved October 12, 2005, from http://www.biausa.org/Pages/types_of_brain_injury.html

Bruns, J., & Hauser, W. A. (2003). The epidemiology of traumatic brain injury: A review. *Epilepsia, 44*(suppl. 10), 2–10.

Bush, B. A., Novack, T. A., Schneider, J. J., & Madan, A. (2004). Depression following traumatic brain injury: The validity of the CES-D as a brief screening device. *Journal of Clinical Psychology in Medical Settings, 11*, 195–201.

Corrigan, J. D., Bogner, J., Lamb-Hart, G., Heinemann, A. W., & Moore, D. (2005). Increasing substance treatment compliance for persons with traumatic brain injury. *Psychology of Addictive Behaviors: Journal of the Society of Psychologists in Addictive Behaviors, 19*, 131–139.

Dixon, M. R. (2002). Setting events. In M. Hersen & W. Sledge (Eds.) *Encyclopedia of psychotherapy.* New York: Academic Press.

Dixon, M. R., Guercio, J., Falcomata, T., Horner, M. J., Root, S., Newell, C., et al. (2004). Exploring the utility of functional analysis methodology to assess and treat problematic verbal behavior in persons with acquired brain injury. *Behavioral Interventions, 19*, 91–102.

Dixon, M. R., Jacobs, E. A., Sanders, S., Guercio, J., Soldner, J., Parker-Singler, S., et al. (2005). Impulsivity, self-control, and delay discounting in persons with acquired brain injury. *Behavioral Interventions, 20*, 101–120.

Durand, V. M., & Crimmins, D. B. (1988). Identifying the variables maintaining self-injurious behavior. *Journal of Autism and Developmental Disorders, 18*, 99–117.

Kurtz, P. F., Chin, M. D., Huete, J. M., Tarbox, R. S. F., O'Connor, J. T., Paclawskyj, T. R., et al. (2003). Functional analysis and treatment of self-injurious behavior in young children: A summary of 30 cases. *Journal of Applied behavior Analysis, 36*, 205–219.

Fyffe, C. T., Kahng, S., Fittro, E., & Russell, D. (2004). Functional analysis and treatment of inappropriate sexual behavior. *Journal of Applied Behavior Analysis, 37*, 401–404.

Gardner, R. M., Bird, F. L., Maguire, H., Carreiro, R., & Abenaim, N. (2003). Intensive positive behavior support for adolescents with acquired brain injury. Long-term outcomes in community settings. *Journal of Head Trauma Rehabilitation, 18*, 52–74.

Granger, C. V., Hamilton, B. B., Keith, R. A., et al. (1986). Advances in functional assessment for medical rehabilitation. *Topics in Geriatric Rehabilitation, 1*, 59–74.

Groden, G. (1989). A guide for conducting a comprehensive behavioral analysis of a target behavior. *Journal of Behavior Therapy and Experimental Psychiatry, 20*, 163–169.

Hanley, G. P., Iwata, B. A., & McCord, B. E. (2003). Functional analysis of problem behavior: A review. *Journal of Applied Behavior Analysis, 36*, 147–185.

Hibbard, M. R., Bogdany, J., Uysal, S., Kepler, K., Silver, J. M., Gordon, W. A., et al. (2000). Axis II psychopathology in individuals with traumatic brain injury. *Brain Injury, 14*, 45–61.

Hibbard, M. R., Uysal, S., Kepler, K., Bogdany, J., & Silver, J. (1998). Axis I psychopathology in individuals with traumatic brain injury. *The Journal of Head Trauma Rehabilitation, 13*, 24–39.

Iwata, B. A., Dorsey, M. F., Slifer, K. J., Bauman, K. E., & Richman, G. S. (1994). Toward a functional analysis of self-injury. *Journal of Applied Behavior Analysis, 27*, 197–209. (Reprinted from *Analysis and Intervention in Developmental Disabilities, 2*, 3–20, 1982).

Jennett, W. B., & Bond, M. R. (1975). Assessment of outcome after severe brain damage. *Lancet, 1*, 480–484.

Mace, F. C. (1994). The significance and future of functional analysis methodologies. *Journal of Applied Behavior Analysis, 27*, 385–392.

Matson, J. L., Bamburg, J. W., Cherry, K. E., & Paclawskyj, T. R. (1999). A validity study on the Questions About Behavioral Functions (QABF): Predicting treatment success for self-injury, aggression, and stereotypies. *Research in Developmental Disabilities, 20*, 142–160.

Matson, J. L., & Vollmer, T. R. (1995). *User's guide: Questions About Behavioral Function (QABF)*. Baton Rouge, LA: Scientific Publishers.

Max, J. E., & Dunisch, D. (1997). Traumatic brain injury in a child psychiatry outpatient clinic: A controlled study. *Journal of the American Academy of Child and Adolescent Psychiatry, 36*, 404–411.

Mayo Clinic. (2004). *Traumatic brain injury.* Retrieved October 12, 2005, from http://www.mayoclinic.com/health/traumatic-brain-injury/DS00552/DSECTION=4&

Mazzini, L., Cossa, F. M., Angelino, E., Campini, R., Pastore, I., & Monaco, F. (2003). Posttraumatic epilepsy: Neuroradiologic and neuropsychological assessment of long-term outcome. *Epilepsia, 44*, 569–574.

Miltenberger, R. G., & Fuqua, R. W. (1985). Evaluation of a training manual for the acquisition of behavioral assessment interviewing skills. *Journal of Applied Behavior Analysis, 18*, 323–328.

National Institute of Neurological Disorders and Stroke. (2005). *Traumatic brain injury: Hope through research.* Retrieved September 5, 2005, from http://www.ninds.nih.gov/disorders/tbi/detail_tbi.htm#42763218

National Institute of Neurological Disorders and Stroke. (2006). *General trauma.* Retrieved January, 28, 2006, from http://www.ninds.nih.gov/disorders/tbi/detail_tbi.htm#56443218

New York Times Company. (2006). *Why men suffer more acquired brain injury.* Retrieved January 28, 2006, from http://menshealth.about.com/library/bltrauma.htm

O'Neill, R. E., Horner, R. H., Albin, R. W., Storey, K., & Sprague, J. R. (1990). *Functional analysis of problem behavior. A practical assessment guide.* Sycamore, IL: Sycamore.

Pace, G. M., Ivancic, M. T., & Jefferson, G. (1994). Stimulus fading as treatment for obscenity in a brain-injured adult. *Journal of Applied Behavior Analysis, 27*, 301–305.

Paclawskyj, T. R., Matson, J. L., Rush, K. S., Smalls, Y., & Vollmer, T. R. (2000). Questions About Behavioral Function (QABF): A behavioral checklist for functional assessment of aberrant behavior, *Research in Developmental Disabilities, 21*, 223–229.

Rappaport, M., Hall, K. M., Hopkins, K., Belleza, T., & Cope, D. N. (1982). Disability rating scale for severe head trauma: Coma to community. *Archives of Physical Medicine and Rehabilitation, 63*, 118–123.

Rowland S. M., Lam, C. S., & Leahy, B. (2005). Use of the Beck Depression Inventory-II (BDI-II) with persons with traumatic brain injury: Analysis of factorial structure. *Brain Injury, 19*, 77–83.

Sigafoos, J., Kerr, M., & Roberts, D. (1994). Inter-rater reliability of the Motivational Assessment Scale: Failure to replicate with aggressive behavior. *Research in Developmental Disabilities, 21,* 333–342.

Sprague, J. R., & Horner, R. H. (1995). Functional assessment and intervention in community sessions. *Mental Retardation and Developmental Disabilities Research Reviews, 1,* 89–93.

Sturmey, P. (1994). Assessing the functions of aberrant behaviors: A review of psychometric instruments. *Journal of Autism and Developmental Disorders, 24,* 293–304.

Sturmey, P. (1996). *Functional analysis in clinical psychology.* New York: Wiley.

Touchette, P. E., MacDonald, R. F., & Langer, S. N. (1985). A scatter-plot for identifying stimulus control of problem behavior. *Journal of Applied Behavior Analysis, 18,* 343–351.

Van Houten, R., & Rolider, A. (1991). Applied behavior analysis. In J. L. Matson & J. A. Mulick (Eds.), *Handbook of mental retardation* (2nd ed., pp. 569–585). New York: Pergamon.

Wikimedia. (2006). *Traumatic brain injury.* Retrieved January 28, 2006, from http://en.wikipedia.org/wiki/Brain_injury

13

A BEHAVIORAL APPROACH TO THE TREATMENT OF SUBSTANCE USE DISORDERS

STEPHEN T. HIGGINS,
SARAH H. HEIL,

AND

STACEY C. SIGMON

Departments of Psychiatry and Psychology
University of Vermont

This chapter describes a behavioral approach to the treatment of substance use disorders (SUDs) based on the principles of behavior analysis and behavioral pharmacology. Those principles support a conceptual framework in which substance use is considered a form of operant responding that is maintained by the primary reinforcing effects of the pharmacological actions of abused drugs in combination with social reinforcement derived from the drug-abusing lifestyle. There is extensive experimental evidence that abused substances can function as highly effective primary reinforcers. Cocaine, other psychomotor stimulants, ethanol, opioids, nicotine, and sedatives serve as reinforcers that are voluntarily self-administered by a wide variety of species (Griffiths, Bigelow, & Henningfield, 1980; Higgins, 1997). Moreover, through respondent and operant conditioning, environmental events (people, places, and things) that are paired with drug use reliably come to occasion drug seeking and use. Physical dependence is not necessary for abused drugs to support

Functional Analysis in
Clinical Treatment

ongoing and stable patterns of voluntary drug seeking and use in otherwise healthy laboratory animals or humans. Moreover, the effects of alterations in drug availability, drug dose, schedule of reinforcement, and other environmental manipulations on drug use are orderly and have functionally similar effects across different species and types of drug abuse (Griffiths et al., 1980; Higgins, 1997). This rich body of empirical evidence about the reliable and orderly reinforcing functions of abused drugs in many different species supports a theoretical position that reinforcement and other principles of learning are fundamental determinants of substance use and related disorders.

Within this conceptual framework, substance use can be thought of as a normal, learned behavior that falls along a frequency continuum ranging from patterns of little use and few problems to excessive use and many untoward effects, including death. The same processes and principles of learning appear to operate across the continuum. Individuals need not have any exceptional or pathological characteristics in order to develop SUDs. All healthy humans are assumed to possess the necessary neurobiological systems to experience drug-produced reinforcement and hence to develop patterns of repeated substance use, abuse, and dependence. Genetic or acquired characteristics such as family history of substance dependence, other psychiatric disorders, and poverty correlate with increases or decreases in the probability of developing SUDs, but are not deemed to be necessary for the emergence or maintenance of SUDs.

Treatment based on this conceptual framework seeks to weaken the behavioral control exerted by drug-produced reinforcement and the drug-abusing lifestyle, and to increase the control exerted by competing, healthier sources of nondrug reinforcement. That basic approach can be applied to any type of SUD, but in this chapter we illustrate how it was used in the development of an effective outpatient treatment for cocaine dependence. The treatment is called the Community Reinforcement plus Vouchers Approach (CRA + Vouchers). A therapist manual containing forms and other relevant details involved in implementing this evidence-based treatment can be downloaded free of cost (Budney & Higgins, 1998; http://www.drugabuse.gov/TXManuals/CRA/CRA1.html). In this chapter we provide an outline of functional assessment and analysis of SUDS-related behavior, description of the CRA + Vouchers treatment, a brief review of the evidence supporting its efficacy, and an illustration of its application in the form of a case study. Before turning to that task, however, we briefly characterize the prevalence of SUDs in the United States to provide a sense of the enormous public health problem that they represent and also review the criteria commonly used in diagnosing SUDs so that our use of the term is clear.

SUD DIAGNOSIS

Substance use diagnoses of abuse or dependence are typically established based on criteria stipulated in the *Diagnostic and Statistical Manual* (4th edition) of the American Psychiatric Association (1994). These criteria represent a cluster of behavioral and physiological signs and symptoms resulting from a pattern of repeated substance use. To receive a diagnosis of substance abuse, a person must satisfy one or more of the following four criteria within a 12-month period, and cannot currently or previously have satisfied a diagnosis of dependence on that same substance: (1) recurrent substance use resulting in failure to fulfill major role obligations; (2) recurrent substance use in situations in which it is physically hazardous; (3) recurrent substance-related legal problems; (4) continued substance use despite having persistent or recurrent social or interpersonal problems caused or exacerbated by the substance. To receive a diagnosis of substance dependence, the person must have exhibited three or more of the following criteria within a 12-month period: (1) tolerance (need for markedly increased amounts to achieve intoxication, markedly diminished effect with continued use of the same amount); (2) withdrawal (substance-specific syndrome due to cessation or reduction in substance use); (3) substance taken in larger amounts or over a longer time than was intended; (4) persistent desire or unsuccessful efforts to cut down or discontinue use; (5) great deal of time spent in activities to obtain or recover from use of the substance; (6) important activities given up or reduced because of substance use; (7) substance use continued despite knowledge of having a recurrent problem that is likely caused or exacerbated by substance use. Note that while tolerance and withdrawal are listed first among the seven criteria, neither is necessary for a dependence diagnosis. One simply must meet any three of the seven criteria listed within a 12-month period to satisfy the diagnosis.

Dependence diagnoses can be qualified as follows. If the client has not met any of the criteria for abuse or dependence for at least 1 month, the dependence can be considered to be in *remission*. The term *early remission* refers to remission of less than 12 months duration and *sustained remission* refers to remission of 12 and more months. The term *full remission* is used when none of the criteria for abuse or dependence have been met during the period of remission and *partial remission* if one or more criteria have been met. The remission term is not used if the period without any dependence signs occurs while the client is on an agonist therapy (e.g., methadone therapy for opiate dependence) or occurred while the client resided in a controlled environment (e.g., prison).

PREVALENCE

SUDs represent an entrenched and costly public health problem in the United States, as they do in virtually all industrialized societies. The most recent estimate of the prevalence of SUDs in the United States comes from the 2004 National Survey on Drug Use and Health (Substance Abuse and Mental Health Services Administration, 1997, 1999) which is conducted annually to estimate the prevalence of substance use among those aged 12 years and older who reside in U.S. households. In the 2004 survey, 50.3% (121 million) were recent users of alcohol, 29.2% (70.3 million) were estimated to be recent users (past 30 days) of tobacco products, and 20.4% (19.1 million) recent users of illicit drugs. An estimated 35.3 million are dependent on tobacco; 15.2 million, dependent on alcohol but not illicit drugs; 3.9 million, on illicit drugs but not alcohol; and 3.4 million, on alcohol and illicit drugs. An estimate was not provided in the survey results, but the majority of those dependent on alcohol or illicit drugs can be expected to be dependent on tobacco as well.

FUNCTIONAL ANALYSIS OF COCAINE AND OTHER DRUG ABUSE

Much of the experimental research functionally analyzing factors controlling human cocaine and other drug abuse has been conducted with drug abusers in controlled laboratory settings where drugs can be administered in accurate doses and the user can be carefully monitored for potential adverse effects. These studies offer key insights into the operant nature of cocaine and other drug abuse that form the cornerstone of the CRA + Vouchers treatment described in this chapter. The research illustrates two important empirical generalizations about substance abuse that follow directly from the recognition that abused drugs function as reinforcers. First, as a form of operant behavior, drug use by definition is sensitive to its consequences. Second, drug use is malleable and dependent on environmental context. In the following text we review research examining the validity of those generalizations.

Experiments examining the influence of alternative, nondrug reinforcers on preference for cocaine use in experienced users illustrate the context-dependent nature of the reinforcing effects of drugs. In one of the seminal experiments on this topic, cocaine users residing in a controlled laboratory setting made a series of choices between 10 mg doses of intranasally administered cocaine and placebo (lactose mixed with a small amount of cocaine) under double-blind conditions and subsequently between cocaine and varying amounts of money (Higgins, Bickel, & Hughes, 1994). The maximum dose that subjects could administer in a single session was

100 mg, which is a psychoactive but modest dose in terms of amounts used in uncontrolled settings. During sessions comparing cocaine versus placebo, subjects exclusively chose cocaine, and they self-administered all of the drug that was available to them. Cocaine functioned as an effective reinforcer in these subjects. In subsequent sessions, subjects chose between cocaine and varying amounts of money. Within that context, choice of cocaine decreased as an orderly, graded function of increasing value of the monetary option. That outcome demonstrated the malleability of cocaine's reinforcing effects, which were robust when the alternative was an inert placebo or little money, but relatively weak as the value of the monetary option increased. The same functional relationship was subsequently demonstrated in studies using smoked and intravenous routes of cocaine administration, relatively higher cocaine doses, and subjects with extensive histories of cocaine abuse and dependence (Foltin & Fischman, 1994; Hatsukami, Thompson, Pentel, Flygare, & Carroll, 1994).

Two other contextual factors not underscored by those studies but essential to understanding substance use disorders is the role of stimulus control and temporal delays in modulating how environmental context influences drug preference. With regard to stimulus control, numerous experimental studies have examined whether stimuli paired with drug use acquire conditioned effects. The typical arrangement is to expose the cocaine abusers to drug-related and drug-neutral stimuli while assessing self-reported desire for cocaine use as well as physiological changes like heart rate and skin temperature. Both reported desire to use, or actual use, in experiments where cocaine use is permitted, increase when exposed to the drug-related compared to drug-neutral stimuli (Modesto-Lowe & Kranzler, 1999).

With regard to temporal delays, it is important to keep in mind that drug abusers typically make choices between using drugs in the present versus abstaining and experiencing a temporally delayed positive consequence—for example, going out and partying on a work night versus staying home and attending work clearheaded the following day. A laboratory study with cigarette smokers illustrates how temporal delays influence the relationships between drug preference and the influence of environmental constraints (Roll, Reilly, & Johanson, 2000). Regular cigarette smokers who had abstained from recent smoking for several hours made repeated choices between puffs on a cigarette available immediately and money that was available at varying values ($0.10–$2.00 per choice) and after varying delays (end of the session, 1 week, and 3 weeks). Preference for the drug option varied as an orderly, graded function of the value of the alternative available consistent with the studies discussed previously, but as the delay interval increased, choice of the alternative decreased.

There is an emerging area of behavioral-economics research suggesting that individuals with substance use disorders discount the value of delayed reinforcement and the magnitude of reinforcement losses to a greater extent than individuals without substance use disorders (Bickel & Marsch, 2001). Moreover, those with substance abuse and other problems, such as depression or gambling, discount the value of delayed consequences more than those with only substance abuse. This characteristic can be summarized as substance abusers showing a greater preference for (1) more immediate, smaller magnitude over more delayed, larger magnitude reinforcement; and (2) more delayed, larger magnitude losses (punishment) over more immediate, smaller magnitude losses. Of course, a shared characteristic of drugs of abuse is their ability to produce relatively immediate, positive reinforcement. Whether the greater discounting rates observed among those with substance use disorders in these studies represents cause or consequence of the disorders is unknown. This is an emerging area of investigation with many questions remaining to be answered. The studies conducted to date provide an impressive amount of evidence of significantly greater discounting of delayed consequences among individuals with substance use disorders. Whether cause or consequence, this characteristic is an important one to consider in efforts to provide a scientific account of substance use disorders, as well as in the more practical effort to treat them.

Other variables also alter the influence of alternative reinforcers on drug self-administration. For example, providing an alternative source of nicotine increases the ability of an alternative monetary reinforcer to decrease the frequency of smoking (Bickel, Madden, & DeGrandpre, 1997), and pretreatment with alcohol (Higgins, Roll, & Bickel, 1996) or cocaine (Donny, Bigelow, & Walsh, 2003) decreases the ability of monetary reinforcement to decrease preference for cocaine reinforcement. There are still other ways that nondrug reinforcers can increase the future probability of drug use. For example, when drug use is associated with increased earnings on a performance task, due to experimenter manipulation and not enhanced abilities, preference for drug use in future sessions increases (Alessi, Roll, Reilly, & Johanson, 2002). In these cases, drug use acquires discriminative or conditioned reinforcing functions. Likewise, when drug use is associated with decreased earnings, future preference for drug use decreases, indicative of discriminative or conditioned punishing effects.

FUNCTIONAL ASSESSMENT

Our clinic is a university-based outpatient center specializing in the treatment of cocaine-dependent adults. All initial clinic contacts are handled by a receptionist who establishes that the treatment seeker reports problems related to cocaine use and is age 18 years or above. Every effort is made to schedule the intake assessment interview within 24 hr of clinic contact, which significantly reduces attrition between the initial clinic contact and assessment interview (Festinger, Lamb, Kirby, & Marlowe, 1996). During the intake assessment, we collect detailed information on cocaine and other substance use; evaluate treatment readiness; and assess psychiatric functioning, employment/vocational status, recreational interests, current social supports, family and social problems, and legal issues. This is all important information that needs to be considered in developing a treatment plan to decrease cocaine use and increase involvement with alternative and healthier sources of reinforcement.

SELF-ADMINISTERED QUESTIONNAIRES

We use several self-administered questionnaires. The *Stages of Change Readiness and Treatment Eagerness Scale* (SOCRATES; Miller & Tonigan, 1996) provides a quantitative index of self-reported commitment to changing substance use, which may be an important indicator of the client's willingness to comply with the treatment plan. We assess readiness to change cocaine, alcohol, and other substances. We use an adaptation of the *Cocaine Dependency Self-Test* (Washton, Stone, & Hendrickson, 1988) as an efficient means to collect specific information regarding the type of adverse effects of cocaine that clients have experienced and the *Michigan Alcoholism Screening Test* (MAST), a widely used brief alcoholism screening instrument (Selzer, 1971), since approximately 60% of those seeking outpatient treatment for cocaine dependence also meet diagnostic criteria for alcohol dependence. The *Beck Depression Inventory* (BDI; Beck, Ward, Mendelson, Mock, & Erbaugh, 1961) is used to screen for depressive symptomatology and readministered on a regular basis to monitor progress with those clients who score in the clinical range at the Intake assessment. Depressed mood is a state that for some clients increases the likelihood of cocaine use. The SCL-90-R (Derogatis, 1983) is also used to screen for psychiatric symptomatology and is helpful in determining whether a more in-depth psychiatric evaluation is warranted. The SCL-90-R also can be easily readministered to monitor progress or change in psychiatric status.

STRUCTURED INTERVIEWS

A brief description (10–15 min) of treatment is provided, including overall duration, the recommended frequency and duration of clinic visits, and the focus on lifestyle changes. Next, a semistructured drug-history interview developed in our clinic is completed to obtain details on current and past drug use to obtain information regarding the duration, severity, and pattern of cocaine and other drug use. Using a calendar as a prompt, clients are asked to recall on a day-by-day basis the number of days they used in the past week and the amount used per occasion (Sobell, Sobell,

Leo & Cancilla, 1988). The same assessment is conducted for the past 3 weeks and as far back in time as needed for diagnostic reasons. This technique results in a good overview of the pattern of cocaine use during the past 30 days. This information is critical for the development of a detailed, individualized treatment plan designed to decrease the frequency of cocaine and other abuse. The *Addiction Severity Index* (ASI) (McLellan et al., 1985) is used to provide reliable, valid, and quantitative assessments of the severity of problems the client is experiencing in major areas of functioning. A *Practical Needs Assessment* developed in our clinic is used to determine if the client has any pressing needs or crises that may interfere with treatment participation such as housing, legal, transportation, or childcare. The probability of engaging and keeping such clients in treatment may be compromised if swift attention is not provided to assist with certain acute crises. We typically use community resources including homeless shelters and resources for battered women in efforts to resolve any such crises.

After completing these assessments, the client meets with a therapist for 15 min or so in order to have the client depart feeling that treatment has begun and with concrete plans for abstaining from cocaine use until the next clinic visit. The session is also used to establish rapport with the client and to provide further rationales for the treatment approach. Clients are oriented to the rigorous urinalysis-testing regimen, and if it appears that disulfiram therapy for alcohol problems is indicated, initial steps are taken toward implementing that protocol.

TREATMENT COMPONENTS

The recommended duration of CRA + Vouchers is 24 weeks of treatment and 6 months of aftercare. The influence of treatment duration has not yet been experimentally examined with this or other efficacious psychosocial treatments for cocaine dependence, but 3 or more months of care is a recommended practice in the treatment of illicit-drug abuse. CRA therapy in this model is delivered in individual sessions, although CRA has also been delivered effectively in group sessions with alcoholics (Azrin, 1976). As the title implies, the treatment involves two main components: CRA and vouchers.

CRA THERAPY

Regarding more general characteristics of CRA therapy, therapists must exhibit good listening skills and express empathy for the difficult challenges that clients face. Active problem solving is a routine part of the therapeutic relationship. Within ethical boundaries, therapists are committed to actively doing what it takes to facilitate lifestyle changes on the part of clients. Therapists take clients to appointments or job interviews, initiate recreational activities with clients, and schedule sessions at different times of day to accomplish specific goals. They have patients make phone calls from their office, and search newspapers for job possibilities or ideas for healthy recreational activities in which patients might be able to participate. In sum, the CRA therapists are empathic, directive, and active in pursuing treatment goals.

CRA is delivered in twice-weekly 1.0–1.5-hr therapy sessions during the initial 12 weeks and once-weekly sessions of the same duration during the final 12 weeks of treatment. Sessions focus on six general topics, depending on the needs of the individual patient. First, patients are instructed in how to recognize antecedents and consequences of their cocaine use—that is, how to functionally assess their cocaine use. They are also instructed in how to use that information to reduce the probability of using cocaine. A twofold message is conveyed to the client: (1) cocaine use is orderly behavior that is more likely to occur under certain circumstances than others, and (2) by learning how to identify the circumstances that affect one's cocaine use, plans can be developed and implemented to reduce the likelihood of future cocaine use. Our approach to teaching functional analysis is based on the work of Miller and Munoz (1982) and McCrady (1986, 1993). Clients are assigned the task of analyzing at least three recent episodes of cocaine use. Learning to analyze one's cocaine use is emphasized during initial treatment sessions, but the exercise is used throughout the treatment process to systematically analyze any instances of cocaine use. To begin this task, clients make a list of places, people, times, and activities where cocaine use is likely and also where use is unlikely. They next list the behavior involved in a typical drug use episode, including how much drug is used, route of administration, what other activities occurred simultaneous with using, etc. Finally, they list all of the positive consequences of use as well as the unpleasant consequences which are typically more delayed. Once clients are oriented to the task of identifying antecedents, behavior, and consequences, they analyze along with their therapist their three most recent episodes of cocaine use. This exercise is repeated each time cocaine use occurs during the course of treatment. In conjunction with functional assessment, clients are taught to develop plans for using the information revealed in the functional analyses to decrease the probability of future cocaine use. Clients are counseled to restructure their daily activities in order to minimize contact with known antecedents of cocaine use, find alternatives to the positive consequences of cocaine use, and to make explicit the negative consequences of cocaine use. A key part of planning that is implemented with most clients is drug-refusal training. We approach this as a special case of assertiveness training using previously reported procedures (McCrady, 1986; Sisson & Azrin, 1989). The therapist and other staff role-play with clients simulated situations involving offers to use cocaine. This permits the therapist to model appropriate responses, clients to practice approximations of those responses, and the therapist to reinforce their performance. Second, the need to develop a new social network that will reinforce a healthier lifestyle and getting involved with recreational activities that are likely to be reinforcing and do not involve cocaine or other drug use is addressed with all clients. Systematically developing and maintaining contacts with drug-free social networks and participation in drug-free recreational activities remain high priorities throughout treatment. These social contacts and new activities are designed to function as substitutes for cocaine use and associated lifestyle. Identifying activities that can be engaged in at times that are high risk for cocaine use is often important. Specific treatment goals are set, and weekly progress on specific goals is monitored. Plans for developing healthy social networks and recreational activities must be individualized depending on the circumstances, skills, and interests of the patient. For those willing to participate, self-help groups such as Alcoholics or Narcotics Anonymous can be an effective way to develop a new social network willing to reinforce a sober lifestyle. Clinic staff will often accompany a client to sample a self-help meeting or two. Self-help involvement is not mandated but is merely one option to increase support for sober behavior. We assist clients with getting involved in a wide variety of healthy activities that can facilitate developing a new social network or reestablishing a prior one that will reinforce a healthy lifestyle. We also teach clients to develop healthy recreational activities and hobbies that can provide alternatives to drug use for obtaining positive reinforcement. Joining the local YMCA, enrolling in continuing education classes, and taking ski lessons are common examples. Clinic staff accompany clients when they try new or reinitiate familiar healthy activities.

Third, various other forms of individualized skills training are provided, usually to address some specific skill deficit that may directly or indirectly influence a client's risk for cocaine use, including time management, problem solving, assertiveness training, social-skills training, and mood management. Essential to success with meeting the treatment goals discussed previously, for example, is some level of time-management skills. All clients are given daily planners to facilitate planning. We implement the *Control Your Depression* protocol (Lewinsohn, Munoz, Youngren, & Zeiss, 1986) with those whose depression continues after discontinuing cocaine use, especially those for whom depressed mood appears to occasion cocaine craving or use. Many clients report problems with insomnia following discontinuation of drug use. With them, we implement a sleep-hygiene protocol (Lacks, 1987). Because clients often have many problems that could benefit from professional assistance, we treat only those that our assessments suggest are functionally related to the probability of initial or longer-term cocaine abstinence. We make referrals for problems that do not appear to be functionally related to cocaine use, but warrant professional attention.

Fourth, unemployed clients are offered Job Club, an efficacious method for assisting chronically unemployed individuals obtain employment (Azrin & Besalel, 1980). A meaningful vocation is fundamental to a healthy lifestyle and typically places demands on one's schedule that are incompatible with ongoing cocaine abuse. We recommend goals directed toward vocational enhancement for all clients.

Fifth, participants with romantic partners who were not drug abusers are offered reciprocal relationship counseling, which is an intervention designed to teach couples positive communication skills and ways to negotiate reciprocal contracts for desired changes in each other's behavior (Azrin, Naster, & Jones, 1973). The rationale here is to improve the quality of the relationship so that it may serve as an attractive alternative to the drug-abusing lifestyle.

Sixth, HIV/AIDS education is provided to all participants in the early stages of treatment, along with counseling directed at addressing any specific needs or risk behavior of the individual patient. We address with all patients the potential for acquiring HIV/AIDS from sharing injection equipment and through sexual activity.

Seventh, all who meet diagnostic criteria for alcohol dependence or report that alcohol use is involved in their use of cocaine are offered disulfiram therapy, which is an integral part of the CRA treatment for alcoholism (Myers & Smith, 1995; Sisson & Azrin, 1989). Disulfiram interferes with the metabolism of alcohol and thereby causes a physically unpleasant reaction if one drinks while on the medication, i.e., punishes drinking. Disulfiram therapy is effective only when implemented with procedures to monitor compliance with the recommended dosing regimen. Clients generally ingest a 250 mg daily dose under clinic staff observation on urine-toxicology test days and when possible under the observation of a significant other on the other days.

Use of substances other than tobacco and caffeine is discouraged as well via CRA therapy. Anyone who meets criteria for opiate dependence is referred to an adjoining service located within our clinic for opioid replacement therapy (Bickel, Amass, Higgins, Badger, Esch, 1997). We recommend marijuana abstinence because of the problems associated with its abuse, but have found no evidence that marijuana use or dependence adversely affects treatment for cocaine dependence (Budney, Higgins, & Wong, 1996). Importantly, we never dismiss or refuse to treat a client due to other drug use. We recommend cessation of tobacco use, but usually not during the course of treatment for cocaine dependence.

Upon completion of the 24 weeks of treatment, participants are encouraged to participate in 6 months of aftercare in our clinic, which involves at least a once-monthly brief therapy session and a urine toxicology screen.

This allows for a gradual rather than abrupt ending of the patient's involvement with the clinic.

Voucher Program

The voucher program is a contingency-management (CM) intervention designed to directly reinforce abstinence from cocaine use. The voucher program is implemented in conjunction with a rigorous urine-toxicology monitoring program. Urine specimens are collected from all clients according to a Monday, Wednesday, and Friday schedule during Weeks 1–12 and a Monday and Thursday schedule during Weeks 13–24 of treatment. Specimens are screened immediately via an onsite Enzyme Multiplied Immunoassay Technique to minimize delays in delivering reinforcement for cocaine-negative specimens. To decrease the likelihood of submitting bogus specimens, all specimens are collected under the observation of a same-sex staff member. All specimens are screened for benzoylecgonine, a cocaine metabolite, and one randomly selected specimen each week also is screened for the presence of other abused drugs. Failure to submit a scheduled specimen is treated as a cocaine-positive. Clients are informed of their urinalysis results within several minutes of submitting specimens.

Urine specimens collected during Weeks 1–12 that test negative for benzoylecgonine earn points that are recorded on vouchers and given to clients. Points are worth the equivalent of $0.25 each. Money is never provided directly to clients. Instead, points are used to purchase retail items in the community. A staff member makes all purchases. The first negative specimen is worth 10 points @ $.25/point or $2.50. The value of vouchers for each subsequent consecutive negative specimen increases by 5 points; e.g., 2nd = 15 points, 3rd = 20 points, etc. To further increase the likelihood of continuous cocaine abstinence, the equivalent of a $10 bonus is earned for each three consecutive negative specimens. Specimens that are cocaine positive or failure to submit a scheduled specimen resets the value of vouchers back to the initial $2.50 value from which they can escalate again according to the same schedule. Submission of five consecutive cocaine-negative specimens following submission of a positive specimen returns the value of points to where they were prior to the reset. The voucher program is discontinued at the end of week 12. During Weeks 13–24, clients receive a single $1.00 Vermont State Lottery ticket per cocaine-negative urinalysis test. Therapists retain veto power over all purchases made with vouchers. Purchases are approved only if therapists deem them to be in concert with individual treatment goals of increasing a healthy lifestyle.

Clinical Supervision

Doctorate-level psychologists who have expertise in behavioral psychology and substance abuse treatment provide supervision in our clinic.

Supervisors provide significant input into treatment plans and selection of targets for behavior change. The supervisor provides guidance about how to monitor progress during weekly sessions that usually last 2–3 hr, during which all cases are reviewed. Therapists update the supervisor and other clinic therapists on each patient's progress at the level of specific treatment goals and whether progress has been made since last supervision meeting. Progress is presented graphically for all goals. Review begins with examination of a graph of the patient's cumulative cocaine urinalysis results from the start of treatment. We also review alcohol or any other drugs that are being targeted for change. Then we review attendance at therapy sessions, primary goals for lifestyle changes, and then secondary goals on the same. Careful attention is paid to whether clients are abstaining from cocaine use and corresponding progress on other goals, which are designed to facilitate short- and longer-term cocaine abstinence. If clients do not abstain from cocaine abstinence or there are other difficulties in achieving goals, the treatment plan is modified as needed. Once those treatment targets have been reviewed and modified as necessary, any recent crises or relevant clinical issues, such as suicidal ideation or newly identified problem behaviors, are discussed.

SUPPORTING RESEARCH

A series of controlled clinical trials support the efficacy of the CRA + Vouchers treatment (Higgins et al., 1991, 1993, 1995, 2003; Higgins, Budney, Bickel, & Badger, 1994; Higgins, Wong, Badger, Haug Ogden, & Dantona, 2000). In the following text we briefly review those studies and related research demonstrating effective clinical practices with the cocaine-dependent population. Interested readers also may want to see two meta-analyses supporting the efficacy of CRA (Roozin et al., 2004) and vouchers (Lussier, Heil, Mongeon, Badger, & Higgins, 2006) in the treatment of SUDs.

The initial two trials conducted with this treatment were comparisons with drug abuse counseling based on the disease-model approach (Higgins et al., 1991, 1993). In both trials, the CRA + Vouchers promoted better retention in outpatient treatment and greater cocaine abstinence than drug abuse counseling. Those two trials demonstrated the efficacy of the CRA + Vouchers intervention. One of those trials also included post-treatment assessments that supported treatment efficacy through 6 months of post-treatment follow-up (Higgins et al., 1995).

Next, a dismantling strategy was implemented. Assessing the efficacy of the voucher component was the first step. Patients were randomly assigned to receive CRA with (N = 20) or without (N = 20) the voucher program (Higgins, Budney, Bickel, et al., 1994). Vouchers significantly improved retention and cocaine abstinence during the 6 months of outpatient treatment. During 6 months of post-treatment follow-up, those treated with vouchers reported greater reductions in cocaine use, and only the vouchers group showed significant reductions in psychiatric symptomatology on the ASI (McLellan et al., 1985). A subsequent trial conducted with 70 cocaine-dependent outpatients demonstrated the contribution of the voucher program to increased cocaine abstinence rates through 1 year of post-treatment follow-up (Higgins et al., 2000).

To examine the efficacy of the CRA component, 100 cocaine-dependent outpatients were randomly assigned to receive the CRA + Vouchers treatment or vouchers only. Because both treatment groups received the abstinence-contingent vouchers component, any outcome differences are attributable to the CRA component. CRA increased retention in treatment, decreased cocaine use during treatment but not follow-up, decreased the frequency of drinking to intoxication during treatment and follow-up, increased the number of days employed during treatment and follow-up, decreased depressive symptoms during treatment, and decreased the number of hospitalizations and arrests for driving while under the influence during the 18-month post-treatment follow-up period.

Monitored disulfiram therapy, a component of the CRA intervention, has been demonstrated to decrease alcohol and cocaine use in clients who abuse both substances in controlled clinical trials (Carroll et al., 1994; Carroll, Nich, Ball, McCance, & Rounsaville, 1998).

Because much of this work was completed in Vermont, Carroll and colleagues determined the generality of those findings to inner-city cocaine-dependent patients. Carroll and colleagues demonstrated the generality of the disulfiram component of CRA to an inner-city population, and a series of well-controlled trials have demonstrated the efficacy of the voucher component in inner-city populations (Kirby, Marlowe, Festinger, Lamb, & Platt, 1998; Piotrowski & Hall, 1999; Silverman et al., 1996).

CASE STUDY

In this section we review a case of someone treated in our clinic using the CRA + Vouchers intervention. The case was chosen because it illustrates well a number of different aspects of using this treatment approach. Outcome was excellent, but certainly not perfect, which is to be expected with this population. The case also illustrates the multifaceted problems with which cocaine-dependent patients present.

Richard was a 30-year-old, single, White male who was referred by his employer to the clinic for help with problems with cocaine use. He currently lived with his parents, neither of whom abused alcohol or drugs, and was employed full time as a truck driver.

PRESENTING COMPLAINT

Richard had recently been charged with possession of cocaine. This was his first serious criminal charge. When he informed his employer about his legal problems, the employer removed him from his driving responsibilities and required him to enter treatment as a way to regain them. Richard reported no previous treatment episodes for alcohol or drug abuse. He reported several prior attempts to stop cocaine use on his own, but with minimal success. He reported being extremely worried about the potential for losing his job and the effect this would have on his financial stability.

FUNCTIONAL ASSESSMENT

Richard reported that his first use of cocaine was at age 21 years, using the intranasal route. Shortly thereafter, he switched to cocaine smoking. Richard's cocaine use had steadily escalated over 9 years to using cocaine an average of 2 days per week, approximately 1 gram per occasion, at the time he sought treatment. His most recent use was 2 days prior to the intake assessment, using 1.5 grams at a bar with friends. He reported this to be his typical pattern of use. At the time of the intake, Richard reported 8 episodes of cocaine use during the prior 30 days, with each episode lasting approximately 12 hr and usually beginning with friends at bars or parties and then continuing at home alone.

His cocaine use was often preceded by working long hours, which left little time to spend with family and friends who were not abusers or to engage in healthy recreational activities. When he did have free time, he played in league billiards games, which primarily took place in bars with friends who also used cocaine. Richard was reluctant to give up this activity, as it represented his only outlet for recreational and social contact outside work. Richard reported a number of serious consequences related to his cocaine use, including the recent encounter with the criminal justice system, financial and employment problems, low energy, sleep problems, and a short temper. He met *DSM-IV* criteria for cocaine dependence.

Richard reported a pattern of relatively light alcohol use, typically drinking 1–2 beers twice per month. He reported no other regular drug use. Richard also reported a history of occasional depressed mood, particularly following periods of cocaine use, but no other current or historic symptoms of emotional or psychological problems or history of suicide attempts or ideation. His BDI score was 8 at intake, which is well below the clinical range.

CONCEPTUALIZATION OF THE CASE

Richard worked long hours as a truck driver and reported minimal involvement in any form of healthy recreational activities. Aside from work, there were few alternative sources of reinforcement to compete with the reinforcing effects of cocaine use. Conceptualized in terms of concurrent operants, reinforcement frequency was higher for drug-related activities than potential healthy alternatives. Since Richard's primary recreational activity and social contact outside work was playing billiards in a bar with other cocaine-abusing friends, this further limited his sources of potential competing reinforcement from healthy, non-drug-abusing sources and placed him in contact with known antecedents for cocaine use. Richard very much enjoyed his job, and his full-time employment likely provided some protection against cocaine becoming more frequent. As such, finding a way to balance both full-time employment and engagement in healthy social and recreational activities would be key to competing effectively with the reinforcement derived from cocaine abuse and thus for Richard's success in treatment.

TREATMENT PLAN

Cocaine abstinence was the first priority in Richard's treatment plan. The voucher program was implemented during the initial 12 weeks of treatment to differentially reinforce cocaine abstinence. Next, we recommended alcohol abstinence or at least reduction and close monitoring of Richard's alcohol consumption to decrease the potential of cocaine use by ensuring that Richard would have no contact with bars, which were a clear antecedent for cocaine use. Removing this source of social contact and recreational activities alone would likely result in relapse, as it would remove both cocaine and the only social reinforcers in Richard's life. Therefore, the plan included other social and recreational activities to provide alternate sources of reinforcement. Thus, high-priority goals were reestablishing a regular pattern of involvement in healthy recreational activities, especially activities that might substitute for cocaine use, and increasing involvement in social activities with family members and healthy others. Since his current job was an important source of healthy behavior and associated reinforcement maintaining that behavior, we assisted Richard in maintaining his current job and working closely with his employer to satisfy his concerns and regain Richard's driving responsibilities.

TREATMENT OUTCOME

Cocaine Abstinence

Richard's only instance of confirmed cocaine use occurred during the first week of treatment. Richard was scheduled to attend the clinic for a counseling session and to provide a urine sample; however, he did not show at the time of his scheduled appointment. Consistent with the CRA focus on active problem solving and outreach, his therapist immediately began efforts to contact Richard. After several calls to the family members and friends whom Richard had identified as contact persons, his therapist learned that Richard was currently at a local bar. The therapist went to the bar and, in an empathic manner, conveyed to Richard that he had been missed at the clinic and asked if he could help in any way. Richard reported that he had run into old friends the previous night, used cocaine, and was embarrassed to come into the clinic the following day for his scheduled appointment since he knew that he would test positive for cocaine use. Richard's therapist reminded him that such challenges are the norm rather than the exception, particularly during the early days of treatment, and that the only way to ensure that a single slip does not develop into a full-blown relapse is to get back into the routine of clinic contact and to actively work with clinic staff on treatment goals. The therapist also pointed out that Richard's pursuit of contact with the bar and drug-using friends simply showed that his treatment goals of increasing engagement in social and recreational activities that can compete with cocaine use were the right place to start. His therapist also pointed out that if Richard stuck with treatment for a little longer, real progress could be made on this front. After this brief but important interaction, Richard said that he would be at the clinic the next day for a counseling session and to continue working with his therapist on the treatment goals. Indeed, he returned to the clinic the following day, he and his therapist implemented the treatment plan as described previously, and Richard remained continuously cocaine abstinent throughout the remainder of treatment.

Alcohol Abstinence

At the beginning of treatment, the therapist worked with Richard to functionally assess his alcohol use. While he was a fairly moderate drinker and had not experienced extensive negative consequences related to drinking, Richard typically drank in the same bar where he and his friends had historically used cocaine. Richard and his therapist reviewed how this context placed him at significant risk for using cocaine, thereby undermining his progress in treatment. They developed a plan for finding alternative ways to relax that did not involve the bar or drinking; in addition they rehearse how to identify and problem-solve the times that he might be tempted to go to the bar and the alternative activities he might engage in instead. Throughout the 24-week treatment, Richard drank only 1 time per month compared to 1–2 times per month prior to treatment. The big change was that Richard avoided the bar when he did drink. On these occasions, he did so at home and not in the presence of the old friends with whom he had used cocaine in the past. This modification in Richard's alcohol use was effective in disrupting the response chain early in the chain of going to the bar, drinking alcohol, and finally using cocaine.

Recreational Activities

An important goal was reestablishing a regular pattern of involvement in healthy recreational activities, especially activities that might substitute for cocaine use by providing social and other forms of reinforcement. Toward this end, Richard and his therapist identified a specific, measurable goal of three recreational activities per week. Richard consistently met this goal during treatment. Activities in which Richard participated included movies, hunting, archery, the gym, and dinners out. He used his vouchers to buy a gym membership, hunting and fishing licenses, archery supplies, and gift certificates to local restaurants. In addition, midway through treatment Richard noted that he would like to try going to the gym, which was an activity that he had enjoyed prior to becoming a cocaine user. Richard's therapist went with Richard to tour the gym facilities, assisted him with using a portion of his earned vouchers to purchase a gym membership, and then established a goal of attending the gym at least once weekly. Richard also consistently met this goal throughout the remainder of treatment.

Family and Social Support

In order to build a social support network that would also provide social reinforcement contingent upon healthy behavior, the therapist discussed with Richard the need for him to expand his social network to include family and other social interaction with non-drug-using people. Richard noted that his family currently represented the only safe, non-drug-using individuals with whom he could enjoy spending time. Thus, Richard and his therapist identified a specific goal of completing two activities per week with family members, which he met consistently throughout treatment. In order to facilitate meeting new, non-drug-using individuals beyond family, Richard and his therapist completed a list of places and activities where Richard could possibly make healthy contacts with new people, including the gym, community activities, and his workplace. While this goal of meeting two new people every week was difficult to implement and objectively measure, Richard and his therapist worked closely to take advantage of any possible activities as an opportunity to expand his social network and meet new people. As a result, Richard increased contact with several safe friends and coworkers during the course of treatment.

Employment and Education

To help Richard maintain his current job and to satisfy his employer's concerns about ongoing cocaine use, his therapist established several specific goals. First, Richard provided consent that allowed his therapist to make weekly calls to his employer, relaying Richard's progress in treatment and fielding any continued concerns on behalf of the employer. Second, the number of hours Richard worked each week was continuously monitored and graphed so that any particularly troubling trends, such as missed days or excessive hours, could be identified and problem-solved. This combination was successful in helping Richard to maintain his job, satisfy his employer, and regain the driving responsibilities.

Psychiatric Monitoring

The occasional mild depression that Richard had reported experiencing subsided following his initiation and maintenance of continuous cocaine abstinence. He had no other psychological or emotional problems. His BDI score remained low throughout treatment.

Summary

Richard made excellent progress during treatment toward establishing a stable record of cocaine abstinence, avoiding contact with bars and drug-using individuals, increasing his involvement in recreational activities, expanding his social-support network, and improving his job and legal situations.

Follow-up

Following completion of the 24-week treatment protocol, Richard participated in aftercare involving brief check-in sessions with his therapist and completed 2 years of follow-up assessments. Thus, we have a relatively extensive picture of his continued progress with drug abstinence and treatment goals. At the 1-year follow-up, 6 months after treatment had ended, Richard reported no cocaine use since treatment termination and provided a urine specimen that tested negative for cocaine and other drug use. Richard reported that he continued to avoid bars and high-risk people. He noted several new, positive events in his life, which included a drug-free girlfriend and hunting for a new apartment of his own. At the 2-year follow-up, he reported that he had become engaged to his girlfriend, was planning an upcoming wedding, and was enjoying living on his own in his new apartment. He continued to report cocaine and other drug abstinence throughout the past year since ending treatment at the clinic.

CONCLUDING COMMENTS

In this chapter we have described an efficacious intervention for cocaine dependence that is based on a functional analysis of the disorder. Most fundamental about this approach is recognition that drug use is a form of operant responding that is sensitive to reinforcement and other environmental consequences. The cornerstone of the CRA + Vouchers treatment approach is recognizing and utilizing the principle of reinforcement in the treatment process. We realize that limitations in resources and other practical constraints will prevent many clinicians from utilizing the treatment practices outlined in this chapter in the exact manner that we have described them, but we hope that the information provided offers insights into the important elements of effective treatment for cocaine dependence and other SUDs.

ACKNOWLEDGMENTS

Preparation of this chapter was supported in part by National Institute on Drug Abuse Research Grants DA09378, DA14028, DA08076, DA018410, and DA017813.

REFERENCES

Alessi, S. M., Roll, J. M., Reilly, M. P., & Johanson, C. E. (2002). Establishment of a diazepam preference in human volunteers following a differential-conditioning history of placebo versus diazepam choice. *Experimental and Clinical Psychopharmacology, 10,* 77–83.

American Psychiatric Association (1994). *Diagnostic and statistical manual of mental disorders* (4th ed.). Washington D.C: Author.

Azrin, N. H. (1976). Improvements in the community-reinforcement approach to alcoholism. *Behaviour Research and Therapy, 14,* 339–348.

Azrin, N. H., & Besalel, V. A. (1980). *Job club counselor's manual.* University Park Press: Baltimore, MD.

Azrin, N. H., Naster, B. J., & Jones, R. (1973). Reciprocity counseling: A rapid learning based procedure for marital counseling. *Behaviour Research and Therapy, 11,* 364–382.

Beck, A. T., Ward, C. H., Mendelson, M., Mock, J., & Erbaugh, J. (1961). An inventory for measuring depression. *Archives of General Psychiatry, 4,* 561–571.

Bickel, W. K., Amass, L., Higgins, S. T., Badger, G. J., & Esch, R. A. (1997). Effects of adding behavioral treatment to opioid detoxification with buprenorphine. *Journal of Consulting and Clinical Psychology, 65,* 803–810.

Bickel, W. K., Madden, G. L., & DeGrandpre R. J. (1997). Modeling the effects of combined behavioral-pharmacological treatment on cigarette smoking: Behavioral-economic analyses. *Experimental & Clinical Psychopharmacology, 5,* 334–343.

Bickel, W. K., & Marsch, L. A. (2001). Toward a behavioral economic understanding of drug dependence; delay discounting processes. *Addiction, 96,* 73–86.

Budney, A. J., & Higgins, S. T. (1998). *The community reinforcement plus vouchers approach. Manual 2: National Institute on Drug Abuse therapy manuals for drug addiction.* NIH publication # 98–4308. Rockville, MD: National Institute on Drug Abuse.

Budney, A. J., Higgins, S. T., & Wong, C. J. (1996). Marijuana use and treatment outcome in cocaine-dependent patients. *Journal of Experimental and Clinical Psychopharmacology, 4,* 1–8.

Carroll, K. M., Nich, C., Ball, S. A., McCance, E., & Rounsaville, B. J. (1998). Treatment of cocaine and alcohol dependence with psychotherapy and disulfiram. *Addiction, 93,* 713–727.

Carroll, K. M., Rounsaville, B. J., Nich, C., Gordon, L. T., Wirtz, P. W., & Gawin, F. H. (1994). One-year follow-up of psychotherapy and pharmacotherapy for cocaine dependence: Delayed emergence of psychotherapy effects. *Archives of General Psychiatry, 51,* 989–997.

Derogatis, L. R. (1983). *SLC-90R: Administration, scoring and procedures manual-II.* Towson, MD: Clinical Psychometric Research.

Donny, E. C., Bigelow, G. E., & Walsh, S. L. (2003). Choosing to take cocaine in the human laboratory: Effects of cocaine dose, inter-choice interval, and magnitude of alternative reinforcement. *Drug & Alcohol Dependence, 69,* 289–301.

Festinger, D. S., Lamb, R. J., Kirby, K. C., & Marlowe, D. B. (1996). The accelerated intake: A method for increasing initial attendance to outpatient cocaine treatment. *Journal of Applied Behavior Analysis, 29,* 387–389.

Foltin, R. W., & Fischman, M. W. (1994). Effect of buprenorphine on the self-administration of cocaine by humans. *Behavioural Pharmacology, 5,* 79–89.

Griffiths, R. R., Bigelow, G. E., & Henningfield, J. E. (1980). Similarities in animal and human drug taking behavior. In N. K. Mello (Ed.), *Advances in substance abuse: Behavioral and biological research* (pp. 1–90). Greenwich, CT: JAI Press.

Hatsukami, D. K., Thompson, T. N., Pentel, P. R., Flygare, B. K., & Carroll, M. E. (1994). Self-administration of smoked cocaine. *Experimental & Clinical Psychopharmacology, 2,* 115–125.

Higgins, S. T. (1997). The influence of alternative reinforcers on cocaine use and abuse: A brief review. *Pharmacology Biochemistry and Behavior, 57,* 419–427.

Higgins, S. T., Bickel, W. K., & Hughes, J. R. (1994). Influence of an alternative reinforcer on human cocaine self-administration. *Life Sciences, 55,* 179–187.

Higgins, S. T., Budney, A. J., Bickel, W. K., Badger, G. J., Foerg, F. E., & Ogden, D. (1995). Outpatient behavioral treatment for cocaine dependence: One-year outcome. *Experimental and Clinical Psychopharmacology, 3,* 205–212.

Higgins, S. T., Budney, A. J., Bickel, W. K., Foerg, F. E., Donham, R., & Badger, G. (1994). Incentives improve outcome in outpatient behavioral treatment of cocaine dependence. *Archives of General Psychiatry, 51,* 568–576.

Higgins, S. T., Budney, A. J., Bickel, W. K., Hughes, J. R., Foerg, F., & Badger, G. (1993). Achieving cocaine abstinence with a behavioral approach. *American Journal of Psychiatry, 150,* 763–769.

Higgins, S. T., Delaney, D. D., Budney, A. J., Bickel, W. K., Hughes, J. R., Foerg, F., et al. (1991). A behavioral approach to achieving initial cocaine abstinence. *American Journal of Psychiatry, 148,* 1218–1224.

Higgins, S. T., Roll, J. M., & Bickel, W. K. (1996). Alcohol retreatment increases preference for cocaine over monetary reinforcement. *Psychopharmacology, 123,* 1–8.

Higgins, S. T., Sigmon, S. C., Wong, C. J., Heil, S. H., Badger, G. J., Donham, R., et al. (2003). Community reinforcement therapy for cocaine-dependent outpatients. *Archives of General Psychiatry, 60,* 1043–1052.

Higgins, S. T., Wong, C. J., Badger, G. J., Haug Ogden, D. E., & Dantona, R. L. (2000). Contingent reinforcement increases cocaine abstinence during outpatient treatment and one year of follow-up. *Journal of Consulting and Clinical Psychology, 68,* 64–72.

Kirby, K. C., Marlowe, D. B., Festinger, D. S., Lamb, R. J., & Platt, J. J. (1998). Schedule of voucher delivery influences initiation of cocaine abstinence. *Journal of Consulting and Clinical Psychology, 66,* 761–767.

Lacks, P. (1987). *Behavioral treatment of persistent insomnia.* Pergamon: New York.

Lewinsohn, P. M., Munoz, R. F., Youngren, M. A., & Zeiss, A. M. (1986). *Control your depression.* New York: Simon & Schuster.

Lussier, J. P., Heil, S. H., Mongeon, J. A., Badger, G. J., & Higgins, S. T. (2006). A meta-analysis of voucher-based reinforcement therapy for substance use disorders. *Addiction, 101,* 192–203.

McCrady, B. S. (1986). *Behavioral marital therapy for alcohol dependence.* Unpublished treatment manual. Rutgers University.

McCrady, B. S. (1993). Alcoholism. In D. H. Barlow (Ed.), *Clinical handbook of psychological disorders: A step-by-step treatment manual* (2nd ed., pp. 362–395). New York: Guilford Press.

McLellan, A. T., Luborsky, L., Cacciola, J., Griffiths, J., Evans, F., Barr, H. L., et al. (1985). New data from the Addiction Severity Index. *The Journal of Nervous and Mental Disease, 173,* 412–423.

Miller, W. R., & Munoz, R. F. (1982). *How to control your drinking.* Albuquerque: University of New Mexico Press.

Miller, W. R., & Tonigan, J. S. (1996). Assessing drinkers' motivation for change: The Stages of Change Readiness and Treatment Eagerness Scale (SOCRATES). *Psychology of Addictive Behaviors, 10,* 81–89.

Modesto-Lowe, V., & Kranzler, H. R. (1999). Using cue reactivity to evaluate medications for treatment of cocaine dependence: A critical review. *Addiction, 94,* 1639–1651.

Myers, R. J., & Smith, J. E. (1995). *Clinical guide to alcohol treatment: The community reinforcement approach.* New York: Guilford Press.

Piotrowski, N. A., & Hall, S. M. (1999). Treatment of multiple drug abuse in the methadone clinic. In S. T. Higgins & K. Silverman (Eds.), *Motivating behavior change among illicit-drug abusers: Research on contingency-management interventions* (pp. 183–202). Washington, D.C.: American Psychological Association.

Roll, J. M., Reilly, M. P., & Johanson, C. E. (2000). The influence of exchange delays on cigarette versus money choice: A laboratory analog of voucher-based reinforcement therapy. *Experimental & Clinical Psychopharmacology, 8,* 366–370.

Roozen, H. G., Boulogne, J. J., van Tulder, M. W., van den Brink, W., De Jong, C. A., & Kerkhof, A. J. (2004). A systematic review of the effectiveness of the community reinforcement approach in alcohol, cocaine and opioid addiction. *Drug and Alcohol Dependence, 74,* 1–13.

Selzer, M. L. (1971). The Michigan Alcoholism Screening Test. *American Journal of Psychiatry, 127,* 89–94.

Silverman, K., Higgins, S. T., Brooner, R. K., Montoya, I. D., Cone, E. J., Schuster, C. R., et al. (1996). Sustained cocaine abstinence in methadone maintenance patients through voucher-based reinforcement therapy. *Archives of General Psychiatry, 53,* 409–415.

Sisson, R., & Azrin, N. H. (1989). The community reinforcement approach. In R. K. Hester & W. R. Miller (Eds.), *Handbook of alcoholism treatment approaches: Effective alternatives* (pp. 242–258). New York: Pergamon Press.

Sobell, L. C., Sobell, M. B., Leo, G. I., & Cancilla, A. (1988). Reliability of a timeline method: Assessing normal drinkers' reports of recent drinking and a comparative evaluation across several populations. *British Journal of Addictions, 83,* 393–402.

Substance Abuse and Mental Health Services Administration (SAMHSA). (1997). *Drug Abuse Warning Network Series: D-3: Year-end preliminary estimates from the 1996 drug abuse warning network.* Rockville MD: National Clearinghouse for Alcohol and Drug Information.

Substance Abuse and Mental Health Services Administration (SAMHSA). (1999). *National Household Survey on Drug Abuse: Population estimates, 1998.* Rockville, MD: National Clearinghouse for Alcohol and Drug Information.

Washton, A. M., Stone, N. S., & Hendrickson, E. C. (1988). Cocaine abuse. In D. M. Donovan & G. A. Marlatt (Eds.), *Assessment of addictive behaviors* (pp. 364–389). New York: Guilford Press.

14

SCHIZOPHRENIA AND OTHER PSYCHOTIC DISORDERS

DAVID A. WILDER

AND

STEPHEN E. WONG

DIAGNOSTIC CRITERIA AND RELATED CHARACTERISTICS

SCHIZOPHRENIA

Schizophrenia and other psychotic disorders are mental disorders marked by the presence of delusions, prominent hallucinations, disorganized speech, or disorganized or catatonic behavior (American Psychiatric Association, 2000). The diagnosis of schizophrenia is assigned when the client meets four criteria. The first is the presence of two or more of the following symptoms ". . . for a significant portion of time during a 1-month period (or less if successfully treated)" (p. 285): (1) delusions; (2) hallucinations; (3) disorganized speech; (4) grossly disorganized or catatonic behavior; and (5) negative symptoms (e.g., flattened affect, poverty of speech). Two additional criteria are a disturbance in social or occupational function, and signs of the disturbance persisting for at least 6 months. Other criteria for schizophrenia include ruling out the competing diagnoses of schizoaffective, mood, substance abuse, and pervasive developmental disorders and general medical conditions.

The prevalence of schizophrenia is estimated to range between 0.5% and 1.5% of the general population. The onset of the disorder typically occurs between 18 and 25 years for men and 25 and 35 years of age for women. Risk of this disorder is higher among urban-than rural-born persons. Like most mental disorders, the probability of developing schizophrenia increases as socioeconomic status decreases (Hudson, 2005; Kessler et al., 1994; Keith, Regier, & Rae, 1991).

OTHER PSYCHOTIC DISORDERS

Other psychoses include schizoaffective, delusional, and brief psychotic disorders. Schizoaffective disorder involves symptoms of schizophrenia plus a depressive or manic episode, or both. Delusional disorder is distinguished on the basis of plausible nonbizarre delusions, as opposed to the clearly implausible bizarre delusions of schizophrenia. Brief psychotic disorder is indicated by a sudden disturbance involving schizophrenic symptoms that lasts at least 1 day but less than 1 month, with an eventual return to full level of premorbid functioning. There is little consistent information on prevalence, course, or demographics of these other disorders, except that they are less prevalent than schizophrenia. For example, the prevalence for delusional disorder is estimated to be around 0.03% (APA, 2000).

For several reasons, distinctions between the previous diagnostic categories may not be especially pertinent to clinicians who will be using functional analytic treatments. For one, during the past 30 years the diagnostic criteria for schizophrenia in particular have been malleable, with key symptoms of the disorders appearing and disappearing in subsequent editions of the *Diagnostic Statistical Manual of Mental Disorders* (DSM). These repeated revisions represent efforts of *DSM* diagnostic subcommittees to raise the initially low inter-rater reliability of many of these categories (Andreasen & Flaum, 1994; Cooksey & Brown, 1998; Kirk & Kutchins, 1992; Kutchins & Kirk, 1997). Second, these diagnoses are mentalistic in nature and depend on self-report of phenomena, such as sensory hallucinations and delusional beliefs. Such phenomena are neither directly observable nor available for independent verification or measurement. In fact, their detection depends on the veracity of client self-report, which in many cases should be suspect given the client's other psychotic behavior. Preoccupation with convoluted and elusive private events is consistent with a cognitive approach that presumes thoughts and beliefs control overt behavior. However, a functional approach that seeks to change behavior by altering environmental stimulus-response relations would not find these subtle differences very useful. Third, *Diagnostic and Statistical Manual 4th edition (text revision)* diagnoses emphasize pathological symptoms of mental disorders, such as disorganized speech and lack of self-initiated activities. In contrast, functional approaches focus on increasing adaptive behavior, such as appropriate language and social and self-care skills, as well as reducing and replacing undesired responses, such as stereotypic behavior. A functional approach is well suited for the training and rehabilitation of adaptive behavior, which is often necessary given the broad spectrum of disabilities associated with these disorders.

THE BIOMEDICAL MODEL

In order to adequately describe the functional analytic model of schizophrenia, the more common biomedical conceptualization of the disorder must first be reviewed. The prevailing biomedical model assumes that psychotic disorders are due to underlying abnormalities in patients' neurochemistry, neuroanatomy, or other nervous system substrata. However, the scientific evidence in support of this model is much less conclusive than is generally believed (Ross & Pam, 1995; Seibert, 1999; Shean, 2001; Valenstein, 1998; Wong, In press). There is currently no biological test used to determine whether or not an individual has the disorder. Assessment is based solely on the extent to which individuals meet the diagnostic criteria described previously, but the alleged underlying abnormalities are rarely, if ever, observed in individual clients.

Currently, psychotropic medications are the core psychiatric treatment for these disorders, including typical antipsychotics (e.g., fluphenazine, haloperidol), atypical antipsychotics (e.g., risperidone, olanzapine), benzodiazepines, and other drugs (APA, 1997). While pharmaceutical companies have vigorously promoted atypical antipsychotics over first-generation antipsychotic drugs, there is little evidence to indicate that the former have superior therapeutic efficacy. Hence, first-generation neuroleptic drugs are still recommended as a mainstay in the armamentarium for schizophrenia (Lehman et al., 2004). Research also suggests that a large proportion of psychiatric patients in the United States are simultaneously prescribed two or more antipsychotic medications, contrary to professional practice guidelines (Faries, Ascher-Svanum, Zhu, Correll, & Kane, 2005).

First- and second-generation antipsychotic drugs have been shown to be helpful in reducing agitation; managing positive symptoms, such as delusions, hallucinations; and preventing rehospitalization in a sizable proportion of patients. However, they have less ameliorative value with negative symptoms, such as blunted affect, poverty of speech, or apathy. In fact, it can be difficult to distinguish the sedating and enervating (i.e., weakening) effects of antipsychotic medication from the negative symptoms of schizophrenia. Limited success in restoring normal social and vocational functioning is another shortcoming of drug treatment of schizophrenia. There is very little data to show that antipsychotic medication improves performance in these domains (Cohen, 1997; Gelman, 1999), and a significant proportion of people with psychotic disorders who are successfully treated, in the sense that they no longer display positive symptoms, continue to have severe disabilities and highly restricted and unsatisfactory lives.

These deficiencies in the biomedical model are serious. They indicate a need for alternative, scientific-based approaches for motivating, teaching, and refining adaptive skills in clients with schizophrenic and other psychotic disorders.

FUNCTIONAL ANALYTIC MODEL

The functional analytic model of schizophrenia focuses on operationally defined specific behaviors that may be exhibited by an individual with a psychotic disorder. That is, as opposed to assessing the presence or absence of the disorder, the functional analytic model focuses on specific behaviors exhibited by the individual, such as bizarre behavior, perseverative or hallucinatory speech, odd facial expressions or body movements, or social skills deficits. Assessment involves determination of the antecedent and consequential environmental events that influence these behaviors. Once the variables controlling the target behaviors have been identified, an intervention that is designed specifically to address the variables responsible for maintenance of each behavior is implemented. The functional analytic model does not make use of biological explanations. The model does not deny that such variables are important; however, instead of focusing on a biological cause and corresponding intervention, the model analyzes the environmental events that may occasion and/or maintain each specific behavior that makes up the individual's diagnosis.

Individuals with a diagnosis of schizophrenia typically exhibit behavioral excesses as well as skill deficits. Behavioral excesses are responses occurring at unusually high rates that disrupt social relations or activities of daily living or both. Behavioral deficits are responses occurring at unusually low rates that are insufficient to maintain independent living. Specific behavioral excesses and skill deficits commonly seen in schizophrenia are described later in this text.

Current practice guidelines recommend combining behavioral programs, such as social skills training, independent living skills training, and token economies, with psychotropic drugs in the treatment of schizophrenia (APA, 1997; Lehman et al., 2004). These behavioral programs are closely related to interventions from a functional analytic model and were applied decades ago by behavior analysts to decrease symptomatic verbal behavior in adults with psychotic disorders (Ayllon & Haughton, 1964) and increase adaptive behavior and skills in this population (Ayllon & Azrin, 1965). The main difference between early behavioral programs such as the token economy described previously and functional analytic approaches is that the former often used arbitrarily selected consequences (e.g., token reinforcement exchangeable for food or privileges and timeout from reinforcement) to override existing environmental contingencies and thereby promote desired behavior or weaken undesired behavior. In contrast to these early behavioral programs, the functional analytic model investigates various hypotheses about contingencies currently maintaining problematic behavior and tests these hypotheses in analogue assessments known as functional analyses. These analyses lead to different combinations of treatment procedures applying *the same reinforcers that originally maintained the problematic behavior* to either decrease or increase the behavior of interest. So, instead of using token reinforcement or token fines to reduce bizarre behavior that was maintained by escape from demands, a functional analytic model would restructure the environment to allow escape from demands for appropriate behavior, such as asking for a break or for a different assignment.

BEHAVIORAL EXCESSES

Some behavioral excesses that have been successfully treated with functional analytic or simpler behavior modification procedures include bizarre behavior, oppositional behavior, and stereotypical behavior.

Bizarre behavior is conduct so unusual that it baffles or bewilders the observer. Bizarre behavior can take myriad forms, in some cases being coherent and centering around religious, somatic, grandiose, or persecutory themes; or in other cases it may be disorganized and seemingly nonsensical. Bizarre behavior can be verbal (e.g., claiming that one is the "Devil's Right Hand"), motoric (e.g., ritualistic gestures or rigid posture), or a combination of both. Rather than focusing on details of response topography, a functional approach examines antecedent and consequent environmental stimuli that maintain the bizarre behavior and that can be altered to modify it.

Oppositional behavior is the failure to respond to verbal requests, direct commands, and other forms of social influence. Oppositional behavior may be passive in nature, such as not reacting to a request and sitting motionless, or may be active or even aggressive in nature, such as screaming profanities when asked to do a household chore or kicking a staff member who tries to get the client out of bed. Oppositional behavior can block the training of new skills as well as interfere with the performance of all manner of daily living activities.

Stereotypic behavior consists of repetitive, nonproductive responses that often occur without any identifiable external reinforcement. Some examples of stereotypic behavior are pacing, posturing, rituals, and self-talk. Although stereotypic behavior appears odd like bizarre behavior, it can be distinguished from the latter in that it usually is less complex and it occurs repeatedly even though it garners little or no attention from others.

BEHAVIORAL DEFICITS

Another category of problematic responses that are targeted by a functional approach is behavioral deficits. Behavioral deficits are adaptive or productive behaviors needed for independent living in the community, but that the client fails to display in acceptable forms, frequently enough, or in required contexts (Wong, Wilder, Schock, & Clay, 2004). Some behavioral deficits that have been successfully rehabilitated with functional analytic or behavioral training procedures are social, self-care, vocational, leisure, and recreational skills.

Social skills include molecular responses (e.g., eye contact, positive comments) needed for effective communication and to develop and sustain interpersonal relationships. More complex social skills can involve interactive strategies for dealing with difficult situations (e.g., appropriate assertiveness, negotiation techniques). Self-care skills are behaviors maintaining personal hygiene (e.g., bathing, grooming, dressing) and one's living environment (e.g., laundry, housekeeping), which are necessary for independent life in the community. Vocational skills are responses needed for finding and keeping gainful employment. Such skills include looking for job ads, filling out job applications, conveying a positive impression during job interviews, following supervisors' instructions, asking for assistance, and working at tasks for extended periods of time. Recreational skills are leisure activities to make productive use of one's free time, such as the development of hobbies like reading, engaging in sporting activities, or collecting things.

The preceding description represents only a partial list of the skill domains that a comprehensive rehabilitation plan can and should address. Schizophrenia and other psychotic disorders are *pervasive disorders* that typically involve multiple problem behaviors and disruption of multiple areas of adaptive functioning. Each of these emergent problems may require systematic behavioral treatment.

FUNCTIONAL ASSESSMENT AND ANALYSIS OF PSYCHOTIC BEHAVIOR

Functional assessment refers to the process of identifying or determining the variable(s) controlling behavior. Three methods of functional assessment exist: informant, descriptive, and experimental methods. Before these methods are described in detail, a brief description of measurement systems and methods of data collection commonly used as part of the functional analytic model is provided.

MEASUREMENT SYSTEMS AND DATA COLLECTION

The functional analytic model of assessment and treatment includes precise and frequent data collection on all target behaviors. Behaviors are operationally defined, and a method of collecting data on the behaviors is determined based on the frequency and/or continuity of the target behaviors. For discrete target behaviors that are low to moderate in frequency, such as an instance of an odd posture or facial expression, a frequency measure is often used. Frequency measures simply involve recording each occurrence of a target behavior. Frequency measures are often converted to rate (frequency per unit of time) when data are graphed or analyzed.

For target behaviors which are not discrete, such as extended episodes of bizarre speech or repetitive stereotypic behavior, interval systems of measurement are typically used. A particular type of interval system, called a partial interval procedure, is often used to record data on behaviors which are not discrete. Partial interval systems involve designating a time interval (e.g., 10s) and recording whether or not a target behavior occurs within the interval. Data are graphed and analyzed based on the percentage of intervals in which the target behavior occurred in a given time period. Interval systems are only estimates of behavior, and as such, they are more subject to errors in data interpretation than are frequency measures. For target behaviors that can only occur when an opportunity arises, such as a social or self-care skill, the percentage of "correct responses" is often used as a measure. This amount is obtained by dividing the number of correct responses by the number of opportunities and multiplying the result by 100%.

INFORMANT METHODS OF ASSESSMENT

Informant methods of assessment involve obtaining information related to the target behavior from other individuals in the environment. Informant methods include asking caregivers, friends, etc., about the occurrence of the behavior, what precedes it, and what comes after it. Informant methods can consist of informal interviews or more structured tools, such as the Motivation Assessment Scale (Durand & Crimmins, 1988).

One aspect of informant assessment that differs when working with individuals with schizophrenia compared to other populations is that the individuals themselves can be asked about the target behavior. When the clinician works with other populations, such as individuals with developmental disabilities, poorly developed language skills may make it difficult or even impossible to obtain information about the target behavior directly from the client. Because most clients with schizophrenia and other psychotic disorders have relatively well-developed language skills, that is often not the case. An advantage of informant assessment is its ease of use.

Unfortunately, since another person provides the information, it can also be an unreliable method of functional assessment. Other methods of assessment should supplement it.

DESCRIPTIVE METHODS OF ASSESSMENT

In contrast to informant methods, descriptive methods of assessment involve directly observing the target behavior as it occurs. Data are collected through scatterplots, antecedent-behavior-consequence (ABC) data sheets, or through narrative recording. The purpose is to describe the context in which the target behavior occurs, what occurs immediately before the target behavior, and what occurs immediately after the target behavior. Observing and collecting data on repeated instances of the target behavior allow the clinician to determine any patterns in the data that may exist. Descriptive assessment data collected for individuals with schizophrenia may be obtained in a variety of settings including board-and-care homes, places of employment, and day training facilities. Descriptive assessment data can be collected wherever target behaviors occur.

An advantage of descriptive assessment is that information is obtained via observation of the target behavior as it occurs, thus increasing the validity of the assessment procedure. One disadvantage is that it can take a long time to observe and record enough instances of the target behavior to obtain useful information, particularly when target behaviors are infrequent. In addition, only correlational information can be obtained with the use of descriptive assessment. That is, observing that an event occurs before and/or after a target behavior does not necessarily mean that the event causes the target behavior. The event and the target behavior may be related to each other only through a third variable.

EXPERIMENTAL METHODS OF ASSESSMENT

Experimental methods of assessment, often called functional analytic methods or simply functional analysis, involve systematic manipulation of independent variables while measuring the target behavior to determine if a relationship exists between the variables manipulated and behavior. Iwata, Dorsey, Slifer, Bauman, and Richman (1982/1994) developed this method and used it to assess self-injurious behavior exhibited by individuals with developmental disabilities. Since then, it has been extended to assess the functions of a wide range of maladaptive behaviors, often in people with developmental disabilities (see Chapter 4, "Functional Analysis Methodology in Developmental Disabilities," by Didden).

In a functional analysis, clients are exposed to various conditions or scenarios in which antecedent and consequence variables are manipulated. Typical conditions employed as part of functional analyses with individuals with schizophrenia include attention conditions, in which attention is delivered contingent upon the target behavior (test for social positive reinforcement); demand conditions, in which a brief break is provided contingent upon the target behavior (test for social negative reinforcement); alone conditions, in which no programmed consequences are provided for the target behavior (test for automatic reinforcement); and control conditions, in which patients are provided with noncontingent attention and no demands are delivered. The details of these conditions are described in the case study section (see section later in chapter). The occurrence of target behaviors is recorded as clients are exposed to each condition. Conditions which produce the most target behaviors relative to the control condition are seen as including the antecedent and consequent events that are likely to influence or maintain the target behavior. See Figure 14.1 for an example.

A number of special modifications have been made to the standard functional analysis method when working with individuals with schizophrenia. These modifications are mainly due to the often well-developed language skills that individuals with schizophrenia possess. One modification involves the demand condition. In typical demand conditions, individuals are asked to perform a basic self-care or academic task such as touching various body parts or stacking wooden blocks. For most individuals with schizophrenia, these tasks are not relevant. Many individuals with schizophrenia are not in school and routinely perform most very basic self-care tasks, such as brushing their teeth. Instead, more age and situation-appropriate tasks are needed. For example, Wilder, Masuda, O'Connor,

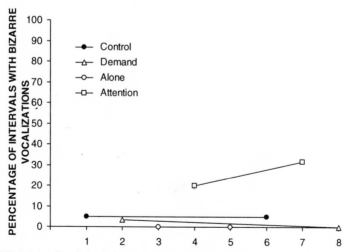

FIGURE 14.1   Graph depicting the results of a functional analysis of the bizarre speech of an adult with schizophrenia.

and Baham (2001) used filling out a job application as the task presented in the demand condition. Another difference between functional analyses conducted with individuals who have well-developed language skills and those who do not involves telling participants the purpose of the assessment procedure. For persons with schizophrenia and other psychotic disorders, telling them that you want to identify the conditions under which they do or do not say odd things or comply with requests may affect the frequency with which they do these things during the assessment procedure. Patients have a right to refuse assessment and treatment procedures, and need accurate information to make a decision about participation. Many individuals with schizophrenia and other psychotic disorders are their own guardians and decide to participate in or refrain from assessment and treatment procedures themselves. In some cases, it may be possible to satisfy the requirement of informed consent while not providing too much information so that the assessment process is made invalid by describing functional analyses in somewhat vague terms. For example, the purpose of the assessment may be described as "determining the ways in which the environment impacts some of the things that you do". This is accurate, and it probably does not provide enough information to negatively impact functional analysis results. The functional analysis conditions themselves might be described in similar terms, with the phrase "you may discontinue participation in the assessment procedure at any time" emphasized.

An advantage of the experimental method of functional assessment is its validity. That is, because it involves testing of each variable posited to maintain behavior in a relatively controlled setting, it is the most reliable way of determining behavioral function. A disadvantage is that it requires more expertise than the previously described methods.

STIMULUS PREFERENCE AND REINFORCER ASSESSMENTS

In addition to assessing the function or reason that a target behavior exhibited by an individual with schizophrenia might occur, it might also be useful to assess what items and activities are preferred and function as reinforcers when developing a treatment program for an individual with schizophrenia. Items identified as preferred and/or as reinforcers might be used as part of a program to increase a behavioral deficit. They might also be used as part of a program to decrease a behavioral excess, particularly when the function of a target behavior cannot be identified.

Stimulus preference assessments are formal methods of assessing preferred items and activities. To date, a number of methods of stimulus preference assessment have been developed, including approach/avoidance or the single item method (Pace, Ivancic, Edwards, Iwata, & Page, 1985), the paired or choice assessment (Fisher et al., 1992), multiple stimulus with

(Windsor, Piche, & Locke, 1994) and without replacement (DeLeon & Iwata, 1996), and the free operant method (Roane, Vollmer, Ringdahl, & Marcus, 1998). Reinforcer assessments consist of procedures used to determine the extent to which items function as reinforcers for a given behavior for an individual. They are sometimes conducted as a supplement to stimulus preference assessments and are often completed in the context of a skill acquisition procedure. They involve an empirical demonstration that some behavior increases in frequency because of a contingency between that behavior and the consequence of interest.

All of these methods were developed for use with children and adults who generally have poor language skills. Although it might be assumed that individuals with schizophrenia who have good language skills can accurately state their preferences for items and activities and accurately describe what might function as a reinforcer for their behavior, this may not always be the case. Indeed, studies using other populations have found that some verbally competent individuals may not be able to accurately describe what functions as a reinforcer for them. Northup, George, Jones, Broussard, and Vollmer (1996) conducted a study with typically developing children with attention deficit hyperactivity disorder in which they evaluated the accuracy of self-report of item/activity preference by comparing a number of different methods of identifying preferred items and subsequently conducting a reinforcer assessment. They found that the accuracy of self-report was poor among this group. In a replication with more participants, Northup (2000) also found similar results.

In order to determine if individuals with schizophrenia can accurately state what will reinforce their behavior and to determine the most effective method of identifying preferred items/activities in this population, Wilder, Ellsworth, White, and Schock (2003) compared three methods of assessing preference for four categories of stimuli in four adults with schizophrenia. The three methods were a survey method, a verbal stimulus choice method, and a pictorial stimulus choice method. After the assessment methods were administered, a coupon system was used to determine which categories of stimuli actually functioned as reinforcers for each participant. Comparisons between the three assessment methods were than made based on the results of the reinforcer assessment. There were few differences in accuracy among the preference assessment procedures: all of the methods were only moderately accurate. Thus, in contrast to the findings of research conducted with typically developing children, self-report does not appear to be any less accurate than other methods of identifying stimuli for adults with schizophrenia.

In the study described in the preceding paragraph, there was no method of preference assessment in which the items were actually present. Thus, it could be that any method that does not involve the items actually being present will be only moderately accurate for adults with schizophrenia. In order to examine this, Wilder, Wilson, Ellsworth, and Heering (2003) compared verbal and tangible methods of identifying preferred items and activities among four adults with schizophrenia. In the verbal assessment, participants were asked "Do you want X or Y?" The items were not present, and the participants made their choice verbally. In the tangible assessment, pairs of items were placed in front of the participant and she/he was asked to choose by approaching the stimuli. The two assessment methods yielded identical highest preference items for three of the four participants and identical lowest preference items for all participants. In addition, the verbal assessment method took less time to complete than the tangible assessment method.

Based on the limited research on this topic thus far, it seems most appropriate to suggest that whatever method is used to identify preferred items/activities for individuals with schizophrenia and other psychotic disorders, assessments should be conducted frequently, particularly for individuals with less-developed language skills. In addition, individual items/activities should be assessed to determine the extent to which they function as reinforcers, even if they are identified as preferred by a formal preference assessment.

FUNCTIONAL ANALYTIC-BASED INTERVENTIONS

INTERVENTIONS FOR BEHAVIORAL EXCESSES

Function-based interventions for behavioral excesses can be divided into three categories: interventions for behavior maintained by social positive reinforcement, social negative reinforcement, and automatic reinforcement. Social positive reinforcement occurs when a behavior results in stimuli being delivered by another person and the frequency of that behavior increases. Examples of specific forms of social positive reinforcement that may maintain or increase target behaviors include the delivery of various forms of attention from others, tangible items, and social activities. Social negative reinforcement occurs when a behavior results in avoidance or removal of stimuli delivered by another person and the frequency of that behavior(s) increases. Examples of specific forms of social negative reinforcement that may maintain or increase target behaviors include the delivery of task demands and various forms of social interaction with others. Automatic reinforcement involves stimulation which is not delivered by another person, but is instead a natural consequence of an immediately preceding behavior and which increases the frequency of that behavior when the individual is in a similar context. Automatic positive reinforcement involves behavior that produces stimulation, while automatic negative reinforcement involves behavior that produces a reduction or alleviation in stimulation. An example of specific forms of positive automatic reinforcement that may maintain or increase target behaviors include inhalation of a cigarette, which may produce a euphoric feeling. An example of negative automatic reinforcement might be striking an odd posture, which may alleviate muscle or joint pain.

Interventions for Behavior Maintained by Social Positive Reinforcement

One approach to intervention for behavior maintained by social positive reinforcement is the manipulation of motivative operations (Laraway, Snycerski, Michael, & Poling, 2003; Michael, 2000). A motivative operation is a stimulus condition or event that (a) establishes or abolishes another stimulus or event as a reinforcer and (b) makes behavior that produces access to this other stimulus or event more (or less) likely to occur. Examples include providing noncontingent access to the reinforcer maintaining the behavior or providing access to the reinforcer maintaining the behavior for extended periods at certain times of the day. In the clinical setting, this might take the form of programmed interaction of a specific duration between staff and clients at frequent times during the day. A staff member might interact with a client in a board-and-care home for 2–3 min every 15 min across the day. If the client's bizarre behavior is maintained by attention, a procedure such as this might reduce the efficacy of attention as a reinforcer for the bizarre behavior. Of course, the reinforcer maintaining the bizarre behavior might be a specific *kind* of attention, in which case that specific kind of attention might have to be delivered on a noncontingent basis to produce reductions in the target behavior. Alternatively, if the client's behavior is maintained by access to a preferred item or activity, providing access to that item or activity on a frequent basis might reduce the effectiveness of that item as a reinforcer for target behavior. Programming specific times that the client has access to the preferred item or activity and scheduling these times so that they occur shortly before or during times that the client is likely to exhibit the target behavior might be another option.

A second approach to intervention for behavior maintained by social positive reinforcement is attention or material extinction. Extinction involves removing the contingency between the response and the reinforcer maintaining that response. In this case, attention or material extinction would involve withholding access to attention or the material(s) maintaining the target behavior contingent upon the occurrence of the target behavior. This might take the form of staff, and perhaps peers, not making eye contact or verbally responding to an individual with schizophrenia when she/he begins to engage in stereotypic behavior, assuming that the reinforcer maintaining stereotypic behavior is some form of attention or access to materials. Of course, in order to be effective, extinction must be implemented with high levels of integrity. Even a single incident of responding to a target behavior can strengthen that behavior, at least in the short term. Extinction may also produce some side effects, such as bursting, or an increase in other inappropriate behavior.

A third intervention option for behavior maintained by social positive reinforcement is differential reinforcement. Differential reinforcement can take many forms, but all forms must incorporate the functional reinforcer to qualify as a function-based intervention. An example of the use of differential reinforcement might consist of providing a lot of the type of attention or tangible item that maintains bizarre speech contingent upon the use of an appropriate phrase.

Most commonly, combinations of two or more of the preceding categories of intervention are used to decrease behavioral excesses. A good example of the combination of two of the preceding interventions is the use of differential reinforcement plus attention extinction. In this case, attention is provided for the use of appropriate phrases, and extinction is employed when bizarre speech occurs. Of course, this is assuming that the target behavior is maintained by attention. An example of the use of this intervention is provided by Wilder et al. (2001). After a brief functional analysis confirmed that attention maintained bizarre speech, an intervention consisting of differential reinforcement of appropriate speech and extinction of bizarre speech was applied. Bizarre speech was nearly eliminated, and appropriate speech increased with the use of this procedure.

Interventions for Behavior Maintained by Social Negative Reinforcement

As with behavioral excesses maintained by social positive reinforcement, excesses maintained by social negative reinforcement can also be addressed using one of three general approaches. First, motivative operations can be manipulated so that reinforcers maintaining client maladaptive behavior are less reinforcing. For behavior maintained by social negative reinforcement, this may involve reducing the aversiveness of the demand or task that the client is being asked to perform. This may be accomplished in a number of ways. One way would be to provide some assistance with the task. Other ways might be to decrease the rate of presentation of the task or portions of the task, or decrease the complexity of the task or to provide frequent breaks during the task. For a client who exhibits bizarre behavior when asked to go out and find a job, a motivative operation intervention might involve breaking the task down into its components and then asking the client to perform just one component while providing as much assistance with the task as possible.

Another intervention option for behavior maintained by negative reinforcement is escape extinction. Escape extinction involves preventing escape when the target behavior occurs. Thus, an example of this might be to require that a client follow through with a task even when a bizarre behavior occurs. The following through might involve using hand-over-hand guidance to assist the client to complete the task. Of course, this should be done only if it safe to do so. For clients who have a history of aggression, using hand-over-hand guidance may not be appropriate. Following through might also take the form of continuous prompting and assistance until the client completes the task. However, the client should be the one performing the task; doing the task for a client does not constitute escape extinction. As with attention and material extinction, the integrity of escape extinction is important. The client must learn that bizarre behavior will not enable avoidance or escape of a task or demand. If avoidance or escape is occasionally obtained, it is unlikely that the procedure will be effective or may even inadvertently strengthen the undesirable behavior.

A third intervention option for behavior maintained by social negative reinforcement is differential negative reinforcement. Differential negative reinforcement involves providing escape/avoidance for an alternative behavior (DNRA) or for the absence of a target behavior for some period of time (DNRO). In practice, this might take the form of allowing a break from a task or demand contingent upon emission of an appropriate request and/or the absence of bizarre behavior.

Interventions for Behavioral Excesses Maintained by Automatic Reinforcement

Behavioral excesses maintained by automatic reinforcement are difficult to treat. The specific stimuli which strengthen behavior maintained by automatic reinforcement, unlike behavior maintained by social variables, are often not able to be identified. Even if they can be identified, manipulating them may be impossible.

If the specific stimuli responsible for the maintenance of automatically reinforced behavior can be identified, intervention becomes more feasible. One option is to manipulate motivative operations. Behavior which produces automatic positive reinforcement might be reduced by either providing a great deal of that or a closely related type of stimulation, or by providing the stimulation maintaining the behavioral excess immediately before the behavior is most likely to occur.

Another option is sensory extinction. Sensory extinction involves breaking the contingency between the behavior and the stimulation produced by the behavior. For behavior maintained by automatic reinforcement, this may be easier said than done. In many cases, there is no way to do so. In other cases, severing the relationship might be possible but is difficult or may involve a medical intervention. For example, at least some cigarette smoking (which is common among persons with schizophrenia) is maintained by the physiological sensation which is produced by the inhalation of the smoke. Medications which block this sensation, or at least reduce it, may exert their behavioral effects through sensory extinction.

A third intervention option for behavior maintained by automatic reinforcement is differential reinforcement. In some cases, it may be possible to teach a client a behavior which produces the same stimulation produced by the target behavior. For example, some odd motor movements exhibited by individuals with schizophrenia may be addressed by teaching an alternate physical response. The alternate response may be particularly likely to reduce the target behavior if it produces the same type of stimulation produced by the target behavior.

Wilder, White, and Yu (2003) provided an example of an intervention used to address a behavioral excess partially maintained by automatic reinforcement in an individual with schizophrenia. A functional analysis suggested that bizarre speech was maintained by attention and automatic reinforcement. Therefore, an intervention consisting of awareness training, competing response training, differential reinforcement of the competing response, and attention extinction was used. Bizarre speech was reduced and appropriate speech was increased.

Interventions for behavior maintained by automatic negative reinforcement are generally medical in nature. They might involve alleviation of some form of aversive stimulation through medication or some other medical procedure. An example might be a woman with schizophrenia who rubs her scalp to the point that she removes hair. She did this only when she awoke each morning. After some investigation, it was determined that her blood pressure was quite low, which probably resulted in a "light-headed" feeling, particularly when she arose from bed each morning. Apparently, rubbing resulted in a reduction of the "light-headed" feeling. Medication to adjust her blood pressure eliminated the behavior.

INTERVENTIONS FOR BEHAVIORAL SKILL DEFICITS

As described earlier, difficulties with social skills are common among individuals with schizophrenia and, although medications are often used to treat behaviors such as hallucinations and delusions, medications generally have little to no beneficial impact on social skills or may sometimes inhibit them. Therefore, these skills need to be taught, and that it is a common task for behavior analysts, psychiatric technicians and aides, and the like. What is the best way to go about teaching these behaviors? The answer to this question may depend on a number of factors, such as the specific social skill being addressed, the learner's history with respect to the skill being taught, and the resources available to the instructor and learner.

Social Skills

The first step in teaching any social skill is to clearly identify what it is that needs to be taught. Social skills can be complex, so any definition must include all aspects of the skill. For example, teaching "appropriate greeting behavior" would involve teaching the appropriate timing and duration of eye contact, appropriate vocal behavior (i.e., what is said and when it is said), appropriate distance between partners, and appropriate hand shaking (i.e., how to do it and how long to hold it), among other things. Once defined, the task should be broken down into its component parts, and these parts should be taught individually. Teaching of these parts will most likely involve a verbal description of the behavior, along with instructor or therapist modeling of the behavior. In addition, the learner should have a number of opportunities to practice the skill being taught while receiving feedback on his/her performance. A criterion for correct performance is often set. Clients should be required to meet this criterion before moving on to the next step. Once the criterion on a subskill is obtained, teaching of the next subskill can begin. Once criteria on all subskills have been met, the skills should be combined to form the composite skill. The learner should have the opportunity to practice the composite skill numerous times and receive feedback from the instructor. Progressively more realistic situations in which to practice the skill should be introduced.

Wilder et al. (2002) provided an example of teaching social skills to an individual with schizophrenia. After the researchers conducted an information-gathering procedure in which they identified asking questions of a conversation partner as the target skill, baseline data collection began. They used a prompting procedure to teach the participant to ask appropriate questions at key points in a conversation. The prompting procedure was evaluated using a multiple baseline design across therapists. The intervention was effective at increasing the number and appropriateness of the questions asked by the participant.

Self-Care

Self-care skills such as grooming, dressing, and personal hygiene skills may also need to be taught to individuals with schizophrenia. Poor self-care can result in health problems and interfere with obtaining and maintaining employment. A few studies have demonstrated the teaching of these skills to individuals with schizophrenia.

As with teaching any other skill, the first step is to identify the skill being taught and precisely define it. Next, the skill should be broken down into smaller, component steps. Each component step should be taught independently before the composite skill is taught in its entirety. The various steps should be first described and then modeled for the learners. After this, learners should have the opportunity to perform the skill and receive feedback based on their performance.

Nelson and Cone (1979) taught grooming skills to a group of patients at a psychiatric hospital unit. They combined the use of token economy with verbal instructions, modeling, and prompts, and found substantial improvements in patient appearance as a result of the teaching procedure. Wong, Wright, Terranova, Bowen, and Zarate (1988) replicated this finding.

Vocational Skills

One of the most common challenges for individuals with psychotic disorders involves vocational activities. Because persons with psychotic disorders often experience social difficulties, and because many vocational activities include social skills, underemployment or unemployment is prevalent among this population. In fact, some studies have reported unemployment to be as high as 80% among individuals with psychotic disorders (Bond & McDonel, 1991).

Behavioral interventions have long been applied to establish vocational skills to individuals with schizophrenia and other psychotic disorders. It is noteworthy that one of the earliest behavioral programs placed emphasis on the teaching of work skills. Ayllon and Azrin's (1968) token economy was used to teach a number of job skills, such as janitorial work, culinary activities, and secretarial work. Behavioral techniques have also been used to teach job interviewing skills. Furman, Geller, Simon, and Kelly (1979) taught clients how to appropriately respond to interview questions and ask questions of the interviewer.

Leisure Skills

Because individuals with schizophrenia and other psychotic disorders often lack appropriate leisure skills, this has also been a focus of intervention. Skinner, Skinner, and Armstrong (2000) increased the reading persistence of an adult with schizophrenia from one page per day during baseline to more than six pages per day during treatment. They also collected follow-up data, which suggested that the client continued to read more per day after the treatment ended.

CASE STUDY

Sara was a 28-year-old woman with a diagnosis of Schizophrenia, Paranoid Subtype. She lived in a board-and-care home with 6 other residents. She did not work but had held a number of previous jobs, and she reported that she would eventually like to be employed in the restaurant industry. She completed 10th grade and dropped out of school at age 18. Sara took a number of medications, including Risperdal, Haldol, and Depakote.

Sara reported seeing and talking to stimuli that are not present. Specifically, she reported that she often saw a small child standing near her. Sara also reported that she had conversations with this girl. Other people who were present when Sara claimed to see the girl did not see or hear anyone. Sara talked to this absent girl two to three times per day, nearly every day. Thus, the target behavior was bizarre speech, defined as phrases or sentences that either referred to stimuli not present or to topics unrelated to the conversation.

After identification of the target behavior, the next step in the functional analysis process was to conduct an informant assessment. The therapist began by asking Sara's housemates and the staff who worked at Sara's home about her bizarre speech. The therapist asked these folks to describe the conditions under which Sara was most likely to say bizarre things. Specifically, the therapist asked about the presence of others, the activity that was occurring or the context in which Sara most often talked to her friend, and what happened when Sara performed the target behavior. The therapist also administered a Motivation Assessment Scale (Durand & Crimmins, 1988) to assist in the informant assessment. Finally, the therapist asked Sara herself to describe the conditions under which she was most likely to talk to the little girl and even asked why she did it.

Next, the therapist began a descriptive assessment in which data were collected on Sara's target behavior as it naturally occurred at her board-and-care home. The therapist collected some of the data but asked staff to collect the majority of the data. The therapist trained staff to record the target behavior, its antecedents, and consequences on a descriptive assessment data sheet. Because the target behavior was not a discrete event (i.e., Sara sometimes talked to the absent girl for 30 min or more), data were collected on episodes of bizarre speech. An episode was defined as bizarre speech that occurred and continued with less than a 60-s break between instances of bizarre speech.

The results of the informant and descriptive assessments were not perfectly clear but suggested that Sara's bizarre speech may be maintained by attention from caregivers. Caregivers typically responded to the speech by asking Sara about the girl, her conversations with the girl, and associated information. Because of the ambiguity in assessment results and because experimental analyses are the most reliable method of assessing function, a functional analysis of bizarre speech was then conducted.

The functional analysis was completed at Sara's board-and-care home. A small, out-of-the-way office was used as the setting for the analysis. Four conditions were employed: alone, demand, social attention, and control. Before the analysis began, Sara was asked if she would be willing to spend some time with the therapist. Sara replied that she was, and Sara's legal guardian also gave consent for Sara to participate in the functional analysis.

Both bizarre speech (target behavior) and appropriate speech were recorded in each session. All sessions were 10 min in duration, and a 10-s partial-interval system of recording was used to record occurrences of both bizarre and appropriate speech. Appropriate speech was defined as any verbal behavior that did not meet the definition of bizarre speech. Bizarre speech was defined as verbal behavior that referred to stimuli which were not present.

In the alone condition, Sara was merely observed. The therapist was not present in the room. In the demand condition, Sara was asked to complete a job application, which was something that she could do but reported that she disliked. The therapist prompted Sara to complete the application in a step-by-step fashion. If Sara did not perform a step, the therapist modeled the correct performance of the step and again asked Sara to complete the step. If Sara still did not complete the step, the therapist gently guided Sara's pencil to write the information required on the application. If Sara engaged in bizarre speech at any time, the therapist immediately said, "This must be stressful for you." told Sara to take a break from work, and removed the application for about 1 min. After 1 min, the application was presented to Sara again. In the attention condition, the therapist sat across a table from Sara. The therapist pretended to be doing paperwork and did not look at or say anything to Sara. However, if Sara engaged in bizarre speech, the therapist immediately looked at Sara and moved a bit closer to her. The therapist maintained eye contact for the duration of Sara's bizarre speech. During the control condition, no demands were presented to Sara, and she and the therapist talked about anything Sara wanted to, as long as it did not involve bizarre speech. Bizarre speech resulted in the therapist briefly looking away from Sara and not verbally responding for about 10 s. Each condition was replicated four times, and the order of conditions was randomized.

The functional analysis showed that Sara talked bizarrely most often during the attention condition. This confirmed the results of the informant and descriptive assessments, both of which hinted at attention as the main-taining variable. Based on these results, a function-based intervention was then developed which consisted of differential reinforcement of alternative behavior and attention extinction. First, Sara was taught some specific conversational skills, including how to appropriately initiate a conversation by approaching another person and saying, "Excuse me," and then asking a question related to stimuli that are present or events that may have recently occurred (e.g., "How was your day?"). She was also taught to wait her turn when talking to others and to appropriately change the topic of conversation by saying, "That reminds me of. . . ." During skills teaching the therapist did not respond to Sara's bizarre speech (i.e., did not make eye contact, did not ask follow-up questions).

The therapist taught each of these skills individually to Sara using behavioral skills training using instruction, modeling, rehearsal, and feedback. The therapist began teaching each skill by first describing the skill. Next, the therapist demonstrated the skill to Sara. Finally, the therapist asked Sara to perform the skill and provided verbal feedback on Sara's performance. Sara had to demonstrate the use of each skill five consecutive times with the therapist before it was considered to be learned.

After the intervention was tested out in the same room in which the functional analysis was conducted and was found to be effective in reducing bizarre speech, the therapist trained the staff at the board-and-care home. Staff and other residents at Sara's home were taught to respond to Sara's newly acquired conversational skills by making direct eye contact and asking follow-up questions about the topic of interest. They were also taught to ignore Sara's bizarre speech.

After 6 months passed, Sara was reevaluated. Although she still occasionally engaged in bizarre speech, it now occurred much less often than it did before the intervention began. Sara was having more appropriate conversations with staff and other residents than ever, and staff reported that she seemed to enjoy her new social abilities.

REFERENCES

American Psychiatric Association. (1997). Practice guidelines for the treatment of patients with schizophrenia. *American Journal of Psychiatry, 154*(Suppl. 4), 1–63.

American Psychiatric Association. (2000). *Diagnostic and statistical manual of mental disorders* (4th ed. text rev.). Washington, D.C.: Author.

Andreasen, N., & Flaum, M. (1994). Characteristic symptoms of schizophrenia. In T. A. Widiger, A. J. Frances, H. A. Pincus, M. B. First, R. Ross, & W. Davis (Eds.), *DSM-IV sourcebook: Volume 1* (pp. 351–379). Washington, D.C.: American Psychiatric Association.

Ayllon, T., & Azrin, N. H. (1965). The measurement and reinforcement of behavior of psychotics. *Journal of the Experimental Analysis of Behavior, 8,* 357–383.

Ayllon, T., & Azrin, N. H. (1968). *The token economy: A motivation system for therapy and rehabilitation.* Englewood Cliffs, NJ: Prentice Hall.

Ayllon, T., & Haughton, E. (1964). Modification of symptomatic verbal behavior of mental patients. *Behavior Research and Therapy, 2,* 87–97.

Bond, G. R., & McDonel, E. C. (1991). Vocational rehabilitation outcomes for persons with psychiatric disabilities: An update. *Journal of Vocational Rehabilitation, 1,* 9–20.

Cohen, D. (1997). A critique of the use of neuroleptic drugs in psychiatry. In S. Fisher & R. P. Greenberg (Eds.), *From placebo to panacea: Putting psychiatric drugs to the test* (pp. 173–228). New York: John Wiley & Sons, Inc.

Cooksey, E. C., & Brown, P. (1998). Spinning on its axes: DSM and the social construction of psychiatric diagnoses. *International Journal of Health Services, 28,* 525–554.

DeLeon, I. G., & Iwata, B. A. (1996). Evaluation of a multiple-stimulus presentation format for assessing reinforcer preferences. *Journal of Applied Behavior Analysis, 29,* 519–532.

Durand, V. M., & Crimmins, D. B. (1988). Identifying the variables maintaining self-injurious behavior. *Journal of Autism and Developmental Disorders, 18,* 99–117.

Faries, D., Ascher-Svanum, H., Zhu, B., Correll, C., & Kane, J. (2005, May). Antipsychotic monotherapy and polypharmacy in the naturalistic treatment of schizophrenia with atypical antipsychotics. *BioMedCentral Psychiatry, 5.* Retrieved September 20, 2005, from http://www.biomedcentral.com/1471-244X/5/26

Fisher, W., Piazza, C. C., Bowman, L. G., Hagopian, L. P., Owens, J. C., & Slevin, I. (1992). A comparison of two approaches for identifying reinforcers for persons with severe and profound disabilities. *Journal of Applied Behavior Analysis, 25,* 491–498.

Furman, W., Geller, M., Simon, S., & Kelly, J. (1979). The use of a behavioral rehearsal procedure for teaching job-interviewing skills to psychiatric patients. *Behavior Therapy, 10,* 157–167.

Gelman, S. (1999). *Medicating schizophrenia: A history.* New Jersey: Rutgers University Press.

Hudson, C. G. (2005). Socioeconomic status and mental illness: Tests of the social causation and selection hypothesis. *American Journal of Orthopsychiatry, 75,* 3–18.

Iwata, B. A., Dorsey, M. F., Slifer, K. J., Bauman, K. E., & Richman, G. S. (1994). Toward a functional analysis of self-injury. *Journal of Applied Behavior Analysis, 27,* 197–209. (Reprinted from *Analysis and Intervention in Developmental Disabilities, 2,* 3–20, 1982).

Keith, S. J., Regier, D. A., & Rae, D. S. (1991). Schizophrenic disorders. In L. N. Robins & D. A. Regier (Eds.), *Psychiatric disorders in America: The epidemiologic catchment area study* (pp. 33–52). New York: The Free Press.

Kessler, R. C., McGonagle, K. A., Zhao, S., Nelson, C. B., Hughes, M., Eshleman, S., et al. (1994). Lifetime and 12-month prevalence of DSM-III-R psychiatric disorders in the United States. *Archives of General Psychiatry, 51,* 8–19.

Kirk, S. A., & Kutchins, H. (1992). *The selling of the DSM: The rhetoric of science in psychiatry.* New York: Aldine De Gruyter.

Kutchins, H., & Kirk, S. A. (1997). *Making us crazy—DSM: The psychiatric bible and the creation of mental disorders.* New York: The Free Press.

Laraway, S., Snycerski, S., Michael, J., & Poling, A. (2003). Motivating operations and terms to describe them: Some further refinements. *Journal of Applied Behavior Analysis, 36,* 407–414.

Lehman, A. F., Kreyenbuhl, J., Buchanan, R. W., Dickerson, F. B., Dixon, L. B., Goldberg, R., et al. (2004). The Schizophrenia Patient Outcomes Research Team (PORT): Updated treatment recommendations 2003. *Schizophrenia Bulletin, 30,* 193–217.

Michael, J. (2000). Implications and refinements of the establishing operation concept. *Journal of Applied Behavior Analysis, 33,* 401–410.

Nelson, G. L., & Cone, J. D. (1979). Multiple baseline analysis of a token economy for psychiatric inpatients. *Journal of Applied Behavior Analysis, 12,* 255–217.

Northup, J. (2000). Further evaluation of the accuracy of reinforcer surveys: A systematic replication. *Journal of Applied Behavior Analysis, 33,* 335–338.

Northup, J., George, T., Jones, K., Broussard, C., & Vollmer, T. R. (1996). A comparison of reinforcer assessment methods: The utility of verbal and pictorial choice procedures. *Journal of Applied Behavior Analysis, 29,* 201–212.

Pace, G. M., Ivancic, M. T., Edwards, G. L., Iwata, B. A., & Page, T. J. (1985). Assessment of stimulus preference and reinforcer value with profoundly retarded individuals. *Journal of Applied Behavior Analysis, 18,* 249–255.

Roane, H. S., Vollmer, T. R., Ringdahl, J. E., & Marcus, B. A. (1998). Evaluation of a brief stimulus preference assessment. *Journal of Applied Behavior Analysis, 31,* 605–620.

Ross, C. A., & Pam, A. (1995). *Pseudoscience in biological psychiatry: Blaming the body.* New York: John Wiley & Sons.

Shean, G. (2001). A critical look at some assumptions of biopsychiatry. *Ethical Human Sciences and Services, 3*, 77–96.

Siebert, A. (1999). Brain disease hypothesis for schizophrenia disconfirmed by all evidence. *Ethical Human Sciences and Services, 1*, 179–189.

Skinner, C., Skinner, A., & Armstrong, K. J. (2000). Analysis of a client-staff-developed program designed to enhance reading persistence in an adult diagnosed with schizophrenia. *Psychiatric Rehabilitation Journal, 24*, 52–57.

Valenstein, E. S. (1998). *Blaming the brain: The truth about drugs and mental health.* New York: The Free Press.

Wilder, D. A., Ellsworth, C., White, H., & Schock, K. (2003). A comparison of stimulus preference assessment methods in adults with schizophrenia. *Behavioral Interventions, 18*, 151–160.

Wilder, D. A., Masuda, A., Baham, M., & O'Connor, C. (2002). An analysis of the training level necessary to increase independent question asking in an adult with schizophrenia. *Psychiatric Rehabilitation Skills, 6*, 32–43.

Wilder, D. A., Masuda, A., O'Conner, C., & Baham, M. (2001). Brief functional analysis and treatment of bizarre vocalizations in an adult with schizophrenia. *Journal of Applied Behavior Analysis, 34*, 65–68.

Wilder, D. A., White, H., & Yu, M. (2003). Functional analysis and treatment of bizarre vocalizations exhibited by an adult with schizophrenia: A replication and extension. *Behavioral Interventions, 18*, 43–52.

Wilder, D. A., Wilson, P., Ellsworth, C., & Heering, P. (2003). A comparison of verbal and tangible stimulus preference assessment methods in adults with schizophrenia. *Behavioral Interventions, 18*, 191–198.

Windsor, J., Piche, L. M., & Locke, P. A. (1994). Preference testing: A comparison of two presentation methods. *Research in Developmental Disabilities, 15*, 439–455.

Wong, S. E. (In press). Behavior analysis of psychotic disorders: Scientific dead end or casualty of the mental health political economy? *Behavior and Social Issues.*

Wong, S. E., Wilder, D. A., Schock, K., & Clay, C. (2004). Behavioral interventions with severe and persistent mental disorders. In H. E. Briggs & T. L. Rzepnicki (Eds.), *Using evidence in social work practice: Behavioral perspectives* (pp. 210–230). Chicago: Lyceum Books.

Wong, S. E., Wright, J., Terranova, M. D., Bowen, L., & Zarate, R. (1988). Effects of structured ward activities on appropriate and psychotic behavior of chronic psychiatric patients. *Behavioral Residential Treatment, 3*, 41–50.

15

MOOD DISORDERS

DEREK R. HOPKO

The University of Tennessee—Knoxville, TN

SANDRA D. HOPKO

Cariten Assist Employee Assistance Program—Knoxville, TN

AND

CARL. W. LEJUEZ

The University of Maryland—College Park, MD

MOOD DISORDERS

Mood disorders include various affective problems such as major depressive disorder (MDD), bipolar disorder, cyclothymic disorder, and dysthymic disorder. The *Diagnostic and Statistical Manual of Mental Disorders, text revision* (American Psychiatric Association, 2000) highlights two primary diagnostic criteria for MDD, depressed mood and loss of interest or pleasure in activities (anhedonia), at least one of which must occur for at least 2 weeks. Secondary symptoms include appetite change and/or weight loss, sleep disturbance, psychomotor agitation or retardation, fatigue or energy loss, feelings of worthlessness or guilt, attentional or concentration difficulties, and recurrent thoughts of death and/or suicide. Of these symptoms, depressed mood, appetite and sleep change, and thoughts of death are most common, while anhedonia and psychomotor change are less common (Weissman, Bruce, Leaf, Florio, & Holzer, 1991).

Although about 30% of American adults report feeling dysphoric at some point during their life span (Weissman et al., 1991), the experience of MDD is less common. The National Comorbidity Survey found a lifetime prevalence of 17.1% and a 1-year prevalence of 10.3% (Kessler, Chiu, Demler, & Walters, 2005; Kessler et al., 1994). The American Psychiatric

Association estimated the lifetime risk of MDD between 10 and 25% for women and 5 and 12% for men, with females twice as likely to develop clinical depression (APA, 2000; Just & Alloy, 1997; Nolen-Hoeksema & Girgus, 1994). Other risk factors include Caucasian ethnicity, separation or divorce, prior depressive episodes, poor physical health and/or medical illnesses, low socioeconomic status, adverse life events (e.g., unemployment, loss of loved one), and family history of depression (cf. Hopko, Lejuez, Armento, & Bare, 2004; Kaelber, Moul, & Farmer, 1995). Although depression may develop at any age, the average age of onset is 15–19 in females and 25–29 for males (APA, 2000; Burke, Burke, Regier, & Rae, 1990).

Functional impairment associated with MDD also is quite extensive, including exacerbation of medical illness (Stevens, Merikangas, & Merikangas, 1995), increased use of medical health services (Simon & Katzelnick, 1997; Simon, Ormel, VonKorff, & Barlow, 1995), maladaptive cognitive processes (Abramson, Metalsky, & Alloy, 1989; Beck, Shaw, Rush, & Emory, 1979), decreased engagement in rewarding behaviors (Hopko, Armento, Chambers, Cantu, & Lejuez, 2003a; Lewinsohn, 1974), and problems with interpersonal relationships (Klerman, Weissman, Rounsaville, & Chevron, 1984). MDD also is highly comorbid with other psychiatric problems such as anxiety disorders (Mineka, Watson, & Clark, 1998) and alcohol abuse (Regier et al., 1990). Finally, the economic costs (e.g., healthcare, medication, lost wages, absenteeism) of treating depressive disorders are enormous (Booth et al., 1997), with several hundred million dollars spent annually to treat MDD (Jonsson & Rosenbaum, 1993; Montgomery, Doyle, Stern, & McBurney, 2005).

Considering the prevalence and functional impairment associated with clinical depression, it is important that continued strides are made toward developing valid and effective assessment and intervention strategies. In pursuit of this objective, several functional analytic models of clinical depression have been proposed. From one perspective, it has been suggested that functional analytic paradigms may be difficult to apply to clinical depression. These arguments are based on the premises that depression largely is experienced in the form of private as opposed to observable public behaviors, and that fundamental behavior analytic principles often are ignored in the current application of functional analytic assessment and intervention strategies (Kanter, Callaghan, Landes, Busch, & Brown, 2004; Moore, 1980). Others have argued that even the most private (schema-based) models of depression may be conceptualized via functional analytic models (Bolling, Kohlenberg, & Parker, 2000) and that several novel functional assessment procedures and functional analytic interventions have been developed that may supplement early behavior analytic formulations of depression (Hayes, Masuda, Bissett, Luoma, & Guerrero, 2004; Hopko, & Hopko, 1999; Hopko, Lejuez, Ruggiero, & Eifert, 2003c). This chapter highlights various functional analytic models of depression, the process of conducting functional assessments with depressed patients, reviews traditional as well as contemporary functional analytic interventions for clinical depression, and concludes with a case illustration depicting the use of functional assessment and treatment methods. Although this chapter could be broadened to include mood disorders in general, the focus will exclusively be on clinical depression given the more comprehensive literature and greater application of behavior analytic models toward assessment and intervention for this disorder.

FUNCTIONAL ANALYTIC MODELS OF DEPRESSION

FERSTER'S BEHAVIOR ANALYTIC MODEL

Ferster (1973, 1981) proposed that depression occurs as a result of sudden environmental changes in which opportunities for positive reinforcement become limited. Leaning on an etiological formulation of depression that involved disrupted early childhood attachment experiences and a relatively fixed ratio of reinforcement contingent on substantial social activity, Ferster suggested that depressed individuals come to experience a low rate of positively reinforced social behaviors that result in a passive lifestyle. This hypothesis was supported in early research examining social correlates of depression (Dykman, Horowitz, Abramson, & Usher, 1991; Gotlib, 1982; Gotlib & Robinson, 1982; MacPhillamy & Lewinsohn, 1982; Rehm, 1988). According to Ferster, it is this passivity or, perhaps more precisely, a response pattern of escape and avoidance behaviors that come to be associated with increased dysphoria and aversive experiences (e.g., negative social encounters). This process is consistent with a matching-law-based model of clinical depression whereby the relative frequency of depressed behavior, compared to nondepressed behavior (i.e., all other types of behavior), is proportional to the relative value of reinforcement obtained for depressed behavior compared to nondepressed behavior (Herrnstein, 1970; Lejuez, Hopko, LePage, Hopko, & McNeil, 2001b; McDowell, 1982). In essence, in response to inadequate environmental reinforcement (historically and/or proximally), a state of deprivation evolves that is accompanied by a narrowing of behavioral repertoires. Depressed or escape and avoidance behaviors (e.g., inactivity, ruminating, pacing) become the primary response. Due to healthy (nondepressed) behaviors being extinguished, depressed behaviors are reinforced on a stronger schedule, and avoidance behaviors (of both aversive and appetitive stimuli) limit opportunities to experience environmental reinforcement.

LEWINSOHN'S MODEL

Closely related to Ferster's behavioral paradigm, Lewinsohn (1974) proposed that the primary cause of depression is low rates of response-contingent positive reinforcement (RCPR) for healthy behavior and a response schedule that serves to elicit depressive behaviors such as passivity, dysphoria, fatigue, and a range of maladaptive cognitions. Decreased RCPR is believed to occur for three primary reasons. First, lower rates of RCPR may be due to a limited number of events that are potentially rewarding to an individual (Lewinsohn & Graf, 1973; Lewinsohn & Libet, 1972). The qualitative and quantitative properties of these events are presumed to be quite variable across individuals and are a function of biological and historical variables. Second, RCPR is a function of the availability and value of reinforcement in the environment. Value is defined as the interaction of several parameters of reinforcement including frequency, magnitude, duration, immediacy, and certainty, and may be affected by any number of variables, including one's learning history, level of deprivation or satiation, and interference created by psychological problems such as anxiety disorders (Barlow, 2002). Depression might also be a function of an increased frequency of punishment that may diminish both the availability and value of reinforcers (Lewinsohn, Antonuccio, Breckenridge, & Teri, 1984). Finally, RCPR is proposed as strongly influenced by instrumental behavior, particularly abilities (or skills) needed to obtain RCPR in the social environment (also see Coyne, 1976). Inadequate behavioral repertoires decrease the likelihood of obtaining (social) reinforcement and are presumed to be a primary antecedent toward eliciting depressive affect.

Although the Lewinsohn (1974) paradigm remains a strong conceptual model, Lewinsohn and colleagues also have proposed a more integrative model of depression that might better address the complexities and variable presentations of major depression (Lewinsohn, Hoberman, Teri, & Hautziger, 1985). In this model, RCPR continues to be a core feature of depressive states, but dispositional factors are given more attention, as are environmentally initiated changes in cognitions. Situational factors continue to be viewed as triggers to depressive episodes, but cognitive experiences are perceived as moderators of the effects of environmental experience.

KANFER AND GRIMM'S MODEL

Kanfer and Grimm's (1977) general model of psychopathology represents an organizational structure of behaviors that are classified on the basis of function as opposed to problem content. This model was not developed to address depression but may readily be applied to this form of psychopathology. Within this model, problematic behaviors are presumed to occur in the context of various response classes, stimulus classes, and contingency relationships. Behavioral and emotional problems are formulated with reference to one or a combination of five primary categories. First, *behavior deficits* are highlighted as an important antecedent to psychological problems. Behavior deficits generally are presumed to occur as a result of an inadequate knowledge base for guiding behavior, skills deficits, inadequate self-directing responses, inability to alter responses in conflict situations, and deficits in self-monitoring and self-reinforcement [also see Lynn Rehm's (1977) self-control model of depression]. Second, *behavior excesses* such as inappropriately conditioned anxiety to objects or events and excessive self-observational activity may be problematic in that these behavioral patterns might be associated with increased depressive and anxious affect as well as inappropriate standards of reinforcement. Third, *problems in environmental stimulus control* may contribute to depressive affect. This functional class would include stimuli that elicit inappropriate emotional reactions, exposure to restrictive environments that do not provide opportunities for reinforcement [similar to Lewinsohn's (1974) formulation], and the inefficient arrangement of controlling stimuli for daily activities (e.g., poor time management). Fourth, depression may be a function of *inappropriate self-generated stimulus control.* This might include tendencies to inaccurately describe one's abilities or behaviors as they pertain to certain contexts as well as inappropriate covert (or verbal) behaviors. Finally, as highlighted in Ferster (1973) and Lewinsohn (1974), *inappropriate contingency arrangement* may result in depressive affect. This might result from a lack of RCPR, environmental maintenance of undesirable behavior (through reinforcement strategies), response noncontingent reinforcement, and even excessive use of reinforcement for desirable behaviors that might result in satiation.

The heuristic framework outlined by Kanfer and Grimm (1977) allows for functional assessment of problematic depressive behaviors. Additionally, it provides direction in selecting the most practical of cognitive-behavioral interventions (see their work for a description of these strategies).

PARADIGMATIC MODEL OF DEPRESSION

The paradigmatic model of depression (Eifert, Beach, & Wilson, 1998; Staats & Heiby, 1985) is a comprehensive formulation of major depression that includes attention to acquired biological risk factors, historical antecedent events, psychological vulnerability in the form of deficient and inappropriate basic behavioral repertoires, current antecedent or precipitating events, and the stimulus properties and consequences of depressive symptoms. According to this model, there are three basic behavioral repertoires, termed *emotional-motivational, language-cognitive,* and *sensory-motor.* Under conditions of inappropriate or deficient learning via interactions with the environment, repertoires may be formed that increase vulnerability to depression, such as negative self-evaluative feedback (emotional-motivational repertoire), depressive attributional style and distorted cognitive processing (language-cognitive repertoire), as well as deficient social skills and avoidance behavior (sensory-motor repertoire). When unlearned or biological vulnerabilities (Malhi, Parker, & Greenwood, 2005) and historical antecedents such as early parental loss, insecure parental attachment, trauma, and noncontingent reward and punishment that affect repertoire development are taken into account, an individual may be particularly predisposed toward developing depressive symptoms. Also contributing to depressive symptoms as manifested within behavioral repertoires, current environmental antecedents are highly functional and may include increased life stress, trauma, or illness; decreased social support or reduction of reinforcement in another life domain; and associated changes in discriminative stimuli that decrease the likelihood of engaging in healthy (nondepressive) behavior and increase the likelihood of behaving in a depressed manner (Lejuez et al., 2001). Taken together, the paradigmatic model is a comprehensive, multilevel framework that is useful toward assessing the function of depressive symptoms from an ideographic perspective.

FUNCTIONAL ASSESSMENT AND ANALYSIS OF DEPRESSIVE SYMPTOMS

As highlighted in earlier works (Hopko et al., 2004; Thorpe & Olson, 1997), functional analytic models of depression are highly compatible with a broad array of assessment strategies that assess depressive symptom patterns and behaviors, including unstructured and structured interviews, self-report measures, observational methods, and functional behavioral assessment. Although a multitude of assessment resources are available, their clinical utility varies greatly across patient and assessment context, as does the level of skill required for appropriate administration (Alexopoulos et al., 2002; Hopko et al., 2004). Importantly, within the framework of functional models presented in the previous section, these assessment strategies all may be used to facilitate a comprehensive behavioral assessment of stimulus-organism-response-consequence variables that are critical toward conceptualizing a depressive episode (Goldfried & Sprafkin, 1976). In relation to depression response classes, target behaviors are formulated based on functional models, but this process generally will include primary attention to the somatic, behavioral, social, and verbal-cognitive symptoms of depression. For a discussion of the functional properties of language, see Skinner (1957) and Hayes, Strosahl, and Wilson (1999). The principal methods of assessing depressive symptom patterns are discussed in this section, as the information obtained from these approaches largely serves as the foundation for functional analytic interventions for depression (Nelson-Gray, 2003).

CLINICAL INTERVIEWS AND SELF-REPORT MEASURES

Functional assessment of depressive symptoms may involve the use of clinical interviews conducted with the patient, family member, spouse, caregiver, or teacher in the case of adolescent or childhood depression. The manner in which interviews are conducted varies substantially, ranging from unstructured and open-ended approaches, to more structured methods that are more restrictive. Insofar as conducting a traditional behavior analytic interview (Kanfer & Grimm, 1977), unstructured clinical interviews are likely of greater utility in determining the stimuli, organismic variables, behavioral responses, and consequences that may be pertinent toward understanding the etiology and maintenance of depressive behaviors. Unstructured interviews also may allow for increased therapist-patient rapport and greater opportunities to explore unique details of a patient's symptom pattern (Haynes, 1998). On the other hand, and with the caveat that structured interviews may be less practical in a variety of clinical settings, given managed care policies, structured interviews may be of some utility as a supplementary source of information. In the realm of clinical depression, these interviewing strategies might include use of the *Structured Clinical Interview for DSM-IV—Patient Version* (SCID-I/P; First, Spitzer, Gibbon, & Williams, 1996), *Anxiety Disorders Interview Schedule* (ADIS-IV; Brown, Di Nardo, & Barlow, 1994), the *Schedule for Affective Disorders and Schizophrenia* (SADS; Endicott & Spitzer, 1978), *Hamilton Rating Scale for Depression* (HRSD; Hamilton, 1960), and the *Diagnostic Interview Schedule* (DIS; Robins, Helzer, Croughan, & Ratcliff, 1981).

Self-report measures of depression also have proven useful in assessing a tremendous range of content areas, including affective, verbal-cognitive, somatic, behavioral, and social symptoms of depression. Several of the most commonly utilized measures include the *Beck Depression Inventories* (BDI; Beck & Steer, 1987; BDI-II; Beck, Steer, & Brown, 1996), *Hamilton Depression Inventory* (HDI; Reynolds & Kobak, 1995), the *Center for Epidemiological Studies' Depression Scale* (CES-D; Radloff, 1977), the *Harvard Department of Psychiatry/National Depression Screening Day Scale* (HANDS; Baer et al., 2000), the *Reynolds Depression Screening Inventory* (RDSI; Reynolds & Kobak, 1998), the *Minnesota Multiphasic Personality Inventory 2 Depression Scale* (MMPI-2-D; Butcher, Dahlstrom, Graham, Tellegen, & Kaemmer, 1989), and the *Personality Assessment Inventory* (PAI; Morey, 1991). The psychometric properties of these structured interviews and self-report instruments generally are quite strong (Nezu, Ronan, Meadows, & McClure, 2000), and their primary benefit in functional assessment may be a precise identification of behavioral responses associated with depression. Nonetheless, these methods also might serve to further generate hypotheses about controlling variables associated with depressive episodes and highlight specific environmental contingencies in need of further assessment through more traditional functional assessment strategies. For example, the *SCID-I/P* and *ADIS-IV* interviews include sections that qualitatively assess the social, medical, and occupational correlates of depression; the relevance of substance use as an antecedent or consequent variable of depressive symptoms; and the environmental context and possible precipitants of depression, including reference to specific triggering events.

OBSERVATIONAL METHODS

More consistent with traditional functional assessment strategies, observational methods of assessing depressive symptoms involve selecting appropriate target behaviors, defining behaviors, deciding on which dimensions of behavior to measure (e.g., frequency, duration, magnitude), and determining the most appropriate method to record data (Foster, Bell-Dolan, & Burge, 1988). Moreover, these observational strategies generally are conducted with reference to the specific environmental context(s) in which they occur. Indeed, direct observational strategies often are used outside the rigors of experimental settings, and in the current context, mental health professionals frequently rely on behavioral observations of patients' significant others to better assess the functionality of depressive behavior. In both formal and informal applications, depressive behaviors may include *excesses* such as crying, increased sleep, irritable/agitated behaviors, increased substance use, and even suicidal behaviors. Depression also may be marked by behavioral *deficits* that include minimal eye contact, psychomotor retardation, decreased recreational and occupational activities, social withdrawal, as well as disruption in sleep, eating, and sexual behaviors. Interestingly, although direct observational assessment of depressive behaviors should intuitively be a primary tool of behavioral therapists and researchers, it is remarkable how little systematic research has been done in this area.

Pertaining to verbal behavior, depressed individuals generally exhibit a slower and more monotonous rate of speech (Gotlib & Robinson, 1982; Libet & Lewinsohn, 1973; Robinson & Lewinsohn, 1973), take longer to respond to the verbal behavior of others (Libet & Lewinsohn, 1973), exhibit more frequent self-focused negative remarks (Blumberg & Hokanson, 1983; Gotlib & Robinson, 1982), and use fewer achievement and "power" words (e.g., committed, strong) in their speech (Andreasen

& Pfohl, 1976). Nonverbal, motoric differences between depressed and nondepressed individuals also are evident. Depressed individuals generally smile less frequently (Gotlib & Robinson, 1982), exhibit reductions in motoric activities (e.g., reading, grooming; Williams, Barlow, & Agras, 1972), make less eye contact during conversation (Field, Healy, Goldstein, & Guthertz, 1990; Gotlib, 1982), hold their head in a downward position more frequently, engage in more self-touching (Ranelli & Miller, 1981), may be more prone to aggression and conflict (Hinchliffe, Hooper, & Roberts, 1978; Suls & Bunde, 2005), and are rated as less competent in social situations (Dykman et al., 1991). When used by mental health personnel, clinician-rated behavior observation systems also may be useful toward identifying individuals with more severe levels of depression (LePage, Mogge, Sellers, & DelBen, 2003).

Behavioral monitoring logs or diaries are another potential means of gathering information about controlling variables and sources of environmental reinforcement. For example, MacPhillamy and Lewinsohn (1971, 1982) developed the Pleasant Events Schedule to assess, monitor, and modify positive activities among individuals with depression. Our research group has used daily diaries to assess the frequency and duration of unhealthy depressive behaviors as part of a functional assessment process and to assist in treatment planning (Lejuez, Hopko, & Hopko, 2001a, 2002). Recent research has indicated that these daily diaries can be useful in assessing both immediate and future reward value of current behaviors, that reward value ratings correlate highly with self-report measures of depressive affect, and that mildly depressed and nondepressed students can be distinguished via response style (Hopko et al., 2003a).

FUNCTIONAL ANALYSIS

Although many different definitions have appeared in the literature (Haynes & O'Brien, 1990; Nelson-Gray, 2003), functional analysis generally refers to the process of identifying important, controllable, and causal environmental factors that may be related to the etiology and maintenance of depressive symptoms. As stated by Jacobson and Gortner:

> The behavior analytic framework emphasizes functional analyses of the environmental events that have impinged upon individual clients to generate depression, and formulates cases in a way that looks outside rather than inside the person for targeting change. That is, instead of emphasizing faulty thinking, [behavior analysis] conceptualizes depression in terms of environmental events that created contextual shifts, which in turn have denied the client access to those reinforcers which normally functioned as anti-depressants. (2000; cited in Martell, Addis, & Jacobson, 2001, p. 112)

In the domain of clinical depression, functional analysis involves the process of operationally defining undesirable depressive behavior(s) such as lethargy, social withdrawal (and more global avoidance behaviors), crying, alcohol abuse, and suicidality, as well as determining establishing operations among environmental variables. Strategies for conducting functional assessments and analyses include interviews, naturalistic observation, and/or the manipulation of specific situations that result in an increase or decrease of target behaviors (Haynes & O'Brien, 1990; O'Neill, Horner, Albin, Storey, & Sprague, 1990). Often incorporating daily monitoring exercises, depressed patients may be asked to record depressive target behaviors, the context (time, place, surroundings) in which they occur, and the consequences that follow. With all functional analytic strategies, the therapist is concerned with identifying the function(s) of an individual's depressed behavior or, put more simply, *why* the depressed behavior occurs. Within this behavioral philosophy, depressed behavior occurs because reinforcement for healthy behavior is minimal and/or healthy behaviors are punished and thus reduced, or because positive and negative reinforcement for depressive behavior is excessive. In other words, depressed behavior may develop following extinction or punishment of healthy behaviors consequent to a decrease in response contingent positive reinforcement and may be maintained via the experience of pleasant consequences (e.g., other people completing responsibilities, attention and sympathy) and/or as a result of the removal of aversive experiences (e.g., unpleasant or stressful activities).

Functional assessment and analysis techniques also may be useful for understanding maladaptive thought processes that more cognitively oriented therapists believe to be critical in eliciting depressive affect (Beck et al., 1979). Indeed, through strategies that include the use of thought-monitoring logs or various thought sampling methods (Csikszentmihalyi & Larson, 1987; Hurlburt, 1997), functional analysis strategies can be used to identify thought patterns elicited by certain environmental events and how these cognitions may correspond with depressive mood states. At the core of cognitive theory, schemas serve as fundamental negative belief patterns that cause public and private depressed behaviors. Within a behavior analytic framework, schemas are viewed as contingency-shaped private behaviors (i.e., rules) that are modified within the context of basic behavioral principles including shaping procedures within the therapeutic relationship (Bolling et al., 2000).

Functional assessment and analysis methods may be useful as an independent procedure or in the context of a broader comprehensive assessment, assisting in developing hypotheses about factors maintaining depressive behaviors and the formulation of a treatment plan (Nelson-Gray, 2003). However, pretreatment functional assessment and analyses are infrequently conducted, and possible controlling variables are rarely verified via empirical assessment (Haynes & O'Brien, 1990; Nelson-Gray, 2003). As highlighted in the following section, traditional (Ferster, 1973) and contemporary behavioral theories and interventions for depression (Hayes et al. 1999; Hopko & Hopko, 1999; Lejuez et al., 2002; Martell et al., 2001; McCullough, 2000; Nezu & Nezu, 1989) to a greater or lesser degree incorporate functional analytic assessment and analysis techniques.

FUNCTIONAL ANALYTIC INTERVENTIONS FOR DEPRESSIVE DISORDERS

The most prominent functional analytic interventions for depression briefly are reviewed in the following section. In reading about these treatment methods, the reader is encouraged to take into account the general contention that clinical depressions often are topographically similar yet functionally quite distinct. In other words, similar symptom patterns may be a product of very different controlling variables. For example, three women with breast cancer, all having recently undergone bilateral mastectomies, may have identical symptoms of depression. One of these women may be depressed primarily due to decreased social reinforcement and punishing experiences in the form of conflictual experiences with her partner. Another woman's symptoms largely may be a product of rule-governed behavior in which she has equated breast surgery with a loss of femininity. In the third case, breast cancer treatment might have resulted in lengthy absences from work and inappropriate termination by an insensitive employer. In these instances, different functional analytic interventions might be indicated. Behavioral activation or cognitive-behavioral analysis system of psychotherapy (CBASP) might be most appropriate for patient 1, FAP or ACT for patient 2, and problem-solving therapy for patient 3. The important point is several treatment options are available, and even though several behavioral therapies are effective for depression, only rarely do well-designed functional analyses dictate choice of treatment modality (Nelson-Gray, 2003). A great deal more research is needed to determine the effectiveness of patient-treatment matching based on functional analysis.

TRADITIONAL BEHAVIORAL INTERVENTIONS

A functional analytic view suggests that depressed behavior develops and persists due to positive and negative reinforcement for depressed behavior and/or a lack of reinforcement or punishment of healthy alternative behaviors (Ferster, 1973; Lewinsohn, 1974). Based on this formulation, conventional behavioral therapy for depression was designed to increase access to pleasant events and positive reinforcers, as well as decrease the intensity and frequency of aversive events and consequences (Lewinsohn & Atwood, 1969; Lewinsohn & Graf, 1973; Lewinsohn, Sullivan, &

Grosscup, 1980). In early investigations assessing the efficacy of these strategies, Lewinsohn and colleagues demonstrated that daily self-monitoring of pleasant/unpleasant events and corresponding intervention methods such as activity scheduling, social skills development, and time management training could be used to reduce depressive symptoms. Indeed, such approaches (i.e., pleasant event scheduling) were found to be as effective in treating depressed outpatients as were cognitive and inter-personal skills training approaches (Zeiss, Lewinsohn, & Munoz, 1979).

These early behavioral interventions were not without their criticisms. Researchers questioned both the adherence of these interventions to behavior analytic principles and the overall empirical support for pure behavioral therapy. For example, the suggestion has been made that Lewinsohn's early interventions were too heavily focused on assessing and treating depression based on decreased response-contingent positive rein-forcement, while neglecting depressive behaviors initiated via the process of negative reinforcement (Kanter, Callaghan, Landes, Busch, & Brown, 2004). Behavioral assessment and intervention strategies were thereby per-ceived as noncomprehensive in that important controlling variables such as negative reinforcers, aversive discriminative stimuli, and deprivation-establishing operations were not taken into account (Kanter et al., 2004). A second criticism was made that evidence supporting the efficacy of pure behavioral interventions was equivocal at best, with more multifaceted interventions that included cognitive restructuring (McLean & Hakstian, 1979) rendering superior outcomes (Blaney, 1981). As a result of these concerns and growing interest in cognitive theory and therapy, pure behav-ioral interventions largely were abandoned in favor of combined cognitive and behavioral approaches (Lewinsohn et al., 1984; Lewinsohn, Munoz, Youngren, & Zeiss, 1986).

CONTEMPORARY BEHAVIORAL INTERVENTIONS

Following the recent publication of a component analysis study indicat-ing that a comprehensive cognitive-behavioral intervention was no more effective than behavioral intervention in treating depression (Jacobson et al., 1996), interest in pure behavioral approaches was revitalized. This is most evident in the development of two novel interventions: *Behavioral Activation* (BA; Martell et al., 2001) and the *Brief Behavioral Activation Treatment for Depression* (BATD; Lejuez et al., 2001a, 2002). Behavioral activation is defined as a therapeutic process that emphasizes structured attempts at engendering increases in overt behaviors that are likely to bring the patient into contact with reinforcing environmental contingencies and produce corresponding improvements in thoughts, mood, and overall quality of life (Hopko et al., 2003c).

Although the BA and BATD treatment protocols utilize somewhat different strategies, both approaches generally are consistent with the earlier etiological formulations of depression. However, these newer protocols may entail important advances with respect to case conceptualization and choice of intervention components. First, the recent activation approaches are more idiographic, giving more attention to the unique environmental contingencies maintaining an individual's depressed behavior (Lejuez et al., 2002; Martell et al., 2001). In line with this philosophy, there has been a movement from targeting pleasant events per se (Lewinsohn & Graf, 1973) to understanding the *functional* aspects of behavior change (Martell et al., 2001). So, rather than indiscriminately increasing exposure to events or activities that are presumed to be rewarding, this functional analytic approach involves a detailed assessment of contingencies maintaining depressive behavior, idiographic assessment of patients' short- and long-term goals, and the subsequent targeting of behaviors that are likely to improve quality of life for that particular patient. Further, in contrast to an *a priori* nomothetic assumption of what is pleasant, the appropriateness of any particular behavioral change is determined by ongoing assessment based on whether the frequency and/or duration of that behavior increases over time and leads to a corresponding reduction in depressive symptoms. Activation strategies are primarily based on the principles of extinction, fading, shaping, and differential reinforcement of incompatible behaviors (Hopko et al., 2003c).

Preliminary data support the utility of BATD among depressed patients in a community mental health center (Lejuez et al., 2001b), an inpatient psychiatric facility (Hopko, Lejuez, LePage, Hopko, & McNeil, 2003b), as a supplemental intervention for patients with coexistent Axis I (Hopko, Hopko, & Lejuez, 2004) and Axis II disorders (Hopko, Sanchez, Hopko, Dvir, & Lejuez, 2003d), and as a treatment for depressed cancer patients in primary care (Hopko, Bell, Armento, Hunt, & Lejuez, 2005). Data also indicate that the more extensive form of behavioral activation (Martell et al., 2001) may be comparable to cognitive therapy and Paroxetine, with the psychosocial interventions associated with longer-term gains and reduced medical costs (Hollon, 2003; Jacobson et al., 1996). Of these studies, three are randomized controlled trials (Hollon, 2003; Hopko et al., 2003b; Jacobson et al., 1996).

FUNCTIONAL ANALYTIC PSYCHOTHERAPY FOR DEPRESSION

Functional analytic psychotherapy (FAP; Kohlenberg & Tsai, 1991) involves the application of functional analysis within the context of the therapeutic relationship. FAP may be utilized as an independent therapeutic modality but is easily integrated into alternative behavioral and cognitive therapy applications, including clinical depression (Hopko & Hopko, 1999; Kohlenberg, Kanter, Bolling, Parker, & Tsai, 2002). Within the FAP framework, the therapist is responsible for (a) identifying problematic behaviors displayed during the session (clinically relevant behaviors or CRB1s), (b) evoking CRB1s for the purpose of developing more appropriate behaviors (CRB2s), (c) facilitating patient observations and interpretations of their behaviors (CRB3s), and (d) reinforcing improvements made during the session while failing to reinforce and occasionally punishing maladaptive behaviors. Accordingly, the therapist takes a very active role in identifying, and on some level manipulating, antecedents and consequences of patient behaviors as they occur during sessions. The FAP therapist works toward shaping more appropriate behaviors via differential and successive reinforcement (Follette, Naugle, & Callaghan, 1996). From a behavior analytic perspective, FAP embraces fundamental concepts such as *contextualism* (i.e., problems are a function of specific environmental variables and are shaped through language and learning) and *generalization* (i.e., the natural and therapeutic environments are functionally similar if they evoke the same behavior). With reference to relational frame theory, a concept described in the following section, client and therapist verbal behaviors and associated contingencies are proposed as the initial mechanism by which the generalization of patient response classes begins to occur. Theoretically, to the degree that the patient is guided toward emitting verbal behaviors with others who exhibit verbal relations similar to that of the therapist, the patient's behavior comes under stimulus control (Follette et al., 1996). It should be noted, however, that direct empirical support for this process is limited.

Applied to depression, therapists are responsible for identifying depressive behaviors as they occur in session, such as passivity, limited eye contact, and verbal deficiencies as outlined earlier. These observations are interpreted within the social (functional analytic) context and communicated to the patient. Alternative nondepressive behaviors are identified and potentially role-played in therapy, and when these preferred behaviors are observed, the patient is rewarded via therapist acknowledgment and praise. In the context of supplementing cognitive therapy for depression, a distinction is made between *cognitive products* (i.e., thoughts and beliefs expressed through language) and the more nonbehavioral *cognitive structures* (i.e., schema). FAP-enhanced cognitive therapy is guided by the degree to which rules or contingencies contribute in the development and maintenance of problem behaviors (Kohlenberg & Tsai, 1994). In the case of rule-governed behavior, antecedents lead to consequences, which may result in certain beliefs (i.e., rule formation). Alternatively, antecedents could lead to beliefs and then to (behavioral) consequences (i.e., rule following). In the case of contingency-shaped behavior, antecedents more directly lead to consequences. If a problem were primarily rule governed, FAP therapists would focus on evoking maladaptive rules via depressive thoughts and verbalizations and take a more traditional cognitive approach to treatment. In the case of more contingency-shaped behaviors, in-session contingencies would be established such that depressed behavior would be identified and extinguished with minimal attention to cognitive processes. For example, if a depressed patient begins to withdraw (consequence) when discussing a recently deceased relative (antecedent), regardless of thought processes that may be occurring, the therapist would illustrate this contingency and teach the patient strategies to cope with the aversive emotional experience (e.g., diaphragmatic breathing, communicating positive memories of the deceased person). When exhibited in future sessions, the therapist would reinforce these more adaptive behaviors. Following a preliminary case study that supported FAP-enhanced cognitive therapy (Kohlenberg & Tsai, 1994), a recent study indicated that relative to traditional cognitive therapy, FAP-enhanced cognitive therapy resulted in significantly greater reductions in depression and increased relationship satisfaction at post-treatment and 3-month follow-up (Kohlenberg et al., 2002). More systematic research is necessary to assess the potential benefits of FAP as applied to individuals with well-diagnosed depression.

ACCEPTANCE AND COMMITMENT THERAPY

Although the intervention was not specifically designed as a treatment for depression, acceptance and commitment therapy (ACT; Hayes et al., 1999) is a functional analytic intervention that recently has been applied to the treatment of emotional disorders (Hayes et al., 2004). ACT is based on the philosophies of *functional contextualism* (Hayes, 1993) and *relational frame theory* (Hayes, 1991). Functional contextualism refers to the idea that human behavior is best understood with reference to both historical and more proximal controlling variables. These variables are external to the behavior of human organisms and are in principle manipulable, which has the potential effect of influencing or modifying behavior (Hayes et al., 1999). Functional contextualism is in many ways similar to functional analysis. Perhaps the most significant point of departure is the increased attention that functional contextualism ascribes to verbal behavior, and how covert and overt verbal behaviors are associated with certain environmental stimuli and serve a predominant role in the development and maintenance of pathological behavior. These rule-governed behaviors are understood with reference to relational frame theory, which is a detailed account of how human language and cognitions come to be associated with human suffering, not necessarily because of the content or frequency of cognitions per se, but more as a function of the environment in which they are experienced. Based on these ideas, depression and other forms of mental illness are perceived as a consequence of environmental contingencies, including behavioral processes outlined earlier (Ferster, 1973; Lewinsohn, 1974). Extending further, however, depression also would involve experiential avoidance, or unhealthy attempts to not only avoid external environmental stimuli associated with negative affect, but also the private experiences of thoughts, memories, and emotions (Hayes, Wilson, Gifford, Follette, & Strosahl, 1996). As these maladaptive cognitive processes occur, one's value system or direction in life becomes impaired, as does the ability to behave in healthy, nondepressed ways that are consistent with this value system.

Given this conceptualization, the primary goal of ACT is to increase psychological acceptance, rather than defend against unwanted cognitions, memories, emotions, and facilitate behavioral change through focusing on more controllable aspects of the external environment. Through interventions that include creative hopelessness, cognitive defusion (diminishing the role of literal thought), decreasing emotional control, and value-defining exercises, ACT has been applied to a number of patient samples (Hayes et al., 2004). Although the theory and applications of ACT may be viable for depressed individuals, thus far only two randomized trials (Zettle & Hayes, 1986; Zettle & Raines, 1989) and two case studies (Hayes, Masuda, & De Mey, 2003; Lopez & Arco, 2002) have been published. These studies provide preliminary support for the use of ACT with depressed individuals, but more research is necessary to fully evaluate the efficacy of this treatment.

PROBLEM-SOLVING THERAPY

Problem-solving therapy (PST; D'Zurilla & Nezu, 2001; Nezu, 1987) is based on the notion that ineffective problem-solving skills are involved in the onset and continuation of depressive symptoms. Supported by an increasing literature, Nezu and colleagues (Nezu, 2004) have outlined how problem-solving skills may moderate the association between negative life events or stressors, negative attributional styles, and depression. Within this framework, problem-solving skills involve five primary components: problem orientation, problem definition and formulation, generation of alternatives, solution implementation, and verification. Although generally not perceived as a functional analytic intervention for depression, PST may be considered in this domain, with specific negative contextual experiences and a corresponding inability to effectively problem solve serving as primary antecedents to depressive symptoms. PST implicates actual or perceived negative consequences (in response to a problematic situation) as having three primary functions: (a) they exacerbate existing problems and increase the likelihood that future problems will occur, (b) they lead to a decrease in perceived and/or actual environmental reinforcement, and

(c) they decrease the likelihood that individuals will engage in future problem solving (Nezu, 1987). Although all three functions may be incorporated within models discussed in the initial section, function (b) has obvious relevance to traditional functional models of depression (Ferster, 1973; Lewinsohn, 1974). Moreover, the idea of inadequate behavioral repertoires as contributing to psychopathology is consistent with formulations of Kanfer and Grimm (1977) and the paradigmatic model of depression (Eifert et al., 1998; Staats & Heiby, 1985).

To a greater degree than all other functional analytic interventions presented in this section, PST has very strong empirical support and generally is considered an efficacious treatment for depression (DeRubeis & Crits-Christoph, 1998). In the context of a number of randomized controlled trials in both academic (Nezu, 1986; Nezu & Perri, 1989) and primary care settings (Wolf & Hopko, 2005), PST has been shown to be superior to wait-list control groups and treatment-as-usual conditions, and in most cases to have comparable efficacy with antidepressant medications (Mynors-Wallis, Gath, Day, & Baker, 2000; Mynors-Wallis, Gath, Lloyd-Thomas, & Tomlinson, 1995; Williams et al., 2000).

COGNITIVE-BEHAVIORAL ANALYSIS SYSTEM OF PSYCHOTHERAPY

The CBASP (McCullough, 2000, 2003) was specifically designed to treat individuals with chronic depression. Based firmly on the developmental theories of Piaget (1923), CBASP is considered a functional analytic intervention in that it is a contingency training program that predominantly is based on principles of negative reinforcement. According to McCullough (2003), "being able to identify the consequences of one's behavior (perceived functionality) and learning to recognize one's stimulus value for others, as well as the stimulus value that others have for the patient, are the essential goals of treatment" (p. 839). So the basic philosophy is that when patients are educated about the negative behavioral consequences elicited by their interpersonal behaviors, an uncomfortable emotional state arises between therapist and patient such that the patient becomes motivated to learn and implement more adaptive interpersonal behaviors to avoid the experiences with the therapist (i.e., negative reinforcement). Primarily through the highly structured contingency procedure of situational analysis, patients present an interpersonal encounter that was experienced as problematic. Through assessing antecedent events, situational behaviors, and both actual and desired outcomes, patients are provided with a functional analytic interpretation of the interpersonal event. In the remediation phase that follows, the therapist and patient focus on the negative interpersonal event and problem-solve with respect to how to achieve a more desirable outcome. This process involves revising irrelevant and inaccurate interpretations of interpersonal behaviors, role playing and modification of interpersonal behaviors, and generalization and transfer of learning (McCullough, 2000). Other more specific procedures such as the interpersonal discrimination exercise are used to assess the functional impact of lengthy relationships (e.g., parents, siblings, intimate friends) on current interpersonal behaviors.

Although the position has been taken that CBASP is an empirically supported intervention for chronic depression (McCullough, 2003; Riso, McCullough, & Blandino, 2003), this suggestion should not be interpreted as indicating CBASP is an empirically validated treatment by contemporary standards (DeRubeis & Crits-Christoph, 1998). Indeed, CBASP represents a very promising intervention, but the efficacy of the treatment has been demonstrated in only one large randomized controlled trial (Keller et al., 2000). In this study, 662 chronically depressed patients were randomized to either CBASP, nefazadone, or the combination treatment (over a 12-week period). Among treatment completers ($n = 519$), response rates were good in the nefazadone (55%) and CBASP (52%) conditions, and exceptional with the combined intervention (85%). Importantly, subsequent data indicated that for treatment responders, monthly exposure to CBASP reduced the likelihood of depression recurrence (Klein et al., 2004). As with the other functional analytic interventions reviewed in this section, a more substantial body of treatment outcome research is necessary to evaluate the efficacy and effectiveness of CBASP relative to other psychosocial and pharmacological interventions.

CASE STUDY

The following case study illustrates the use of functional assessment and functional analytic intervention with a patient diagnosed with clinical depression.

CLIENT DESCRIPTION

"Kim" was a 36-year-old married Caucasian woman with two young children. She had a doctorate in engineering and had been a professor for 5 years at a large university. Kim presented with depressive symptoms that included depressed mood, decreased sleep and appetite, anhedonia, concentration difficulties, and feelings of guilt and low self-worth. She also reported several symptoms of generalized anxiety that included excessive worry in a number of life domains (e.g., personal and family health, career, finances); difficulty controlling worry; and psychosomatic symptoms such as increased muscle tension, insomnia, and general restlessness. Physiological symptoms of anxiety included periodic experiences of tachycardia, shortness of breath, and nausea. Cognitive symptoms included a pronounced fear of failure with respect to both her professional and familial responsibilities (i.e., work/family conflict). In addition to substance use behaviors, behavioral symptoms included periodic gambling, overt anger episodes, and social withdrawal. Such behaviors frequently resulted in marital conflict and an associated increase in depressive and anxiety symptoms.

HISTORY OF THE DISORDER

Kim indicated that she had experienced minor depression for as long as she could remember, with the most severe symptoms manifesting over the past 6 years. She reported positive peer and parental relationships, a history of alcohol abuse, and had no significant medical history other than a chronic case of asthma that developed early in childhood. She did report a substantial family history of depression and anxiety problems, as well as polysubstance abuse. Kim reported that the past several years had been particularly difficult, following the initiation of her academic appointment and birth of her two children. Kim reported no history of inpatient or outpatient psychological treatment, but had been taking moderate doses of Klonopin and paroxetine for about 5 years. Kim reported that she periodically engaged in binge drinking (i.e., approximately three times per month, 6–10 beers each occasion), which usually was followed by a visit to the local casino where she played the slot machines for several hours.

PSYCHOLOGICAL ASSESSMENT

Kim initially was given a brief unstructured clinical interview followed by the Anxiety Disorders Interview Schedule (ADIS-IV; Brown et al., 1994). Results suggested that Kim met *DSM-IV-TR* (APA, 2000) criteria for major depressive disorder and generalized anxiety disorder. Further supporting the diagnosis of major depressive disorder, Kim scored 23 on the Hamilton Rating Scale for Depression (HRSD; Hamilton, 1960) and 34 on the Beck Depression Inventory (severe depression; Beck & Steer, 1987). Kim's profile on the Personality Assessment Inventory (PAI; Morey, 1991) was valid and interpretable, and elevations were evident on the depression (T = 75) and anxiety (T = 72) clinical scales. On the Quality of Life Inventory (QOLI; Frisch, 1994), which assesses life satisfaction in various life domains (e.g., health, relationships, money), Kim scored in the "low" range of life satisfaction (QOLI total = –5).

As a pretreatment assessment strategy and part of a BATD (Lejuez et al., 2001a, 2002), Kim completed a daily diary for 1 week (Hopko et al., 2003c). Daily monitoring revealed that Kim was leading a very active lifestyle, predominantly characterized by employment-related activities, but also involving frequent household chores, television viewing, and transportation of children. Binge drinking and gambling also were evident. When queried about the reinforcement value of such activities, Kim indicated that minimal pleasure was being experienced, other than the peace and solitude that she associated with alcohol use.

Following daily monitoring, the assessment process involved an identification of Kim's values and goals within life domains. These were family, social, and intimate relationships; education; employment/career; hobbies/ recreation; volunteer work/charity; physical/health issues; and spirituality (Hayes et al., 1999). Based on this evaluation, it was clear that Kim strongly valued her roles as an academician and mother, but also her identification with her Christian religion.

Functional analytic procedures were conducted via unstructured interviews with the patient and husband to identify environmental factors that may be serving to maintain depressive and anxious behaviors. These interviews revealed that Kim's depressive and anxiety-related behaviors were substantially maintained by the consequences that followed. For example, Kim's compulsive work behaviors were negatively reinforced in that she avoided persistent fears of failure. Unfortunately, these behaviors also resulted in less frequent contact with her husband and children, thus compromising important aspects of her life. In response to the negative affect elicited by these circumstances, Kim engaged in binge drinking and gambling behaviors that were perceived as experiential avoidance activities. Thus, binge drinking and gambling were negatively reinforced. Anger-related behaviors such as emotional outbursts directed at husband were largely a function of Kim's perception that she was being asked to increase responsibility for childcare, which she perceived as unreasonable given her substantial work demands and perceived role as family provider.

TARGETS SELECTED FOR TREATMENT

The first treatment goal was to systematically increase response contingent positive reinforcement by facilitating increased exposure to behaviors that were consistent with Kim's value/goal assessment. To accomplish this objective, Kim engaged in a BATD (Lejuez et al., 2001a), whereby she moved through a constructed behavioral hierarchy in a progressive manner, moving from the easier behaviors to the more difficult. Activities included in this hierarchy involved increasing time spent with her children and spouse, increased time in spiritual activities (e.g., solitary prayer, attending church), and health-related behaviors such as physical fitness activities and designated nutritional meal plans. Anxiety-reducing exercises also were included such as progressive-muscle relaxation and walks in a nearby park, which also functioned to increase family time. For each activity, Kim and the clinician collaboratively determined what the *final goal* would be in terms of the frequency and duration of activity per week. These goals were recorded on the master activity log that was kept in the possession of the therapist. *Weekly goals* were recorded on the behavioral checkout form that Kim brought to therapy each week. At the start of each session, the behavioral checkout form was examined and discussed, with new weekly goals being established as a function of Kim's success or difficulty.

A second component of this treatment was ACT-based exercises described earlier (Hayes et al., 1999). As Kim's depression and anxiety symptoms (e.g., drinking, gambling) clearly involved experiential avoidance, or unhealthy attempts to avoid private experiences, Kim's ability to live according to her value system was impaired. Thus, through cognitive defusion exercises and strategies aimed at decreasing emotional control, Kim was taught to be more accepting of unwanted cognitions and emotions, and to focus more extensively on controllable aspects of the environment through BATD. Certain cognitions, including those focused on a fear of failure, were presented as being a function of distal life experiences, including experiences with a strongly evaluative father during childhood and adolescence. Through understanding the development of such thoughts, and how they became functionally linked to more proximal contexts such as work, Kim was able to deliteralize these cognitions, or see them merely as cognitive experiences that potentially were nonvalid. Kim was in fact the most productive member of her department. Finally, through utilization of the BATD- and ACT-based strategies, Kim's engagement in family activities and perceptions of family responsibilities were altered such that the primary antecedent to expressed anger was removed (i.e., verbal behaviors of spouse).

Kim made fairly robust improvements during the 15-week BATD-ACT–based intervention, with a documented decrease in her HRSD (8) and BDI (11) scores and a notable increase in quality of life (1). Drinking and gambling behaviors also were substantially minimized, as were anger-related behaviors directed toward her husband.

REFERENCES

Abramson, L. Y., Metalsky, G. I., & Alloy, L. B. (1989). Hopelessness depression: A theory-based subtype of depression. *Psychological Review, 96*, 358–372.

Alexopoulos, G. S., Borson, S., Cuthbert, B. N., Devanand, D. P., Mulsant, B. H., Olin, J. T., et al. (2002). Assessment of late life depression. *Biological Psychiatry, 52*, 164–174.

American Psychiatric Association. (2000). *Diagnostic and statistical manual of mental disorders* (4th ed., text rev.). Washington, D.C.: Author.

Andreasen, N. J. C., & Pfohl, B. (1976). Linguistic analysis of speech in affective disorders. *Archives of General Psychiatry, 33*, 1361–1367.

Baer, L., Jacobs, D. G., Meszler-Reizes, J., Blais, M., Fava, M., Kessler, R., et al. (2000). Development of a brief screening instrument: The HANDS. *Psychotherapy and Psychosomatics, 69,* 35–41.

Barlow, D. H. (2002). *Anxiety and its disorders: The nature and treatment of anxiety and panic* (2nd ed.). New York: Guilford Press.

Beck, A. T., Shaw, B. J., Rush, A. J., & Emery, G. (1979). *Cognitive therapy of depression.* New York: Guilford Press.

Beck, A. T., & Steer, R. A. (1987). *Beck Depression Inventory: Manual.* San Antonio, TX: The Psychiatric Corporation.

Beck, A. T., Steer, R. A., & Brown, G. K. (1996). *Manual for the BDI-II.* San Antonio, TX: The Psychological Corporation.

Blaney, P. H. (1981). The effectiveness of cognitive and behavioral therapies. In L. P. Rehm (Ed.), *Behavior therapy for depression: Present status and future directions* (pp. 1–32). New York: Academic Press.

Blumberg, S. R., & Hokanson, J. E. (1983). The effects of another person's response style on interpersonal behavior in depression. *Journal of Abnormal Psychology, 92,* 196–209.

Bolling, M. Y., Kohlenberg, R. J., & Parker, C. R. (2000). Behavior analysis and depression. In M. J. Dougher (Ed.), *Clinical behavior analysis* (pp. 127–152). Reno, NV: Context Press.

Booth, B. M., Zhang, M., Rost, K. M., Clardy, J. A., Smith, L. G., & Smith, G. R. (1997). Measuring outcomes and costs for major depression. *Psychopharmacological Bulletin, 33,* 653–658.

Brown, T. A., DiNardo, P. A., & Barlow, D. H. (1994). *The Anxiety Disorder Interview Schedule for DSM-IV.* Center for Stress and Anxiety Disorders. Albany: State University of New York.

Burke, K. C., Burke, J. D., Regier, D. A., & Rae, D. S. (1990). Age at onset of selected mental disorders in five community populations. *Archives of General Psychiatry, 47,* 511–518.

Butcher, J. N., Dahlstrom, W. G., Graham, J. R., Tellegen, A., & Kaemmer, B. (1989). *Minnesota Multiphasic Personality Inventory-2 (MMPI-2): Manual for administration and scoring.* Minneapolis, MN: University of Minnesota Press.

Coyne, J. C. (1976). Toward an interactional description of depression. *Psychiatry, 39,* 28–40.

Csikszentmihalyi, M., & Larson, R. (1987). Validity and reliability of the experience sampling method. *Journal of Nervous and Mental Disease, 175,* 526–536.

DeRubeis, R. J., & Crits-Christoph, P. (1998). Empirically supported individual and group psychological treatments for adult mental disorders. *Journal of Consulting and Clinical Psychology, 66,* 37–52.

Dykman, B. M., Horowitz, I. M., Abramson, L. Y., & Usher, M. (1991). Schematic and situational determinants of depressed and nondepressed students' interpretation of feedback. *Journal of Abnormal Psychology, 100,* 45–55.

D'Zurilla, T. J., & Nezu, A. M. (2001). Problem-solving therapies. In K. S. Dobson (Ed.), *The handbook of cognitive-behavioral therapies* (2nd ed., pp. 211–245). New York: Guilford Press.

Eifert, G. H., Beach, B. K., & Wilson, P. H. (1998). Depression: Behavioral principles and implications for treatment and relapse prevention. In J. J. Plaud & G. H. Eifert (Eds.), *From behavior research to behavior therapy* (pp. 68–97). Needham Heights, MA: Allyn & Bacon.

Endicott, J., & Spitzer, R. L. (1978). A diagnostic interview: The schedule for affective disorders and schizophrenia. *Archives of General Psychiatry, 35,* 837–844.

Ferster, C. B. (1973). A functional analysis of depression. *American Psychologist, 28*, 857–870.

Ferster, C. B. (1981). A functional analysis of behavior therapy. In L. P. Rehm (Ed.), *Behavior therapy for depression: Present status and future directions* (pp. 181–196). New York: Academic.

Field, T., Healy, B., Goldstein, S., & Guthertz, M. (1990). Behavior-state matching and synchrony in mother-infant interactions of nondepressed versus depressed dyads. *Developmental Psychology, 26*, 7–14.

First, M. B., Spitzer, R. L., Gibbon, M., & Williams, J. (1996). *Structured Clinical Interview for DSM-IV Axis I Disorders—Patient Edition (SCID-I/P, Version 2.0)*. New York: Biometrics Research Department, New York Psychiatric Institute.

Follette, W. C., Naugle, A. E., & Callaghan, G. M. (1996). A radical behavioral understanding of the therapeutic relationship affecting change. *Behavior Therapy, 27*, 623–641.

Foster, S. L., Bell-Dolan, D. J., & Burge, D. A. (1988). Behavioral observation. In A. S. Bellack & M. Hersen (Eds.), *Behavioral assessment: A practical handbook* (3rd ed., pp. 119–160). New York: Pergamon Press.

Frisch, M. B. (1994). *Manual and treatment guide for the Quality of Life Inventory (QOLI)*. Minneapolis, MN: National Computer Systems.

Goldfried, M. R., & Sprafkin, J. N. (1976). Behavioral personality assessment. In J. T. Spence, R. C. Carson, & J. W. Thibaut (Eds.), *Behavioral approaches to therapy* (pp. 295–321). Morristown, NJ: General Learning Press.

Gotlib, I. H. (1982). Self-reinforcement and depression in interpersonal interaction: The role of performance level. *Journal of Abnormal Psychology, 91*, 3–13.

Gotlib, I. H., & Robinson, L. A. (1982). Responses to depressed individuals: Discrepancies between self-report and observer-rated behavior. *Journal of Abnormal Psychology, 91*, 231–240.

Hamilton, M. (1960). A rating scale for depression. *Journal of Neurology, Neurosurgery, and Psychiatry, 23*, 56–62.

Hayes, S. C. (1991). A relational control theory of stimulus equivalence. In L. J. Hayes & P. N. Chase (Eds.), *Dialogues on verbal behavior* (pp. 19–40). Reno, NV: Context Press.

Hayes, S. C., (1993). Analytic goals and the varieties of scientific contextualism. In S. C. Hayes, L. J. Hayes, H. W. Reese, & T. R. Sarbin (Eds.), *Varieties of scientific contextualism* (pp. 11–27). Reno, NV: Context Press.

Hayes, S. C., Masuda, A., Bissett, R., Luoma, J., & Guerrero, L. F. (2004). DBT, FAR and ACT: How empirically oriented are the new behavior therapy technologies? *Behavior Therapy, 35*, 35–54.

Hayes, S. C., Masuda, A., & De Mey, H. (2003). Acceptance and commitment therapy and the third wave of behavior therapy. *Gedragstherapie (Dutch Journal of Behavior Therapy), 2*, 69–96.

Hayes, S. C., Strosahl, K. D., & Wilson, K. G. (1999). *Acceptance and commitment therapy: An experiential approach to behavior change*. New York: Guilford Press.

Hayes, S. C., Wilson, K. G., Gifford, E. V., Follette, V. M., & Strosahl, K. (1996). Emotional avoidance and behavioral disorders: A functional dimensional approach to diagnosis and treatment. *Journal of Consulting and Clinical Psychology, 64*, 1152–1168.

Haynes, S. N. (1998). The assessment-treatment relationship and functional analysis in behavior therapy. *European Journal of Psychological Assessment, 14*, 26–35.

Haynes, S. N., & O'Brien, W. H. (1990). Functional analysis in behavior therapy. *Clinical Psychology Review, 10*, 649–668.

Herrnstein, R. J. (1970). On the law of effect. *Journal of the Experimental Analysis of Behavior, 13*, 243–266.

Hinchliffe, M., Hooper, D., & Roberts, F. J. (1978). *The melancholy marriage.* New York: Wiley.

Hollon, S. D. (2003). *Behavioral activation, cognitive therapy, and antidepressant medication in the treatment of major depression.* Symposium presented at the 37th annual convention of the Association for the Advancement of Behavior Therapy, Boston, MA.

Hopko, D. R., Armento, M., Chambers, L., Cantu, M., & Lejuez, C. W. (2003a). The use of daily diaries to assess the relations among mood state, overt behavior, and reward value of activities. *Behaviour Research and Therapy, 41,* 1137–1148.

Hopko, D. R., Bell, J. L., Armento, M. E. A., Hunt, M. K., & Lejuez, C. W. (2005). Behavior therapy for depressed cancer patients in primary care. *Psychotherapy: Theory, Research, Practice, Training, 42,* 236–243.

Hopko, D. R., & Hopko, S. D. (1999). What can functional analytic psychotherapy contribute to empirically validated treatments? *Clinical Psychology and Psychotherapy, 6,* 349–356.

Hopko, D. R., Hopko, S. D., & Lejuez, C. W. (2004). Behavioral activation as an intervention for co-existent depressive and anxiety symptoms. *Clinical Case Studies, 3,* 37–48.

Hopko, D. R., Lejuez, C. W., Armento, M. E. A., & Bare, R. L. (2004). Depressive disorders. In M. Hersen (Ed.), *Psychological assessment in clinical practice: A pragmatic guide* (pp. 85–116). New York: Taylor & Francis.

Hopko, D. R., Lejuez, C. W., LePage, J., Hopko, S. D., & McNeil, D. W. (2003b). A brief behavioral activation treatment for depression: A randomized trial within an inpatient psychiatric hospital. *Behavior Modification, 27,* 458–469.

Hopko, D. R., Lejuez, C. W., Ruggiero, K. J., & Eifert, G. H. (2003c). Contemporary behavioral activation treatments for depression: Procedures, principles, and progress. *Clinical Psychology Review, 23,* 699–717.

Hopko, D. R., Sanchez, L., Hopko, S. D., Dvir, S., & Lejuez, C. W. (2003d). Behavioral activation and the prevention of suicide in patients with borderline personality disorder. *Journal of Personality Disorders, 17,* 460–478.

Hurlburt, R. T. (1997). Randomly sampling thinking in the natural environment. *Journal of Consulting and Clinical Psychology, 65,* 941–949.

Jacobson, N. S., Dobson, K. S., Truax, P. A., Addis, M. E., Koerner, K., Gollan, J. K., et al. (1996). A component analysis of cognitive-behavioral treatment for depression. *Journal of Consulting and Clinical Psychology, 64,* 295-304.

Jonsson, B., & Rosenbaum, J. (1993). *Health economics of depression.* New York: Wiley.

Just, N., & Alloy, L. B. (1997). The response styles theory of depression: Tests and an extension of the theory. *Journal of Abnormal Psychology, 106,* 221–229.

Kaelber, C. T., Moul, D. E., & Farmer, M. E. (1995). Epidemiology of depression. In E. E. Beckham & W. R. Leber (Eds.), *Handbook of Depression* (2nd ed., pp. 3–35). New York: Guilford Press.

Kanfer, F. H., & Grimm, L. G. (1977). Behavioral analysis: Selecting target behaviors in the interview. *Behavior Modification, 1,* 7–28.

Kanter, J. W., Callaghan, G. M., Landes, S. J., Busch, A. M., & Brown, K. R. (2004). Behavior analytic conceptualization and treatment of depression: Traditional models and recent advances. *The Behavior Analyst Today, 5,* 255–274.

Keller, M. B., McCullough, J. P., Klein, D. N., Arnow, B., Dunner, D. L., Gelenberg, A. J., et al. (2000). A comparison of nefazadone, the cognitive-behavioral analysis system of psychotherapy, and their combination for the treatment of chronic depression. *New England Journal of Medicine, 342,* 1462–1470.

Kessler, R. C., Chiu, W. T., Demler, O., & Walters, E. E. (2005). Prevalence, severity, and comorbidity of 12-month DSM-IV disorders in the National Comorbidity Survey Replication. *Archives of General Psychiatry, 62,* 617–627.

Kessler, R. C., McGonagle, K. A., Zhao, S., Nelson, C. B., Hughes, M., Eshleman, S., et al. (1994). Lifetime and 12-month prevalence of DSM-III-R psychiatric disorders in the United States: Results from the National Comorbidity Survey. *Archives of General Psychiatry, 51*, 8–19.

Klein, D. N., Santiago, N. J., Vivian, D., Arnow, B. A., Blalock, J. A., Dunner, D. L., et al. (2004). Cognitive-behavioral analysis system of psychotherapy as a maintenance treatment for chronic depression. *Journal of Consulting and Clinical Psychology, 72*, 661–688.

Klerman, G. L., Weissman, M. M., Rounsaville, B. J., & Chevron, E. S. (1984). *Interpersonal psychotherapy of depression*. New York: Basic Books.

Kohlenberg, R. J., Kanter, J. W., Bolling, M. Y., Parker, C., & Tsai, M. (2002). Enhancing cognitive therapy for depression with functional analytic psychotherapy: Treatment guidelines and empirical findings. *Cognitive and Behavioral Practice, 9*, 213–229.

Kohlenberg, R. J., & Tsai, M. (1991). *Functional analytic psychotherapy: Creating intense and creative therapeutic relationships*. New York: Plenum Press.

Kohlenberg, R. J., & Tsai, M. (1994). Improving cognitive therapy for depression with functional analytic psychotherapy: Theory and case study. *The Behavior Analyst, 17*, 305–319.

Lejuez, C. W., Hopko, D. R., & Hopko, S. D. (2001a). A brief behavioral activation treatment for depression: Treatment manual. *Behavior Modification, 25*, 255–286.

Lejuez, C. W., Hopko, D. R., & Hopko, S. D. (2002). *The Brief Behavioral Activation Treatment for Depression (BATD): A comprehensive patient guide*. Boston, MA: Pearson Custom Publishing.

Lejuez, C. W., Hopko, D. R., LePage, J., Hopko, S. D., & McNeil, D. W. (2001b). A brief behavioral activation treatment for depression. *Cognitive and Behavioral Practice, 8*, 164–175.

LePage, J. P., Mogge, N. L., Sellers, D. G., & DelBen, K. (2003). Line staff use of the behavioral observation system: Assessment of depression scale validity and cut scores. *Depression and Anxiety, 17*, 217–219.

Lewinsohn, P. M. (1974). A behavioral approach to depression. In R. M. Friedman & M. M. Katz (Eds.), *The psychology of depression: Contemporary theory and research* (pp. 157–178). New York: Wiley.

Lewinsohn, P. M., Antonuccio, D. O., Breckenridge, J. S., & Teri, L. (1984). *The "Coping with Depression" course*. Eugene, OR: Castalia.

Lewinsohn, P. M., & Atwood, G. E. (1969). Depression: A clinical-research approach. *Psychotherapy: Theory, Research, & Practice, 6*, 166–171.

Lewinsohn, P. M., & Graf, M. (1973). Pleasant events and depression. *Journal of Consulting and Clinical Psychology, 2*, 261–268.

Lewinsohn, P. M., Hoberman, H., Teri, L., & Hautzinger, M. (1985). An integrative theory of depression. In S. Reiss & R. Bootzin (Eds.), *Theoretical issues in behavior therapy* (pp. 331–359). New York: Academic Press.

Lewinsohn, P. M., & Libet, J. (1972). Pleasant events, activity schedules, and depressions. *Journal of Abnormal Psychology, 3*, 291–295.

Lewinsohn, P. M., Munoz, R. F., Youngren, M. A., & Zeiss, A. M. (1986). *Control your depression*. New York: Prentice Hall Press.

Lewinsohn, P. M., Sullivan, J. M., & Grosscup, S. J. (1980). Changing reinforcing events: An approach to the treatment of depression. *Psychotherapy: Theory, Research, and Practice, 47*, 322–334.

Libet, J., & Lewinsohn, P. M. (1973). The concept of social skill with special reference to the behavior of depressed persons. *Journal of Consulting and Clinical Psychology, 40*, 304–312.

Lopez, S., & Arco, J. L. (2002). ACT as an alternative for patients that do not respond to traditional treatments: A case study. *Analisis Modificacion de Conducta, 28*, 585–616.

MacPhillamy, D. J., & Lewinsohn, P. M. (1971). *Pleasant Events Schedule*. Eugene: University of Oregon.

MacPhillamy, D. J., & Lewinsohn, P. M. (1982). The Pleasant Events Schedule: Studies on reliability, validity, and scale intercorrelations. *Journal of Consulting and Clinical Psychology, 50*, 363–380.

Malhi, G. S., Parker, G. B., & Greenwood, J. (2005). Structural and functional models of depression: From sub-types to substrates. *Acta Psychiatrica Scandinavica, 111*, 94–105.

Martell, C. R., Addis, M. E., & Jacobson, N. S. (2001). *Depression in context: Strategies for guided action*. New York: W. W. Norton.

McCullough, J. P. Jr. (2000). *Treatment for chronic depression: Cognitive behavioral analysis system of psychotherapy*. New York: Guilford Press.

McCullough, J. P. Jr. (2003). Treatment for chronic depression using cognitive behavioral analysis system of psychotherapy. *Journal of Clinical Psychology, 59*, 833–846.

McDowell, J. J. (1982). The importance of Herrnstein's mathematical statement of the law of effect for behavior therapy. *American Psychologist, 37*, 771–779.

McLean, P. D., & Hakstian, A. R., (1979). Clinical depression: Comparative efficacy of outpatient treatments. *Journal of Consulting and Clinical Psychology, 47*, 818–836.

Mineka, S., Watson, D., & Clark, L. A. (1998). Comorbidity of anxiety and unipolar mood disorders. *Annual Review of Psychology, 49*, 377–412.

Montgomery, S., Doyle, J. J., Stern, L., & McBurney, C. R. (2005). Economic considerations in the prescribing of third-generation antidepressants. *Pharmacoeconomics, 23*, 477–491.

Moore, J. (1980). On behaviorism and private events. *The Psychological Record, 30*, 459–475.

Morey, L. C. (1991). *The Personality Assessment Inventory professional manual*. Odessa, FL: Psychological Assessment Resources.

Mynors-Wallis, L., Gath, D., Day, A., & Baker, F. (2000). Randomized controlled trial of problem solving treatment, antidepressant medication, and combined treatment for major depression in primary care. *British Medical Journal, 320*, 26–30.

Mynors-Wallis, L., Gath, D., Lloyd-Thomas, A., & Tomlinson, D. (1995). Randomized controlled trial comparing problem solving treatment with amitriptyline and placebo for major depression in primary care. *British Medical Journal, 310*, 441–445.

Nelson-Gray, R. O. (2003). Treatment utility of psychological assessment. *Psychological Assessment, 15*, 521–531.

Nezu, A. M. (1986). Efficacy of a social problem-solving therapy approach for unipolar depression. *Journal of Consulting and Clinical Psychology, 54*, 196–202.

Nezu, A. M. (1987). A problem-solving formulation of depression: A literature review and proposal of a pluralistic model. *Clinical Psychology Review, 7*, 121–144.

Nezu, A. M. (2004). Problem solving and behavior therapy revisited. *Behavior Therapy, 35*, 1–33.

Nezu, A. M., & Perri, M. G. (1989). Social problem-solving therapy for unipolar depression: An initial dismantling investigation. *Journal of Consulting and Clinical Psychology, 57*, 408–413.

Nezu, A. M., Ronan, G. F., Meadows, E. A., & McClure, K. S. (2000). *Practitioner's guide to empirically based measures of depression*. New York: Kluwer Academic/Plenum Publishers.

Nezu, C. M., & Nezu, A. M. (1989). Unipolar depression. In A. M. Nezu & C. M. Nezu (Eds.), *Clinical decision making in behavior therapy: A problem-solving perspective* (pp. 117–156). Champaign, IL: Research Press.

Nolen-Hoeksema, S., & Girgus, J. S. (1994). The emergence of gender differences in depression during adolescence. *Psychological Bulletin, 115,* 424–443.

O'Neill, R. E., Horner, R. H., Albin, R. W., Storey, K., & Sprague, J. R. (1990). *Functional analysis of problem behavior: A practical assessment guide.* Sycamore, IL: Sycamore Publishing Company.

Piaget, J. (1923). *The language and thought of the child.* New York: Harcourt, Brace.

Radloff, L. (1977). The CES-D scale: A self-report depression scale for research in the general population. *Applied Psychological Measurement, 1,* 385–401.

Ranelli, C. J., & Miller, R. E. (1981). Behavioral predictors of amitriptyline response in depression. *American Journal of Psychiatry, 138,* 30–34.

Regier, D. A., Farmer, M. E., Rae, D. S., Locke, B. Z., Keith, S. J., Judd, L. L., et al. (1990). Comorbidity of mental disorders with alcohol and other drug abuse. *Journal of the American Medical Association, 264,* 2511–2518.

Rehm, L. P. (1977). A self-control model of depression. *Behavior Therapy, 8,* 787–804.

Rehm, L. P. (1988). Assessment of depression. In A. S. Bellack & M. Hersen (Eds.), *Behavioral assessment: A practical handbook* (3rd ed., pp. 313–364). New York: Pergamon Press.

Reynolds, W. M., & Kobak, K. A. (1995). *Hamilton Depression Inventory (HDI): Professional manual.* Odessa, FL: Psychological Assessment Resources.

Reynolds, W. M., & Kobak, K. A. (1998). *Reynolds Depression Screening Inventory: Professional manual.* Odessa, FL: Psychological Assessment Resources.

Riso, L. P., McCullough, J. P., & Blandino, J. (2003). Focus on empirically-supported methods: The cognitive-behavioral analysis system of psychotherapy: A promising treatment for clinical depression. *Scientific Review of Mental Health Practice, 2,* 61–68.

Robins, L. N., Helzer, J. E., Croughan, J. L., & Ratcliff, K. S. (1981). National Institute of Mental Health Diagnostic Interview Schedule: Its history, characteristics, and validity. *Archives of General Psychiatry, 38,* 381–389.

Robinson, J. C., & Lewinsohn, P. M. (1973). Behavior modification of speech characteristics in a chronically depressed man. *Behavior Therapy, 4,* 150–152.

Simon, G. E., & Katzelnick, D. J. (1997). Depression, use of medical services and cost-offset effects. *Journal of Psychosomatic Research, 42,* 333–344.

Simon, G. E., Ormel, J., VonKorff, M., & Barlow, W. (1995). Health care costs associated with depressive and anxiety disorders in primary care. *American Journal of Psychiatry, 152,* 352–357.

Skinner, B. F. (1957). *Verbal behavior.* Acton, MA: Copley Publishing Group.

Staats, A. W., & Heiby, E. M. (1985). Paradigmatic behaviorism's theory of depression: Unified, explanatory, and heuristic. In S. Reiss & R. R. Bootzin (Eds.), *Theoretical issues in behavior therapy* (pp. 279–330). New York: Academic Press.

Stevens, D. E., Merikangas, K. R., & Merikangas, J. R. (1995). Comorbidity of depression and other medical conditions. In E. E. Beckham & W. R. Leber (Eds.), *Handbook of depression* (2nd ed., pp. 147–199). New York: Guilford Press.

Suls, J., & Bunde, J. (2005). Anger, anxiety, and depression as risk factors for cardiovascular disease: Problems and implications of overlapping affective dispositions. *Psychological Bulletin, 131,* 260–300.

Thorpe, G. L., & Olson, S. L. (1997). *Behavior therapy: Concepts, procedures, and applications.* Boston, MA: Allyn & Bacon.

Weissman, M. M., Bruce, M. L., Leaf, P. J., Florio, L. P., & Holzer, C. (1991). Affective disorders. In L. N. Robins & D. A. Regier (Eds.), *Psychiatric disorders in America: The Epidemiological Catchment Area Study* (pp. 53–80). New York: Free Press.

Williams, J., Barrett, J., Oxman, T., Frank, E., Katon, W., Sullivan, M., et al. (2000). Treatment of dysthymia and minor depression in primary care: A randomized controlled trial in older adults. *Journal of the American Medical Association, 284,* 1519–1526.

Williams, J. G., Barlow, D. H., & Agras, W. S. (1972). Behavioral measurement of severe depression. *Archives of General Psychiatry, 27,* 330–333.

Wolf, N. J., & Hopko, D. R. (2005). *Psychosocial and pharmacological interventions for depressed adults in primary care.* Manuscript submitted for publication.

Zeiss, A. M., Lewinsohn, P. M., & Munoz, R. F. (1979). Nonspecific improvement effects in depression using interpersonal cognitive, and pleasant events focused treatments. *Journal of Consulting and Clinical Psychology, 47,* 427–439.

Zettle, R. D., & Hayes, S. C. (1986). Dysfunctional control by client verbal behavior: The context of reason giving. *The Analysis of Verbal Behavior, 4,* 30–38.

Zettle, R. D., & Raines, J. C. (1989). Group cognitive and contextual therapies in treatment of depression. *Journal of Clinical Psychology, 45,* 438–445.

16

THE FEAR FACTOR: A FUNCTIONAL PERSPECTIVE ON ANXIETY

PATRICK C. FRIMAN

Father Flanagan's Boys' Home
University of Nebraska School of Medicine

Danger lurks in every part of human life, and fear is a ubiquitous human emotion. But human fear is not confined to true danger. The range of fear extends to almost all human experience. People are afraid of sex, lack of sex, affection, lack of affection, attention, lack of attention, being with others, being alone, flying, missing a flight, riding, missing a ride, walking, not being able to walk, getting lost, being found, eating too much, eating too little, and other examples too numerous to list. Humans are also afraid of more abstract phenomena such as the past, future, unknown, freedom, or restriction. Many people do not know why they are afraid nor what of, but they do experience chronic fear. Currently, the most widely used categorical term for fear that does not involve true danger is *anxiety*, and this chapter will address it from a functional perspective.

DEFINING ANXIETY

The *Diagnostic and Statistical Manual of Mental Disorders* (*DSM-IV;* American Psychiatric Association, 1994) distinguishes 12 anxiety disorders. Although the section on anxiety is one of the longest in the *DSM*, no definition of anxiety is provided. Each anxiety disorder is defined in terms typical of the *DSM* system by listing the clinical features, prevalence, course, and differential diagnosis, but no attempt is made to define the construct of anxiety itself. Searching major authoritative sources on anxiety reveals the likely reason for this. Anxiety commands one of the largest literatures in all of behavior science, and yet the millions of published words devoted to it appear to have brought no precision to the term nor consensus as to its definition (Friman, Wilson, & Hayes, 1998). Whether anxiety is merely a metaphor (Friman et al., 1998; Sarbin, 1964, 1968), a lay construct without a unique set of referents (Hallam, 1985), an indefinable term because of unlimited possible variations (Levitt, 1967; Sidman, 1964), a synonym for fear (Barrios & Hartmann, 2001; Ohman, 2000), or a technical term for a psychobiological condition that resembles fear but is distinct from it in terms of biological and psychological essence (Lang, Cuthbert, & Bradley, 1998) is not firmly established, nor may it ever be. A current, comprehensive, and influential book on anxiety discusses it for the first 100 out of 704 pages before attempting a definition (Barlow, 2002). The definition then supplied includes more than 100 words, among which are a variety of other hypothetical constructs, such as apprehension and failed coping mechanisms, that are themselves not technically defined.

Central to the *DSM* descriptions of various anxiety disorders, as well as descriptions provided in the numerous books on the topic, are perceptions of peril, cognitions about that peril, physiological activation (e.g., increased heart rate), and overt and covert avoidance and escape responses. In other words, the defining features of anxiety disorders are virtually identical with the defining features of fear. The majority of attempts to distinguish anxiety from fear focus on the specificity, immediacy, and perilous nature of the events that set the occasion for responses of concern. Events that are specific, immediate, and actually perilous are said to produce fear, and events that are nonspecific, removed in time, and/or nonperilous are said to produce fear-like responses, the general term for which is anxiety (Barlow, 2002; Lang et al., 1998). As reflected in the writings of authors skeptical about the concept of anxiety, the attempt to distinguish anxiety from fear on this basis does not survive serious scrutiny (Friman et al., 1998; Hallam, 1985; Levitt, 1967; Sarbin, 1964, 1968). For example, specific phobias are classified as anxiety disorders (APA, 1994) but typically involve specific, immediate, and potentially perilous events (e.g., some spiders are poisonous, dogs bite, planes crash, falls from a great height can kill). Thus, phobic responses may not be just fear-like but actually fearful, and subcategorizing them in terms of anxiety can easily be made to seem superfluous. The arbitrariness of conventional attempts to define and categorize phobias in terms of anxiety rather than fear, while ignoring the concept of function altogether, is revealed by comparing them with sexual aversion. If a person is fearful of, and therefore persistently avoids, snakes, blood, or heights, he/she is diagnosed with an anxiety disorder. If a person is fearful of, and therefore persistently avoids, sexual activity, however, he/she is diagnosed with sexual aversion, which is not an anxiety disorder. In a functional account, both would be classified in the same category, although each would be occasioned by different circumstances and represented by different topographies.

Some investigators attempting to determine the essential basis of anxiety, distinct from fear, bypass specific phobias and focus more on generalized disorders such as post-traumatic stress disorder (PTSD) or generalized anxiety disorder (GAD; Barlow, 2002; Lang et al., 1998; McTeague & Lang, 2005). That phobias can be explicated cogently using functional concepts is well established, and doing so often requires only elementary behavior analytic concepts (Jones & Friman, 1999). For this reason, phobias will be used here to introduce the functional perspective on anxiety, and the case example will involve specific phobia. However, a functional account of less specific conditions such as PTSD can be explicated just as cogently, although not as self-evidently, because doing so requires the use of complex functional concepts such as derived relational responding and experiential avoidance (see related sections later in this chapter).

PREVALENCE OF ANXIETY DISORDERS

Regardless of how anxiety is defined, the phenomena the term is used to categorize are abundant in everyday life. In fact, anxiety is reported to co-occur so prevalently with psychological distress that it has been referred to as the psychological equivalent of fever (Carson, 1997). Even when the term is confined to *DSM-IV* defined disorders, anxiety is highly prevalent. Epidemiological studies estimate a lifetime prevalence of any anxiety disorder, regardless of gender, at 25% with ranges in specific disorders from 3.5% for panic to 13.3% for social phobia (Barlow, 2002).

FEAR IS A MORE FUNCTIONALLY RELEVANT TERM

A functional account of the concept of anxiety involves assessment and analysis of the behaviors that compose it, and the term *fear* is actually a more useful term in classifying these activities, for at least three reasons. First, it is more parsimonious. As indicated previously, anxiety researchers have yet to provide a clear, widely accepted definition of anxiety. All definitions include fear-like responses, and most definitions implicitly or explicitly allude to the presence of something else that is usually defined vaguely (Freud, 1917/1966; Barlow, 2002), metaphorically (Friman et al., 1998; Sarbin, 1964, 1968) or not at all (APA, 1994). At present, however, the concept of fear can be used to account for virtually all of the known facts (for one possible exception—although not an unanswerable one— see McTeague & Lang, 2005). Second, as indicated earlier, there is no consensus that the two are different (Barrios & Hartmann, 2001; Friman et al., 1998; Ohman, 2000), and the most authoritative investigators who espouse a technical distinction between anxiety and fear have yet to provide a widely persuasive argument (Barlow, 2002; Lang et al., 1998; McTeague & Lang, 2005). Third, almost all attempts to define anxiety emphasize the vagueness of functional environmental variables or even doubt about their very existence (Barlow, 2002; Friman et al., 1998). The always implicit, and often explicit, assumption is that the source of anxiety is within the person rather than in his/her environment. This assumption would seem to be the source of the monumental search of physiological systems for the source of anxiety that has been occurring for decades (Barlow, 2002; Lang et al., 1998; McTeague & Lang, 2005). In other words, the term *anxiety* does not direct investigators to the environment, the preferred source of functional variables in behavior analysis, whereas the term *fear* does. People in fear are presumed to be afraid of something. So natural language leads to a search for that something (e.g., "What are you afraid of?") and substituting the term *anxiety* in similar constructions would usually be awkward or even nonsensical (e.g., "What are you anxious of?").

Furthermore, functional accounts require documentation of at least two clearly defined variables that are functionally related. In behavior analysis, the documented relations involve stimuli and responses that are observable in principle (Friman, In press). The most illustrative example involves experimental functional analysis, arguably the most active area of the science of applied behavior analysis. The prototypical experimental functional relationship in applied behavior analysis is between precisely defined environmental events and observable behaviors with applied significance (Iwata, Dorsey, Slifer, Bauman, & Richman, 1982/1994; Neef & Iwata, 1994). Thus, conducting an experimental functional analysis of anxiety itself is not actually possible because it is not a behavior, but rather is a hypothetical construct and cannot be observed directly (Friman, In press; Friman et al., 1998). Fear is also a construct but its status is not nearly as hypothetical as that of anxiety, and most of the particulars that compose the construct of fear are observable in principle and, thus, can be analytic units in experimental attempts to establish functional relations. At an elementary level, a functional analysis of a "fearful behavior" would merely involve precisely defining the behavior in observable terms, measuring it over time and assessing the extent to which changes in any of its dimensions (rate, duration) correspond with changes in experimentally manipulated, consequential events (e.g., Jones & Friman, 1999). Therefore, in the remainder of this chapter, the compound term *fear/anxiety* will be substituted for the term *anxiety* in all but a small number of instances. Before further discussion of a functional analysis of fear/anxiety is attempted, however, a brief description of two additional concepts—derived relational responding and experiential avoidance—is needed.

FUNCTIONAL RELEVANCE OF DERIVED
RELATIONAL RESPONDING

Derived relations can be used to explain how fear-like responses to perilous events can generalize so broadly, not only to nonperilous events that are formally similar to the originating events, but to events that bear no formal relationship of any kind with the perilous events. In other words, derived relations can be used to explain how nonspecific, temporally distal, and harmless events can set the occasion for the fearful responses that compose the particulars of fear/anxiety disorders. Note only a brief description will be provided here; several much more complete accounts are available (Friman et al., 1998; Hayes, Barnes-Holmes, & Roche, 2001; Sidman, 1994). The most widely documented derived relation is equivalence, but several others have also been shown (e.g., more than, less than). Some investigators assert that derived relational responding is actually at the heart of language development (Hayes et al., 2001) and, pertinent to this chapter, also of common forms of psychopathology such as anxiety disorders (e.g., Friman et al., 1998).

A large body of research shows that when language-able humans learn certain types of stimulus relations directly, they also derive, or learn indirectly, a number of other relations—even though these other relations were not among those in the direct contingency relationships arranged in the learning context. As an elementary example of equivalence, if a person is trained that A is equivalent to B and B to C, he/she comes to "know" a number of relations that were not trained but rather derived from the initial training (e.g., B is equivalent to A, C to B, A to C, and C to A). As another example, when children are trained to select the word *dog* from an array of words when shown a picture of a dog, they do not need to be trained to select a picture of a dog from an array of pictures when shown the word *dog*. Selecting the word *dog* is trained directly, and selecting the picture is derived or learned indirectly. If the children are then trained to say the word *dog* when an actual dog is seen, even more derived relationships emerge (e.g., they point to picture of a dog when a dog is seen, look for a dog when the word *dog* is heard, etc.). The relevance to fear is reflected by the fact that if a child has a painful experience with a dog, subsequently not only dogs, but pictures of dogs and the written and spoken word *dog* can evoke fear, even though the pictures and the word were not part of the painful experience.

This process emerges as early as 23 months (Devany, Hayes, & Nelson, 1986; Lipkens, Hayes, & Hayes, 1993) and appears to last through the life span. As indicated previously, the prototypical relations involve stimulus equivalence (Sidman, 1994), but a rapidly growing body of research shows that many relations other than equivalence are also derivable, such as greater than, less than, opposition, difference. The ease with which these relations are formed, along with the extent of responses that emerge as they form, leads to networks of stimulus relations of almost indescribable complexity (Dymond & Barnes, 1995, 1996; Hayes et al., 2001; Steele & Hayes, 1991). Additionally, a large, multidisciplinary literature shows that generalized responding to stimuli with discriminative functions easily spreads to stimuli that are formally similar via stimulus generalization. For example, eating contaminated meat is usually followed by extremely unpleasant gastrointestinal distress that can, in turn, lead to apprehension and avoidant responding not just in the presence of similar meat, but all meat as well as a broad range of stimuli that bear a formal resemblance to the meat (e.g., any food substance with a similar taste, smell, texture, or appearance). Furthermore, stimulus classes formed via derived relations can merge with classes formed via stimulus generalization, resulting in extremely large, relational categories of responses (Fields, Reeve, Adams, & Verhave, 1991; Hernstein, 1984; Rosche & Mervis, 1975). So persons who have eaten contaminated meat may avoid not only stimuli that are formally similar to the meat, but also thoughts about those stimuli, words that depict those stimuli, or pictures of places (e.g., grocery stores, restaurants) where those stimuli might be found. There are innumerable other examples. The fundamental point is that events in derived relational classes, which are often very large in and of themselves, can spread not just through derived relations, such as equivalence, but also through the effects of stimulus generalization to become several orders of magnitude larger. That is, large, relational classes of responses created via derived relations can merge with large, relational, fuzzy classes of responses created via stimulus generalization if a member of one enters into an equivalence relation with a member of the other. Adding to the magnitude of learning generated by derived relations and stimulus generalization, psychological functions of elements in these relational networks can pass to other elements in accordance with the underlying derived stimulus relation. Functions are merely transferred in equivalence relations but can actually be transformed in other types or relations (Dymond & Barnes, 1995; Roche & Barnes, 1997). This passage of function has been shown with conditioned reinforcing functions (Hayes, Brownstein, Devany, Kohlenberg, & Shelby, 1987; Hayes, Kohlenberg, & Hayes, 1991), discriminative functions of public (Hayes et al., 1987) and private (DeGrandpre, Bickel, & Higgins, 1992) stimuli, elicited conditioned emotional responses (Dougher, Augustson, Markham, Greenway, & Wulfert, 1994), extinction functions (Dougher et al., 1994), and sexual responses (Roche & Barnes, 1997). Or, using the meat example, the apprehension, nausea, and gastrointestinal motility responses that follow eating contaminated meat can also follow exposure not only to stimuli that are formally similar to the meat, but also to stimuli that bear no formal similarity to the meat but that are in a derived relationship with it (e.g., thoughts, written descriptions, etc.).

FUNCTIONAL RELEVANCE OF
EXPERIENTIAL AVOIDANCE

Among the earliest and most powerfully learned behaviors in humans is a large and versatile repertoire of strategies for avoiding unpleasant events such as withdrawing and vigilance. This avoidance repertoire was central to Skinner's early attempt to describe anxiety in operant terms. His account also anticipated the construct of experiential avoidance by integrating the emotional dimensions of exposure to aversive events with the response avoidance that was generated by the exposure. Jointly, the two dimensions composed anxiety (Skinner, 1953). Experiential avoidance is a more comprehensive construct because it incorporates a verbal dimension. As development progresses, verbal repertoires that are functionally related to the emotional and avoidance dimensions also emerge and expand rapidly. The parallel development of the three dimensions and their response repertoires leads to derived relations between them (e.g., equivalence) and, subsequently, persons not only avoid the events but also the thoughts and feelings about them, resulting in a pattern of behavior characterized as experiential avoidance (Hayes, Wilson, Gifford, Follette, & Strosahl, 1996). Briefly, experiential avoidance involves public and private behavior whose function is avoidance of a broad range of private events such as unpleasant bodily sensations, emotions, thoughts and memories. More specifically, experientially avoidant behavior is reinforced by diminished contact with or alteration in the form or frequency of these events and the contexts that occasion them (Hayes et al., 1996). Behavior classified as fear/anxiety is one of the most prevalent examples of experiential avoidance. That is, fear/anxiety disorders have a functional unity: The behaviors that compose them are reinforced by the elimination, minimization, or reduction in the form, frequency, or situational sensitivity of various public and private events that occasion the use of the term *anxiety*.

DISORDERS WITH NONSPECIFIC FEAR/
ANXIETY-EVOKING EVENTS

Investigators attempting to distinguish anxiety from fear typically bypass specific phobia and focus upon disorders that involve much less specific fear-evoking stimuli (Barlow, 2002; Lang et al., 1998; McTeague & Lang, 2005). Foremost among those disorders is PTSD, a condition that is inaugurated by terrifying or traumatic events, but that ultimately involves highly generalized responding to events that bear no evident formal or functional relationship to the inaugural events. This type of responding taxes functional accounts that rely on direct contingency relationships between observable events and overt behaviors.

The expanding research on derived relations, stimulus generalization, and experiential avoidance, however, provides an empirically derived basis for explaining how specific terrifying or traumatic events and subsequent public and private responses to them can lead to highly generalized, chronic fear/anxiety disorders, such as PTSD. For example, as described earlier in the section on derived relations, both private and public events can become part of the same equivalence class (DeGrandpre et al., 1992), and functions such as respondent elicitation can transfer through such classes (Dougher et al., 1994). In addition, functions can be transformed, not just transferred, when the underlying stimulus relation is not one of equivalence (Dymond & Barnes, 1995; Roche & Barnes, 1997). The process of stimulus generalization can expedite, and tremendously expand, membership in the response classes formed by derived relations (Fields et al., 1991; Hernstein, 1984; Rosche & Mervis, 1975). Couple the facts that growing classes can include public and private responses and that the functions of members can be transferred or transformed, depending on the relation involved, with the facts that experiential avoidance generates a powerful source of negative reinforcement and that fearful responses also often generate a range of positive reinforcers (i.e., secondary gain), and a cogent functional perspective on the generalized responding in disorders such as PTSD emerges. In other words, the mysterious, seemingly nonfunctional nature of responding that occurs with such disorders that has led to a vigorous, essentialistic, intraorganismic search for something other than fear can be plausibly accounted for using only functional concepts. The extraordinary complexity of responding that can be generated through the combined effects of stimulus generalization, derived relations, and stimulus equivalence can help explain why a single fearful episode can lead to chronically impairing, psychological conditions involving stimuli that were not part of the episode nor formally similar to them (see Friman et al., 1998, for a more complete discussion).

A Functional Perspective on PTSD

PTSD is the most recent and the most widely applied diagnostic category for trauma-related emotional disturbance. Previous descriptors were more colloquial and specific, but appeared to refer to the same emotional phenomenon such as shell shock and rape trauma syndrome. PTSD involves direct exposure to a traumatic event and the subsequent emergence of three clusters of symptoms: reexperience (nightmares, flashbacks), avoidance in active (avoiding trauma-related stimuli) and passive (numbing; disassociation) forms, and increased arousal (e.g., insomnia, hypervigilance, exaggerated startle responses; APA, 1994; Foa & Meadows, 1997; Foa & Riggs, 1995).

The value of the experiential avoidance approach to classifying PTSD is evident in its symptom clusters. The first cluster involves the unpleasant experience and reexperience of the traumatic event, the avoidance or escape from which reinforces responses in the second cluster, numbing, active avoidance, and disassociation. The third cluster involves persistent arousal that has an apparently elicited basis and, in part, occurs as a function of formal similarities between the traumagenic and current environments. The pervasiveness of the arousal, however, would seem to unduly tax respondent conditioning as its sole explanation because the hyperarousal often extends to conditions that bear no formal similarity to the original traumagenic settings. Situations that are only verbally or metaphorically related, such as unknown or unpredictable situations, may also generate hyperarousal. Although conceivable that such responses could be explained by higher-order respondent conditioning (Forsyth & Eifert, 1996a, 1996b), the incorporation of derived relational responding provides a more flexible account with considerably more scope. For example, respondent accounts do not explain why avoidance should lead to increases in hyperarousal. From a strictly respondent perspective, decreases in arousal would be expected. From a derived relations perspective, however, the elevated, reactivating influence of the avoidance cluster could be explained by a derived relationship between avoidance responses, accompanied by verbal behavior that increases sensitivity to emotional reactions (e.g., "Oh no, it is happening again"), and perceived direct dire consequences resulting in more reactions, such as imagined, verbally specified stimuli. This dynamic could then lead to avoidance responses, a major portion of which would involve experiential avoidance. And the literature shows that the presence, severity, and chronicity of PTSD are better predicted by signs of avoidance (e.g., numbing, avoidance, disassociation) than by the level of fear and horror at the time of the inaugural traumatic events (Foa & Riggs, 1995; Orsillo & Batten, 2005). The condition worsens as the emotional reactions and verbal accompaniments produce increasingly aberrant and impairing avoidance and escape responses, such as diminished activity, excess sleep, drugs, alcohol, sensation seeking, and suicidal gestures. Not surprisingly, a common element across several successful treatments for PTSD involves exposure to public and private events that set the occasion for public and private avoidance. Although the literature appears not to include a data-based account of successful PTSD treatment based on derived relational theory, a recent paper describes how one such treatment, Acceptance and Commitment Therapy, could be used and the paper includes a case report of a successful application (Orsillo & Batten, 2005).

FUNCTIONAL ASSESSMENT AND ANALYSIS

Despite the difficulties in establishing a consensual definition of fear/anxiety, most investigators agree that the response clusters that compose the fear/anxiety disorders generally have the same three dimensions that compose fear: physiological activity, cognitive activity, and behavioral activity (APA, 1994; Barlow, 2002). Note the assumption, implicit in distinguishing these three dimensions, that physiological and cognitive activity is not behavior. Because the erroneous assumption is virtually ubiquitous outside the field of behavior analysis, especially as it regards cognition, this chapter will allow it, but only for the sake of continuing the discussion and not to endorse a technical distinction between the three—in a functional account, physiological and cognitive responses are merely different kinds of behavior to be explained and changed.

Consistent with the conventional tridimensional perspective, in a typical clinically significant episode of phobia, a person encounters a stimulus event (e.g., boards a plane), experiences physiological arousal (e.g., changes in respiration, heart rate, salivary production, etc.), engages in cognitive activity which includes the semantic and behavioral function that initiates avoidance or escape (e.g., "I have to get off this plane), and exhibits such behavior (e.g., leaves plane). The sequence of these responses is not always consistent with the one used here—or at least the affected person's awareness of the sequence may not correspond with it. For example, behavioral activity may occur before awareness of the physiological or cognitive activity. Additionally, even in phobias, the stimulus events setting the occasion for the subsequent phobic responses are not always as apparent as the preceding example. For example, a person who is phobic about flying may exhibit phobic responses when in the presence of any stimulus event whose topography is formally and/or functionally related to flying, such as discussing a proposed trip or seeing a small bag of peanuts. In less specific fear/anxiety disorders such as Post Traumatic Stress Disorder (PTSD) and Generalized Anxiety Disorder (GAD), the range of events that set the occasion for related responses can seem limitless.

In addition to the physiological, cognitive, and behavioral activity dimensions of fear, there is a fourth. It includes events and behaviors that are not obviously related to fear or fear-pertinent stimuli, but are drawn into the relevant functional classes through environmental events occasioned by fear-related behavior. In more psychodynamic accounts of fear/anxiety, the fourth domain is often referred to as secondary gain (Morrison, 2002). The term *secondary* refers to its indirect relationship to fear. For example, children who are phobic about school exhibit behavior whose relationship to the phobia is intuitively evident, such as obsessing about grades, crying on the way to school, hyperventilating in school, or complaining of physical ailments with no physiological basis on school days. These behaviors may result in school avoidance or escape. But they may also result in expressions of sympathy by parents, teachers, and peers; access to the nurse's office at difficult times in class; reduced academic expectations; teacher intercession in difficult encounters with peers; and a range of other responses that function to broaden the response class beyond

TABLE 16.1   Functional Assessment of Fear/Anxiety

1. Assess material events associated with the condition (e.g., school, flying, being out in the open).
2. Assess cognitive events associated with the condition (e.g., thoughts of danger, lack of preparedness, disastrous outcomes).
3. Assess emotional events associated with the conditions (e.g., feelings associated with or occasioned by the material and cognitive events).
4. Assess physiological events associated with the condition (e.g., heart rate, respiration, muscle tension).
5. Assess all possibilities for secondary gain (e.g., excused absences, expressions of sympathies, reduced expectations).
6. Assess all domains of impairment resulting from the condition (e.g., lost jobs, missed social events, relationship difficulties, disrupted work).

(For a much fuller description, see assessment and treatment manuals—e.g., Bourne, 2005.)

events that are obviously fear related. Table 16.1 describes the steps necessary to conduct a behavioral assessment of fear/anxiety.

FURTHER DESCRIPTION OF THE FOUR FUNCTIONAL DIMENSIONS OF FEAR/ANXIETY

A functional assessment of anxiety/fear is employed to inform and develop functional treatments. Therefore, the exigencies of clinical outcome guide the assessment; its breadth is determined by the amount of information needed to design an effective treatment. As indicated previously, there are four functional dimensions of fear/anxiety, and any or all could be the focus of the assessment. Although it is possible that effective treatments could be based on data obtained from only one or two dimensions (Jones & Friman, 1999; Friman & Lucas, 1996; Swearer, Jones, & Friman, 1997), a paper providing a functional perspective should provide information on all four; following is an abbreviated description of each dimension.

Physiological Activity

In most anxious events, and especially those involving panic, clinically significant physiological responding, such as increased blood flow, oxygen intake, and heart rate acceleration, is often present. Relief from the experience of this type of responding and, subsequently, avoidance of events that bring it about supply the major portion of the reinforcement for anxious behavior. Therefore, the extent of the physiological dimensions of the anxious behavior is important in a functional account because it can lead to functionally related treatment. The central rationale for functional assessment of clinically significant behavior is that knowledge of the reinforcing consequences for the behavior leads to treatment that allows those consequences to be obtained in more adaptive ways. Relatedly, knowledge of the physiological dimensions of the behavior of a fearful/anxious person can inform the design of treatments that establish adaptive, nonclinically significant behavior that produces relief from or avoidance of the clinically relevant physiology.

Another related target for functional assessment that addresses physiology is the physiological state of the person when he/she encounters feared stimuli. Thresholds for avoidance responses are lowered by motivating events (more conventionally bodily states) such as fatigue, illness, hunger, or pain. In behavioral terms, such events heighten the reinforcing properties of avoidance and escape as well as increase the rate of behaviors previously reinforced by both. In more colloquial terms, afflicted persons affected by these bodily states are less prepared to confront and manage aversive situations. Finally, any physiological condition that affects breathing, such as asthma or emphysema, can increase the probability of anxious responses in predisposed individuals. In fact, interference with oxygen flow is such a potent instigator of anxious responding that laboratory preparations that affect oxygen intake, such as CO_2, lactate infusion, or hyperventilation provocation, are virtually standard in the experimental study of fear/anxiety (Barlow, 2002; Forsyth & Eifert, 1996a).

Cognitive Activity

There are also cognitive dimensions, and these should also be the targets of assessment. From a functional perspective, it is more parsimonious to refer to this dimension as verbal behavior. In response to either physiological states or exposure to events that instigate them, afflicted individuals typical engage in highly pessimistic verbal behavior. The behavior usually involves judgments that some stimulus events are potentially harmful and that the person is not capable of adaptively confronting them. The extent of the verbal behavior can range from a few fleeting private statements, such as those that might occur as a phobic stimulus is encountered and quickly escaped, to an intensive private dialogue that can dominate virtually all adaptive forms of verbal behavior, such as extreme obsessions. Through operant and respondent processes, fear-related verbal behavior can become absorbed into the functional stimulus class that sets the occasion for perpetuated anxious responding (Friman et al., 1998; Forsyth & Eifert, 1996a; Hayes et al., 1996). Afflicted persons engage in behavior that not only generates relief and avoidance of fear/anxiety provoking stimuli, they also engage in behavior that generates relief and avoidance from language about those stimuli (Friman et al., 1998; Hayes et al., 1996).

Behavioral Activity

The behavioral dimension of fear/anxiety is the one most likely to lead to a referral because it involves responses reinforced by escape and avoidance, and these are often more overt than physiological and verbal responses. It is also the dimension that contributes the most to the impairment aspect of a fear/anxiety diagnosis. In the *DSM* system, merely exhibiting symptoms is not sufficient to meet criteria for diagnosis. There must be an accompanying level of impairment brought about by the symptoms. Thus, experiencing physiological arousal and engaging in fear-related verbal behavior in the presence of phobic stimuli (e.g., airplanes, podiums, classrooms) is not sufficient to warrant a diagnosis of phobia. There must also be an adverse effect of the phobic person's life (e.g., refusal to fly, speak in public, or go to school). In a typical behavior assessment, anxious persons are queried about all behaviors whose occasion is set by fear-related stimuli and events. These queries include behaviors that are exhibited (e.g., taking a bus instead of a plane) and inhibited (e.g., refusing to fly). If children are the targets of assessment, the queries are posed to them as well as to caretakers and sometimes teachers.

Secondary Gain

Assessment of this dimension merely involves a comprehensive search for all possible sources of reinforcement that symptoms of fear/anxiety produce outside those sources that technically describe the condition. For example, school avoidance is a defining component of school phobia, and the behaviors that produce the avoidance are reinforced by it. But other secondary sources of reinforcement may be available at home, such as access to leisure activity, recreation, increased contact with a parent, and sympathy.

FUNCTIONAL ANALYTIC-BASED INTERVENTIONS

As indicated earlier, functional analytic interventions supply the functional consequences produced by target behaviors in more adaptive ways. Empirical evidence supporting these interventions for fear/anxiety disorders is limited because so few treatment studies employ functional analysis and assessment. One of the few studies to do so will be described in the case study described in a subsequent section (Jones & Friman, 1999). Another example involves a case of social phobia in a 14-year-old boy. Although the referral concerns involved extremely disruptive behavior, a diagnostic interview revealed the presence of social phobia, and a search of the literature revealed no published papers on the comorbid presence of social phobia and any disruptive behavior disorder. The core symptoms of social phobia involve avoidance of any kind of publicly delivered attention, including both praise and criticism. A cross-situational functional assessment revealed that the boy's high frequency and intensity outbursts were always preceded by public delivery of corrective feedback by authority figures in the boy's life (e.g., teachers, house parents). The functionally related intervention involved sustaining the rate of corrective feedback, but arranging for it to be delivered in a private context. Almost immediately, disruptive episodes were reduced to zero levels and maintained for an entire year (Friman & Lucas, 1996).

In another functionally pertinent treatment study, a socially anxious 15-year-old boy was referred for treatment because he bit the insides of his mouth so frequently and intensely in social situations that his mouth would fill with blood. The assessment indicated the biting resulted in reduced physiological arousal and escape from social encounters. The functionally related treatment to reduce physiological arousal appropriately included relaxation exercises and a recommendation that the boy always carry gum and chew it whenever socially engaged. Almost immediately following treatment, cheek biting was reduced to zero levels and maintained throughout the study and at long term follow-up (Swearer et al., 1997).

A final example involves a study of 7 school-avoidant children that used both conventional diagnostic instruments to assess presenting problems (Kearny & Silverman, 1990) and a functional measure, the *School Refusal Assessment Scale* (SRAS; Kearny & Silverman, 1988), to assess conditions that can motivate school refusal. The instrument yields four functional constructs that are analogous to the four conditions usually included in an experimental functional analysis: social, attention, tangible, and fear (which would correspond with automatic reinforcement). Treatments were individualized on the basis of SRAS and resulted in 6 of 7 children beginning to attend school on a full-time basis. As an illustration of the ubiquity of the nonfunctional, nonbehavioral perspective that dominates the literature on fear/anxiety, the school attendance results were reported as secondary outcomes. The primary outcomes reported involved change scores on the diagnostic instruments even though these did not inform treatment, nor did they add anything of substance to the observable clinical outcome. The clinical problem, requiring assessment in cases of school avoidance, is the avoidance, and the desired clinical outcome is attendance and appropriate classroom performance—at least from a functional perspective. As indicated previously, however, the vast majority of published case accounts of fear/anxiety includes data not just obtained from standardized measures of anxiety, but also emphasizes those measures over and above behavioral outcomes with some notable exceptions (Jones & Friman, 1999; Swearer et al., 1997).

Rather than describe additional case studies employing functional assessment/analysis of fear/anxiety—there are very few to choose from—it may be more informative to note that virtually all the standard, empirically supported treatments for fear/anxiety disorders include two core components: exposure and response prevention (Barlow, 2002; Bourne, 2005).

These two components have such an abundance of empirical support across the entire spectrum of fear/anxiety disorders that their prescription now occurs almost as automatically as prescriptions of antibiotics for bacterial infections. Although the design of related treatments is rarely informed by functional assessment, at least in the published literature, it is plausible to characterize them as functionally derived treatments. The overarching functions of all fear/anxiety disorders are escape and avoidance. The exposure-based component of standard treatment involves being in the presence of fear/anxiety-provoking stimuli, whether they be private, as with obsessions, or public as with specific phobia. The response prevention component involves preventing escape and inhibiting avoidance-based responses. Thus, from a behavior analytic perspective, exposure and response prevention merely involves establishing contact between client and fear/anxiety-provoking stimuli and establishing escape and avoidance extinction. Although this characterization invokes an operant process, respondent extinction processes are also involved (Barlow, 2002; Forsyth & Eifert, 1996b). The following section describes a case study of specific phobia that included functional assessment and a treatment derived from that assessment. See Table 16.2.

CASE STUDY

The *DSM-IV* defines phobia as persistent fear that is excessive or unreasonable, cued by the presence or anticipation of an object or situation (e.g., spiders, flying). It is the least complex of the anxiety disorders, and a functional perspective of it is, if not explicitly described, at least implicitly apparent in most current conceptual accounts (Mineka & Zinbarg, 2006; Ohman, 2000). The primary difference between specific phobia and more complex disorders involves the stimuli that evoke those responses; those

TABLE 16.2   Functional Treatment for Fear/Anxiety

1. Make improvements in domains of impairment the primary target of treatment (e.g., return to school, complete tasks in the presence of feared stimuli).
2. Gradually and systematically expose the individual to material events associated with the condition (e.g., diminishing distance from feared events).
3. Address the logical problems with the cognitive events and also employ acceptance-based strategies for fearful/anxious thoughts (e.g., Hayes, 2005).
4. Address the emotional events associated with the condition with acceptance-based strategies.
5. Address the physiological events associated with condition with arousal-reducing exercises (e.g., relaxation, focused breathing, meditative practice, yoga).
6. Eliminate or neutralize all identified sources of secondary gain.

in phobia are more specific and identifiable than those in the more complex disorders such as PTSD.

METHOD

Participant

Mike was a 14-year-old male enrolled in middle school at a residential care program. His school principal referred him because the presence of insects in his classroom, and taunts about insects, seriously disrupted his academic performance. Mike reported that he had difficulty concentrating and working when he thought insects might be nearby and that he was often teased by peers (e.g., "Mike, there is a bug under your chair!"). His responses included ignoring his work, pulling the hood of his jacket over his head and, occasionally, yelling. Mike identified crickets, spiders, and ladybugs as the insects he feared most. For a more complete account, see Jones and Friman (1999).

Measurement

Phobic stimuli may influence behavior in various ways; however, we focused on academic performance because it was the primary referral concern reported by his school principal. The dependent measure was Mike's work completion rate in the presence of crickets purchased from a local pet store. Two or three 4-min math probes were administered each session, during which Mike sat at a desk in a 7 × 7 m work room with one of 30 alternate-form, third-grade math sheets on the desk. Mike was instructed to complete as many problems as possible. His response rate was the mean number of correct digits per 4-min probe. Twenty math sheets (26%) were independently scored by the therapist and another person.

Assessment Procedures

The assessment targeted the effects of insects present and absent and verbal statements about insect presence on Mike's academic response rates. Between administration of math probes, Mike and his therapist engaged in 15–20 min of casual conversation about sports, grades, and friends. There were three assessment conditions: Insects Present, Statements About Insects, and Insects Absent. During *Insects Present* condition following instructions, the therapist released three live crickets in the center of the floor and left the room. In the *Statements About Insects* condition, the therapist removed the crickets and examined the room to make certain there were no insects, with Mike just outside. Then the therapist brought Mike back in and said, "There are bugs somewhere in this room." Finally, in the *Insects Absent* condition, the therapist told Mike, "There are no bugs anywhere in this room."

Treatment Procedures

Two treatment conditions were implemented: (a) graduated exposure and (b) graduated exposure plus reinforcement. During *graduated exposure*, Mike engaged in 15–20 min of graduated exposure exercises immediately before each math probe. These exercises included a hierarchy of behavioral approach tasks, ranging from holding a jar of crickets to holding a cricket in each hand for 1 min. Mike selected the initial exposure level for each session and continued until he refused to proceed with the next step. Mike completed six steps with assistance during the first session and independently completed nine steps by the final session of the exposure alone phase. Thereafter, time requirements were increased (e.g., holding a cricket for 40 or 60 s). The *graduated exposure plus reinforcement* was identical to the exposure condition, except Mike earned points for each correct digit. These points were exchanged at the end of each week for items from a reinforcement menu, including Blockbuster gift certificates, videos, candy, and Legos.

EXPERIMENTAL DESIGN

A multielement design was used to evaluate the effects of the three assessment conditions. An A-B-BC-A-BC design was used to compare the effect of the experimental conditions. Mike's performance during the initial "bugs" condition served as the initial baseline phase.

RESULTS AND DISCUSSION

Assessment data in the first panel of Figure 16.1 show higher rates of correct digits in the *no bugs* condition relative to the other conditions, initially low but increasing rates in the *statements* condition, and low rates in the *insects present* condition. Treatment data indicate no improvement in the *exposure* condition and increasing trends within both *exposure plus reinforcement* phases. A reversal phase resulted in a modest decline in scores, with the last two sessions yielding lower numbers of digits correct than any single session of either combined treatment phase.

Although teachers reported taunts a primary concern for Mike, assessment results indicated that performance deficits were sustained only in the presence of actual crickets. Performance problems in the *insects present* condition resolved during assessment when this verbal stimulus was repeatedly presented in the absence of actual insects. The results demonstrate the value of targeting adaptive behavior directly affected by the phobic stimuli rather than mere approach and/or indirect measures of fear or fear/anxiety (Friman et al., 1998). Lastly, the results suggest that programmed rewards, contingent on adaptive responding, may sometimes be

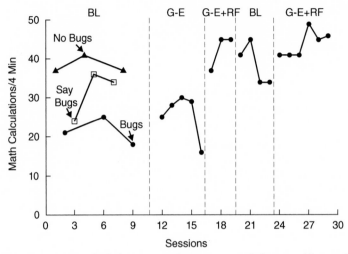

FIGURE 16.1 Math calculations across the presence and absence of insects ("bugs") and statements about insects in baseline (BL) as well as across treatment conditions of graduated exposure (G-E) and graduated exposure with reinforcement (G-E+RF).

needed to increase the level of adaptive behavior to that similar to typical peers.

CONCLUSION

The vast literature on anxiety has yet to produce a technical definition of its central topic that commands wide acceptance. Despite the reluctance to define the nominative target of the investigations in this vast literature, there is a widespread assumption that anxiety is distinct from fear in some essential way. The reasoning behind the assumed distinction is mostly elliptical, although sophisticated attempts have been made to fill in the ellipses with either hypothetical cognitive (Barlow, 2002) or biological constructs (Lang et al., 1998; McTeague & Lang, 2005). The related cognitive research directs investigators and clinicians to nonmaterial, hypothetical constructs that are unobservable, abstract, internal, and difficult to manipulate (e.g., cognitive schema) and lead to circular explanations. The related biological research directs investigators and clinicians to variables that are beyond the range of manipulation by individuals outside the medical field. Furthermore, no persuasive biological evidence distinguishing fear from anxiety has been supplied. Functional perspectives on anxiety do not require it to be distinguished from fear. In fact, in a functional account, using either term can quickly appear superfluous (Friman & Jones, 1999; Swearer et al., 1997). For ease of communication, however, a term is often needed, and the position of this chapter is that, when this is the case, *fear* is preferred because its linguistic and semantic properties direct attention outside the individual; thus, it is much better suited for a functional account. The fundamental advantage of functional accounts is that they direct investigators and clinicians to actual events that are, in principal, observable, manipulable, and directly influence fear-related behavior. Until recently, functional perspectives were seen as useful primarily for specific phobia. For less specific, more complex disorders such as PTSD and GAD, hypothetical conceptual equipment was needed. This chapter not only underscores the value of a functional perspective on specific phobia, but also provides, using the advanced functional concepts of derived relations and experiential avoidance, a plausible functional perspective on the less specific conditions.

REFERENCES

American Psychiatric Association. (1994). *Diagnostic and statistical manual of mental disorders* (4th ed.). Washington, D.C.: Author.

Barlow, D. H. (2002). *Anxiety and its disorders* (2nd ed.). New York: Guilford.

Barrios, B. A., & Hartmann, D. H. (2001). Fears and anxieties. In E. J. Mash & L. B. Terdal (Eds.), *Assessment of childhood disorders, third edition* (pp. 230–237). New York: Guilford.

Bourne, E. J. (2005). *The anxiety and phobia workbook* (4th ed.). Oakland, CA: New Harbinger.

Carson, R. C. (1997). Costly compromises: A critique of the DSM. In S. Fisher & R. Greenberg (Eds.), *Psychopharmacology scientifically reappraised* (pp. 98–114). New York: Wiley.

DeGrandpre, R. J., Bickel, W. K., & Higgins, S. T. (1992). Emergent equivalence relations between interoceptive (drug) and exteroceptive (visual) stimuli. *Journal of the Experimental Analysis of Behavior, 58,* 9–18.

Devany, J. M., Hayes, S. C., & Nelson, R. O. (1986). Equivalence class formation in language-able and language disabled children. *Journal of the Experimental Analysis of Behavior, 46,* 243–257.

Dougher, M. J., Augustson, E., Markham, M. R., Greenway, D. E., & Wulfert, E. (1994). The transfer of respondent eliciting and extinction functions through stimulus equivalence classes. *Journal of Experimental Analysis of Behavior, 62,* 331–351.

Dymond, S., & Barnes, D. (1995). A transformation of self-discrimination response functions in accordance with the arbitrarily applicable relations of sameness, more-than, and less-than. *Journal of the Experimental Analysis of Behavior, 64,* 163–184.

Dymond, S., & Barnes, D. (1996). A transformation of self-discrimination response functions in accordance with the arbitrarily applicable relations of sameness and opposition. *The Psychological Record, 46,* 271–300.

Fields, L., Reeve, K. E., Adams, B. J., & Verhave, T. (1991). Stimulus generalization and equivalence classes: A model for natural categories. *Journal of the Experimental Analysis of Behavior, 55,* 305–312.

Foa, E. B., & Meadows, E. A. (1997). Psychosocial treatments for posttraumatic stress disorder: A critical review. *Annual Review of Psychology, 48,* 449–480.

Foa, E. B., & Riggs, D. S. (1995). Posttraumatic stress disorder following assault: Theoretical considerations and empirical findings. *Current Directions in Psychological Science, 4,* 61–65.

Forsyth, J. P., & Eifert, G. H. (1996a). The language of feeling and the feeling of anxiety: Contributions of the behaviorisms toward understanding the function altering effects of language. *The Psychological Record, 46,* 607–649.

Forsyth, J. P., & Eifert, G. H. (1996b). Systematic alarms in fear conditioning I: A reappraisal of what is being conditioned. *Behavior Therapy, 27,* 441–462.

Freud, S. (11966). Anxiety (lecture XXV). In J. Strachey (Ed.), *The complete introductory lectures on psychoanalysis* (pp. 392–411). New York: Norton. (Original work published 1917).

Friman, P. C. (2006). Behavior assessment. In D. Barlow, F. Andrasik, F., & M. Hersen. *Single case experimental designs.* Boston: Allyn & Bacon.

Friman, P. C., & Lucas, C. (1996). Social phobia obscured by disruptive behavior disorder: A case study. *Clinical Child Psychology and Psychiatry, 1,* 401–409.

Friman, P. C., Wilson, K., & Hayes, S. C. (1998). Behavior analysis of private events is possible, progressive, and nondualistic: A response to Lamal. *Journal of Applied Behavior Analysis, 31,* 707–708.

Hallam, R. S. (1985). *Anxiety.* London: Academic Press.

Hayes, S. C. (2005). *Get out of your mind and into your life.* Oakland, CA: New Harbinger.

Hayes, S. C., Barnes-Holmes, D., & Roche, B. (2001). *Relational frame theory.* New York: Plenum.

Hayes, S. C., Brownstein, A. J., Devany, J. M., Kohlenberg, B. S., & Shelby, J. (1987). Stimulus equivalence and the symbolic control of behavior. *Mexican Journal of Behavior Analysis, 13,* 361–374.

Hayes, S. C., Kohlenberg, B. S., & Hayes, L. J. (1991). Transfer of consequential functions through simple and conditional equivalence classes. *Journal of the Experimental Analysis of Behavior, 56,* 119–137.

Hayes, S. C., Wilson, K. G., Gifford, E. V., Follette, V. M., & Strosahl, K. (1996). Experiential avoidance and behavior disorders: A functional dimensional approach to diagnosis and treatment. *Journal of Consulting and Clinical Psychology, 64,* 1152–1168.

Herrnstein, R. J. (1984). Objects, categories, and discriminative stimuli. In H. L. Roitblatt, T. G., Bever, & H. S. Terrace (Eds.), *Animal cognition* (pp. 233–261). Hillsdale, NJ: Erlbaum.

Iwata, B. A., Dorsey, M. F., Slifer, K. J., Bauman, K. E., & Richman, G. S. (1994). Toward a functional analysis of self-injury. *Journal of Applied Behavior Analysis, 27,* 197–209. (Reprinted from *Analysis and Intervention in Developmental Disabilities, 2,* 3–20, 1982).

Jones, K. M., & Friman, P. C. (1999). Behavior assessment and treatment of insect phobia: A preliminary case study. *Journal of Applied Behavior Analysis, 32,* 95–98.

Kearny, C. A., & Silverman, W. K. (1988). *Measuring the function of school refusal behavior: The School Refusal Assessment Scale (SRAS).* Paper presented at the meeting of the Association for Advancement of Behavior Therapy, New York.

Kearny, C. A., & Silverman, W. K. (1990). A preliminary analysis of a functional model of assessment and treatment for school refusal behavior. *Behavior Modification, 14,* 340–366.

Lang, P. J., Cuthbert, B. N., & Bradley, M. M. (1998). Measuring emotion in therapy: Imagery, activation, and feeling. *Behavior Therapy, 29,* 655–674.

Levitt, E. E. (1967). *The psychology of anxiety.* Indianapolis, IN: Bobbs-Merril.

Lipkens, G., Hayes, S. C., & Hayes, L. (1993). Longitudinal study of derived relations in an infant. *Journal of Experimental Child Psychology, 56,* 201–239.

McTeague, L. M., & Lang, P. J. (2005). *The psychophysiology of fear and anxiety: Same or different? Hyper- or Hypo-Reactivity?* Symposium presented at the 39th annual convention of the Association for Advancement of Behavior Therapy, Washington, D.C.

Mineka, S., & Zinbarg, R. (2006). A contemporary learning theory perspective on the etiology of anxiety disorders. *American Psychologist, 61,* 10–26.

Morrison, A. K. (2002). Somatoform disorders. In M. Hersen & W. Sledge (Eds.), *The encyclopedia of psychotherapy* (pp. 679–685).

Neef, N. A., & Iwata, B. A. (1994). Current research on functional analysis methodologies: An introduction. *Journal of Applied Behavior Analysis, 27,* 211–214.

Ohman, A. (2000). Fear and anxiety: Evolutionary, cognitive, and clinical perspectives. In M. Lews & J. M. Haviland-Jones (Eds.), *Handbook of emotions* (pp. 573–593). New York: Guilford.

Orsillo, S. M., & Batten, S. J. (2005). Acceptance and commitment therapy in the treatment of posttraumatic stress disorder. *Behavior Modification, 29,* 95–129.

Roche, B., & Barnes, D. (1997). A transformation of respondently conditioned stimulus function in accordance with arbitrarily applicable relations. *Journal of the Experimental Analysis of Behavior, 67,* 275–301.

Rosche, E., & Mervis, C. B. (1975). Family resemblance: Studies in the internal structure of categories. *Cognitive Psychology, 7,* 573–605.

Sarbin, T. R. (1964). Anxiety: The reification of a metaphor. *Archives of General Psychiatry, 10,* 630–638.

Sarbin, T. R. (1968). Ontology recapitulates philology: The mythic nature of anxiety. *American Psychologist, 23,* 411–418.

Sidman, M. (1964). Anxiety. *Proceedings of the American Philosophical Society, 108,* 478–481.

Sidman, M. (1994). *Equivalence relations: A research story.* Boston, MA: Authors Cooperative.

Skinner, B. F. (1953). *Science and human behavior.* New York: Macmillan.

Steele, D. L., & Hayes, S. C. (1991). Stimulus equivalence and arbitrarily applicable relational responding. *Journal of the Experimental Analysis of Behavior, 56,* 519–555.

Swearer, S. M., Jones, K. M., & Friman, P. C. (1997). Relax and try this instead: Abbreviated habit reversal for oral self-biting. *Journal of Applied Behavior Analysis, 30,* 697–700.

17

SEXUAL DISORDERS

JOSEPH J. PLAUD, PH.D., BCBA

Applied Behavioral Consultants, Inc., and Brown University

DIAGNOSTIC ISSUES

The common factors involved in sexual disorders, or paraphilias, include distressing and repetitive sexual fantasies, sexual urges, or overt sexual behaviors. Further, these sexual fantasies, sexual urges, or overt sexual behaviors must occur for a significant period of time and must interfere with either appropriate sexual relations or the person's daily functioning for a diagnosis to be made. The *Diagnostic and Statistical Manual of Mental Disorders, fourth edition* (text revision; *DSM-IV-TR*) also specifies that these areas of sexual expression are usually accompanied by a sense of distress from the person (American Psychiatric Association, 2000). These sexual disorders include exhibitionism (intense sexually arousing fantasies, urges, or behaviors in which the individual exposes his or her genitals to an unsuspecting stranger); voyeurism (intense sexually arousing fantasies, urges, or behaviors in which the individual observes an unsuspecting stranger who is naked, disrobing, or engaging in sexual activity); fetishism (intense sexually arousing fantasies, urges, or behaviors in which the individual uses an inanimate object, such as a woman's shoe in a sexual manner); frotteurism (intense sexually arousing fantasies, urges, or behaviors in which the individual touches or rubs against a nonconsenting person in a sexual manner); pedophilia (intense sexually arousing fantasies, urges, or behaviors involving sexual activity with a prepubescent child, typically age 13 years or younger); sexual masochism (intense sexually arousing fantasies, urges, or behaviors in which the individual is humiliated, beaten, bound, or made to suffer in some way); sexual sadism (intense sexually arousing fantasies, urges, or behaviors in which the individual is sexually aroused by causing humiliation or physical suffering of another person); and transvestic fetishism (intense sexually arousing fantasies, urges, or behaviors involving cross-dressing, which generally refers to a male wearing female clothing).

FUNCTIONAL CONCEPTIONS OF SEXUAL DISORDERS

Behavior analytic approaches to sexual disorders have as their bases the relationships among habituation, classical and operant conditioning, and overt sexual behavior (Plaud, 2005). Given the wide applicability of conditioning principles (Plaud & Vogeltanz, 1993), a common theoretical assumption is that sexual responses are learned (O'Donohue & Plaud, 1994). Learning-based accounts of sexual behavior focus on the importance of both conditioning and habituation of sexual arousal. However, there are differences in the extent to which different behavioral theories emphasize the role of respondent conditioning-related factors (O'Donohue & Plaud, 1994; Plaud, Muench-Plaud, Kolstoe, & Orvedal, 2000; Skinner, 1969; Watson, 1925). Further, theories of human behavior which are not explicitly based on the principles of learning also rely on the assumption that, at least to some extent, sexual behavior and arousal are learned. For example, theories derived from anthropology and sociology have also claimed that human beings learn the range of sexual behavior through a variety of societal and cultural mechanisms (Davenport, 1987; DeLamater, 1987).

In applied behavioral contexts, the acquisition, maintenance, and modification of unusual sexual behavior have been explained by theories of learning (Barbaree & Marshall, 1991; Laws & Marshall, 1991; LoPiccolo & Stock, 1986; Masters & Johnson, 1970). For example, Laws, Meyer, and Holmen (1978) and O'Donohue and Plaud (1994) hypothesized that paraphilias develop due to early conditioning experiences and are maintained by their association with orgasm from masturbation to deviant fantasies or through positive reinforcement. Many behavior interventions used to treat sexual disorders also rely on learning principles in the modification of sexually deviant behavior (Kelly, 1982).

O'Donohue and Plaud (1994) conducted an extensive review and analysis of the literature on behavior analytic approaches to the conditioning of human sexual behavior, focusing on habituation/sensitization, classical conditioning, and the operant conditioning of sexual behavior. In order to understand the importance of each of these areas of behavior analysis, empirical research should first demonstrate that the principles of learning and behavior are involved in sexual behavior. Additionally, outcome studies that use the principles of learning to modify patterns of sexual arousal and deviant sexual behavior should demonstrate that behavioral interventions are effective.

HABITUATION AND SENSITIZATION

Habituation and sensitization are among the simplest forms of learning (Domjan & Burkhard, 1986). These processes are regarded as fundamental because they involve a two-term relation between stimuli and responses (S-R), rather than a four-term relation found in classical conditioning (conditioned stimulus, CS; unconditioned stimulus, US; unconditioned response, UR; and conditioned response, CR), or a three-term relation thought to be involved in operant conditioning (discriminative stimulus, S^D; response, R; and reinforcing consequence, S^{r+}). Habituation and sensitization are also thought to be more elementary than classical and operant conditioning because these forms of learning occur in organisms that are phylogenetically simple. Further, it is hypothesized that they are a precondition for the occurrence of other conditioning processes. For example, emotional responding of experimental animals to the experimental chamber must habituate before operant conditioning can take place (Skinner, 1938).

O'Donohue and Plaud (1991) documented that there are five possible patterns of responding to repeated presentations of a constant eliciting stimulus: (1) response magnitude can systematically decrease; (2) response magnitude can systematically increase; (3) response magnitude can remain constant; (4) changes in response magnitude can vary unsystematically; and (5) changes in response magnitude can vary between these previous four possibilities in any complex variation. When a systematic decrease in response magnitude is not due to physiological fatigue or response adaptation (the responsive adjustment of a sense organ to varying conditions), then habituation is said to occur. When a systematic increase in response magnitude is observed, sensitization is said to occur.

Plaud, Gaither, Amato-Henderson, and Devitt (1997) investigated the long-term habituation of male sexual arousal. They defined long-term habituation when (a) short-term (intra-session) habituation occurred; (b) habituated arousal spontaneously recovered; (c) the magnitude of spontaneous remission decreased across habituations sessions; and (d) the number of trials to habituation decreased across sessions. Plaud et al. (1997) replicated O'Donohue and Plaud's (1991) study in that they observed a short-term habituation of sexual arousal. They observed spontaneous recovery in all subjects who showed intrasession habituation. Finally, subjects also generally met the other criteria for long-term habituation.

CLASSICAL CONDITIONING

The earliest behavior analytic theories of sexual disorders were based on a classical conditioning paradigm. Binet (1888), Jaspers (1963), and Rachman (1961) believed unusual sexual behavior was the result of an accidental pairing. However, an accidental pairing of an unusual stimulus with sexual arousal or ejaculation, though feasible, leaves much to be explained. For instance, why do only some people whose first sexual experience involved a partner of the same sex practice homosexuality in adulthood, while many others do not? Or why would a stimulus such as a woman's stocking gain such erotic value as to become a fetish, while other objects that were also present, such as a pillow, do not become fetishistic objects?

O'Donohue and Plaud (1994) reviewed studies of classical conditioning and sexual behavior using the following criteria: (1) Was the CS presented alone in order to test for familiarity with the CS? (2) Were any novel conditioned stimuli used in order to test for the unconditioned effects of the CS? (3) Was the US presented alone to test for any prior sensitization or habituation to the US? (4) Was backward conditioning investigated by presenting the US prior to the CS in order to test for any effects of temporal order? (5) Was a truly random control procedure utilized by presenting the CS, US, each programmed entirely independently, in order to test for all nonassociative effects? and, (6) Could other factors, such as subject awareness of the experimental procedures, account for the findings as plausibly as a conditioning explanation?

An issue that confounds conditioning-based explanations of sexual arousal and overt sexual behavior, either classical or operant, concerns the issue of voluntary control of sexual behavior. Laws and Rubin (1969) provided support for the fact that males have voluntary control of penile tumescence—the engorgement of the penis by blood, which leads to an erection. They found that 4 of 7 male subjects developed full erections when exposed to erotic motion pictures. When the experimenters instructed the subjects to inhibit their erections when viewing the erotic films, every subject showed at least a 50% reduction in erections. This effect lasted as long as the instructions were in effect and ceased once the verbal instructions to inhibit responding were lifted. They also found that when asked to develop an erection in the absence of erotic stimuli, every subject did so, each reaching a peak of approximately 30% of maximum tumescence. This study has been criticized because subjects might have inhibited sexual responding by not orienting to the erotic stimuli. Henson and Rubin (1971) controlled for this possibility by requiring subjects to verbally describe the stimuli aloud or press a button in response to a random pattern of lights. They replicated Laws and Rubin's (1969) results. These studies suggest that research concerning conditioning in humans may be difficult due to the ability of humans to use verbally based strategies to inhibit or produce arousal in the absence of any explicit environmental contingencies. In other words, self-instructive verbal behavior may mediate the physiological process of responding to sexual stimuli.

Rachman (1966) conducted one of the earliest studies on the classical conditioning of sexual arousal and sexual fetishism by pairing a visual stimulus of a pair of black boots with visual stimuli of attractive, nude women. Rachman defined a conditioned response as five successive penile responses to the conditioned stimulus (black boots). However, Rachman did not define the minimum size of penile responding necessary for criterion conditioned responding. Rachman also assessed stimulus generalization after criterion responding to the CS by presenting stimuli of other types of boots and shoes. He found that all 3 subjects showed criterion conditioned responding, extinction of conditioned responding, and stimulus generalization. Consistent with behavioral accounts of unusual sexual behavior, Rachman concluded that sexual behavior can be conditioned to previously neutral stimuli, providing a basis for arguing that sexual behavior can be conditioned to unusual stimuli. However, there are several problems with Rachman's conclusions. Although he did pretest for initial levels of sexual arousal to black boots as well as the other stimuli, this study did not have a control for pseudoconditioning (which refers to the development of association between an unconditioned stimulus and a previously neutral stimulus that is not paired with the unconditioned stimulus in time but, due to chance, at times appears close to it). In other words, it may have been the US itself that led to sexual responding to the supposed CS and similar stimuli.

Rachman and Hodgson (1968) replicated this earlier study to rule out any pseudoconditioning confound and attempted to control for subject awareness by not informing the subjects about the nature of the study. The control condition consisted of a backward conditioning procedure; that is, the CS was presented after the US. They found the acquisition of a classically conditioned arousal response and failed to find conditioning in the control group. Although Rachman and Hodgson employed a backward conditioning condition, the sample size was small, no mention was made concerning any pretesting of the CS alone, no novel stimuli were used in either the forward or backward conditioning conditions to test for unconditioned effects of the CS, and, although they used many different unconditioned stimuli, they did not administer a random control procedure.

Plaud and Martini (1999) conducted the only laboratory-based classical conditioning study of male sexual arousal employing appropriate control conditions. Nine subjects participated in three sessions. There were 15 stimulus periods and 15 detumescence periods in each session. Each subject preselected sexually explicit visual stimuli which were used as unconditioned stimuli (US). The experimenter used a slide of a penny jar as the CS. In the first procedure, short delay conditioning, the CS was presented for 15 s, followed immediately by the US for 30 s with a CS/US overlap interval of 1 s. Interspersed in the 15 trials were 5 probe trials in which the CS was presented alone. Following each trial a 2-min detumescence period permitted a return to baseline. The second procedure was a backward conditioning procedure, in which the US was presented before the CS, to control for any effects of temporal ordering. In the third procedure, a random control condition, the experimenter presented the CS and US in a random order to test for nonassociative effects such as pseudoconditioning as defined above. The subjects showed systematic maximum increases in penile tumescence from baseline in the short delay conditioning procedure, but not in the other two control procedures, indicating that classical conditioning of the sexual response had in fact occurred.

Another issue related to classical conditioning of sexual arousal concerns the hypothesis that natural selection has favored the associations between certain kinds of stimuli and responses (Garcia & Koelling, 1966; Seligman, 1970). For example, Gosselin and Wilson (1980) found that male fetishists evidenced sexual arousal mainly to stimuli that were pink, black, smooth, silky, and shiny. McConaghy (1987) argued that the stimuli are similar to the female vulva, suggesting a dimension of biological preparedness. A theory such as biological preparedness can be used to plausibly explain why it is uncommon to find persons sexually aroused by pillows even though they are often paired with orgasm and other reinforcers. One explanation is that these items have not been prepared in evolution for association because they are recent artifacts, are not biologically significant, and are not directly related to sexual behavior.

OPERANT CONDITIONING

Skinner (1969, 1988) argued that past and present contingencies of survival and past and present contingencies of reinforcement shape sexual behavior. By "contingencies of survival," Skinner referred to the selective action of the environment on the species' gene pools. In Skinner's behavioral account, sexual contact has come to function as a powerful primary reinforcer through the contingencies of survival. Skinner also argued that the ontogeny and phylogeny of sexual behavior are closely related. According to Skinner, the contingencies of survival shaped global behavioral patterns exhibited by different organisms. However, natural selection is also complemented during the organism's lifetime through selection of behavior by its consequences. The capacity to learn through operant conditioning was an adaptive mutation. Organisms that were responsive to immediate environmental consequences that biological evolution could not prepare them for survived temporally unstable shifts in prevailing environmental features. Skinner termed this selection of behavior *operant conditioning* and the behaviors selected through consequences were called *operants*.

While Skinner focused only on natural selection in the evolution of human sexuality, Darwin (1871) suggested that evolution operated through *two* selection mechanisms: natural selection and sexual selection. According to Darwin (1871), sexual selection involves male competition for females and female choice among males. Darwin thought that the "law of battle" or male competition for mates accounted for the evolution of male aggressiveness, the male's greater size, and in some species the male's anatomical weapons for fighting. Furthermore, Darwin believed that females would choose males based on factors such as quality of their adornment, their courtship display, and the quality of the resources they control. Sexual selection is currently an area of intense interest among evolutionary biologists and behavioral geneticists and ought to be considered in any account of the phylogenetic heritage of an organism (O'Donohue & Plaud, 1994).

A central question in operant studies of sexual arousal is whether operant or classical conditioning has been demonstrated. O'Donohue and Plaud (1994) used Millenson and Leslie's (1979) criteria to evaluate studies claiming to demonstrate operant conditioning of sexual behavior. These criteria were (1) breaking of the contingency produces a short-term extinction burst in operant but not classical conditioning; (2) intermittent reinforcement produces greater resistance to extinction in operant conditioning, but this effect is not seen in classical conditioning; (3) complex skeletal behavior involving striated muscle is readily conditioned in operant but not classical conditioning; (4) autonomic behavior is readily conditioned in classical but not operant conditioning; (5) the conditioned response is not usually a component of behavior elicited by the reinforcer in operant conditioning, in contrast to classical conditioning; and (6) in operant but not classical conditioning the experimenter usually specifies the nature of the conditioned response within the often broad constraints of biological preparedness.

Quinn, Harbisan, and McAllister (1970) studied operant conditioning of sexual arousal with one homosexual subject. They placed the subject on a water-deprivation schedule in which presentation of liquid, its delivery signaled by a light, was contingent on increases in penile tumescence to a slide of an adult female. In contrast to baseline, the subject emitted increased levels of tumescence when tumescence (operant response) to a slide of a female adult (discriminative stimulus) was explicitly reinforced by a cold lime juice concentration. Although this study found a direct contingent relationship between tumescence and consequences, the researchers did not attempt to break the contingency in order to investigate whether an extinction burst would occur, and they did not investigate whether penile tumescence was sensitive to the effects of intermittent reinforcement.

Using an aversive conditioning procedure, Rosen and Kopel (1977) scheduled a contingent relationship between penile tumescence and the loudness of an alarm clock buzzer. As penile tumescence increased to video stimuli of the subject's initial sexual preference, the alarm sound's loudness also increased. They found a reduction in tumescence while this contingency was in effect. Different schedules of punishment were not used, and the potential physiological effects of fatigue or habituation were not ruled out.

Rosen, Shapiro, and Schwartz (1975) found that subjects demonstrated increased penile tumescence to a discriminative stimulus when monetarily reinforced for such responding. In contrast a yoked control group, which received noncontingent reinforcement in the presence of the same discriminative stimulus, showed no increased penile tumescence to the same stimulus. However, Rosen et al. did not investigate the effects of contingent versus noncontingent reinforcement and did not investigate the direct effects of intermittent reinforcement.

Schaefer and Colgan (1977) also studied whether sexual reinforcement increased penile responding. The researchers used sexually explicit scripts and neutral scripts which were read by the subjects. They found that penile responding continued to increase over trials when the sexually explicit scripts were followed by sexual reinforcement, a sexually explicit stimulus. The control subjects demonstrated decreased responding over trials.

Cliffe and Parry (1980) conducted the most direct study of the operant conditionability of sexual arousal by using sexual stimuli to test the matching law (Plaud, 1992). Originally formulated by Herrnstein (1970), the matching law predicts that when concurrent variable-interval schedules of reinforcement are in effect, there exists a one-to-one or matching relation between relative overall number of responses and the overall relative number of reinforcement presentations. A variable-interval schedule is one in which a reinforcer is presented after the first response that occurs after a variable amount of time has passed since the previously reinforced response. Concurrent schedules exist when two or more schedules are simultaneously in effect. Cliffe and Parry studied the sexual behavior of a male pedophile. Three concurrent variable-interval schedules were available. The first concurrent choice involved pressing keys to view either slides of women or men. The second concurrent choice was between slides of men or children. The third concurrent choice was between slides of women or children. The matching law accurately described the subject's behavior in all three conditions. Thus, these studies found support for the operant conditioning of sexual arousal. However, like classical conditioning, the data base at present is limited by the methodologies employed. Further, there exists a paucity of operant studies of female sexual arousal.

FUNCTIONAL ASSESSMENT AND ANALYSIS

Functional assessment of sexual disorders should identify the possible respondent and operant mechanisms that may lie behind problematic sexual behavior in order to identify alternate appropriate, satisfying forms of sexual behavior that service the same function as the problematic sexual

TABLE 17.1   A Multipurpose Framework for Looking at Sexuality

1. Sexually relevant behavior and experiences are present across virtually all ages and developmental levels.
2. Adults experience great diversity in forms of sexual expression and preferences regardless of whether intellectual functioning is within or below "normal" limits.
3. These diverse forms of sexual expression reflect genetic, constitutional, physiological, developmental, and environmental influences, or may also reflect "accidents" of conditioning history, based on actual life experiences.
4. Sexual arousal and orgasm constitute a potent reinforcer; the pairing of heightened sexual arousal and orgasm with some object, event, fantasy, activity, or person produces powerful conditioned associations, attachments, and emotions.
5. Once this association occurs, it may promote the development of strong and unusual preferences.
6. Because sexual reinforcement is so powerful, it is difficult to alter preferences for certain types of sexual behavior except under the following conditions: (1) there is some acceptable alternative sexual behavior which is more reinforcing; (2) the specific sexual behavior produces more negative consequences than the problematic sexual behavior.

behavior. Additionally, therapists should identify appropriate forms of sexual behavior, either in the client's current or past behavioral repertoire, as well as appropriate forms of sexual expression that are presently not in the client's repertoire. Thus, as a result of functional assessment, interventions should be individualized to each client in all sexual and general curriculum areas. Functional approaches eschew self-report data. In many cases of problematic sexual behavior, there may be a long history of reinforcement for inaccurate self-reporting because of negative consequences. Thus, responses are best if they are directly measured, for example, through plethysmography. These and other data should graphed in order to identify potential stimulus control of problematic and appropriate forms of sexual behavior. The therapist should develop operational definitions of problematic and appropriate forms of sexual behavior. For example, terms such as *stalking* or *approaching a child* should be operationally defined in order that therapist and client can agree on which behaviors are to be changed and in order to prevent subtle client noncompliance with intervention. Finally, the therapist should identify environmental stimuli that control the client's appropriate and inappropriate forms of sexual behavior. Table 17.1 presents a multipurpose behavior analytic framework for assessing human sexuality.

ETHICAL ISSUES

Before proceeding to define behaviorally based intervention strategies designed to eliminate sexual deviancy and promote prosocial sexual behavior, it is important to discuss the professional ethics of doing so. Many clients receiving behavioral intervention for sexual disorders will also be involved in the legal system, often serving a criminal sentence or under some legal jurisdiction such as probation or parole. Behavior analysts may be required to work directly with law enforcement or other criminal justice officials in providing professional services and should be aware of legal consequences on an offender's participation in behavioral intervention services (Glaser, 2005).

Beyond the potential limitations in behavioral intervention services that can result as a function of the legal system's direct involvement as a third party in cases involving clients with sexual disorders, it is also important to recognize that the client as well as societal safety can be better guaranteed if the offender does not have to sacrifice excessively in his/her life to lead a law-abiding lifestyle and, in particular, that the offender is able to have the same legal rights and privileges of any other citizen (Glaser, 2005). There is no agreed-upon professional code of ethics or practice standards that adequately address the many special issues that confront those who treat individuals with sexual disorders. Ethical codes and practice standards advocated by some organizations that focus on assessment and treatment of sexual disorders rely too heavily on the influence of law enforcement and the legal system. Behavior analysts working in this specialty area should be aware that such practice standards may compromise the delivery of appropriate services to those with sexual disorders (Glaser, 2005). Moreover, many behavioral techniques designed to treat sexual disorders involve the application of aversive contingencies or punishing consequences in addition to more positively based behavioral skills training. As such, it is critical that the client give full, informed consent to participating in any behavioral regimen designed to treat sexual disorders and that behavioral interventions should be employed only by trained and supervised professionals familiar with the behavioral techniques described in this chapter.

Providing for full, informed consent and understanding the potentially conflicting roles of providing behavioral interventions in context of ongoing legal issues are two major ethical concerns confronting behavior analysts who treat persons with sexual disorders. Another significant ethical consideration can be stated thus: Just because a behavioral intervention *can* be employed with an individual does not mean that it *should* be employed, even if the intervention has been empirically validated. For example, before homosexuality was removed as a mental disorder by the American Psychiatric Association in 1972, behavioral interventions were attempted to change the sexual orientation of homosexual men, a practice that has been shown to be unethical (Davison, 1976; Haldeman, 1994; Martin, 1984). Behavioral interventions in this domain should not be employed, and behavior analysts should take special caution when applying any intervention designed to change sexual behavior, including basic physiological sexual arousal. A basic ethical principle for professionals who work with clients in this area is that the client's welfare, as well as societal safety, must both be served through the implementation of behavioral interventions. This principle applies to both clients who have a diagnosed sexual disorder as well as clients who do not have a clinical diagnosis or any legal involvement, yet verbalize problems in their sexual functioning.

FUNCTIONAL-ANALYTIC BASED INTERVENTIONS

Like some other disorders, there is an established corpus of behavior therapy approaches to the treatment of sexual disorders, such as Masters and Johnson's sensate focus, social skills and assertiveness training for people with sexual problems, and various aversive conditioning approaches.

FUNCTIONAL ANALYTIC-BASED POSITIVE INTERVENTIONS

Skills training and contingency management approaches to the treatment of sexual behavior can also form an important component of intervention for sexual disorders. Positive interventions include individual and group behavioral skills training to provide opportunities for the social reinforcement of appropriate social and sexual interactive behaviors (Plaud, et al., 2000). Specialized behavioral programming skills should be offered on an ongoing basis for clients based on individual behavioral assessment results and upon behavioral research to date. In other words, excesses and deficits in sexual behavior may either cause, or at least maintain, sexual deviations. For instance, a lack of social skills or communication skills in dealing with appropriate adult sexual partners may lead a person to seek out individuals with whom these skills are not required (e.g., children). Thus, behavioral skills training techniques such as social skills training, assertiveness training, communication skills training, relaxation training, and systematic desensitization have been included in more comprehensive treatment approaches of sexual disorders, at least as adjunctive interventions to the other behavioral techniques described here.

The importance of including behavioral skills training in this area cannot be understated. Clients should benefit from precision teaching in areas of sexual knowledge, "cognitive distortions," and behavioral skills training (e.g., social skills, communication skills, assertiveness skills), direct social reinforcement intervention strategies (individual and group) linking in a direct fashion appropriate sexual behavior in social contexts (employing role playing and *in vivo* behavioral strategies), with the addition of conditioning-based intervention programs described later in this chapter.

ORGASMIC RECONDITIONING

According to Marquis (1970), orgasmic reconditioning "attaches sexual arousal and rehearses sexual behavior in response to socially acceptable stimuli" (p. 267). He further stated that it "extinguishes sexual responses to the deviant stimulus by preventing them from being paired with orgasm and eventually decreasing to zero the amount of arousal with which they are paired" (p. 267). Thus, during orgasmic reconditioning a nondeviant conditioned stimulus is paired with orgasm, while the sexual behavior in response to the deviant stimulus or fantasy is respondently extinguished. Orgasmic reconditioning involves substituting a nondeviant stimulus for the deviant stimulus, which initially elicited sexual arousal.

This technique entails having a subject masturbate while imagining a deviant fantasy. When the subject is near orgasm, he/she should change to a nondeviant fantasy so that the nondeviant fantasy will be paired with orgasm. After several sessions, the subject will be able to move to the point at which he/she switches to the nondeviant fantasy earlier until he/she is aroused by the nondeviant fantasy alone. Several researchers have reported successes using orgasmic reconditioning (Davison, 1968; Jackson, 1969; LoPiccolo, Stewart, & Watkins, 1972; Marshall, 1973; Thorpe, Schmidt, Brown, & Castell, 1964; Thorpe, Schmidt, Brown & Castell, 1964), although it has been combined with other techniques very often.

MASTURBATORY EXTINCTION

In masturbatory extinction the subject is instructed to masturbate to ejaculation while engaging in nondeviant fantasizing and then to continue masturbating for an extended period of time while engaging in deviant fantasizing (Alford, Morin, Atkins, & Shoen, 1987). Thus, perhaps nondeviant fantasies are reinforced with ejaculation, while deviant fantasies are not and are perhaps extinguished.

MASTURBATORY SATIATION

In masturbatory satiation or satiation therapy, a subject masturbates continuously for 1 hr or more, while verbalizing his/her deviant fantasies aloud (Marshall, 1979). If the subject ejaculates, he/she is to continue masturbating until the time period is over. Thus, the deviant fantasy no longer ends with a reinforcing ejaculation. Therefore, the subject will lose interest in the deviant fantasies, because they are paired with habituated sexual stimuli. According to Marshall (1979), this method is especially beneficial because it capitalizes on a naturally occurring phenomenon (i.e., masturbating to a deviant fantasy). The effectiveness of this technique has been reported by several researchers (Johnson, Hudson, & Marshall, 1992; Marshall, 1979; Marshall & Barbaree, 1978).

AVERSIVE CLASSICAL CONDITIONING

A number of intervention methods have been developed in which some aversive stimulus, such as an aversive covert image, aversive olfactory stimulus, or contingent shock, is paired with a problematic conditioned stimulus, such as a fetishistic image or object or reports of private images of problematic situations, such as approaching children or rape. Consonant with matching law, these approaches often combine this approach with reinforcement of alternate, acceptable forms of sexual behavior. These approaches are less preferred than skills teaching and other positive approaches but may have an important role to play when other approaches are not possible, when the client consents to this form of treatment, or when the inappropriate sexual behavior is dangerous or illegal and response to alternative treatment is likely to be ineffective or too slow. Functional assessment has an important role in this group of interventions in accurately identifying the presumptive conditioned stimuli to be paired with the aversive stimuli and identifying functionally equivalent forms of acceptable behavior to be reinforced.

Covert Sensitization

Covert sensitization may be a form of conditioning in which behaviors and their precipitative events are paired with an aversive image in order to promote avoidance of the precipitative events and to decrease the undesirable behaviors. Cautela and Kearney (1990) wrote

> Covert conditioning refers to a family of behavioral therapy procedures which combine the use of imagery with the principles of operant conditioning. Covert conditioning is a process through which private events such as thoughts, images, and feelings are manipulated in accordance with principles of learning, usually operant conditioning, to bring about changes in overt behavior, covert psychological behavior (i.e. thoughts, images, feelings, and/or physiological behavior (e.g. glandular secretions). (p. 86)

In covert sensitization, the aversive image usually consists of an anxiety- or nausea-inducing scene in which the therapist presents images to the client. The aversive scene is created individually and specifically to suit each client. Covert sensitization has been used to treat sexual disorders effectively both singly and in combination with other intervention methods (Dougher, Crossen, Ferraro, & Garland, 1987; Enright, 1989; Haydn-Smith, Marks, Buchaya, & Repper, 1987; Hayes, Brownell, & Barlow, 1978;

King, 1990; Marshall, Eccles, & Barbaree, 1991; Moergen, Merkel, & Brown, 1990).

Other Aversive Therapies

Olfactory aversive and electrical aversive therapies pair a problematic behavior and its precipitating events with a noxious stimulus. These procedures can be used in the treatment of sexual deviance, helping to change deviant sexual arousal and/or deviant fantasies. For example, Abel, Levis, and Clancy (1970) used contingent shock therapy to treat paraphilia in 6 men aged 21–31 years. The researchers paired mild electrical shocks with sexual arousal to a deviant audiotape for 5 of the subjects, while the 6th subject received noncontingent shocks. The 5 subjects receiving contingent shock therapy showed suppressed penile responding to the deviant stimuli at an 18-week follow-up assessment. The subject who had received noncontingent shocks exhibited less suppression.

THERAPY COMBINATIONS

Many early studies of sexual deviance took the simplistic view that deviance consisted only of an excess of arousal to deviant objects or behaviors (Abel & Blanchard, 1976). Thus, early procedures focused only on aversive procedures to reduce deviance. However, many clients also reported behavioral deficits, such as inadequate arousal to appropriate objects or behaviors (Abel & Blanchard, 1976). Therefore, behavior therapy techniques are frequently used in combinations (Enright, 1989; King, 1990; Marshall et al., 1991; McNally & Lukach, 1991; Rangaswamy, 1987; Wolfe 1992) to reduce a subject's deviant arousal and another method used to increase his/her nondeviant arousal.

REHABILITATION

Behaviorally based clinical programs that are designed to treat sexual disorders should also address the community reintegration. Once a client makes satisfactory progress in a program, the client should be placed in a transition program that will serve as the first step to community reintegration and client aftercare of the client. The first two years after release from an in-patient treatment program or correctional facility are the most critical in terms of recidivism (Gaither, Rosenkranz, & Plaud, 1998). Active criteria should also be developed to initiate this transition process. Sample criteria are shown in Table 17.2. A behaviorally based sexual offender treatment program should always be pledged for societal safety and the protection of the rights of sexual offenders relative to effective, objective, and empirically validated assessment and treatment, to provide the necessary resources to assess sexual arousal and behavior.

TABLE 17.2 Sample Behavioral Criteria for Transitional Programming

1. Successful and documented participation in the behavioral treatment regimens in the treatment program, with a 3-month minimum absence of any major target or major disruptive behaviors.
2. Psychophysiological plethysmographic evaluation indicating a relative absence of disordered patterns of sexual arousal. Other forms of psychological/behavioral assessment will also be integrated into a transition risk management assessment protocol.
3. An absence of verbalizations of denial (covert/verbal) with corroborating overt behavior to support and document acknowledgment and acceptance of the sexually offending behavior patterns in question.
4. Identification of the chain of behaviors that put the client at risk for reoffending (sometimes referred to as the "cycle," which involves identifying the precursors that may lead to the ultimate commission of a sexual offense).
5. Demonstration of relapse prevention/safety plan skills while a client in the program (behavioral rehearsal plans will be integrated into behavioral training).
6. Documentation of successful supervised visits with family/social networks while a client in the program.

Professionals who assess and treat those with sexual disorders who have legal involvement should be knowledgeable of the proliferation of laws in the past decade that apply to those with sexual offending histories, such as probation/parole involving electronic monitoring through global position satellite devices, sexual offender registration, prohibitions to live or work in certain jurisdictions, and day-to-life civil commitment of sexual offenders. Given the proliferation of such legal interventions in the daily lives of sexual offenders, careful planning is required in any societal transition program.

CASE STUDY

REFERRAL

A local human service center psychologist referred a 24-year-old male for a penile plethysmographic evaluation due to his failure to progress in group treatment at the human service center. This ultimately led to his termination from the group, which had focused on psychoeducational issues relating to human sexuality, consent and victim empathy issues, appropriate and inappropriate sexual behavior, and disclosure to other members of the group. The client did not participate actively in any phase of the group treatment.

The client had an extensive history of sexually abusive behavior. He earlier pleaded guilty to a charge of sexual assault and was serving probation at the time of the initiation of therapy services.

FUNCTIONAL ASSESSMENT

The client's penile responses during the course of therapy were recorded by a penile plethysmograph utilizing a Type A mercury-in-rubber penile strain gauge. During the original assessment, penile tumescence was continually monitored as the client listened to sexually explicit audiotapes. Eighteen standard audio scripts were presented during the initial assessment. These were descriptions of 2 adult homosexual interactions, 2 adult heterosexual interactions, 2 acts of adult female exhibitionism, 2 adult female rapes, 1 male child physical aggression, 1 female child physical aggression, 1 male child nonphysical coercion, 1 female child nonphysical coercion, 3 male child fondling, and 3 female child fondling. The client's subjective reports of sexual arousal were assessed by having him rate how aroused he felt using a 10-point Likert scale. The client was aroused by adult females. However, he also displayed arousal toward stimuli depicting sexual activities with a male child, specifically anal intercourse. Based on these data, the therapist noted three problematic stimulus categories which elicited the greatest levels of sexual arousal: fondling a male child (FMC), coercing a female child into sexual activity (CFC), and fondling a female child (FFC.) The therapist recommended that the client participate in eight sessions of assisted covert sensitization in addition to being readmitted to group treatment at the local human service center.

INTERVENTION

Shortly after the client's initial assessment, the therapist implemented an assisted covert sensitization protocol. The client consented in writing, and the therapist gave a full explanation of the procedure and answered all questions concerning the procedure. The therapist scheduled the initial assisted covert sensitization session for the following week. During the week the therapist developed audiotapes containing 3-min descriptions of the three classes of problematic sexual activity. The description of each problematic sexual activity was followed by a description of possible negative consequences including either legal (e.g., being beaten up by the father of the child and then being arrested) or physiological (e.g., feeling very nauseous and vomiting) consequences.

When the client arrived for the first session, the therapist conducted an abbreviated assessment to obtain baseline measurements of his sexual arousal to problematic as well as mutually consenting heterosexual and mutually consenting homosexual activity.

Following a 10-min break the therapist presented 10 MFC stimuli. At the end of the session, the therapist gave the client a copy of the tape and instructed him to listen to and visualize the sexual activity as well as the aversive consequences being delivered five times per day. The therapist conducted the remaining five sessions at 1-week intervals apart beginning with Session 1.

During Session 2, the therapist presented the client with the same 10 MFC stimuli from the previous session, and again instructed him to listen to the tape five times per day until the next session. In Sessions 3 and 4, the therapist followed the same procedures, with the exception that the therapist presented the MPF stimuli only twice and the CFC stimuli the other times. The therapist again provided the client with a copy of the new tape and instructed him to listen to it five times per day between sessions as before. In sessions the therapist presented 5 and 6 FFC stimuli six times, FMC two times, and CFC two times each.

OUTCOME

After completion of Session 6, the client returned to the clinic for a 30-day and 3-month follow-up assessment. The therapist used the same stimuli from the baseline assessment to determine patterns of sexual arousal. The client's physiological data for the initial assessment, pretreatment assessment, 30-day follow-up, and 90-day follow-up were calculated and converted to percentages of full erection by subtracting his minimum penile circumference for an entire assessment period from his maximum penile circumference for each trial (the presentation of one audiotaped stimulus represents a trial) and dividing this number by 3. Three centimeters is thought to reflect the circumference change most males undergo from flaccidity to complete engorgement. This number was then multiplied by 100% to give a percentage of full erection. Thus, percentage of full erection data give an indication of *absolute levels of arousal*. In other words, the client's response to each stimulus is viewed in this manner independently of the other stimuli presented in the session. The stimuli elicited less arousal each time the client was assessed during the assisted covert sensitization procedure.

The client's physiological data for the assessments were next converted to z-scores, in which the mean of the distribution equals zero (0) and the standard deviation is 1.0. Client's sexual preferences are expressed as positive z-scores, while negative z-scores reflect sexual aversions. Z-scores give an indication of *relative arousal* or preferences and aversions among a group of stimuli. In the initial assessment, four of the five categories, including the three which were treated, had positive z-scores above 0.50. When the therapist looked across the assessments for each of the deviant categories, it was clear that the client's arousal to these decreased across time, although his arousal to adult mutually consenting sexual activity was clearly his most preferred stimulus in all assessments, except for the 30-day follow-up in which mutually consenting heterosexual activity was the most preferred stimulus. The client reported similar results using the 10-point Likert scale ratings mentioned previously.

The client clearly showed progress in both his physiological and self-report of arousal toward sexually deviant stimuli that were the main areas of concern, utilizing the assisted covert sensitization procedure. After this intervention was completed, the local human service center psychologist received follow-up that the client's participation in group therapy improved, and the client was later designated by his treatment program for inclusion in a lesser risk category of sexual offense recidivism. Recall that the underlying behavior principle of covert sensitization is most often theorized to be a combination of classical and operant conditioning. Given decrements in physiological arousal and self-report normally observed in covert sensitization procedures, we may presume that the aversive image associated with deviant sexual arousal (the UCS) becomes a CS by virtue of its being contingently paired with the UCS. Also, it is logical and theoretically coherent to conclude that both the conditioned response (CR) and the unconditioned response (UCR) consist of a negative reaction which may be emotional (e.g. fear), physiological (e.g. nausea), or in some other way repulsive, which further serves to negatively reinforce avoidance or escape behavior (operant conditioning).

REFERENCES

Abel, G. G., & Blanchard, E. B. (1976). The measurement and generation of sexual arousal in male sexual deviates. In M. Hersen, R. M. Eisler, & P. M. Miller (Eds.), *Progress in behavior modification, volume 2* (pp. 99–136). New York: Academic Press.

Abel, G. G., Levis, D. J., & Clancy, J. (1970). Aversion therapy applied to taped sequences of deviant behavior in exhibitionism and other sexual deviations: A preliminary report. *Journal of Behavior Therapy and Experimental Psychiatry, 1*, 59–66.

Alford, G. S., Morin, C., Atkins, M., & Schoen, L. (1987). Masturbatory extinction of deviant sexual arousal: A case study. *Behavior Therapy, 18*, 265–271.

American Psychiatric Association. (2000). *Diagnostic and statistical manual of mental disorders* (4th ed., Refs rev.). Washington, D.C.: Author.

Barbaree, H. E., & Marshall, W. L. (1991). The role of male sexual arousal in rape: Six models. *Journal of Consulting and Clinical Psychology, 59*, 621–630.

Binet, A. (1888). *Le fetichisme dans l'amour.* Paris: Octave Doin Editeur

Cautela, J. R., & Kearney, A. J. (1990). Behavior analysis, cognitive therapy and covert conditioning. *Journal of Behavior Therapy and Experimental Psychiatry, 21*, 83–90.

Cliffe, M. J., & Parry, S. J. (1980). Matching to reinforcer value: Human concurrent variable-interval performance. *Quarterly Journal of Experimental Psychology [A], 32*, 557–570.

Darwin, C. (1871). *The descent of man and selection in relation to sex.* London: Murray.

Davenport, W. H. (1987). An anthropological approach. In J. Geer & W. O'Donohue (Eds.), *Theories of human sexuality* (pp. 197–236). New York: Plenum Press.

Davison, G. C. (1968). Elimination of a sadistic fantasy by a client-controlled technique: A case study. *Journal of Abnormal Psychology, 73*, 84–90.

Davison, G. C. (1976). Homosexuality: The ethical challenge. *Journal of Consulting and Clinical Psychology, 44*, 157–162.

DeLamater, J. (1987). A sociological approach. In J. Geer & W. O'Donohue (Eds.), *Theories of human sexuality* (pp. 237–256). New York: Plenum Press.

Domjan, M., & Burkhard, B. (1986). *The principles of learning and behavior.* Belmont, CA: Brooks/Cole.

Dougher, M. J., Crossen, J. R., Ferraro, D. P., & Garland, R. (1987). The effects of covert sensitization on preference for sexual stimuli. *Journal of Behavior Therapy and Experimental Psychiatry, 18*, 337–348.

Enright, S. J. (1989). Pedophilia: A cognitive/behavioral treatment approach in single case. *British Journal of Psychiatry, 155*, 399–401.

Gaither, G. A., Rosenkranz, R. R., & Plaud, J. J. (1998). Sexual disorders. In J. J. Plaud & G. H. Eifert (Eds.), *From behavior theory to behavior therapy* (pp. 152–171). Boston, MA: Allyn & Bacon.

Garcia, J., & Koelling, R. (1966). Relation of cue to consequence in avoidance learning. *Psychonomic Science, 4*, 123–124.

Glaser, W. (2005). An ethical paradigm for sex offender treatment: A response to Levenson and D'Amora. *Western Criminology Review, 6*, 154–160.

Gosselin, C., & Wilson, G. (1980). *Sexual variations.* London: Faber & Faber.

Haldeman, D. (1994). The practice and ethics of sexual orientation conversion therapy. *Journal of Consulting and Clinical Psychology, 62*, 221–227.

Haydn-Smith, P., Marks, I., Buchaya, H., & Repper, D. (1987). Behavioral treatment of life threatening masochistic asphyxiation: A case study. *British Journal of Psychiatry, 150*, 518–519.

Hayes, S. C., Brownell, K. D., & Barlow, D. H. (1978). The use of self-administered covert sensitization in the treatment of exhibitionism and sadism. *Behavior Therapy, 9*, 283–289.

Henson, D. E., & Rubin, H. B. (1971). Voluntary control of eroticism. *Journal of Applied Behavior Analysis, 4*, 37–44.

Herrnstein, R. J. (1970). On the law of effect. *Journal of the Experimental Analysis of Behavior, 13*, 243–266.

Jackson, B. (1969). A case of voyeurism treated by counter conditioning. *Behaviour Research and Therapy, 7*, 133–134.

Jaspers, K. (1963). *General psychopathology.* Manchester: Manchester University Press.

Johnson, P., Hudson, S. M., & Marshall, W. L. (1992). The effects of masturbatory reconditioning with non-familial child molesters. *Behaviour Research and Therapy, 30*, 559–561.

Kelly, R. J. (1982). Behavioral reorientation of pedophiliacs: Can it be done? *Clinical Psychology Review, 2*, 387–408.

King, M. B. (1990). Sneezing as a fetishistic stimulus. *Sexual and Marital Therapy, 5*, 69–72.

Laws, D. R., & Marshall, W. L. (1991). A conditioning theory of the etiology and maintenance of deviant sexual preference and behavior. In W. L. Marshall, D. R. Laws, & H. E. Barabaree (Eds.), *Handbook of sexual assault* (pp. 209–229). New York: Plenum Press.

Laws, D. R., Meyer, J., & Holmen, M. L. (1978). Reduction of sadistic sexual arousal by olfactory aversion: A case study. *Behaviour Research and Therapy, 16*, 281–285.

Laws, D. R., & Rubin, H. B. (1969). Instructional control of an autonomic sexual response. *Journal of Applied Behavior Analysis, 2*, 93–99.

LoPiccolo, J., Stewart, R., & Watkins, B. (1972). Treatment of erectile failure and ejaculatory incompetence of homosexual etiology. *Journal of Behavior Therapy and Experimental Psychiatry, 3*, 233–236.

LoPiccolo, J. & Stock, W. E. (1986). Treatment of sexual dysfunction. *Journal of Consulting and Clinical Psychology, 54*, 158–167.

Marquis, J. N. (1970). Orgasmic reconditioning: Changing sexual object choice through controlling masturbation fantasies. *Journal of Behavior Therapy and Experimental Psychiatry, 1*, 263–271.

Marshall, W. L. (1973). The modification of sexual fantasies: A combined treatment approach to the reduction of deviant sexual behavior. *Behaviour Research and Therapy, 11*, 557–564.

Marshall, W. L. (1979). Satiation therapy: A procedure for reducing deviant sexual arousal. *Journal of Applied Behavior Analysis, 12*, 377–389.

Marshall, W. L., & Barbaree, H. E. (1978). The reduction of deviant arousal: Satiation treatment for sexual aggressors. *Criminal Justice and Behavior, 5*, 294–303.

Marshall, W. L., Eccles, A., & Barbaree, H. E. (1991). The treatment of exhibitionists: A focus on sexual deviance versus cognitive and relationship features. *Behavior Research and Therapy, 29*, 129–135.

Martin, A. D. (1984). The emperor's new clothes: Modern attempts to change sexual orientation. In E. S. Hetrick & T. S. Stein, (Eds.), *Psychotherapy with homosexuals* (pp. 24–57). Washington, D.C.: American Psychiatric Association.

Masters, W., & Johnson, V. (1970). *Human sexual inadequacy.* Boston: Little, Brown.

McConaghy, N. (1987). A learning approach. In J. Geer & W. O'Donohue (Eds.), *Theories of human sexuality* (pp. 287–334). New York: Plenum Press.

McNally, R. J., & Lukach, B. M. (1991). Behavioral treatment of zoophilic exhibitionism. *Journal of Behavior Therapy and Experimental Psychiatry, 22*, 281–284.

Millenson, J. R., & Leslie, J. C. (1979). *Principles of behavioral analysis.* New York: Macmillan.

Moergen, S. A., Merkel, W. T., & Brown, S. (1990). The use of covert sensitization and social skills training in the treatment of an obscene telephone caller. *Journal of Behavior Therapy and Experimental Psychiatry, 21*, 269–275.

O'Donohue, W., & Plaud, J. J. (1991). The long-term habituation of human sexual arousal. *Journal of Behavior Therapy and Experimental Psychiatry, 22*, 87–96.

O'Donohue, W., & Plaud, J. J. (1994). The conditioning of human sexual arousal. *Archives of Sexual Behavior, 23*, 321–344.

Plaud, J. J. (1992). The prediction and control of behavior revisited: A review of the matching law. *Journal of Behavior Therapy and Experimental Psychiatry, 23*, 25–31.

Plaud, J. J. (2005). Covert sensitization conditioning. In M. Hersen & J. Rosquist (Eds.), *Encyclopedia of behavior modification and cognitive behavior therapy, volume 1: Adult clinical applications* (pp. 235–241). London: Sage Publications.

Plaud, J. J., Gaither, G. A., Amato-Henderson, S., & Devitt, M. K. (1997). The long-term habituation of sexual arousal in human males: A crossover design. *The Psychological Record, 47*, 385–398.

Plaud, J. J., & Martini, J. R. (1999). The respondent conditioning of male sexual arousal. *Behavior Modification, 23*, 254–268.

Plaud, J. J., Muench-Plaud, D., Kolstoe, P. D., & Orvedal, L. (2000). Behavioral treatment of sexually offending behavior. *Mental Health Aspects of Developmental Disabilities, 3*, 54–61.

Plaud, J. J., & Vogeltanz, N. D. (1993). Behavior therapy and the experimental analysis of behavior: Contributions of the science of human behavior and radical behavioral philosophy. *Journal of Behavior Therapy and Experimental Psychiatry, 24*, 119–127.

Quinn, J. T., Harbisan, J. J., & McAllister, H. (1970). An attempt to shape human penile responses. *Behaviour Research and Therapy, 8,* 213–216.

Rachman, S. (1961). Sexual disorders and behavior therapy. *American Journal of Psychiatry, 18,* 35–240.

Rachman, S. (1966). Sexual fetishism: An experimental analogue. *Psychological Record, 16,* 293–296.

Rachman, S., & Hodgson, R. J. (1968). Experimentally-induced "sexual fetishism": Replication and development. *Psychological Record, 18,* 25–27.

Rangaswamy, K. (1987). Treatment of voyeurism by behavior therapy. *Child Psychiatry Quarterly, 20*(3–4), 73–76.

Rosen, R. C., & Kopel, S. A. (1977). Penile plethysmography and bio-feedback in the treatment of a transvestite-exhibitionist. *Journal of Consulting and Clinical Psychology, 45,* 908–916.

Rosen, R. C., Shapiro, D., & Schwartz, G. (1975). Voluntary control of penile tumescence. *Psychosomatic Medicine, 37,* 479–483.

Schaefer, H. H., & Colgan, A. H. (1977). The effect of pornography on penile tumescence as a function of reinforcement and novelty. *Behavior Therapy, 8,* 938–946.

Seligman, M. E. P. (1970). On the generality of the laws of learning. *Psychological Review, 77,* 406–418.

Skinner, B. F. (1938). *The behavior of organisms.* New York: Appleton-Century-Crofts.

Skinner, B. F. (1969). *Contingencies of reinforcement: A theoretical analysis.* New York: Appleton-Century-Crofts.

Skinner, B. F. (1988). The phylogeny and ontogeny of behavior. In A. C. Catania & S. Harnad (Eds.), *The selection of behavior: The operant behaviorism of B. F. Skinner* (pp. 382–400). Cambridge: Cambridge University.

Thorpe, J., Schmidt, E., Brown, P., & Castell, D. (1964). Aversion-relief therapy: A new method for general application. *Behaviour Research and Therapy, 2,* 71–82.

Watson, J. B. (1925). *Behaviorism.* New York: Norton.

Wolfe, R. W. (1992). Video aversive satiation: A hopefully heuristic single-case study. *Annals of Sex Research, 5,* 181–187.

18

EATING DISORDERS

RICHARD F. FARMER

Oregon Research Institute

AND

JANET D. LATNER

University of Hawaii

In traditional behavioral assessment, emphasis is placed on within-person variability of behavior and the contextual factors that influence this variability. In this respect, behavioral assessment is idiographic, or person centered. With rare exception, research on the functional aspects of eating-disorder-related behaviors has been based on between-groups designs, group trend analyses, or correlational methods, and largely centered on variables (e.g., diagnostic group membership). Such nomothetic approaches, however, are useful for illuminating common patterns among groups of persons that can, in turn, suggest useful hypotheses to evaluate in relation to the individual case. Therefore, in the review that follows, we first describe a functional analytic model that will serve as a conceptual framework for the subsequent literature review. Based on available theory and research, we then present a nomothetic functional analysis of response topographies (or forms of behavior) that define the eating disorders, which is then followed by a survey of behavioral assessment methods that can aid in the functional assessment of the client. Three different forms of eating disorders will be emphasized: anorexia nervosa (AN), bulimia nervosa (BN), and binge eating disorder (BED). Based on our nomothetic functional formulation, we then describe treatment components to address functional aspects of eating disorder behavior. This is then followed by an illustrative case study.

Functional Analysis in Clinical Treatment

FUNCTIONAL ANALYTIC MODELS OF EATING DISORDERS

Behavioral assessments and therapies influenced by radical behavioral (operant) theory regard the principal unit of analysis to be the whole person and functional aspects of his or her behavior in particular environmental contexts (Follette, Naugle, & Linnerooth, 2000). The focus of assessment and treatment is on *why* people behave as they do (Farmer & Nelson-Gray, 2005). Correspondingly, the functional analytic approach in behavioral assessment involves the identification of functional relationships between clinically relevant behaviors and the environmental variables that select, influence, and maintain them (Sturmey, 1996).

The functional assessment framework utilized in this chapter is Goldfried and Sprafkin's (1976) SORC model, which is an acronym for Stimuli–Organism variables–Responses–Consequences. In this model, an individual's responses (physiological, emotional, verbal, cognitive, motor) are thought to be a joint function of contextual factors (establishing operations, antecedent stimuli, and consequences) and organismic variables that mediate or moderate environment–behavior relations (e.g., heritable biological endowments, past learning history). In addition to illuminating the functional contexts of clinically relevant behavior, other goals associated with such an analysis include the selection of an appropriate intervention and the provision of the methods and means to monitor treatment progress and evaluate treatment effectiveness (Follette et al., 2000; Sturmey, 1996). Procedural steps for developing a functional understanding of eating disorder behavior, elaborated in the following sections, are summarized in Table 18.1.

CENTRAL BEHAVIORS THAT DEFINE EATING DISORDERS

Behaviors that define contemporary eating disorders, such as self-starvation, binge eating, and purging, have been manifested and documented throughout history. However, the covarying clusters of behaviors

---

TABLE 18.1　Recommendations for the Functional Assessment of Eating Disorders

- Assess the forms of eating disorder behaviors (e.g., vomiting, exercise, social avoidance), as well as the associated history, pattern, and course of these behaviors.
- With nomothetic findings on the functional nature of eating disorder behaviors in mind, conduct an SORC functional analysis of a client's eating disorder behavior.
- From the resultant SORC analyses, generate hypotheses concerning contextual and organismic factors that influence current problematic eating behavior.

that define modern forms of AN, BN, and BED began to receive significant research and clinical attention only in the last half of the 20th century. This growing attention parallels the relatively recent and rapid rise in extreme dieting behaviors and self-starvation tendencies found in many contemporary Western societies where thinness is increasingly idealized. This idealization is particularly evident in the media, as revealed by the thinning of fashion models (Morris, Cooper, & Cooper, 1989) and Miss America contestants (Rubinstein & Caballero, 2000). Correspondingly, the disapproval of excess weight among girls and boys has increased in the past 40 years (Latner & Stunkard, 2003). This cultural acceptance and valuation of the thinness ideal have been implicated as risk factors for eating disturbances and body dissatisfaction (Stice, 2002).

In the *Diagnostic and Statistical Manual of Mental Disorders* (4th edition; *DSM-IV*), Criterion A of AN is the "Refusal to maintain body weight at or above a minimally normal weight for age and height" (American Psychiatric Association [APA], 2000, p. 589). Associated with this are disturbances characterized by fear of fat and dysfunctional attitudes toward one's body image. Criteria B and C involve "Intense fear of gaining weight or becoming fat, even though underweight" and "Disturbance in the way in which one's body weight or shape is experienced, undue influence of body weight or shape on self-evaluation, or denial of the seriousness of the current low body weight." Criterion D, amenorrhea, applies to postmenarcheal females only. The lifetime prevalence of AN is approximately 0.5%, with the typical age of onset being middle to late adolescence.

"Recurrent binge eating episodes," characterized by eating a large quantity of food and experiencing a subjective sense of loss of control over eating, comprise Criterion A of the *DSM-IV* BN diagnosis (APA, 2000, p. 594). Criterion B, "Recurrent inappropriate compensatory behavior in order to prevent weight gain," often follows binge episodes, and commonly involves self-induced vomiting, laxative misuse, fasting, or excessive exercise. For a diagnosis, binge episodes and compensatory behaviors must occur at least twice a week for 3 months, on average (Criterion C). Similar to AN, "self-evaluation is unduly influenced by body shape and weight" in BN (Criterion D). Body-related checking and avoidance behaviors often co-occur with excessive shape and weight concerns, and are frequently displayed by persons with eating disorders. Checking behaviors can include reassurance seeking, frequent self-weighing, pinching one's flesh, and looking at particular body parts in the mirror. Body-related avoidance behaviors include avoiding of form-fitting clothing, shunning activities that require body exposure (e.g., shopping, exercise), or refusing to view oneself in the mirror. The lifetime prevalence of BN is about 1–3% and the typical age of onset is late adolescence to early adulthood. About 10% of persons with AN or BN are male.

Similar to BN, *DSM-IV* research criteria for BED include frequent binge eating, at an average rate of 2 days a week for 6 months (APA, 2000, p. 787). These episodes must be associated with three of the following: eating more rapidly than normal; eating until uncomfortably full; eating large amounts when not physically hungry; eating alone because of embarrassment by how much one is eating; and feeling disgusted, depressed, or guilty after overeating. "Marked distress regarding binge eating" must also be present. BED is common among individuals seeking professional weight-control treatment (15–50%); the lifetime prevalence in the general population is 0.7–4%. Like BN, the onset of BED is typically in late adolescence or the early 20s. Unlike the other eating disorders, about 40% of individuals with BED are male (APA, 2000).

THE CONTEXT OF BEHAVIOR

A core assumption of behavioral theories is that behavior varies according to antecedent stimuli that precede behavior, establishing operations (e.g., reinforcer deprivation or satiation), and the consequences that follow. These constitute the context of behavior.

Antecedent Stimuli

Antecedent or discriminative stimuli (or S^D) signal to the individual the likelihood of reinforcement or punishment following some type of response. The informational value associated with an S^D is based on the person's previous experiences when in the presence of those or related stimuli. The sights and smells of food, for example, are often potent antecedents associated with potential reinforcement for food consumption. Rules are one important subgroup of S^D. A rule is a verbal expression of a behavioral contingency that describes a specific form of behavior, an outcome or consequence associated with that behavior when enacted, and/or an antecedent condition in the presence of which the behavior will produce the specified outcome (Anderson, Hawkins, Freeman, & Scotti, 2000). Rules are often communicated via instructions, moral injunctions, culturally defined values, "if . . . , then . . ." statements, advice, and modeled behavior (Baum, 1994). They often alter the functions of the environment, either by strengthening or weakening its influence. Examples of rules are, "If I purge, then I'll get rid of all of the calories from when I binged" and "If I'm thin people will like me more." As Skinner (1969) acknowledged, much of our behavior is rule governed and not directly influenced by actual experiences with environmental contingencies.

Establishing (or Motivational) Operations (EOs)

EOs refer to the impact that environmental events, operations, or conditions have on behavior by temporarily altering the reinforcing or punishing properties of consequences. In so doing, EOs increase or decrease the likelihood of behavior (Michael, 1982). For example, if a person is deprived of food, this deprivation (the EO) can enhance the reinforcing value of other associated stimuli (e.g., highly caloric foods), thereby increasing the intensity and frequency of behaviors previously reinforced by the deprived stimulus (e.g., the speed and volume of food consumption). Other behaviors that have been in the past reinforced by access to food, such as thinking about and seeking food, will also increase in frequency and intensity. Similarly, emotional states and thoughts that arise from certain environmental events may be construed as establishing operations for a variety of behaviors (Miltenberger, 2005). Studies reviewed in the following sections suggest food deprivation, certain forms of thinking, and mood may increase behaviors related to eating disorders as well as the influence of food as a reinforcer.

Consequences

A basic pretreatment assessment goal is to determine the consequences that maintain problematic behavior. Behavior maintained by positive reinforcement is influenced by the presentation of rewarding events or the rewarding stimulation it produces, whereas behavior maintained by negative reinforcement is influenced by the effect it has on avoiding or terminating aversive events or stimulation. Avoidance behavior is usually negatively reinforced behavior and can include escape from ongoing aversive stimuli or avoidance of aversive stimuli before they are presented. Another consideration in relation to behavior consequences is whether they are immediate or delayed. In the case of problematic behaviors, short-term reinforcing consequences typically maintain such behaviors, whereas long-term aversive consequences make them problematic (Nelson-Gray & Farmer, 1999).

Commonly Identified Contexts for Eating-Disorder-Related Behaviors

Common Antecedents

A number of antecedents have been identified that influence the amount of food consumed. They include the packaging of food items, portion and plate size, ambient lighting, and the presence of a social context for eating (Wansink, 2004). Some of these factors may also be S^D that occasion binge eating. Other studies have identified being home and/or alone (Greeno, Wing, & Shiffman, 2000; Stickney & Miltenberger, 1999; Stickney, Miltenberger, & Wolff, 1999; Waters, Hill, & Waller, 2001), negative environmental events (Sherwood, Crowther, Wills, & Ben-Porath, 2000), and the availability of large quantities of binge foods (McManus & Waller, 1995) as frequent antecedents to binge episodes.

Cultural practices and values as they relate to feminine concepts of physical beauty also have been observed to have a significant role in the development of eating-disorder-related behaviors (Fairburn, Shafran, & Cooper, 1998). Stice (2001), for example, found that the acceptance of thinness as an ideal standard predicted body dissatisfaction that, in turn, elevated susceptibility to subsequent dieting behavior and negative affect. These factors in combination increased the likelihood of engagement in bingeing and purging. Similarly, Stice (1998) found that the onset of binge-ing and purging was associated with a history of social reinforcement for the thin ideal by family and friends. These findings suggest that cultural overvaluation of the thin ideal contributes to the development of unhealthy behavior patterns to achieve this ideal.

Common EOs

Several studies also suggest a common set of establishing operations that precede binge episodes. They include negative mood states (Greeno et al., 2000; Johnson, Schlundt, Barclay, Carr-Nangle, & Engler, 1995; Stickney & Miltenberger, 1999) and limited food or calorie intake and associated verbal reports of hunger or food cravings (Davis, Freeman, & Garner, 1988; Davis & Jamieson, In press; Redlin, Miltenberger, Crosby, Wolff, & Stickney, 2002; Schlundt, Johnson, & Jarrell, 1985). In some instances, sleep deprivation or poor sleep quality, as well as negative thoughts related to self, weight, and shape, also serve as establishing operations that occa-sion binge episodes (Redlin et al., 2002; Stickney et al., 1999).

Common Consequences

The consequences that appear to exert their greatest influence on the maintenance of binge and purge behavior are related to the immediate impact that these behaviors have on altering aversive antecedent condi-tions. For example, during the binge episode, negative mood states and food cravings often become attenuated, while positive moods become more pronounced (Deaver, Miltenberger, Smyth, Meidinger, & Crosby, 2003; Redlin et al., 2002; Stickney et al., 1999). Shortly after the binge episode, however, negative moods often resurface (Deaver et al., 2003; Redlin et al., 2002; Stickney & Miltenberger, 1999; Wegner et al., 2002). Davis and Jamieson (2005), for example, administered an adjective check-list to a large sample of treatment-seeking women with BN and evaluated the endorsement of typical mood states immediately before and during binge episodes. Ratings of several adjectives associated with negative mood (e.g., angry) decreased during binge episodes, whereas ratings of several positive affective adjectives (e.g., contented) increased during binges. A smaller number of negative somatic states (e.g., bloated) and mood states (e.g., disgusted), however, were more frequently endorsed during the binge episode than immediately beforehand (cf. Stickney & Miltenberger, 1999; Stickney et al., 1999). Interestingly, individuals who displayed the largest absolute change scores for the two measurement occasions were those most likely to report a longer eating disorder history and the least likely to report an absence of binge and purge episodes 4 months following initial assessment and subsequent treatment completion. The type and degree of negative moods experienced after binge episodes may also vary as a function of the type of eating disorder. Mitchell et al. (1999), for example, reported that persons with BED were less anxious and distressed than those with BN following binge episodes.

Consistent with the preceding research, several theories on binge eating suggest that it is a form of escape from aversive private behavior, such as thoughts related to one's inability to realize high personal standards (Heatherton & Baumeister, 1991) or negative mood states (Deaver et al., 2003). Such models emphasize the role of negative reinforcement in the maintenance of binge eating, although accompanying positive reinforcers also likely serve a maintaining function. They include the presence of others in the social context of eating (Johnson et al., 1995) and the taste of food consumed during binge episodes (Mitchell et al., 1999). However, Mitchell et al. (1999) suggested that the degree to which the taste and consumption of food might be reinforcing varies as a function of the type of eating disorder. In this study, those with BED reported greater enjoyment associated with eating during binges compared to those with BN. Purging behavior among those with BN and AN may also be maintained by immediate positive as well as negative reinforcers. Abraham and Joseph (1987), for example, demonstrated that purging is sometimes followed by a release of endogenous opiates that, in turn, can potentiate euphoria and increase pain tolerance. Increases in positive moods immediately after purging have also been reported (Cooper et al., 1988; Hsu, 1990).

Positive and negative reinforcers associated with overly rigid and excessive dieting among those with AN and BN are largely delayed. Positive reinforcement processes might include self-reinforcement for weight loss, for exercising or self-control over eating, or for acting in accordance with one's moral values or beliefs; reinforcement through experiencing coercive control over others; and reinforcement from others via recognition for thinness (Slade, 1982; Vitousek, Watson, & Wilson, 1998). Delayed negative reinforcers can include avoidance of possible sexual encounters, concerns over menstruation, and the pressures associated with maturity and independence (Vitousek et al., 1998). The delayed nature of many of these potential reinforcers suggests that extreme dieting behaviors are more likely to be rule governed than influenced by direct contingencies.

Whereas the immediate consequences of bingeing and purging tend to be reinforcing and explain their maintenance over time, the delayed consequences of these behaviors as well as self-starvation are often aversive. For individuals who repeatedly binge eat without compensatory behaviors, more calories are typically consumed than are expended, with obesity being a common outcome. A number of health-related problems are associated with obesity, such as diabetes, hypertension, and cardiovascular disease. For those who use purging compensatory methods following binges or who engage in severe dietary restriction, a number of health-related problems are likely to emerge over time. These problems include electrolyte abnormalities, impaired renal functions, and esophageal weakening or rupture (Garner, 1997). A 0.5% risk of death per year among those with AN has also been reported (Isager, Brinch, Kreiner, & Tolstrup, 1985).

Organismic Variables

Organismic variables include both biological characteristics and the effects of past learning, and constitute part of the person's physiological make-up. Such variables can potentially exert a mediating or moderating effect on environment–behavior relations. *Biological characteristics* include genetic predispositions, temperament, neurotransmitter and hormonal activities, physical appearance, and other physiological characteristics, such as the effect of disease or aging. *Learning history* refers to behavioral shaping over one's lifetime. Knowledge of past learning history can suggest the likelihood that a given individual will respond to a particular set of antecedent conditions or EOs with eating disorder behavior.

Biological Characteristics

Maturational timing predicts body perception, body dissatisfaction, and subsequent eating disorder behavior. Girls who mature earlier than the majority of their peers tend to be at greater risk of body dissatisfaction and disordered eating (Heinberg, 1996). Numerous abnormalities in appetitive and metabolic functioning may also influence problematic eating patterns in individuals with BN and BED (Latner & Wilson, 2000).

Temperament and associated processes also have relevance as potential mediators or moderators of eating disorder behavior (Farmer, 2005). Temperament is a heritable and temporally stable feature of persons that is modifiable through basic maturational processes and environmental experience. Anxious and impulsive temperament, respectively, have been associated with individual differences in sensitivity to punishment and reinforcement (Gray, 1987). In relation to the eating disorders, high levels of trait anxiety are common and precede the onset of eating disorders (Pearlstein, 2002). Among persons with BN, high levels of impulsivity are also often observed (Pearlstein, 2002). For those with BED, anxiety is often elevated but not to the extent as those with AN and BN. Impulsivity levels are also elevated, although less so compared to those with BN (Pearlstein, 2002; Santonastaso, Ferrara, & Favaro, 1999). Consistent with these descriptive observations, those with either AN or BN are generally inclined to engage in passive and active avoidance behavior and demonstrate a heightened sensitivity to the influence of punishment. The latter is perhaps most evident in rule statements that signify a need to achieve a thin body shape in order to avoid negative evaluation or social criticism (Chiodo, 1987). In the case of BN, however, more frequent approach-avoidance conflicts are expected given avoidance tendencies associated with elevated levels of anxiety and increased reward sensitivity linked to higher levels of impulsivity (Farmer, Nash, & Field, 2001; Kane, Loxton, Staiger, & Dawe, 2004). In contrast, those with BED, who are lower in levels of anxiety and impulsivity, might have less intense or frequent approach-avoidance conflicts concerning food (Mitchell et al., 1999), which perhaps accounts for why members of this subgroup do not characteristically purge following binge episodes.

The long-term effects of eating-disorder-related behavior on organismic functioning are also important to consider. Many of the symptoms and features of AN and BN can be accounted for by the common effects of starvation (Garner, 1997). Consequently, an undernourished individual's behavioral and psychological functioning is likely to be impaired compared to when weight and nutrition are returned to more normative levels (Garner, 1997).

Learning History

The observation that eating disorders often persist over years, if not decades (Pike, 1998), attests to the long-standing learning history associated with these behaviors. Correspondingly, the longer the history of eating disorder behaviors, the worse the prognosis for treatment success (Pike, 1998). Because of the frequent long-standing nature of eating disorders and the high volume of positive and negative reinforcers that maintain associated behaviors, problematic eating behaviors are often not viewed as especially troublesome by those who engage in them (Vitousek et al., 1998).

Distal factors in one's learning history, such as exposure to societal values concerning thinness or being teased about one's shape and weight (Heinberg, 1996), might contribute to the development of problematic eating behavior and body dissatisfaction. However, factors that contributed to the initial establishment of a set of problematic behaviors may have little or no relevance in the maintenance of those behaviors over time. Consequently, the extent to which initiating events continue to exert influence on eating-disorder-related behaviors would need to be idiographically determined on a case-by-case basis.

FUNCTIONAL ASSESSMENT AND ANALYSIS

To establish the functional context of problematic behavior, functional assessment and analysis (Sturmey, 1996) are ideal assessment methods. Interview or questionnaire assessments can be used to develop a functional understanding of a client's behavior, as can direct observation, self-monitoring, and other behavioral assessment methodologies. Semistru-

TABLE 18.2　Steps in the functional formulation of eating-disorder-related behaviors

---

- Use questionnaire and interview measures to obtain information about the form and function of eating disorder behaviors.
- In initial assessments and during therapy, use ongoing self-monitoring to record target behaviors and their associated context.
- Generate hypotheses about the functions of eating disorder behaviors and explore the validity of these with the client.
- In collaboration with the client, develop a therapeutic program that specifically takes into account the functional aspects of the client's eating-disorder-related behaviors.

---

ctured interviews and questionnaires that assess behavioral topographies or diagnostic criteria may also be useful for clarifying the nature and form of the problematic behaviors. In this section, we briefly outline a number of topographical and functional measures used to assess eating-disorder-related behaviors. Table 18.2 summarizes the process of functional assessment and analysis. Before reviewing these areas, however, we wish to acknowledge the importance of a physician's evaluation of the client's medical status prior to initiating therapy for eating disorders given the potentially serious health-related problems that may be present.

ASSESSMENT OF THE TOPOGRAPHY OF EATING DISORDER BEHAVIOR

Many of the assessment interviews and questionnaires for eating disorders focus on the topography of eating disorder behavior, including symptoms or features of clinical syndromes. The Eating Disorders Examination (Fairburn & Cooper, 1996) is the most widely used interview for this purpose. A commonly used self-report measure for assessing the broad range of behaviors associated with eating disorders is the Eating Disorders Inventory (Garner, Olmstad, & Polivy, 1983). While measures such as these are useful for obtaining retrospective information on behavioral topographies associated with eating disorders, they do not provide information about the functional context within which such behavior occurs. Given the focus of the present chapter, the following sections emphasize instruments or methods designed to illuminate the functional aspects of eating disorder behavior.

FUNCTIONAL ASSESSMENTS OF EATING DISORDER BEHAVIOR

Checklists, Questionnaires, and Interviews

Several behavioral assessment methods have been developed to assess the functional context of eating-disorder-related behavior including interview and self-report measures. One of these is the Binge Eating Interview (BIN; Stickney et al., 1999), a semistructured interview designed to identify antecedents and consequences of binge eating. The Binge Eating Questionnaire (Redlin et al., 2002; Stickney et al., 1999) is identical to the BIN except that it is in a questionnaire format. The Function of Binge Eating Scale (Stickney & Miltenberger, 1999), subsequently modified and renamed the Conditions Associated with Binge Eating Questionnaire (Redlin et al., 2002; Stickney et al., 1999), consists of 15 Likert-scaled items that assess the degree to which certain mood (e.g., anxiety, boredom) and cognitive events (e.g., worry, focus on food) were present prior to, during, and immediately after binge-eating episodes. The Binge Eating Adjective Checklist (Davis & Jamieson, In press) is a retrospective self-report measure designed to assess mood and somatic states prior to and during binge episodes, and consists of 103 mood-related adjectives that can be used to describe a variety of private experiences that might arise in the context of a binge-eating episode.

The Antecedent Checklist (Stickney & Miltenberger, 1999) assesses overt and covert antecedent conditions associated with binge episodes. Antecedent events assessed by this measure are considered to be temporally remote rather than proximal, as events are assessed for their occurrence during the day that a binge episode occurred, not necessarily immediately before the binge took place. Redlin et al. (2002) modified this measure to include a rating to accompany each checked event that indicates the desire to binge eat in relation to that event.

The Description of Binge Episode Questionnaire (Stickney et al., 1999; Redlin et al., 2002) is a self-monitoring measure that inquires about specific binge episodes. The completion of this questionnaire is event triggered; that is, it is completed within 15 minutes of a binge episode. This measure consists of five open-ended questions that assess both private and environmental antecedents and consequences of binge episodes.

Self-Monitoring

Self-monitoring cards and diaries of various types have been used in research studies and the development of treatment programs to identify the forms and functional features of eating-disorder-related behaviors. Most often, antecedents and consequences associated with binge or purge episodes or engagement in excessive exercise and food-related rituals are monitored for those with BN and BED. In the case of AN, the types and amounts of foods and liquids consumed might also be monitored, as well as the occasions when consumption occurred. An analysis of environmental variables related to periods of not eating might also be considered.

Ecological momentary assessments (EMA) are sampling or monitoring strategies used to assess phenomena of interest in the natural environments where they occur. Such an approach is thought to maximize ecological validity while minimizing biases commonly associated with retrospective reports. When employed as a sampling methodology, a pager, handheld computer, or some other signaling device is used that, when activated, indicates to the individual to immediately record target behaviors and associated environmental contexts (e.g., Johnson & Larson, 1982; Wegner et al., 2002). Such assessments are often performed on a quasi-random basis and are signal contingent, whereby the participant has little or no *a priori* indication that a momentary assessment is forthcoming. EMA assessments can also be event-contingent, whereby participants record information following the display of target behaviors. Stein and Corte (2003), for example, used this method in a study of the functional contexts within which problematic eating behavior occurred. Handheld computers were supplied to aid in the self-monitoring, and participants were asked to bring the computer with them as they conducted their daily activities and to enter data related to their target behaviors (e.g., binge-eating episodes, instances of self-induced vomiting) as they occurred.

FUNCTIONAL ANALYTIC-BASED INTERVENTIONS

Functional analyses often result in a set of scientific hypotheses about the behavior of interest (Sturmey, 1996). Based on functional assessments, a therapeutic intervention can be devised that specifically addresses the antecedents, establishing operations, and consequences of target behaviors. Therapies for eating disorders will typically involve altering antecedent conditions that set the occasion for problematic behavior to occur, extinction of the problematic behavior(s), and differential reinforcement of behaviors that are functionally similar to the target behaviors but are not as problematic (Miltenberger, 2005). Direct instruction in the use of new behaviors (e.g., coping skills), challenges to inaccurate rules that influence problematic behavior, therapist instruction in and provision of acceptance and validation, and the incorporation of relapse prevention procedures are often useful additions to these basic strategies (see Table 18.3).

TABLE 18.3   Recommendations for the Functional Therapy of Eating Disorders

---

- Target motivation for change early in therapy if low or absent.
- Alter the antecedent and consequent conditions that support problematic eating behavior.
- Teach new behavioral skills.
- Balance behavior change strategies with acceptance and validation strategies.
- Anticipate lapses and teach relapse prevention strategies.

MOTIVATIONAL INTERVIEWING

During pretreatment assessments, it is often important to assess the degree to which clients desire to change their behavior. Individuals with eating disorders, especially those with AN, often exhibit no or low motivation for change (Vitousek et al., 1998). For such clients, motivational enhancement strategies (Miller & Rollnick, 1991) might help to increase clients' awareness of longer-term negative consequences of problematic behaviors and reduce the salience of immediate reinforcing consequences. From a behavioral perspective, effective motivational enhancement interventions alter EOs and behavioral rules that occasion problematic eating behavior. In addition to motivational interviewing and standard cognitive-behavior therapy (CBT; Garner, Vitousek, & Pike, 1997; Fairburn, Marcus, & Wilson, 1996), alternative therapies that explicitly target ambivalence about change such as acceptance and commitment therapy (Hayes, Strosahl, & Wilson, 1999) might be considered.

DISESTABLISHING OPERATIONS

In the case of long-standing and severe restricted eating patterns common in AN, initial therapeutic emphasis must usually be placed on refeeding and weight gain so as to reduce acute risk associated with marked malnourishment and to increase the likelihood that the client will be able to adequately participate in and benefit from additional therapeutic interventions (Garner, 1997). For those with BN, extended periods of food abstinence or restraint are often EOs for binge eating. Binge-eating episodes, in turn, are frequently regarded as evidence of a failure in self-regulation and in meeting personal and dietary standards. Because food abstinence appears to be an important EO that occasions binge eating among those with BN, excessive food restriction and dietary restraint should be targeted in order to remove this influence. Several effective treatments for BN, for example, emphasize the elimination of restrictive eating patterns and the normalization of eating habits and behavior (Fairburn et al., 1996). Rigid food rules might also be targeted in treatment, particularly if the presence of such rules functions to inhibit normal eating behavior.

As our review has noted, negative emotions and the events that occasion them often serve as potent establishing operations for binge eating and subsequent purging. Therefore, some persons who receive treatment for eating disorders might also benefit from therapies that target emotional experiences. Examples of such interventions include relaxation training, meditation practices, exposure therapy, behavioral activation, and cognitive therapy. Preliminary evidence indicates that a modified form of dialectical behavior therapy (Linehan, 1993) that fosters the development and use of mindfulness and emotion regulation skills to cope with negative affect is effective in reducing binge and purge frequency for BN (Safer, Telch, & Agras, 2001) and binge frequency for BED (Telch, Agras, & Linehan, 2001).

ANTECEDENT EVENTS

Modification of Antecedents

When problematic eating behavior routinely occurs in the presence of specific S^D, steps might be taken to eliminate or modify these stimuli, particularly during high-risk periods. For example, if being at home alone when upset reliably occasions binge episodes, steps might then be taken to avoid being in that environment when emotionally aroused, and instead to seek out environmental contexts that lessen the likelihood for binge eating (e.g., visiting a friend at his or her home).

The elimination of antecedents is potentially most useful when the presence of S^D reliably occasions problematic eating behavior. Such S^D include large amounts of desired foods that are the frequent object of binge episodes (Stickney et al., 1999). Stimulus control procedures are sometimes used to alter one's eating habits or increase one's monitoring of the act of eating. Such procedures can be particularly helpful for individuals whose eating behavior is less influenced by internal cues associated with periods of food deprivation and more strongly associated with other stimuli (e.g., the sights and smells of food, the television, being alone). Examples of stimulus control techniques include placing food items out of view (e.g., in a cupboard) rather than out in the open, reducing the amount of food stored in the home, eating meals at the same times and places, preparing one food portion at a time (e.g., if a person is preparing to eat crackers, a small amount might be poured onto a dish as an alternative to eating them out of a box), increasing the availability and salience of healthy foods, and utilizing small plates, bowls, and utensils when eating so as to reduce the influence that these cues have on the quantity of food consumed (Fairburn et al., 1996). Wansink (2004) provides additional ideas for reducing the volume of food consumed through environmental manipulation.

Exposure to Stimulus Cues

As noted previously, bingeing and purging are frequently associated with the reduction of several negative mood and somatic states as well as increases in positive mood states. Binge eating may serve an avoidance function. Therefore, therapies that block avoidance and promote a willingness to experience aversive private events might be particularly helpful. Similarly, for those who avoid viewing one's body or engage in body checking, exposure-based exercises might be employed as part of body image therapy to challenge individuals to exhibit and view their bodies while blocking avoidance and ritualistic behavior. Examples might include wearing form-fitting clothing, participation in activities such as swimming, and mindfully viewing oneself in a mirror (Key et al., 2002; Wilson, 2004). Graded mirror exposure, for example, has been found to decrease body dissatisfaction and body avoidance among women with AN (Key et al., 2002).

Some approaches to treatment for BN simultaneously expose clients to feelings of fullness while blocking the purge response (i.e., exposure and response prevention [ERP]). The nonreinforced exposure to feelings of fullness promotes the respondent extinction of anxiety to these somatic cues, thus eliminating the reinforcing value of purging. Several studies that have evaluated ERP have found this method to be effective for a subgroup of persons with BN (Johnson, Corrigan, & Mayo, 1987; Rosen & Leitenberg, 1985).

Modification of Rules

Clients who purge often do so as a result of faulty rules that suggest purging is an effective escape behavior that eliminates calories from binge episodes. However, research suggests that the effectiveness of such methods for this purpose is modest at best. Only about 12% of consumed calories are eliminated through laxative misuse (Bo-Linn, Santa Ana, Morawski, & Fordtran, 1983), and less than 50% of calories consumed during a binge, on average, are expelled as a result of self-induced vomiting (Kaye, Weltzin, Hsu, McConaha, & Bolton, 1993). Providing clients with more accurate rules about the effectiveness of purging, along with increasing the salience of possible distal and negative health-related consequences of such behavior, may lessen the likelihood of binge eating among purging clients.

If relevant for a given client, therapy might also target rules that refer to societal values concerning the desirability of thinness. For example, rules such as "I must be thin in order to be accepted by others" might be challenged via standard CBT techniques, or examined in terms of the utility of holding such thoughts and behaving in a manner consistent with them (Vitousek et al., 1998). Reducing the acceptance of beliefs associated with the thin ideal can reduce body dissatisfaction, bulimic symptomatology, and negative affect (Stice, Mazotti, Weibel, & Agras, 2000). Similarly, acceptance-based interventions (Wilson, 2004) might also help the client validate his/her experience, promote self-acceptance and valuation, lessen the tendency toward critical and negative self-evaluation, and provide a useful balance to other change-oriented strategies employed in a multicomponent behavioral intervention. In the case of AN, this may be particularly important given that self-starvation and the goal for thinness are often experienced as a challenge in self-control, a triumph over weakness, or as a moral obligation, behavior that is thus consistent with one's values, goals, and aspirations (Vitousek et al., 1998).

MODIFY BEHAVIOR

Self-Monitoring

When monitoring of food consumption is inhibited or interrupted, the volume of food consumed is greater (Wansinck, 2004). Similarly, self-monitoring of food intake and binge eating has been associated with substantial reductions in binge frequency among women with BN and BED (Latner & Wilson, 2002). To lessen the likelihood that larger amounts of food might be consumed, particularly during high-risk occasions (e.g., when emotionally upset), the environment might be modified in order to facilitate self-monitoring. Pliner and Iuppa (1978), for example, found that the presence of a mirror when eating reduced food consumption among obese persons. Mindfulness skills (e.g., Linehan, 1993) can also be taught to promote fuller awareness of processes and experiences associated with eating and to lessen the likelihood of automatic or dissociative eating.

Increasing Nonharmful Alternative Behavior

Functionally similar behaviors are those that have a similar function, or produce a similar effect, as the problematic target behaviors. If a client learns different and nonharmful behaviors that produce similar consequences, these alternative behaviors will be more likely used as replacements for the target behaviors (Miltenberger, 2005). For example, eating to relieve negative emotions could be replaced with functionally similar, alternative pleasant activities (Lewinsohn & Gotlib, 1995). Participation in these activities is likely to be maintained over time given the natural consequences that such activities provide.

Problem Solving

Consistent with Linehan's (1993) view that some clinically relevant behaviors represent efforts to solve life's problems, problematic eating behavior can be regarded as "the problem," with a major task of therapy being the identification and implementation of alternative "solutions" to the problems that eating-disorder-related behavior may temporarily solve for the individual. In addition to the development of a firm understanding of the problem resolution qualities of eating disorder behavior, such an approach would also involve teaching clients basic problem-solving skills that, in turn, would assist them in identifying alternative behaviors that are less personally harmful but have some of the same beneficial functional properties of the problematic behavior.

CONTINGENCY MANAGEMENT

Among clients with AN, the provision of rewards contingent on calorie intake for increases in weight often facilitates short-term weight gain;

however, such procedures alone generally are not effective in maintaining weight over time (Touyz & Beumont, 1997). Although such procedures may be useful for acutely restoring weight, additional procedures would likely need to be employed to promote the maintenance of weight in the long term.

MANAGEMENT OF RELAPSE

Relapse prevention is designed to help the individual anticipate and cope with relapse situations (Marlatt & Gordon, 1985). As applied to the eating disorders, Garner (1997) has offered the "four Rs" for responding to lapses: (a) reframe a lapse as a "slip" rather than a "full blown relapse"; (b) renew a commitment to maintain beneficial changes, which might include a commitment to refrain from certain behaviors (e.g., purging) and reduce harm associated with overeating (e.g., alter the types of foods consumed in a binge, such as fruit rather than sweets); (c) return to regular eating without engaging in compensatory behavior; and (d) reinstitute behavioral controls, such as adaptive coping skills, to interrupt future episodes. To this, we would add the use of relaxation strategies and appropriate exercise for self- and emotion-regulation purposes, an anticipation of future behavior lapses in recognition that they are a natural part of the habit change process, the promotion of acceptance when lapses occur in an effort to reduce the likelihood of a full-blown relapse, and planned follow-up sessions.

CASE STUDY

Jill, an 18-year-old woman, had been struggling with bulimic behaviors over the past 4 years. With the help of self-monitoring records that Jill completed during the pretreatment assessment phase, the therapist assessed the functional context within which the last five binge and purge episodes had taken place. In each instance, Jill engaged in highly restrictive eating for a period of 2 days, typically before a large social gathering, such as a party. Each time, she wanted to lose a couple of pounds, "so I would look my best." The importance of Jill's appearance in her self-evaluation was captured in a number of statements she made about the reinforcing qualities of being thin (e.g., "only thin, attractive people get noticed by others"). However, as the social event approached, Jill became preoccupied with feelings of inadequacy, especially in relation to her appearance. As these preoccupations intensified, she became more socially isolative (e.g., staying at home alone, a negatively reinforced avoidance behavior) and depressed. Binge episodes would typically occur when she reached peak levels of hunger, depression, and rumination. Binge eating quickly alleviated her sensation of being hungry, low mood, and her negative thoughts about herself (negative reinforcement for binge eating). Jill typically binged on foods that she enjoyed the taste of, most notably ice cream (positive reinforcement for binge eating). After a delay of about 5 minutes following binge eating, however, Jill often experienced intensely negative somatic sensations (e.g., bloated), distressing thoughts (e.g., "I'm going to get fat now"), and a different set of negative emotions (e.g., disgust, anxiety). Based on previous experiences, Jill had learned that she could terminate these negative experiences by vomiting (negative reinforcement for vomiting). Shortly after vomiting, however, she reported feeling somewhat demoralized and "freakish," although these feelings were of relatively low intensity. Figure 18.1 illustrates the functional formulation of Jill's binge and purge episodes.

Jill's therapist formulated that excessive dietary restraint was an important EO for her binge episodes, as food deprivation heightened the reinforcing qualities of food and increased her uncomfortable feelings of hunger. Similarly, large quantities of ice cream stored at home also increased the likelihood of a binge. Social isolation contributed to Jill's feelings of sad mood that, in turn, increased the likelihood of bingeing and purging.

Therapeutic strategies to target these antecedents included the normalization of eating patterns (i.e., three planned meals and two planned snacks), limiting the quantities of binge foods in the household at any one time, and increasing pleasurable activities and social contact through behavioral activation strategies. Given that bingeing and purging were, in part, used as coping behaviors to alter antecedents and EOs, the therapist worked with Jill on problem-solving skills and emotion regulation skills. Her therapist also addressed Jill's rules associated with thinness ideal (e.g., "if I'm thin, I'm in") and those associated with the effectiveness of self-induced vomiting as a means of preventing weight gain. These rules were specifically targeted as they influenced Jill's purging and strict dieting. The therapist discussed the relative ineffectiveness of vomiting for preventing the absorption of calories. Rules related to the thinness ideal were challenged via standard cognitive strategies, such as Socratic questioning, with particular attention to Jill's all-or-none thinking related to the consequences associated with being thin versus a healthy weight.

Therapy also included graded exposure to binge food items. This involved eating small to standard portions of a given food item and inhibiting purging, while simultaneously identifying and expressing private experiences (e.g., thoughts about what she had eaten, physical sensations such as fullness, and emotional states such as anxiety) until their intensity subsided. Other exposure-based activities also facilitated body image acceptance and reduced body checking. These included wearing more form fitting clothes and systematically and deliberately viewing her body in the mirror. Within the last couple of sessions, the therapist discussed with Jill

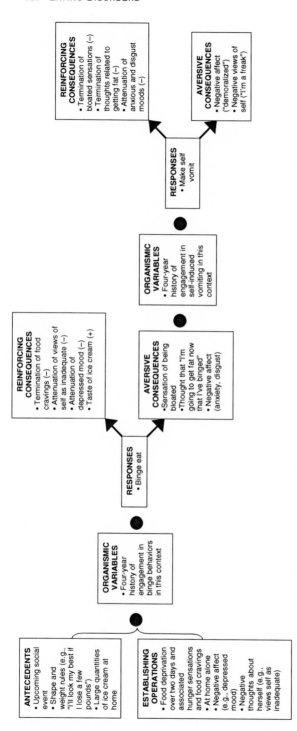

FIGURE 18.1 Functional analysis of Jill's binge and purge episodes. Following Follette et al. (2000) and Farmer and Nelson-Gray (2005), the dots in between components in the response chain indicate a probability function, whereby each dot represents a probability that the preceding component will influence the following component. The dot that follows the *Antecedents* and *Establishing Operations* components and precedes the *Organismic Variables* component indicates that in a given environmental context there is a probability that an individual is sensitive to the environment's influence based on relevant biological variables and/or learning history. The dots between the *Organismic Variables* and *Responses* components indicate that in certain environmental contexts and in conjunction with past learning history and biological factors, there is a probability that a particular form of behavior will follow. The arrows between the *Responses* and *Consequences* components indicate the effect that behavior has on altering antecedents and/or establishing operations. The dot between *Aversive Consequences* and *Organismic Variables* in the second half of the diagram indicates that aversive response consequences associated with binge eating probabilistically function as antecedents and establishing operations for subsequent purge episodes given Jill's learning history. Items listed within the *Reinforcing Consequences* components that are followed by (+) indicate positive reinforcing consequences, whereas components followed by (−) indicate negative reinforcing consequences, each of which function to maintain or strengthen preceding behavior.

the high probability of lapses, which were framed as a normal part of the habit change process and which can be reduced by preparing for probable antecedents.

Over the course of therapy, Jill's eating habits had normalized and she had learned and used alternative behaviors for responding to problematic situations and coping with negative private behavior. She also realized that the pursuit of the thinness ideal through eating-disorder-related behaviors has not been especially effective, as her behaviors thus far have not provided the consequences she assumed (e.g., weight loss, social approval). Finally, she also learned that she could eat regular meals and snacks, and standard-sized portions of desired foods, and experience no significant adverse consequences as a result. Any discomfort that she experienced as a result of doing so (e.g., feelings of guilt) would diminish over time without her needing to vomit. Coinciding with these behavior changes were significant reductions in binge and purge behavior over the course of therapy, with complete abstinence from bingeing and purging 1 month following treatment completion.

REFERENCES

Abraham, H. D., & Joseph, A. B. (1987). Bulimic vomiting alters pain tolerance and mood. *International Journal of Psychiatry in Medicine, 16*, 311–316.

American Psychiatric Association. (2000). *Diagnostic and statistical manual of mental disorders* (4th ed., text rev.). Washington, D.C.: Author.

Anderson, C. M., Hawkins, R. P., Freeman, K. A., & Scotti, J. R. (2000). Private events: Do they belong in a science of human behavior? *The Behavior Analyst, 23*, 1–10.

Baum, W. M. (1994). *Understanding behaviorism: Science, behavior, culture.* New York: Harper Collins.

Bo-Linn, G. W., Santa Ana, C., Morawski, S., & Fordtran, J. (1983). Purging and calorie absorption in bulimic patients and normal women. *Annals of Internal Medicine, 99*, 14–17.

Chiodo, J. (1987). Bulimia: An individual behavior analysis. *Journal of Behavior Therapy and Experimental Psychiatry, 18*, 41–49.

Cooper, J. L., Morrison, T. L., Bigman, O. L., Abramowitz, S. I., Levin, S., & Krener, P. (1988). Mood changes and affect disorders in the bulimic binge-purge cycle. *International Journal of Eating Disorders, 7*, 469–474.

Davis, R., Freeman, R. J., & Garner, D. M. (1988). A naturalistic investigation of eating behavior in bulimia nervosa. *Journal of Consulting and Clinical Psychology, 56*, 273–279.

Davis, R., & Jamieson, J. (2005). Assessing the functional nature of binge eating in the eating disorders. *Eating Behaviors, 6*, 345–354.

Deaver, C. M., Miltenberger, R. G., Smyth, J., Meidinger, A., & Crosby, R. (2003). An evaluation of affect and binge eating. *Behavior Modification, 27*, 578–599.

Fairburn, C. G., & Cooper, Z. (1996). The Eating Disorder Examination (12th ed.). In C. G. Fairburn & G. T. Wilson (Eds.), *Binge eating: Nature, assessment, and treatment* (pp. 317–332). New York: Guilford Press.

Fairburn, C. G., Marcus, M. D., & Wilson, G. T. (1996). Cognitive-behavioral therapy for binge eating and bulimia nervosa: A comprehensive treatment manual. In C. G. Fairburn & G. T. Wilson (Eds.), *Binge eating: Nature, assessment, and treatment* (pp. 361–404). New York: Guilford Press.

Fairburn, C. G., Shafran, R., & Cooper, Z. (1998). A cognitive behavioural theory of anorexia nervosa. *Behaviour Research and Therapy, 37*, 1–13.

Farmer, R. F. (2005). Temperament, reward and punishment sensitivity, and clinical disorders: Implications for behavioral case formulation and therapy. *International Journal of Behavioral Consultative Therapy, 1*, 56–76.

Farmer, R. F., Nash, H. M., & Field, C. E. (2001). Disordered eating behaviors and reward sensitivity. *Journal of Behavior Therapy and Experimental Psychiatry, 32*, 211–219.

Farmer, R. F., & Nelson-Gray, R. O. (2005). *Personality-guided behavior therapy.* Washington, D.C.: American Psychological Association.

Follette, W. C., Naugle, A. E., & Linnerooth, P. J. N. (2000). Functional alternatives to traditional assessment and diagnosis. In M. J. Dougher (Ed.), *Clinical behavior analysis* (pp. 99–125). Reno, NV: Context Press.

Garner, D. M. (1997). Psychoeducational principles in treatment. In D. M. Garner & P. E. Garfinkel (Eds.), *Handbook of treatment for eating disorders* (2nd ed., pp. 145–177). New York: Guilford Press.

Garner, D. M., Olmstead, M. P., & Polivy, J. (1983). Development and validation of a multidimensional eating disorder inventory for anorexia nervosa and bulimia. *International Journal of Eating Disorders, 2*, 15–34.

Garner, D. M., Vitousek, K. M., & Pike, K. M. (1997). Cognitive-behavioral therapy for anorexia nervosa. In D. M. Garner & P. E. Garfinkel (Eds.), *Handbook of treatment for eating disorders* (2nd ed., pp. 94–144). New York: Guilford Press.

Goldfried, M. R., & Sprafkin, J. N. (1976). Behavioral personality assessment. In J. T. Spence, R. C. Carson, & J. W. Thibaut (Eds.), *Behavioral approaches to therapy* (pp. 295–321). Morristown, NJ: General Learning Press.

Gray, J. A. (1987). *The psychology of fear and stress* (2nd ed.). New York: Cambridge University Press.

Greeno, C. G., Wing, R., & Shiffman, S. (2000). Binge antecedents in obese women with and without binge eating disorder. *Journal of Consulting and Clinical Psychology, 68*, 95–102.

Hayes, S. C., Strosahl, K. D., & Wilson, K. G. (1999). *Acceptance and commitment therapy: An experiential approach to behavior change.* New York: Guilford Press.

Heatherton, T. E., & Baumeister, R. F. (1991). Binge eating as escape from self-awareness. *Psychological Bulletin, 110*, 86–108.

Heinberg, L. J. (1996). Theories of body image disturbance: Perceptual, developmental, and sociocultural factors. In J. K. Thompson (Ed.), *Body image, eating disorders, and obesity: An integrative guide for assessment and treatment* (pp. 27–47). Washington, D.C.: American Psychological Association.

Hsu, L. K. G. (1990). Experiential aspects of bulimia nervosa: Implications for cognitive behavioral therapy. *Behavior Modification, 14*, 50–65.

Isager, T., Brinch, M., Kreiner, S., & Tolstrup, K. (1985). Death and relapse in anorexia nervosa: Survival analysis of 151 cases. *Journal of Psychiatric Research, 19*, 515–521.

Johnson, C., & Larson, R. (1982). Bulimia: An analysis of moods and behavior. *Psychosomatic Medicine, 44*, 341–351.

Johnson, W. G., Corrigan, S. A., & Mayo, L. L. (1987). Innovative treatment approaches to bulimia nervosa. *Behavior Modification, 11*, 373–388.

Johnson, W. G., Schlundt, D. G., Barclay, D. R., Carr-Nangle, R. E., & Engler, L. B. (1995). A naturalistic functional analysis of binge eating. *Behavior Therapy, 26*, 101–118.

Kane, T. A., Loxton, N. J., Staiger, P. K., & Dawe, S. (2004). Does the tendency to act impulsively underlie binge eating and alcohol use problems? An empirical investigation. *Personality and Individual Differences, 36*, 83–94.

Kaye, W. H., Weltzin, T. E., Hsu, G., McConaha, C. W., & Bolton, B. (1993). Amount of calories retained after binge eating and vomiting. *American Journal of Psychiatry, 150*, 969–971.

Key, A., George, C. L., Beattie, D., Stammers, K., Lacey, H., & Waller, G. (2002). Body image treatment within an inpatient program for anorexia nervosa: The role of mirror exposure in the desensitization process. *International Journal of Eating Disorders, 31*, 185–190.

Latner, J. D., & Stunkard, A. J. (2003). Getting worse: The stigmatization of obese children. *Obesity Research, 11*, 452–456.

Latner, J. D., & Wilson, G. T. (2000). Cognitive-behavioral therapy and nutritional counseling in the treatment of bulimia nervosa and binge eating. *Eating Behaviors, 1*, 3–21.

Latner, J. D., & Wilson, G. T. (2002). Self-monitoring and the assessment of binge eating. *Behavior Therapy, 33*, 465–477.

Lewinsohn, P. M., & Gotlib, I. H. (1995). Behavioral theory and treatment of depression. In E. E. Beckham & W. R. Leber (Eds.), *Handbook of depression* (2nd ed., pp. 352–375). New York: Guilford Press.

Linehan, M. M. (1993). *Cognitive-behavioral treatment of borderline personality disorder.* New York: Guilford Press.

Marlatt, G. A., & Gordon, J. R. (Eds.) (1985). *Relapse prevention: Maintenance strategies in the treatment of addictive behaviors.* New York: Guilford Press.

McManus, F., & Waller, G. (1995). A functional analysis of binge eating. *Clinical Psychology Review, 15*, 845–863.

Michael, J. (1982). Distinguishing between the discriminative and motivational functions of stimuli. *Journal of the Experimental Analysis of Behavior, 37*, 149–158.

Miller, W. R., & Rollnick, S. (1991). *Motivational interviewing: Preparing people to change addictive behavior.* New York: Guilford Press.

Miltenberger, R. G. (2005). The role of automatic negative reinforcement in clinical problems. *International Journal of Behavioral Consultative Therapy, 1*, 1–11.

Mitchell, J. E., Mussell, M. P., Peterson, C. B., Crow, S., Wonderlich, S. A., Crosby, R. D., et al. (1999). The hedonics of binge eating in women with bulimia nervosa and binge eating disorder. *International Journal of Eating Disorders, 26*, 165–170.

Morris, A., Cooper, T., & Cooper, P. J. (1989). The changing shape of female fashion models. *International Journal of Eating Disorders, 8*, 593–596.

Nelson-Gray, R. O., & Farmer, R. F. (1999). Behavioral assessment of personality disorders. *Behaviour Research and Therapy, 37*, 347–368.

Pearlstein, T. (2002). Eating disorders and comorbidity. *Archives of Women's Mental Health, 4*, 67–78.

Pike, K. M. (1998). Long-term course of anorexia nervosa: Response, relapse, remission, and recovery. *Clinical Psychology Review, 18*, 447–475.

Pliner, P., & Iuppa, G. (1978). Effects of increasing awareness of food consumption in obese and normal weight subjects. *Addictive Behaviors, 3*, 19–24.

Redin, J. A., Miltenberger, R. G., Crosby, R. D., Wolff, G. E., & Stickney, M. I. (2002). Functional assessment of binge eating in a clinical sample of obese eaters. *Eating and Weight Disorders, 7*, 106–115.

Rosen, J. C., & Leitenberg, H. (1985). Exposure plus response prevention in the treatment of bulimia. In D. Garner & P. Garfinkel (Eds.), *Handbook of psychotherapy for anorexia nervosa and bulimia* (pp. 193–209). New York: Guilford Press.

Rubinstein, S., & Caballero, B. (2000). Is Miss America an undernourished role model? *Journal of the American Medical Association, 283,* 1569.

Safer, D. L., Telch, C. F., & Agras, W. S. (2001). Dialectical behavior therapy for bulimia nervosa. *American Journal of Psychiatry, 158,* 632–634.

Santonastaso, P., Ferrara, S., & Favaro, A. (1999). Differences between binge eating disorder and nonpurging bulimia. *International Journal of Eating Disorders, 25,* 215–218.

Schlundt, D. G., Johnson, W. G., & Jarrell, M. (1985). A naturalistic functional analysis of eating behavior in bulimia and obesity. *Advances in Behaviour Research and Therapy, 7,* 149–162.

Sherwood, N. E., Crowther, J. H., Wills, L., & Ben-Porath, Y. S. (2000). The perceived function of eating for bulimic, subclinical bulimic, and non-eating disordered women. *Behavior Therapy, 31,* 777–793.

Skinner, B. F. (1969). *Contingencies of reinforcement: A theoretical analysis.* New York: Appleton-Century-Crofts.

Slade, P. (1982). Toward a functional analysis of anorexia nervosa and bulimia nervosa. *British Journal of Clinical Psychology, 21,* 167–179.

Stein, K. F., & Corte, C. M. (2003). Ecologic momentary assessment of eating-disordered behaviors. *International Journal of Eating Disorders, 34,* 349–360.

Stice, E. (1998). Modeling of eating pathology and social reinforcement of the thin-ideal predict onset of bulimic symptoms. *Behaviour Research and Therapy, 36,* 931–944.

Stice, E. (2001). A prospective test of the dual pathway model of bulimic pathology: Mediating effects of dieting and negative affect. *Journal of Abnormal Psychology, 110,* 124–135.

Stice, E. (2002). Risk and maintenance factors for eating pathology: A meta-analytic review. *Psychological Bulletin, 128,* 825–848.

Stice, E., Mazotti, L., Weibel, D., & Agras, W. S. (2000). Dissonance prevention program decreases thin-ideal internalization, body dissatisfaction, dieting, negative affect, and bulimic symptoms: A preliminary experiment. *International Journal of Eating Disorders, 27,* 206–217.

Stickney, M. I., & Miltenberger, R. G. (1999). Evaluating direct and indirect measures for the functional assessment of binge eating. *International Journal of Eating Disorders, 26,* 195–204.

Stickney, M. I., Miltenberger, R. G., & Wolff, G. (1999). A descriptive analysis of factors contributing to binge eating. *Journal of Behavior Therapy and Experimental Psychiatry, 30,* 177–189.

Sturmey, P. (1996). *Functional analysis in clinical psychology.* New York: Wiley.

Telch, C. F., Agras, W. S., & Linehan, M. M. (2001). Dialectical behavior therapy for binge eating disorder. *Journal of Consulting and Clinical Psychology, 69,* 1061–1065.

Touyz, S. W., & Beumont, P. J. V. (1997). Behavioral treatment to promote weight gain in anorexia nervosa. In D. M. Garner & P. E. Garfinkel (Eds.), *Handbook of treatment for eating disorders* (2nd ed., pp. 361–371). New York: Guilford Press.

Vitousek, K., Watson, S., & Wilson, G. T. (1998). Enhancing motivation for change in treatment-resistant eating disorders. *Clinical Psychology Review, 18,* 391–420.

Wansink, B. (2004). Environmental factors that increase the food intake and consumption volume of unknowing consumers. *Annual Review of Nutrition, 24,* 455–479.

Waters, A., Hill, A., & Waller, G. (2001). Internal and external antecedents of binge eating episodes in a group of women with bulimia nervosa. *International Journal of Eating Disorders, 29,* 17–22.

Wegner, K. E., Smyth, J., Crosby, R., Wittrock, D., Wonderlich, S., & Mitchell, J. (2002). An evaluation of the relationship between mood and binge eating in the natural environment using ecological momentary assessment. *International Journal of Eating Disorders, 32*, 352–361.

Wilson, G. T. (2004). Acceptance and change in the treatment of eating disorders: The evolution of manual-based cognitive-behavioral therapy. In S. C. Hayes, V. M. Follette, & M. M. Linehan (Eds.), *Mindfulness and acceptance: Expanding the cognitive-behavioral tradition* (pp. 243–260). New York: Guilford Press.

19

PERSONALITY DISORDERS

PRUDENCE CUPER

Duke University

RHONDA MERWIN

Duke University Medical Center

AND

THOMAS LYNCH

Duke University

The *Diagnostic and Statistical Manual of Mental Disorders* (4th edition text revision; *DSM-IV-TR*) provides diagnostic criteria for 10 specific Personality Disorders, as well as for Personality Disorder Not Otherwise Specified (American Psychiatric Association, 2000). To meet criteria for any personality disorder, an individual must demonstrate an enduring pattern of culturally abnormal behavior, including cognitions, affectivity, interpersonal functioning, and/or impulse control, traceable to early adulthood. The pattern must be inflexible and present across a wide range of situations, and it must result in distress or impaired functioning. The *DSM* takes a categorical approach to diagnoses; thus, behavior patterns are considered distinct and indicative of different personality disorders. Table 19.1 includes names and brief descriptions of the disorders.

Psychologists from diverse traditions have challenged the utility of the *DSM*'s categorical approach, citing poor convergent and discriminant validity. This is particularly problematic for the personality disorders, which have a high rate of comorbidity. Some critics suggest that personality disorders are better understood in dimensional terms than as distinct categories. Though dimensional models vary, many theorists assert that personality dimensions such as extraversion sit at the top of a hierarchy composed of behaviors, constructs, and traits, and that each dimension is a continuum, with problematic behaviors occurring when an individual

Functional Analysis in
Clinical Treatment

TABLE 19.1   A summary of *DSM-IV* Personality Disorders including the three main
clusters, disorders, and brief description and prevalence rates, as cited in
DSM-IV-TR (APA, 2000)

Cluster	Disorder Name and Brief Description	Prevalence Rate(s), General Population
Cluster A: Odd/Eccentric	Paranoid PD *Behavior marked by distrust and suspiciousness of others*	0.5–2.5%
	Schizoid PD *Restricted emotions and indifference toward close relationships*	"uncommon"
	Schizotypal PD *Cognitive or perceptual distortions and difficulties in close relationships*	3%
Cluster B: Dramatic/Erratic	Antisocial PD *Deceit, manipulation, and disregard for the rights of others*	3% males; 1% females
	Borderline PD *Interpersonal and affective instability; impulsive behaviors*	2%
	Histrionic PD *Excessive attention seeking*	2–3%
	Narcissistic PD *Overestimation of abilities; need for admiration from others*	<1%
Cluster C: Anxious/Avoidant	Avoidant PD *Pervasive feelings of inadequacy; inhibition in social situations*	0.5–1%
	Dependent PD *Submissive behaviors; excessive need for caretaking from others*	"among the most frequently reported"
	Obsessive-Compulsive PD *Excessive orderliness and need for control*	1%
NOS	Personality Disorder Not Otherwise Specified *Distress resulting from a mix of symptoms from various PD categories*	Not applicable

falls too far at either end of the continuum (Clark, Livesley, & Morey,
1997).

From a behavior analytic perspective, the concepts of traits and personality are not as useful as identifying the contextual factors maintaining an
individual's behavior. Viewing behaviors as reflective of traits can be problematic for a number of reasons (Koerner, Kohlenberg, & Parker, 1996).
First, trait concepts do not provide information regarding the variables that

generate and maintain a behavior. As a result, interventions are not likely to be focused on factors that are directly observable and manipulable. Second, viewing a set of behaviors as reflective of a trait may result in a failure to discern the different functions that a behavior serves. Third, traits have sometimes been treated as opposing ends of a continuum, implying that an individual cannot simultaneously possess both traits. This is inconsistent with behavioral approaches that maintain that an individual could possess two seemingly opposite traits, and that the presence of behaviors characteristic of either would depend on the situation.

In most cases, the criterion behaviors listed in the *DSM* provide a formal account that describes the form or topography of behavior, illustrating how someone presenting with one of these disorders might appear. To develop a functional account of personality disorders, one must go beyond formal descriptions and identify the function of the patterns of behavior that characterize the various disorders.

A FUNCTIONAL APPROACH TO PERSONALITY DISORDERS

A functional analytic account makes use of the principles of learning to explain how particular behavior patterns become pervasive, inflexible, and long standing. Behaviors are pervasive if they occur across a variety of contexts. This suggests that the contexts in which the behaviors occur contain similar discriminative stimuli (S^Ds), or that many S^Ds exist for a given maladaptive behavior. This was noted by Skinner (1953) when he stated: "[A] personality may be tied to a particular type of occasion ... when a system of responses is organized around a given discriminative stimulus. ... Responses which lead to a common reinforcement, regardless of the situation, may comprise a functional system" (p. 285). The task for the clinician is to determine what S^D is common across various situations. For example, consider an individual with a tendency to demonstrate considerable anger and aggression. Careful analysis may reveal that this behavior occurs in the presence of family members that serve as S^Ds because they signal the availability of reinforcement for emotional escalation.

Behaviors become inflexible or stereotypical as the result of repeated reinforcement. Basic studies of the acquisition of operant behavior demonstrate that operant behavior is stereotypical in terms of topography, effort, and sequencing (Leslie, 1999). When schedules requiring novel responses are implemented, novel and varied behavior emerges (Lee, McComas, & Jawor, 2002; Lee & Sturmey, 2006; Page & Neuringer, 1985). Responses may become inflexible even if reinforcement is short term and the behavior has long-term negative consequences. For example, consider the long-standing and inflexible pattern of anger and aggression described previously. Emotional escalation may function to activate others to respond to one's wants or needs. As a result, such behavior would be continuously or intermittently reinforced, resulting in response invariability. Reinforcement would maintain the behavior despite damage to important relationships. As maladaptive responses to functionally similar situations are reinforced across time, opportunities for other, more adaptive, behaviors to be developed are directly reduced resulting in (1) a limited number of responses available to a stimulus event and (2) particular responses (those that have been repeatedly reinforced) occurring at high strength. Again, a basic property of the acquisition of operant behavior is the reduction of many other forms of behavior (Leslie, 1999).

FUNCTIONAL ASSESSMENT OF PERSONALITY DISORDERS

Frequent assessment serves to structure, evaluate, and improve an intervention. The goal of an assessment is to understand the interaction between the client and his/her environment and to illuminate the contextual variables that are influencing the target behavior. From a radical behavioral perspective, private events such as thoughts, feelings, bodily states, and behavioral urges are also considered behavior and, thus, may be targeted and functionally analyzed in the same way as overt behavior. Due to the nature of personality disorders, it is often necessary to arrange target behaviors in a hierarchy and address behaviors that are the most problematic or distressing first. Table 19.2 lists behaviors commonly targeted in personality disorder treatment.

Nelson-Gray and Farmer (1999) suggested an assessment approach based on Goldfried and Sprafkin's (1976) Stimuli-Organism variables, Responses, Consequences (SORC) model. They suggested that problematic responses, or target behaviors, be drawn in part from the criterion behaviors given in the *DSM*, as well as from in-session behavior, self-monitoring (discussed later in this chapter), and interviews. According to Nelson-Gray and Farmer (1999) efforts should focus on contexts in which a behavior becomes inflexible for an individual. Organism variables of interest include those within a client's history (modeling and reinforcement) and physiological variables. Finally, they suggested evaluating both the short- and long-term consequences of the problem behavior.

By using such a method to develop a case formulation, a clinician can tailor an intervention to suit the specific needs of a client and can monitor progress with ongoing assessments. If more than one behavior serves the same function, these behaviors can be grouped into a response class, and the entire response class can be monitored as treatment progresses. For

TABLE 19.2   Common behavioral targets in Personality Disorder treatment Adapted from Lynch, Cheavens, Cukrowicz, and Linehan (2006)

Avoidant Personality Disorder	*Decrease:* Maladaptive thoughts about rejection, inferiority, and failure Avoidance of social interactions Avoidance of intimacy and vulnerability Distraction from emotion Efforts to avoid receiving attention from others Ignoring problems to avoid affect
Dependent Personality Disorder	*Decrease:* Asking others for input when making decisions Avoidance of being alone Seeking accompaniment before going places Suppressing opinions that might counter the thoughts of others Soliciting help from others before taking on responsibility
Obsessive-Compulsive Personality Disorder	*Decrease:* Maladaptive thoughts about mistakes, imperfections, and morals Checking the details after completing a task Spending prolonged amounts of time on simple tasks Allowing others autonomy when delegating tasks Discounting the opinions of people with different values Hoarding behavior Excessive work behaviors Willful urges (e.g., the urge to tell another what to do)
Antisocial Personality Disorder	*Decrease:* Lying to others or failing to keep promises Cheating (on exams, taxes, games, etc.) Stealing from or harming others *Increase:* Expressions of empathy Perspective taking Thoughts about consequences (for self and others)
Narcissistic Personality Disorder	*Decrease:* Use of comparative or superlative language Expectations of special treatment and privileges Disregard for rules; thoughts that rules are for others Extreme selectivity in friendships *Increase:* Empathy and perspective taking Openness to criticism or judgment
Histrionic Personality Disorder	*Decrease:* Dramatic behaviors, such as overexpression of private events Flirtatious or provocative behavior Avoidance of being alone *Increase:* Neutral behaviors in groups Tolerance of boredom

(Continues)

TABLE 19.2 (*Continued*)

Schizoid and Schizotypal Personality Disorder	*Decrease:* Maladaptive thoughts about freedom and independence Maladaptive thoughts about close relationships Disregard for the goals and standards of others *Increase:* Interactions with other people Confiding in others Discussion of personal information in therapy
Paranoid Personality Disorder	*Decrease:* Maladaptive thoughts about the motives of others Avoidance of situations in which positive emotion might occur Expectations about "repairs" for grievances Holding grudges *Increase:* Participation in the community Forming close relationships Confiding in others Prosocial behaviors ("chit-chat," saying "thank you" to praise)

example, in dependent personality disorder, both volunteering for unpleasant tasks and requiring reassurance from a partner about appearance may serve to obtain positive attention from others. These behaviors could be targeted simultaneously.

Self-monitoring via recording forms can serve to illuminate the contextual factors that maintain a problem behavior. The client completes an entry in the form whenever he/she engages in the problem behavior. The form includes space to record vulnerability factors, events preceding the problem behavior, and events following the behavior. An alternative to recording forms of this nature is a diary card in combination with an in-session behavioral or chain analysis. For example, if a client with borderline personality disorder (BPD) were to record a cutting episode on a diary card, the therapist and client could perform a chain analysis and intervene by addressing inadequate or inappropriate stimulus control by (a) providing skills to cope with intense emotions that acted as an antecedent, (b) decreasing vulnerability factors that act as establishing operations for problem behaviors, such as strategizing to increase compliance with medications, and (c) using contingency management strategies (i.e., highlighting long-term negative effects of the behavior).

If the appropriate contextual factors have been identified, one should be able to predict and influence target behaviors with precision. Behavior analysts often use single case designs such as multiple baseline or ABA designs to determine whether there is a functional relationship between an independent and dependent variable. Wolpe and Turkat (1985) called this "clinical experimentation," and Maisto (1985) gave an example of its use in investigating the drinking behaviors in a client with a personality disorder. An ABA design can be difficult with personality disorders, because the target behavior may be adaptive or maladaptive, depending on the situation. An additional challenge is the dangerous nature of some target behaviors such as dangerous forms of self-injury.

RULE-GOVERNED BEHAVIOR

A functional analytic approach to psychological difficulties would not be complete without a discussion of the positive and negative impact of verbal abilities on direct environmental contingencies. On the positive side, the human capacity for language makes it possible to respond to contingencies that would be ineffective in controlling the behavior of nonhumans. For example, verbal behavior allows for possible outcomes to be detected and for lengthy behavioral sequences to be performed in anticipation of particular consequences, despite the fact that they may not have been previously experienced (Barnes-Holmes, Hayes, & Dymond, 2001). It also allows for appropriate behavior in novel situations (Follette, Naugle, & Linnerooth, 2000). In addition, the ability to verbally discriminate one's own behavior and the contingencies controlling it alters the psychological functions of these events (see Wilson & Blackledge [1999] for a more extensive discussion) and allows for greater self-control and behavior change (Barnes-Holmes et al., 2001; Hayes & Wilson, 1993).

However, verbal abilities also interfere with contingencies that would shape more adaptive behavior. For example, studies have demonstrated that individuals who are verbally instructed on a task do not change strategies when contingencies change. Rather, they persist in a given strategy because it is consistent with a rule that verbally specifies contingencies (Hayes, Brownstein, Haas, & Greenway, 1986; Shimoff, Catania, & Matthews, 1981). The fact that rule-governed behavior is less sensitive to environmentally available information is relevant to personality disorder behaviors that result from literal belief in thoughts and rigid adherence to inaccurate self-generated rules. For example, an individual diagnosed with dependent PD might have a verbally constructed rule that says, "If I let others make decisions for me, I will avoid negative events," when in many cases letting others decide will contribute to life problems.

THREE DOMAINS OF DIFFICULTY

Personality disorders are diverse; thus, it may be useful to discern primary domains of difficulty that are common among them. Behavior patterns associated with personality disorders include difficulties in the following areas: (1) self, (2) private events, and (3) interpersonal relations.

SELF ISSUES

Problems related to the self are a central feature of personality disorders, although the nature of the disturbance can vary considerably from one PD to another. Issues range from an unstable or absent sense of self (e.g., Borderline PD; Dependent PD) to egocentric beliefs and overt behaviors (e.g., Paranoid PD; Narcissistic PD). Historically, behavioral accounts of the self have focused primarily on self-knowledge and awareness. Skinner (1974) defined such as discriminating one's own responding and suggested that this behavior is learned in the context of social interactions:

> Self-knowledge is of social origin . . . it is only when a person's private world
> becomes important to others that it is made important to him. . . . [A person] made
> "aware of himself" by the questions he has been asked . . . is in a better position
> to predict and control his own behavior. (p. 31)

Skinner (1945) described the process by which discriminations of private events are learned. Specifically, according to Skinner (1945), the verbal community infers private stimuli from co-occurring public stimuli and collateral nonverbal responses and reinforces appropriate verbal behavior. For example, the social-verbal community assumes stomach pain (private event) and reinforces the utterance "my stomach hurts" when a child is hit in the stomach by a ball (public stimulus event that accompanies pain) or is seen with his arms wrapped tightly around his midsection (collateral response). In this way, a child learns how to label and appropriately respond to private events.

Contemporary behavioral accounts of the self emphasize the importance of perspective taking and elaborate on the role of verbal processes (Koerner et al., 1996). These accounts emerge from the theoretical and empirical work of functional analytic psychotherapy (FAP; Kohlenberg & Tsai, 1991) and relational frame theory (RFT; Hayes, Barnes-Holmes, & Roche, 2001).

PERSPECTIVE TAKING

The ability to speak from a consistent perspective allows one to experience the self as constant and present across contexts. The FAP (Kohlenberg & Tsai, 1991) approach to perspective taking and self is a direct application of Skinner's analysis of verbal behavior. It focuses on the use of "I" and the specification of stimuli controlling this verbal response. RFT expands on this analysis by incorporating recent research on derived relational responding. According to RFT, individuals are asked a number of questions that require that they take perspective (e.g., "What are you doing now?"). Each time the questions are asked, aspects of the physical environment are different, but the relational properties of I versus YOU, HERE versus THERE, and NOW versus THEN remain the same. Over time, these perspectives are abstracted and become an important component of all verbal activity (Barnes-Holmes et al., 2001).

FAP APPROACH TO SELF DISTURBANCES

FAP maintains that mild to moderate disturbances of self occur when "I [blank]" responses are largely under public, rather than private, control (Kohlenberg & Tsai, 1991). As a result, one's sense of self is affected by others, creating difficulties in knowing what one wants or feels, extreme sensitivity to criticism or opinions of others, and similar problems. This is particularly relevant to behavior patterns characteristic of avoidant and dependent personality disorders.

Public, rather than private, control of the "I [blank]" response might occur as the result of a number of learning histories. One such history is frequent aversive control. Under such conditions behavior does not occur as "free" or originating from the self. The implication of this is a lack of access to one's "true self" and/or spontaneity (Kohlenberg & Tsai, 1991). Waltz and Linehan (1999) described the results of a particular type of aversive environment. They maintained that in an invalidating environment, self-generated behavior is punished, resulting in an individual not engaging in behaviors that would be reinforcing. As a result of frequent punishment of self-generated responses, one has less access to self-relevant information.

RFT APPROACH TO SELF DISTURBANCES

RFT is a post-Skinnerian account of complex human behavior that expands early analyses of verbal behavior and the self (Hayes et al., 2001). According to RFT, derived relational responding is a learning process integral to language and cognition. While the particulars of this theory and research program are beyond the scope of this chapter, the basic premises can be outlined using a simple, concrete example of stimulus equivalence:

> A child is reinforced for saying the word *car* in the presence of a toy car. As a result, he *derives* (in the absence of direct reinforcement) the corresponding relation between the toy car and the word *car*, such that when the verbal stimulus *car* is presented in the appropriate context, he *responds relationally* (e.g., he points to the toy car; mutual entailment). Later, the child learns that the written word *c-a-r* also

corresponds to "car," and due to a history of reinforcement for relational responding, he *combines the relations* and discerns that *c-a-r* and a toy car are also equivalent (combinatorial entailment).

Perhaps the most important aspect of relational responding is the resulting bidirectional transfer or transformation of psychological functions that occur among stimulus events (Barnes & Keenan, 1993; Dymond & Barnes, 1995). For example, as the result of derived equivalence relations, some of the functions of the toy car (e.g., appetitives, play time) are present in the spoken word *car*. Figure 19.1 illustrates how derived equivalence relations and the resulting transfer of function(s) are related to self.

RFT elucidates three variants of "self" that result from these processes that may be useful in conceptualizing self-disturbances characteristic of the personality disorders. They include (1) self as content of verbal relations, (2) self as ongoing process of verbal relations, and (3) self as context of verbal relations (Barnes-Holmes et al., 2001). *Self-as-content* (the conceptualized self) refers to an individual's verbal construction of self that is formed through predictions, explanations, or interpretations of one's own behavior. According to RFT, when one interacts verbally with respect to his/her own behavior, one derives relations between one's ongoing stream of activity and various discrete categories. The content of an individual's verbally constructed self may be more or less positive, rigid, and multilayered (Barnes-Holmes et al., 2001; Barnes-Holmes, Stewart, Dymond, & Roche, 1999). The domination of one's verbal construction of self in the organization of behavior can be problematic if it decreases sensitivity to environmentally available information or if self rules are poorly constructed and do not accurately reflect contingencies (Follette et al., 2000).

Self-as-process refers to ongoing awareness and description of one's thoughts, feelings, bodily states, and behavioral predispositions (Barnes-Holmes et al., 2001; Hayes, Strosahl, & Wilson, 1999). The ability to

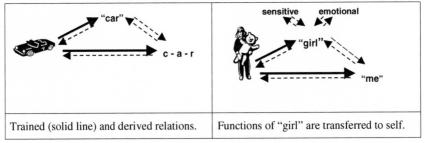

| Trained (solid line) and derived relations. | Functions of "girl" are transferred to self. |

FIGURES 19.1 An illustration of trained and derived relations that may occur in stimulus equivalence training (left panel) and how transfer of function related to personality disorder may occur within a stimulus class (right panel).

verbally report one's private events is learned early by the social-verbal community that assumes private experience given public correlates (Skinner, 1945) and is expanded on via relational conditioning processes. How one behaves in a given situation is highly dependent on this process and the training one received regarding the social implications of particular private events (Barnes-Holmes et al., 2001). It is easy to imagine deviant learning histories that would result in a weak or inappropriate self-as-process, such as learning histories in which conventional verbal discriminations were not made. The most readily accessible example is the case of physical, sexual, or emotional abuse. However, less extreme situations, such as invalidating environments (Linehan, 1993b), may also produce deficiencies in this sense of self. For example, in the context of an invalidating environment, a child may be told "you are not sad" despite the fact that he/she is crying. This would result in confusion regarding when it is appropriate to make such a statement and what private events are associated with this emotion label.

Self-as-context (the observing self) refers to the sense of self that is constant and stable across time and place. It is best understood as the context from which one experiences thoughts, emotions, and other private events. Deficiencies in one's observing self may be related to a lack of "sense of self," dissociation, and/or feelings of emptiness. This suggests the need to train clients to notice and attend to the continuity of perspective and to make a distinction between one's thoughts, feelings, and behaviors and the "I" that is discriminating these events (Barnes-Holmes et al., 2001; Hayes, 1984).

MANAGING PRIVATE EVENTS

Individuals meeting criteria for a personality disorder also have difficulty managing private events such as emotions, thoughts, and impulses. Mismanagement of private events can take a number of forms, including impulsive action, maladaptive attempts to reduce aversive thoughts or feelings, problems with overt expression of emotion, and other difficulties related to a lack of development of self-control skills. Impulsive behavior may be understood as an inability to respond effectively to impulsive urges and counter immediate environmental contingencies in order to regulate one's own behavior. (See Chapter 5, "Attention Deficit Hyperactivity Disorder," by Neef and Northup on ADHD and the analysis and development of self-control.) Self-control skills are necessary when behaviors have delayed detrimental consequences or when the contingencies that would support more adaptive behaviors are far removed from the current situation (Follette et al., 2000; Skinner, 1953). Addressing these competing contingencies is necessary to decrease PD behaviors characterized by impulsive action.

MALADAPTIVE ATTEMPTS TO REDUCE AVERSIVE
PRIVATE EVENTS

PD behaviors may also be developed and maintained by the effect they have on immediately reducing aversive thoughts, feelings, or bodily states. This has been discussed as thought suppression (Wegner & Gold, 1995), emotion regulation (Gross, 1998; Linehan, 1993a), and experiential avoidance (Hayes, Wilson, Gifford, Follette, & Strosahl, 1996). However, like impulsive action, such behaviors tend to be effective only in the short term, and research suggests that avoidance of aversive private events may have detrimental long-term consequences. Research examining thought suppression indicates that this strategy of managing private events often leads to an increase in the frequency of unwanted thoughts (Gross, 1998; Wegner & Gold, 1995), which may be relevant to a functional analysis of various PD criterion behaviors. For example, someone meeting criteria for paranoid PD might try to suppress paranoid thoughts but actually increase the thoughts. He/she might then engage in a number of problem behaviors in an attempt to reduce arousal, such as questioning a partner about faithfulness or prematurely ending a relationship.

Many researchers have turned their attention to emotion regulation strategies that, in some contexts, are problematic and may support PD behaviors. Gross (1998) delineated five emotion regulation strategies: (1) situation selection (approaching or avoiding specific people, places, or objects); (2) situation modification (actively attempting to modify a situation and thus change its emotional impact); (3) attentional deployment (focusing attention); (4) cognitive change (e.g., reappraisal): and (5) response modulation (directly influencing experiential, physiological, or behavioral responding). Emotion regulation is central to the conceptualization and treatment of BPD in dialectical behavior therapy (DBT; Linehan, 1993a). DBT maintains that BPD is primarily a disorder of emotion dysregulation, and the BPD criteria reflect the direct result of this dysregulation and that maladaptive attempts to regulate or alter affect and research thus far have been supportive. For example, studies have found that thought suppression mediates the relationship between affect intensity and BPD criterion behaviors (Cheavens et al., 2005; Rosenthal, Cheavens, Lejuez, & Lynch, 2005).

Experiential avoidance is a general term that encompasses thought suppression, problematic emotion regulation strategies, and the like. Experiential avoidance is defined as a functional class of behaviors designed to alter the form or frequency of all types of aversive private experiences and the contexts that occasion them (Hayes et al., 1996). Research has demonstrated that experiential avoidance, as measured by the Acceptance and Action Questionnaire (Hayes et al., 2004b), is associated with a number of negative outcomes, including BPD behaviors such as self-harm (e.g., Gratz & Roemer, 2004; Hayes, Luoma, Bond, Masuda, & Lillis, 2006).

OVERT EXPRESSION OF PRIVATE EVENTS

One way to organize the various PDs is to consider them on a dimension of emotional expressiveness. While histrionic and borderline PD are characterized by an overexpression of emotion, schizoid and obsessive-compulsive PD are characterized by underexpression. This is likely a function of different reinforcement histories for emotional expression and resulting inadequate or inappropriate discriminative, antecedent, or consequential stimulus control. Factors maintaining overexpression may include displays of intense emotion that in the past may have resulted in greater nurturance or self-validation for one's level of pain (Linehan, 1993b). This history combined with an inability to discriminate when it is and is not appropriate would result in emotional displays in situations in which this behavior is not interpersonally effective.

INTERPERSONAL DIFFICULTIES

PD behaviors that are interpersonally ineffective may be developed and maintained in several ways. First, similar to the previous example of emotional expression, PD behaviors that interfere with social functioning may be the product of reinforcement coupled with a lack of appropriate discriminative stimulus control. For example, an individual diagnosed with Narcissistic PD may have received reinforcement for being demanding in a situation that necessitated such behavior and then failed to discriminate contexts in which this behavior is desirable. Demanding behavior may have then been subsequently reinforced with compliance. Second, problem behaviors may result from an individual deriving an inaccurate association between a given behavior or property of behavior and its positive outcome. For example, an individual with Obsessive-compulsive PD might be praised by his/her supervisor for a presentation. This praise may inadvertently serve to reinforce the meticulous way in which the individual composed slides and graphs, rather than the intended behavior of including relevant information in his/her presentation. Third, problematic interpersonal behavior may result from a poverty of reinforcement that leads to an escalation in behavior until that reinforcement is delivered. For example, consider a history in which attempts to appropriately communicate needs were not rewarded; however, curt, loud tones and aggressive behavior were responded to immediately. Over time, the former response would be shaped out of the individual's behavioral repertoire, and the latter would become the dominant response. Fourth, interpersonal difficulty may arise from deficits in one of the other primary domains such as self issues and difficulty managing private events. For example, if an individual is not able to appropriately label, express, and respond to private events, then it will be exceedingly difficult for him/her to get needs met. This may lead to significant interpersonal conflict and a number of behaviors that negatively impact social relations. Finally, interpersonal difficulties may be due to social skill deficits. Skills deficits may be the function of a lack of opportunities to practice—the absence of either an appropriate S^D or reinforcement for appropriate social behavior, or an inability to generalize or execute skills one has to relevant situations.

FUNCTIONAL ANALYTIC THERAPIES FOR PERSONALITY DISORDERS

Behavior therapies can be divided into three waves. In the first wave, clinicians focused on first-order change, directly targeting problem behaviors such as anxiety symptoms or avoidance of social situations. These therapies relied on the direct application of principles derived in the laboratory to human psychological problems. In the second wave, cognitive content became another target of intervention, although strong models for cognition had yet to be developed. Clinicians focused on the detection and correction of dysfunctional or maladaptive thoughts, schemas, and information-processing strategies (Hayes, 2004; O'Leary & Wilson, 1987). Third-wave behavioral treatments include dialectical behavior therapy (DBT; Linehan, 1993a), acceptance and commitment therapy (ACT; Hayes et al., 1999), and functional analytic psychotherapy (FAP; Kohlenberg & Tsai, 1991), which build on the previous two waves but also reach back through the decades to draw on such sources as Skinner's concept of radical behaviorism, functional analysis, and applied behavior analysis. Common to the new therapies and their anachronistic influences are emphases on context and the importance of controlling variables. Third-wave behavior therapies tend to emphasize context and second-order change strategies. According to a functional contextualistic philosophy, which informs ACT and similar therapies, psychological events are ongoing interactions between a whole organism and historical and situational contexts. Second-order change targets not the form or frequency of a problematic thought or emotion, but its function and context. Thus, interventions are closely tied to a functional assessment of the problem behaviors. Functional analyses can help to elucidate the variables controlling a problematic behavior, which in turn will help the clinician develop and implement an intervention aimed at altering the function of a stimulus event by changing the contexts in which they occur. The context of a private event may be changed using acceptance and mindfulness-based strategies.

Several functional analytic treatment options exist, and they can be applied to the interpersonal, regulatory, and self problems that have been identified as characteristic of the personality disorders. Three treatments—DBT, ACT, and FAP—are summarized in the following sections.

DIALECTICAL BEHAVIOR THERAPY

DBT is a specific treatment, designed for individuals diagnosed with BPD. Evidence from several randomized clinical trials (Koons et al., 2001; Linehan et al., 1991, 1999, 2002, 2006; Turner, 2000; Verheul et al., 2003) suggests that DBT can be efficacious in treating the symptoms of BPD, as well as interpersonal aggression and sensitivity among older adults meeting a diagnosis for depression comorbid with a personality disorder (Lynch et al., 2006). Linehan reorganized the *DSM-IV* criteria for BPD and conceptualized the disorder as a set of maladaptive behavioral patterns, including emotional vulnerability, self-invalidation, unrelenting crises, inhibited grieving, active passivity, and apparent competence. However, as noted by Waltz and Linehan (1999), these descriptions are best used for heuristic purposes; clinicians should always carefully assess each client's specific behavioral patterns. For example, adaptations of DBT for treating older adult personality disorders resulted in differing sets of behavioral targets specific to this population (e.g., behavioral rigidity; see Lynch et al., 2006).

Dialectics is the world view that underlies DBT. This world view emphasizes the wholeness and interrelatedness of systems. It assumes that reality is continually changing and that it is composed of opposing forces—thesis and antithesis—which, when integrated into a synthesis, produce a new set of such forces. Dialectical dilemmas are present in the behavioral patterns of the client and the therapist. The therapy itself requires dialectical thinking. The primary dialectical tension in DBT is between acceptance and change. Each interaction between a DBT therapist and a client involves a balance between validation of the truth in the client's behavior (acceptance) and problem-solving strategies that promote therapeutic change.

Linehan's biosocial theory of BPD is the basis for her treatment. Biosocial theory posits that three factors contribute to the development and maintenance of BPD: (1) emotional vulnerability, (2) an invalidating environment, and (3) emotional dysregulation, due to maladaptive emotion regulation strategies. "Emotional vulnerability" refers to a high sensitivity to emotional stimuli, extreme emotional reactions, and a slow return to baseline following an emotionally arousing episode. An "invalidating environment" is "one in which communication of private experiences is met by erratic, inappropriate, and extreme responses" (p. 50). Linehan explained that when parents disregard, punish, or trivialize the painful emotional experiences of an emotionally reactive child, they are providing an environment that does little to help the child learn to recognize and regulate his/her own emotions and associated behaviors. "Emotional dysregulation" refers to the use of ineffective emotion regulation strategies in the face of overwhelming negative emotions. Such strategies might include performing self-mutilating acts, attempting suicide, or attempting to inhibit affect.

DBT systematically prioritizes target behaviors for intervention. The first behaviors targeted for reduction are suicide-related behaviors, including suicide crisis behaviors, parasuicidal acts, suicidal ideation or communications, suicide-related expectancies, and suicide-related affect. Second on the list of targets are the therapy-interfering behaviors of both client and therapist. After these come quality-of-life-interfering behaviors and post-traumatic-stress-related behaviors, and associated increases in behavioral skills and respect for self. DBT proceeds through four stages, which move the client from acquiring basic emotion regulation and interpersonal skills through dealing with trauma histories and problems with emotions and eventually to skills for living a fuller and more joyful life. This multimodal therapy consists of weekly individual therapy sessions, weekly skills training group sessions, and telephone coaching for the client, as well as consultation team meetings for clinicians. The skills training groups teach clients distress tolerance skills, emotion regulation skills, mindfulness skills, and interpersonal effectiveness skills.

ACCEPTANCE AND COMMITMENT THERAPY

ACT (Hayes et al., 1999) is a third-wave behavior therapy based on behavior analysis and an operant theory and a basic experimental program of derived stimulus relations known as RFT. From an ACT/RFT perspective, psychological difficulties are the result of the way in which verbal processes interact with direct contingencies to generate narrow and inflexible behavioral repertoires. ACT targets core language processes that foster rigidity and attempts to bring these processes under contextual control. The primary targets for intervention are behaviors that interfere with an individual's ability to participate fully in valued life domains. ACT interventions aim to modify the social-verbal contexts that support problematic emotional control, literality, and reason giving. This is achieved via mindfulness and acceptance skills and commitment and behavior change skills. Components of ACT include acceptance, defusion, self-as-context, contact with the present moment, values, and committed action. Hayes et al. (1999) and Hayes and Strosahl (2005) provided book-length treatments of these topics. Hayes et al. (2006) provided a concise description of ACT theory and intervention components, as well as a review of the current state of the process and outcome data.

ACT is a functional contextualistic intervention directly tied to a functional analysis and the applied implications of that analysis, particularly with regards to the acceptance aspects of treatment. Thus, there is an explicit emphasis on the function of behavior, as well as the situational, historical, and socio-cultural contexts in which it occurs. While ACT is a principle-driven intervention in which specific protocols may vary, typical steps taken in assessment and treatment can be identified. These steps include (1) identifying a functional class of avoided events; (2) identifying a functional class of avoidant repertoires that are negatively reinforced by the attenuation of these events; (3) identifying valued directions and corresponding overt behaviors that would be pursued if psychological obstacles were removed; (4) making experiential contact with the paradox of control in the context of valued living; (5) conducting exposure to avoided events and defusing/deliteralizing verbal barriers to valued living; and (6) facilitating engagement in committed action and conducting ongoing exposure and defusion as needed for an individual to move in a chosen valued direction. ACT has not been applied to personality disorders specifically. However, research examining ACT suggests that it might be useful for treating this population. For example, Bach and Hayes (2002) compared a brief ACT intervention to treatment as usual for psychosis. Individuals who received the ACT intervention reported a significant decrease in the belief in hallucinations/delusions and a reduced rate of rehospitalization. Gaudiano and Herbert (2006) reported a similar outcome. These two studies suggest the potential utility of ACT in the treatment of Cluster A personality disorder behaviors. ACT has also been demonstrated to be effective in the treatment of disorders that are often comorbid with personality disorders, such as substance abuse (Gifford et al., 2004; Hayes, Masuda, Bissett, Luoma, & Guerrero, 2004a), depression (Zettle & Raines, 1989), and chronic pain (Dahl, Wilson, & Nilsson, 2004). Furthermore, research examining a combined ACT/DBT intervention for self-harm has yielded large effect sizes (Gratz & Gunderson, 2006). Despite these promising results, much more work needs to be done to determine how to apply this treatment to personality disorders.

FUNCTIONAL ANALYTIC PSYCHOTHERAPY

FAP is theoretically based on applied behavior analysis. Treatment proceeds through the delivery of nonarbitrary reinforcement for certain behaviors termed "clinically relevant behaviors" (CBR) as they occur within a therapy session. CBR are those that are targets of change for the client and which tend to occur both in and out of session. Because environments that are functionally similar tend to evoke the same behavior, behavior change that occurs in a therapy environment that evokes problem behaviors should generalize to functionally similar contexts outside therapy. Kohlenberg and Tsai (1991) define three types of CRBs. Any problematic client behavior that occurs during session is labeled a "CRB1." CRB1s are typically controlled by aversive stimuli and may include avoidance behavior. "CRB2" describes a client improvement that occurs in session. "CRB3" refers to a client's interpretation of his/her own behavior. At their best,

CRB3s will incorporate the client's observations of behaviors as they occur; descriptions of behaviors; and hypotheses about the associated eliciting, discriminative, and reinforcing stimuli.

Kohlenberg and Tsai structured their explanation of therapeutic technique into five rules: (1) Watch for CRBs, because careful observation ensures consistent reinforcement; (2) evoke CRBs, through homework assignments, hypnosis, behavioral tasks, and withholding acceptance comments; (3) reinforce CRBs, both directly and indirectly, focusing on wide response classes and matching expectations to the client's skill level; (4) observe the potentially reinforcing effects of clinician behavior in response to CRBs; and (5) give interpretations of variables affecting a client's behavior, to help clients develop a verbal repertoire that includes the form "[discriminative stimulus] leads to [response] leads to [reinforcement]." Clinicians can enhance their awareness of CRBs by furthering their knowledge of verbal behavior. Kohlenberg and Tsai used Skinner's (1957, as cited in Kohlenberg & Tsai, 1991) classification system to classify verbal behavior according to the type of stimulus controlling it.

FAP can be useful in treating a disturbed sense of self. As described previously, an individual with a disturbed or unstable self requires external cues to make sense of private stimuli. To assess a disturbed sense of self, Koerner et al. (1996) suggested that clinicians be alert to clients acting warily, overly attentive, or overly concerned with therapists' opinions. The authors of FAP suggested that one method for treating an unstable sense of self is to use the classic psychoanalytic approach of free association, though not to the same end. By asking a client to freely speak about current thoughts and emotions, a clinician requires that the client describe the private stimuli associated with self. If a client is unable to perform this task, the clinician should shape the behavior by first asking the client to free associate for a very short period of time and gradually increasing the amount of time in which the client describes private events.

FAP is also particularly well suited to address the interpersonal difficulties characteristic of personality disorders, due to its focus on the interpersonal interactions in the therapy room. Because FAP is designed to be used in conjunction with other treatments, a clinician could address impulsive behaviors and emotion dysregulation directly by adding components of other treatments, such as skills training from DBT.

CASE STUDY

The client (C) is a 27-year-old Caucasian woman who had been diagnosed with BPD and bulimia and who shows signs of alcohol dependence. Here, she met with her individual DBT therapist (T) for their 24th weekly session. The session began with the therapist checking the client's diary card on which the client recorded episodes of purging and self-harm. The therapist proceeded by performing a chain analysis, examining controlling variables, and looking for places where the client might have used DBT skills to achieve a different outcome:

T: I see that you had one purging episode and one self-harming episode, both on Sunday.

C: Yeah, actually, they were back-to-back.

T: Okay, so if we started our chain here [*draws target behavior circle on white board*], then at the end of the chain we would have . . . [*draws more circles*] . . . you purged and then you self-harmed? [*The therapist is determining the specific problem behaviors and the sequence of their occurrence.*]

C: Yeah, I purged, and that just didn't help, I still felt really bad. . . . so I burned myself. [*Client links behavior to anticipated consequence; indicates that behavior is operant.*]

T: Okay, and what time was that?

C: It was about 11:00 or so that I purged, and probably about 11:15 that I burned myself.

T: We'll put that on the end of the chain. Did you feel badly all day, or was there a point when you noticed your mood changing? [*Therapist is determining contextual factors.*]

C: Well, I first started noticing my mood being different when I was watching TV. I remember noticing that I didn't think the shows were as funny as usual.

T: So you picked up on the fact that something was different than it ordinarily is. Do you remember any thoughts in particular that you were having during that time?

C: Well, the show that I was watching was a holiday episode. It made me think about how this is my first holiday season alone. I've been with my boyfriend for the last five years. Last Christmas we got engaged, so this Christmas being alone is just really tough.

T: Okay, so what emotion did you notice going along with that?

C: I was feeling sad and lonely.

T: All right, so "sad" goes right here in the chain. How long would you say that lasted, and what was the next thought, emotional change, or behavior that occurred?

C: I watched the shows until around 9:45. For a couple of hours I was trying to distract myself, but I still felt sad and I was thinking about him. After the shows were over I noticed that I started to get more sad, because there wasn't anything to distract me.

T: Okay, so that was 9:45, and you said you purged around 11. So what happens next?

C: At 9:45 I started getting really tearful. I went to the bathroom, and then I went to get a glass of water. I was getting sadder as that was going on, so I decided to call a friend.

T: What happened in this phone call? Did you notice any emotions or thoughts?

C: I called this friend of mine, who knows my ex. She told me that she and her boyfriend were getting engaged. That was hard because I'd just been thinking about being engaged myself. She also told me that she saw my ex with another girl. That made me a *lot* sadder.

T: How long were you on the phone with her?

C: Until about 10:30. She went on and on about getting married, and I got feeling worse.

T: You're noticing that feelings of sadness are increasing. Are you noticing anything else?

C: I guess I was feeling jealous of her, for getting engaged. And I was angry, too, because I called her to try to feel better and she didn't even notice that I was feeling lousy.

T: What did you do after that? You said you got off the phone about 10:30 or so?

C: Yeah, that's right. Then, I called him. I called my ex.

T: Oh. You called your ex? What thought prompted you to call him?

C: I don't know if it was a thought. I think I just reacted. I just wanted to ask him, "Who are you with?" I wanted to know if he missed me. Looking back, it was pretty impulsive.

T: Tell me about that phone call.

C: As soon as he picked up the phone he said, "You know you're not supposed to call me. You know your parents will get mad at me." Then he hung up.

T: Okay, so he hung up on you. I'm guessing that probably increased some of the emotions you were already feeling, correct?

C: Yeah, I got a lot angrier. I couldn't believe he'd hung up on me! I hadn't gotten any answers, and I was angry at my parents for making it so hard for me to talk to him.

T: On a scale of 1 to 10, how intense were your emotions at that point? [*Therapist gets subjective rating of emotional intensity. This information may be used as an indicator of high risk for self-harm by the client or therapist.*]

C: The sadness and loneliness were at about a nine.

T: So now we're at about 10:30. Walk me through from 10:30 to 11:00. What happens?

C: I was having some thoughts about purging, but I really didn't want to do that, so I watched TV to try to distract myself. I watched TV until about 11. [*Stimulus control*]

T: What did you watch?

C: I watched something funny. I was trying to get my mind off of it.

T: That's good! You tried to create an opposite emotion. That's something we talked about. Unfortunately, it sounds like it didn't work. [*Therapist praises skill use.*]

C: No, it didn't. The TV was on, but I was in my head.

T: Okay, so now we're at about 11:00, and you're watching TV. What are your thoughts? [*Therapist continues to assess contextual factors with a focus on cognition.*]

C: I couldn't stop thinking about purging, so I drank a couple of shots of vodka. Then I felt even worse because I know I shouldn't do that. It increased my desire to throw up.

T: Okay, so it sounds like having the alcohol made it even easier to purge. [*Therapist is defining contingencies for alcohol use.*]

C: Yeah, I would say that's true.

T: Just after 11:00 you go into the bathroom and get sick. Then what happens?

C: I felt even worse because it wasn't helping. I still felt terrible, and I was still crying a lot. That's when I first started thinking about hurting myself. I was hoping that burning myself would do what the purging had not, which was make me feel better.

T: But there was some time between when you purged and when you self-harmed. Any other thoughts in that time? About other things you could do?

C: No, not at the time. I could think of some now.

T: We'll go back and look at some spots in this chain where you could have done things differently. So then, at 11:30 you burned yourself. What happened after that?

C: I actually felt better for a few minutes. I had this one-track mind about burning myself and after I did it, there was relief from feeling miserable. [*Client receives short-term reinforcement for problem behavior.*]

T: Was some of the relief because you'd been able to get that thought to stop for a little bit?

C: Yeah, I guess that was part of the relief.

T: Okay, so you feel better for a few minutes and then what happens? [*Therapist begins highlighting negative consequences as a form of contingency management.*]

C: And then I started feeling *so* guilty. I just felt horrible. Sad and angry and guilty.

T: So let's talk about that. What are some of the consequences as far as you can tell?

C: For one thing, now I have this burn on my arm.

T: Any thoughts?

C: My first thought was that I messed up, I let you down, and that this is a major setback.

T: All right, we can talk about that. Were there other consequences?

C: Well, my Mom saw it, so there was a big fight. I also just felt a lot of shame.

T: So there were lots of consequences to having done this. Let's take a look back at this chain and see if there are any places where you could have done something differently. [*Therapist is looking for places to intervene and disrupt the functional relationships that support the problem behavior.*]

C: I know you're gonna say that I should have called you.

T: Well, you're right. The role of coaching calls is to interrupt this process before it gets to you burning yourself. So where on this chain do you think you might have called me?

C: I guess I could have called you when I was watching TV, after I talked to my ex.

T: That would have been a *great* place to call. I think another place to call might have been right before you called your ex. Are there any other skills that you can think of practicing, that wouldn't have necessitated a call to me?

C: I guess she probably would have told me to do that dive reflex thing. [*This is a DBT skill designed to change body temperature during emotional crisis by splashing ice water on the face or by having the client place his/her face in ice water.*]

T: Where do you think a good place to practice the dive reflex might have been?

C: Right after I got off the phone with my friend, or when I got off the phone with my ex.

T: Those would have been great places. Anything else you can think of? Remember to think of distress tolerance skills.

C: Oh, so you mean things like sucking on a lemon wedge or taking a bath? [*These are skills that include use of stimuli that might cause strong sensation or self-soothing.*]

T: Yes, those are good ideas. This is a nice chain, and I appreciate your willingness to do this. I should point out that another consequence of you engaging in self-harm is having to spend so much of our session doing one of these chains. On the one hand, this is good, because you're developing the skill of being able to look, moment-by-moment, at what's happening with your thoughts and emotions. On the other hand, we have less time to talk about other things that happened in your week. The synthesis is that you can take these skills out of here and apply them to other events in your life. Okay, good work.

REFERENCES

American Psychiatric Association. (2000). *Diagnostic and statistical manual of mental disorders* (4th ed., text rev.). Washington, D.C.: Author.

Bach, P., & Hayes, S. C. (2002). The use of acceptance and commitment therapy to prevent rehospitalization of psychotic patients: A randomized controlled trial. *Journal of Consulting and Clinical Psychology, 70,* 1129–1139.

Barnes, D., & Keenan, M. (1993). A transfer of functions through derived arbitrary and no-arbitrary stimulus relations. *Journal of Experimental Analysis of Behavior, 59,* 61–81.

Barnes-Holmes, D., Hayes, S. C., & Dymond, S. (2001). Self and self-directed rules. In S. C. Hayes, D. Barnes-Holmes, & B. Roche (Eds.), *Relational frame theory: A post Skinnerian account of human language and cognition* (pp. 119–140). New York: Kluwer Academic/Plenum Publishers.

Barnes-Holmes, D., Stewart, I., Dymond, S., & Roche, B. (1999). A behavior-analytic approach to some of the problems of self: A relational frame theory analysis. In M. J. Dougher (Ed.), *Clinical behavior analysis* (pp. 47–74). Reno, NV: Context Press.

Cheavens, J. S., Rosenthal, M. Z., Daughters, S. B., Nowak, J., Kosson, D., Lynch, T. R., et al. (2005). An analogue investigation of the relationships among perceived parental criticism, negative affect, and borderline personality disorder features: The role of thought suppression. *Behaviour Research and Therapy, 43,* 257–268.

Clark, L. A., Livesley, W. J., & Morey, L. (1997). Personality disorder assessment: The challenge of construct validity. *Journal of Personality Disorders, 11,* 205–231.

Dahl, J., Wilson, K. G., & Nilsson, A. (2004). Acceptance and commitment therapy and the treatment of persons at risk for long-term disability resulting from stress and pain symptoms: A preliminary randomized trial. *Behavior Therapy, 35,* 785–801.

Dymond, S., & Barnes, D. (1995). A transformation of self-discrimination response functions in accordance with the arbitrarily applicable relations of sameness, more-than, and less-than. *Journal of Experimental Analysis of Behavior, 64,* 163–184.

Follette, W. C., Naugle, A. E., & Linnerooth, P. J. N. (2000). Functional alternatives to traditional assessment and diagnosis. In M. J. Dougher (Ed.), *Clinical behavior analysis* (pp. 99–125). Reno, NV: Context Press.

Gaudiano, B. A., & Herbert, J. D. (2006). Acute treatment of inpatients with psychotic symptoms using acceptance and commitment therapy. *Behaviour Research and Therapy, 44,* 415–437.

Gifford, E. V., Kohlenberg, B. S., Hayes, S. C., Antonuccio, D. O., Piasecki, M. M., Rasmussen-Hall, M. L., et al. (2004). Acceptance theory-based treatment for smoking cessation: An initial trial of acceptance and commitment therapy. *Behavior Therapy, 35,* 689–706.

Goldfried, M. R., & Sprafkin, J. N. (1976). Behavioral personality assessment. In J. T. Spence, R. C. Carson, & J. W. Thibault (Eds.), *Behavioral approaches to therapy* (pp. 295–321). Morristown, NJ: General Learning Press.

Gratz, K. L., & Gunderson, J. G. (2006). Preliminary data on an acceptance-based emotion regulation group intervention for deliberate self-harm among women with borderline personality disorder. *Behavior Therapy, 37,* 25–35.

Gratz, K. L., & Roemer, L. (2004). Multidimensional assessment of emotion regulation and dysregulation: Development, factor structure, and initial validation of the difficulties in an emotion regulation scale. *Journal of Psychopathology and Behavioral Assessment, 26,* 41–54.

Gross, J. J. (1998). The emerging field of emotion regulation: An integrated review. *Review of General Psychology, 2,* 271–299.

Hayes, S. C. (1984). Making sense of spirituality. *Behaviorism, 12,* 99–110.

Hayes, S. C. (2004). Acceptance and commitment therapy and the new behavior therapies. In S. C. Hayes, V. M. Follette, & M. M. Linehan (Eds.), *Mindfulness and acceptance: Expanding the cognitive-behavioral tradition* (pp. 1–29). New York: Guilford Press.

Hayes, S. C., Barnes-Holmes, D., & Roche, B. (Eds.). (2001). *Relational frame theory: A post-Skinnerian account of human language and cognition*. New York: Kluwer Academic/Plenum Publishers.

Hayes, S. C., Brownstein, A. J., Haas, J. R., & Greenway, D. E. (1986). Instructions, multiple schedules, and extinction: Distinguishing rule-governed behavior from schedule controlled behavior. *Journal of the Experimental Analysis of Behavior, 46*, 137–147.

Hayes, S. C., Luoma, J. B., Bond, F. W., Masuda, A., & Lills, J. (2006). Acceptance and commitment therapy: Model, processes, and outcomes. *Behaviour Research and Therapy, 44*, 1–25.

Hayes, S. C., Masuda, A., Bissett, R., Louma, J., & Guerrero, L. F. (2004a). DBT, FAP, and ACT: How empirically oriented are the new behavior therapy technologies? *Behavior Therapy, 35*, 35–54.

Hayes, S. C., & Strosahl, K. D. (Eds.). (2005). *A practical guide to acceptance and commitment therapy*. New York: Springer-Verlag.

Hayes, S. C., Strosahl, K. D., & Wilson, K. G. (1999). *Acceptance and commitment therapy: An experiential approach to behavior change*. New York: Guilford Press.

Hayes, S. C., Strosahl, K. D., Wilson, K. G., Bissett, R. T., Pistorello, J., Toarmino, D., et al. (2004b). Measuring experiential avoidance: A preliminary test of a working model. *The Psychological Record, 54*, 553–578.

Hayes, S. C., & Wilson, K. G. (1993). Some applied implications of a contemporary behavior-analytic account of verbal events. *The Behavior Analyst, 16*, 283–301.

Hayes, S. C., Wilson, K. G., Gifford, E. V., Follette, V. M., & Strosahl, K. (1996). Experiential avoidance and behavioral disorders: A functional dimensional approach to diagnosis and treatment. *Journal of Consulting and Clinical Psychology, 64*, 1152–1168.

Koerner, K., Kohlenberg, R. J., & Parker, C. R. (1996). Diagnosis of personality disorder: A radical behavioral alternative. *Journal of Consulting and Clinical Psychology, 64*, 1169–1176.

Kohlenberg, R. J., & Tsai, M. (1991). *Functional analytic psychotherapy*. New York: Plenum Press.

Koons, C. R., Robins, C. J., Tweed, J. L., Lynch, T. R., Gonzalez, A. M., Morse, J. Q., et al. (2001). Efficacy of dialectical behavior therapy in women veterans with borderline personality disorder. *Behavior Therapy, 32*, 371–390.

Lee, R., McComas, J. J., & Jawor, J. (2002). The effects of differential reinforcement on varied verbal responding by individuals with autism. *Journal of Autism and Developmental Disabilities, 35*, 391–402.

Lee, R., & Sturmey, P. (2006). The effects of lag schedules and preferred materials on variable responding in students with autism. *Journal of Autism and Developmental Disabilities, 36*, 421–428.

Leslie, J. (1999). *Behavior analysis: Foundations and applications to psychology*. Amsterdam, Netherlands: Routledge.

Linehan, M. M. (1993a). *Skills training manual for borderline personality disorder*. New York: Guilford Press.

Linehan, M. M. (1993b). *Cognitive behavioral treatment of borderline personality disorder*. New York: Guilford Press.

Linehan, M. M., Armstrong, H., Suarez, A., Allmon, D., & Heard, H. L. (1991). Cognitive-behavioral treatment of chronically parasuicidal borderline patients. *Archives of General Psychiatry, 48*, 1060–1064.

Linehan, M. M., Comtois, K. A., Murray, A., Brown, M. Z., Gallop, R. J., Heard, H. L., et al. (2006). Two-year randomized trial + follow-up of dialectical behavior therapy vs. therapy by experts for suicidal behaviors and borderline personality disorder. *Archives of General Psychiatry, 18,* 303–312.

Linehan, M. M., Dimeff, L. A., Reynolds, S. K., Comtois, K. A., Welch, S. S., Heagerty, P., et al. (2002). Dialectical behavior therapy versus comprehensive validation therapy plus 12-step for the treatment of opioid dependent women meeting criteria for borderline personality disorder. *Drug & Alcohol Dependence, 67,* 13–26.

Linehan, M. M., Schmidt, H., Dimeff, L., Craft, J. C., Kanter, J., & Comtois, K. (1999). Dialectical behavior therapy for patients with borderline personality disorder and drug-dependence. *American Journal on Addictions, 8,* 279–292.

Lynch, T. R., Cheavens, J. S., Cukrowicz, K. C., & Linehan, M. M. (2006). *Dialectical behavior therapy for depression with co-morbid personality disorder: An extension of standard DBT with a special emphasis on the treatment of older adults.* Unpublished manuscript.

Lynch, T. R., Cheavens, J. S., Cukrowicz, K. C., Thorp, S., Bronner, L., & Beyer, J. (In press). Treatment of older adults with co-morbid personality disorder and depression: A dialectical behavior therapy approach. *International Journal of Geriatric Psychiatry.*

Maisto, S. A. (1985). Behavioral formulation of cases involving alcohol abuse. In I. D. Turkat (Ed.). *Behavioral case formulation* (pp. 43–86). New York: Plenum Press.

Nelson-Gray, R. O., & Farmer, R. F. (1999). Behavioral assessment of personality disorders. *Behaviour Research and Therapy, 37,* 347–368.

O'Leary, K. D., & Wilson, G. T. (1987). *Behavior therapy: Application and outcome* (2nd ed.). Englewood Cliffs, NJ: Prentice-Hall.

Page, S., & Neuringer, A. (1985). Variability is an operant. *Journal of Experimental Psychology: Animal Behavior Processes, 11,* 429–452.

Rosenthal, M. Z., Cheavens, J. S., Lejuez, C. W., & Lynch, T. R. (2005). Thought suppression mediates the relationship between negative affectivity and borderline personality disorder symptoms. *Behaviour Research and Therapy, 43,* 1173–1185.

Shimoff, E., Catania, A. C., & Matthews, B. A. (1981). Uninstructed human responding: Sensitivity of low rate performance to schedule contingencies. *Journal of the Experimental Analysis of Behavior, 36,* 207–220.

Skinner, B. F. (1945). The operational analysis of psychological terms. *Psychological Review, 52,* 270–277.

Skinner, B. F. (1953). *Science and human behavior.* New York: Macmillan.

Skinner, B. F. (1974). *About behaviorism.* New York: Knopf.

Turner, R. (2000). Naturalistic evaluation of dialectical behavior therapy-oriented treatment for borderline personality disorder. *Cognitive & Behavioral Practice, 7,* 413–419.

Verheul, R., van den Bosch, L. M., Louise, M. C., Koeter, M. W. J., de Ridder, M. A. J., Stinjen, T., et al. (2003). Dialectical behavior therapy for women with borderline personality disorder: 12-month, randomized clinical trial in The Netherlands. *British Journal of Psychiatry, 182,* 135–140.

Waltz, J., & Linehan, M. M. (1999). Functional analysis of borderline personality disorder behavioral criterion patterns: Links to treatment. In J. J. L. Derksen, C. Maffei, & H. Groen (Eds.), *Treatment of personality disorders* (pp. 183–206). New York: Plenum Publishers.

Wegner, D. M., & Gold, D. B. (1995). Fanning old flames—Emotional and cognitive effects of suppressing thoughts of a past relationship. *Journal and Personality and Social Psychology, 68,* 782–792.

Wilson, K. G., & Blackledge, J. T. (1999). Recent developments in the behavioral analysis of language: Making sense of clinical phenomena. In M. J. Dougher (Ed.), *Clinical behavior analysis* (pp. 27–46). Reno, NV: Context Press.

Wolpe, J., & Turkat, I. D. (1985). Behavioral formulation of clinical cases. In I. D. Turkat (Ed.), *Behavioral case formulation* (pp. 5–36). New York: Plenum Press.

Zettle, R. D., & Raines, J. C. (1989). Group cognitive and contextual therapies in treatment of depression. *Journal of Clinical Psychology, 45*, 438–445.

20

IMPULSE CONTROL
DISORDERS

MARK R. DIXON

AND

TAYLOR E. JOHNSON
Southern Illinois University

OVERVIEW OF IMPULSE CONTROL
DISORDERS

Impulse control disorders are characterized by an individual's inability to control impulses that may be harmful to oneself or to others. The lack or loss of control is often distinguished by increased feelings of tension or anxiety prior to engaging in a behavior and feelings of relief or gratification afterward. While many psychological disorders may coexist with feelings of lack of control, such as substance abuse or obsessive compulsive disorder, impulse control disorders are defined mainly by this lack of control. In other words, an individual with another psychological disorder may demonstrate a marked inability to control impulses, but this inability is part of a larger pattern of maladaptive behavior within some other diagnosis.

Impulse Control Disorders Not Elsewhere Classified are listed in the *Diagnostic and Statistical Manual of Mental Disorders* (*DSM*) under Axis I Clinical Disorders and include intermittent explosive disorder (IED), kleptomania, pathological gambling, pyromania, and trichotillomania. Initially, this wide range of disorders may appear too diverse for a cohesive functional analysis, but upon critical investigation of behavioral functions, many psychologists will see similarities across the disorders.

INTERMITTENT EXPLOSIVE DISORDER

IED is the inability to control recurring aggressive impulses. Individuals with this disorder demonstrate discrete violent outbursts significantly out of proportion with respect to the given situation. For example, an individual with IED may exhibit destructive aggression toward objects such as mutilating a family pet in response to small psychological stressors like being scolded by a parent for wearing muddy shoes in the house. Following such an outburst, the individual may express feelings of remorse or guilt, but there may also be feelings of tension reduction or release. Physiological sensations such as tingling or built-up pressure have been reported to coincide with such a sense of release. For example, McElroy, Soutullo, Beckman, Taylor, and Keck (1998) questioned 24 subjects with IED regarding such feelings of tension. Eighty-eight percent reported the experience of tension prior to their aggressive impulses, 75% reported experiences of relief following their outbursts, and 46% reported feelings of pleasure connected with the aggressive acts. However, with IED, unlike many of the other impulse control disorders, feelings of tension and relief are not required for a diagnosis.

According to the *DSM-IV*, IED is "characterized by discrete episodes of failure to resist aggressive impulses resulting in serious assaults or destruction of property"(American Psychiatric Association [APA], 1994, p. 609). The specific diagnostic criteria are "(a) Several discrete episodes of failure to resist aggressive impulses that result in serious assaultive acts or destruction of property. (b) The degree of aggressiveness expressed during the episodes is grossly out of proportion to any precipitating psychosocial stressors (APA, 1994, p. 612). Additionally, other mental disorders should be excluded.

IED was first identified as a disorder in 1980 with its inclusion in the *DSM-III* (APA, 1980). The *DSM-III* and *DSM-III-R* definitions included an additional criterion for diagnosis. Not only did the individual need to exhibit several aggressive outbursts far in excess of the given situation, but the individual should also exhibit "no signs of generalized impulsiveness or aggressiveness between the episodes" (APA, 1987, p. 322). This additional requirement made the true diagnosis of IED extremely infrequent, and the *DSM-IV* broadened the category by excluding this criterion. However, even under the *DSM-IV*, the diagnosis is relatively narrow. Because there cannot be another diagnosis which better explains the aggressive behavior, IED appears to be quite rare. For example, out of 842 cases of episodic violent behavior reviewed for the *DSM-IV*, only 17 met IED criteria (Bradford, Geller, Lesieur, Rosenthal, & Wise, 1994). IED is thought to be predominantly exhibited in males, but it has been associated with menstrual cycles in females (McElroy et al., 1998). There is very limited research on the life course of intermittent explosive disorder, but age of onset appears to be from late adolescence to the mid-30s (APA, 1994).

Despite *DSM* inclusion, some clinicians feel that IED should be considered a symptom of other psychological disorders (McElroy et al., 1998). Other clinicians feel that the *DSM-IV* criteria are too narrow and therefore preclude diagnosis of patients with impulsive aggression problems (Coccaro, Kavoussi, Berman, & Lish, 1998). In fact, these researchers have suggested a modified diagnostic category, Intermittent Explosive Disorder-Revised. This revision broadens the criteria found in the *DSM-IV*. For example, out of the 76 subjects studied who met the revised criteria, only 19 would have qualified for IED as defined by the *DSM-IV*. The following is the suggested diagnostic criteria for Intermittent Explosive Disorder-Revised (Coccaro et al., 1998):

A. Recurrent incidents of verbal or physical aggression towards other people, animals, or property.
B. The degree of aggressive behavior is out of proportion to the provocation.
C. The aggressive behavior is generally not premeditated (e.g., is impulsive) and is not committed in order to achieve some tangible objective (e.g., money, power etc.).
D. Aggressive outbursts occur twice a week, on average, for at least a period of 1 month.
E. Aggressive behavior is not better accounted for by mania, major depression, or psychosis. It is not solely due to the direct physiological effect of a substance (e.g., drug of abuse) or of a general medical condition (e.g., closed head trauma, Alzheimer's).
F. The aggressive behavior causes either marked distress (in the individual) or impairment in occupational or interpersonal functioning. (p. 369)

Individuals with IED have high rates of comorbidity with other psychiatric disorders, such as mood disorders, substance abuse, anxiety disorders, and eating disorders (McElroy, 1999; Olvera, 2002). The functional significance of this comorbidity is difficult to understand without a case-by-case analysis of which disorder preceded and perhaps caused the other disorder. Studies have suggested that individuals with this disorder may respond to serotonin reuptake inhibitors (SRIs) and mood stabilizers (McElroy, 1999; McElroy et al., 1998, Olvera, 2002). A case study of a young woman with autism, mental retardation, IED, and bipolar mood disorder showed clinically significant reductions in problem behavior while on risperidone (Yoo et al. 2003). However, response rates and time to task completion during a matching task were reduced as well. These undesirable reductions in response rate were lessened in comparison to a nonreinforcement condition by a continuous schedule of tangible reinforcement. This suggests that behavior producing strong reinforcement may be less influenced by pharmacological treatment. Thus, it appears that clinicians should consider functional consequences for behavioral challenges and not rush toward applying or continuing medications that may have been administered based on nonfunctional assessments.

KLEPTOMANIA

Kleptomania is characterized by the recurrent inability to resist the impulse to steal items that are not desired for financial gain or individual use. For example, persons with kleptomania may steal a candy bar and simply throw it away afterward, while having enough money in their pocket to pay for the item if they wanted to. According to the *DSM-IV*, kleptomania is "characterized by the recurrent failure to resist impulses to steal objects not needed for personal use or monetary value" (APA, 1994, p. 609). The specific diagnosis criteria are (a) recurrent failure to resist impulses to steal objects that are not needed for personal use of for their monetary value, (b) increasing sense of tension immediately before committing the theft, (c) the stealing is not committed to express anger or vengeance and is not in response to a delusion or a hallucination, and (d) the stealing is not better accounted for by conduct disorder, a manic episode or antisocial personality disorder (APA, 1994, p. 613).

Persons with kleptomania, like other impulse control disorders, will often describe a feeling of tension prior to taking the item, followed by a sense of release or gratification after the act of stealing has been completed. If an individual with kleptomania reports such feelings, automatic reinforcement may be considered as a possible maintaining reinforcer for this individual. Kleptomania is quite separate from shoplifting, in which individuals steal items because of the monetary value or personal gain, and in shoplifting the act is often preplanned. Individuals with kleptomania are aware the act is wrong and often feel guilt following stealing. They also frequently feel anxiety regarding possible apprehension. The *DSM-IV* describes three typical courses of the disorder: (1) "sporadic with brief episodes and long periods of remission;" (2) "episodic with protracted periods of stealing and periods of remission;" and (3) "chronic with some degree of fluctuation" (APA, 1994, p. 613). According to the *DSM-IV*, other disorders are commonly comorbid with kleptomania. These disorders include mood disorders, anxiety disorders, eating disorders, and personality disorders.

Kleptomania is thought to be quite rare and to have a much higher incidence in women (APA, 1994). According to the *DSM-IV*, kleptomania occurs in less than 5% of identified shoplifters. There is little experimental research on psychopharmacology and kleptomania. However, case studies have indicated that SSRIs may be beneficial. In fact, Dannon (2002) reported positive results in 19 out of 30 cases.

PATHOLOGICAL GAMBLING

Pathological gambling is the recurrent, irrepressible urge to gamble. Individuals with this impulse control disorder persist in gambling despite devastating cost to relationships and employment. The diagnosis is not based on the amount of money spent or lost; it is determined by the exhibited lack of control and the effect on quality of life. It is often characterized by preoccupation with betting and feelings of tension, followed by feelings of excitement and release. Although, as noted with IED, these feelings of tension and relief are not required for a diagnosis.

Pathological gambling was first included in the *DSM-III* in 1980. The *DSM-IV* defines pathological gambling as the inability to resist gambling to such a level that it interrupts major life activities. For example, a pathological gambler may destroy marital relationships by opening lines of credit in the spouse's name in order to raise more money for gambling. The specific diagnosis criteria are as follows:

A. Persistent and recurrent maladaptive gambling behavior as indicated by five (or more) of the following:
1) Is preoccupied with gambling (e.g., preoccupied with reliving past gambling experiences, handicapping or planning the next venture, or thinking of ways to get money with which to gamble)
2) Needs to gamble with increasing amounts of money in order to achieve the desired excitement
3) Has repeated unsuccessful efforts to control, cut back, or stop gambling
4) Is restless or irritable when attempting to cut down or stop gambling
5) Gambles as a way of escaping from problems or of relieving a dysphoric mood (e.g., feelings of helplessness, guilt, anxiety, depression)
6) After losing money gambling, often returns another day to get even ("chasing" one's losses)
7) Lies to family members, therapist, or others to conceal the extent of involvement with gambling
8) Has committed illegal acts such as forgery, fraud, theft, or embezzlement to finance gambling
9) Has jeopardized or lost a significant relationship, job or educational or career opportunity because of gambling
10) Relies on others to provide money to relieve a desperate financial situation caused by gambling.
B. The gambling behavior is not better accounted for by a Manic Episode. (APA, 1994, p. 618)

Pathological gamblers have a higher incidence of mood disorders; attention deficit hyperactivity disorder; substance abuse or dependence; and antisocial, narcissistic, and borderline personality disorders (APA, 1994).

The *DSM-IV* reports that pathological gambling may be prevalent in 1–3% of the adult population. Shaffer, Hall, and Vander Bilt (1999) conducted a meta-analysis of 119 studies. They estimated that 3.9% of the population in North America would be considered at-risk gamblers and that 1.6% would be considered pathological gamblers. Pathological gambling usually begins in early adolescence for men and much later in life for women. The National Research Council (1999) has conceptualized four levels of gambling, which are gaining in popularity as a means by which to examine gambling severity. Level 0 is no gambling. Level 1 describes individuals who gamble socially or recreationally. Level 2 applies to individuals who have developed some gambling-related problems, and Level 3 is used to classify those individuals with pathological gambling. These levels are usually assessed via self-report or interview.

There is very little experimental research on pharmacological approaches to treating pathological gamblers. SSRIs, mood stabilizers, and opioid antagonists have all been used with somewhat promising results, but large, placebo-controlled studies are needed. As of yet, no medication has been approved by the Food and Drug Administration for the treatment of pathological gamblers (Hollander, Kaplan, & Pallanti, 2004; Petry, 2005).

PYROMANIA

Pyromania is the uncontrollable impulse to set fires in which the gratification results from the fire itself and not for any other motive such as vandalism, hiding a crime, or financial gain. Individuals with pyromania are fascinated by fire and often actively participate in the aftermath of the fires they set by helping the firefighters and talking to victims, etc. As with other impulse disorders, the individual reports feelings of tension prior to setting the fire, followed by a sense of relief or pleasure after the act. However, unlike individuals with other impulse control disorders, individuals with pyromania may prepare for the fire setting in advance. For example, a person with pyromania may scope out an abandoned warehouse weeks before the actual act of setting a fire. Therefore, the clinician should seek to understand not only the functional relations of the fire setting event itself, which may be somewhat too late in terms of intervention, but also the preparatory behaviors such as scoping out the warehouse.

The *DSM-IV* defines pyromania as failure to resist urges to purposely start fires. Such individuals experience preoccupation with fire, the consequences of fire, and activities related to fires. Setting the fires may supply relief or gratification related to tension felt prior to the act. The specific diagnosis criteria are as follows:

A.   Deliberate and purposeful fire setting on more than one occasion.
B.   Tension or affective arousal before the act.
C.   Fascination with, interest in, curiosity about, or attraction to fire and its situational contexts (e.g., paraphernalia, uses, consequences).
D.   Pleasure, gratification, or relief when setting fires, or when witnessing or participating in their aftermath.

E. The fire setting is not done for monetary gain, as an expression of sociopoliti-cal ideology, to conceal criminal activity, to express anger or vengeance, to improve one's living circumstances, in response to a delusion or a hallucina-tion, or as a result of impaired judgment (e.g., in Dementia, Mental Retarda-tion, Substance Intoxication).

F. The fire setting is not better accounted for by Conduct Disorder, a Manic Episode, or Antisocial Personality Disorder. (APA, 1994, p. 615)

The *DSM-IV* simply notes that "pyromania is apparently rare" (APA, 1994, p. 614). This lack of prevalence data could be due in part to the little research differentiating pyromania from general fire setting. For example, in 2002, children playing with fire resulted in 13,900 fires, 210 deaths, and $339 million in damages (Hall, 2005), yet pyromania in childhood is con-sidered to be rare and is usually linked with conduct disorder, attention deficit hyperactivity disorder, or adjustment disorder rather than pyroma-nia (APA, 1994). Pyromania occurs more often in males, and fire-setting incidents may be episodic and infrequent (APA, 1994).

TRICHOTILLOMANIA

Trichotillomania is the inability to control the impulse to pluck out one's own hair. It is characterized by significant, visible hair loss; feelings of tension prior to the act; and feelings of release or gratification after. Also, the plucked hairs are often repeatedly manipulated. For example, the individual may twist the hair around fingers or rub the hair against the skin. The individual may also eat the hair (trichophagia), which may lead to vomiting, abdominal pain, and constipation. Hair is typically pulled from the head, but eyebrow and eyelash plucking is also common. Hair may also be pulled from the arms, legs, or pubic area. See Chapter 8, "Tic Disorders and Trichotillomania", by Miltenberger, Woods, and Himle for a detailed discussion of hair pulling.

FUNCTIONAL ASSESSMENT AND ANALYSIS

The understanding of behavioral function of an impulse control disor-der should be approached the same way a clinician would attempt to understand the function of any other psychological disorder. Efforts should be made to minimize hypothesizing, speculation, and construction of infer-ences about the internal causes of behavior. Instead, the clinician should strive for objective definitions of the targeted behavior and look to external events in the environment which may be responsible for the emission of the disorder. Although client-specific personalities, histories, and mental states may play a role in the manifestation of an impulse control disorder, the clinician should look to the external environment as much as possible when identifying function. Too much emphasis on internal mechanisms that may underlie the behavior will not be useful when attempting to produce behavior change as readily as if external agents are of central focus.

When the clinician seeks to identify external agents or environmental stimuli that may be functionally related to the impulse control disorder of interest, that clinician should examine to what degree the following functional relationships exist. First, does the behavior have an attention or social reinforcement function to it? For example, does the pathological gambler tend to gamble only with a group of friends on Friday night and never by him/herself any other time of the week? If this appears to be the case, perhaps the behavior is maintained by the social reinforcement of the card game itself, and not necessarily the amount of money that is won by the client. Second, does the behavior have an escape or removing oneself from demands placed upon function? If this appears correct, then the clinician may learn that the pathological gambler he/she is treating is only relapsing into gambling episodes upon being given longer hours at work, which result in anxiety and stress, and gambling reduces these feelings. Third, does the behavior have an automatic or sensory function which seems to be maintaining it? Here, a clinician may discover that the pathological gambler finds him/herself very much physiologically aroused when entering into a casino. The gambler speaks of his/her heart racing and enjoying the thrill of the game. Fourth, does the behavior occur to gain access to other items or places? In other words, does the behavior seem to have a tangible component maintaining it? In the case of the problem gambler, the clinician may discover that the behavior occurs in response to seeking out free buffet tickets, offers for drawings for a new car, and in response to wanting to go on vacation to somewhere warm, such as Las Vegas. While the noted functions are often considered independently, and perhaps only one may maintain an impulse control disorder, it is very possible that a combination of more than one type of function could sustain the disorder. In such a case, the clinician will need to identify the various functional relations and design treatment strategies to address each of these functions independently.

The rather infrequent emission of the targeted behaviors described under impulse control disorders results in very difficult means by which the clinician can attempt to understand functional relations. As opposed to other types of psychological disorders, impulse control disorders involving the reoccurrence of severe problem behaviors (i.e., starting a fire, or gambling an entire family estate) may be very destructive even with just one instance. Therefore, the clinician needs to quickly and accurately gain an understanding of the antecedents, contextual stimuli, and consequences which surround the problematic behavior or chains of behavior that lead up to the target behavior. Direct observation of the behavior may or may not be possible, and thus the clinician often needs to use less objective measures in searching for function.

INDIRECT ASSESSMENT

Questionnaires

A fair number of possible techniques and indirect assessments can be used with impulse control disorders. Many questionnaires are designed to help assist the clinician better understand the nature of behavior for their specific client. These questionnaires may take the form of "Yes" or "No" responses (e.g., Lesieur, & Blume, 1987), Likert-type ratings on scales from 1 to n (e.g., DeCaria et al., 1998), or open-ended questions regarding severity and potential function (e.g., Kolko & Kazdin, 1989b). For example, the South Oaks Gambling Screen (SOGS; Lesieur & Blume, 1987) is a questionnaire with 20 items based on the *DSM* criteria for pathological gambling. However, if the clinician's primary objective is to assess function of the behavior, most questionnaires will need to be supplemented with additional assessments, as in general, they do more to establish severity or history, and less to shed light on behavioral function.

In addition to the diagnosis-specific questionnaires, a few published items, although initially developed for persons with severe mental retardation, hold promise as potentially useful for persons with impulse control disorders. The *Motivation Assessment Scale* (MAS; Durand & Crimmins, 1988) is a 16-item behavior rating scale scored on a 7-point scale anchored by the options "Never" and "Always." The MAS identifies function in four categories of behavior: attention, tangible, escape, and sensory. The questions involve such inquiries as Does the behavior occur when no one else is around, or does the behavior occur when the person is asked to do something that he/she may not like? Each type of question is designed to address a potential behavior function. In the previous examples, the first type of question may, if answered as "Never," be able to rule out a potential sensory function that the behavior might serve, and the latter type of question would, if answered similarly, potentially rule out an escaping from demand function of the behavior. The MAS may be completed by the therapist, the clients themselves, or by other people who are well acquainted with the behavior in question. The *Questions About Behavior Function* (QABF; Matson & Volmer, 1995) is a questionnaire initially designed for people with mental retardation which evaluates the maintaining effects of attention, escape, tangible, physical discomfort, and nonsocial reinforcement. Questions and response options are similar to that of the MAS, yet the QABF items are scored on a 5-point scale. The QABF has successfully identified the function of self-injurious behavior, aggression, and stereotypic movements in 84% of subjects in a study of nearly 400 people with mental retardation (Matson, Bamburg, Cherry, & Paclawaskyj, 1999).

Questions from both of these scales are rather similar and seek to examine what possible primary function a behavior may serve. For example, "Does the behavior occur while you (or client) are alone?" A response to this question with the extreme of "Always" or "Often" may lead a clinician to deduce that perhaps the target behavior in question is maintained by something other than the social reinforcement it might have if emitted in the presence of peers or loved ones that repeatedly attended to the emission of the behavior.

Additional questionnaires that were designed for various other clinical populations or to assess various personality traits or states may also be useful with persons with impulse control disorders. For example, persons with kleptomania have been noted to score significantly higher on novelty-seeking scores and harm-avoidance scores than matched control subjects on the Tridimensional Personality Questionnaire (Grant & Kim, 2002), and pathological gamblers tend to show higher than normal scores for depression on the Diagnostic Inventory for Depression (Black & Moyer, 1998). Comorbidity within the diagnostic class of impulse control disorders (i.e., pathological gambling and kleptomania [Kim & Grant, 2001]) and across diagnosis classes (i.e., pathological gambling and bipolar disorder [Linden, Pope, & Jonas, 1986]) is common and should be assessed by the clinician as much as possible.

Self-Report

Self-report is one additional means by which function is assessed in impulse control disorders. Often, the secretive and infrequent nature of many of the behaviors associated with impulse control disorders make retrospective self-report a very practical, although limited, way to assess function. It is practical because often the individual is the only person present when the target behavior occurs, but it is limited because there are some concerns regarding the accuracy of self-report. One way to offset this potential limitation is to ensure that self-reports of decreasing levels of the target behavior are not differentially reinforced. Rather, the individual's accurate self-reporting should be reinforced. Furthermore, incorporating self-report-dependent measures into other assessment methods may hold added benefit for the client. As the client begins to learn to discriminate the antecedent conditions which may be responsible for the targeted behavior's emission, he/she may learn to act in ways to potentially minimize those conditions or find functional replacement behaviors.

Interviews

While the clinical interview technique is something that will vary across therapists and will be rooted in each clinician's theoretical orientation, a number of structured or guided interview methods exist and can be helpful in gaining a greater functional understanding of the behavior problem of interest. It is very common for two therapists to disagree on the occurrence or nonoccurrence of a psychological disorder, in part due to inadequateness of diagnostic criteria and in part to the lack of standardized questions offered to patients (Segal & Falk, 1998). The behavioral interview for gaining a functional assessment of the behavior in question will focus heavily on understanding the antecedents, establishing operations, setting events, and history of the problem, as well as the consequences which the behavior may serve. To do this, the clinician must guide questions to the client along the lines of objective definitions and environmental events. This may require reframing common questions which occur in traditional clinical interviews. For example, instead of asking the client *"How do you feel when starting fires?"* the clinician should seek to discover functional relationships via questioning such as *"What seems to happen immediately before you want to start a fire?"* or *"What happens to you immediately after a fire has been set?"* Questions may also assess the magnitude of the disorder such as *"How long has this behavior occurred?" "Has it increased in frequency?"* and so on. Again, the focus of these questions during the interview should be along the lines of objective measures by which to evaluate function.

The structure of the interview, regardless of specific impulse control disorder, should take a general format of a brief introduction, the assessment, and a brief closing. During the introduction, the clinician should recap the problem behavior which has brought the client into treatment. The clinician should summarize all of the information about the client that he/she currently has and ask the client to provide missing pieces to that initial information. Detailed questioning about functional relations should be left until initial rapport has been developed between parties, and reassurance should be given that the clinician is there to help. During the assessment period of the interview, the clinician should seek out means by which the client can describe the antecedents and consequences of his/her behavior of concern. At this time the therapist should explore severity, intensity, history, and triggers of the problem behavior, but also as carefully explore the conditions under which the problem behavior does not occur. In other words, what are some of the places, people, and events that result in the problem not occurring? Such information will be useful when attempting to discover methods and strategies that can be used for treatment. During the closing of the behavioral interview, the clinician should summarize the information gathered during the interview, including initial analyses of the functional relationships that exist for the problem behavior. The clinician should stress the importance of seeing the problem as not an internal flaw of personality of the person, but rather as an example of how situations in that person's life can arise which lend themselves to the problem behavior occurring. The clinician should console the patient and inform him/her that there are others with similar disorders, and while difficult, behavior change can occur, and this will be the focus of subsequent therapy sessions.

A number of guided behavioral interviews have been developed for the functional assessment of impulse control disorders. With respect to childhood pyromania, Kazdin and Kolko developed a series of interviews that may detect severity and function of the disorder. They include the *Fire Setting History Screen* (Kolko & Kazdin, 1988), whereby the clinician would interview both child and parent or caregiver; the *Fire Setting Risk Interview* (Kolko & Kazdin, 1989a), which is designed primarily for the caregiver; and the *Children's Fire Setting Interview* (Kolko & Kazdin, 1989b), which is targeted directly at the suspected child. Such screening interviews, while designed for children with the disorder, may be worthy of attempts at revising to serve the adult population.

The *Gambling Behavior Interview* (GBI; Stinchfield, Govoni & Frisch, 2005) and the Structured Clinical Interview for Pathological Gambling (SCI-PG; Grant, 2001) both have high inter-rater reliability and content validity with the *South Oaks Gambling Screen*. The latter also has discriminant validity with anxiety and depression. It is also quite common for clinicians to utilize more general psychological interviews, while not specific for impulse control disorders, that may hold insight to eventual diagnosis (e.g., Baylé, Caci, Millet, Richa, & Olié, 2003).

DIRECT ASSESSMENT

Direct Observation

Direct observation provides a number of advantages over the more indirect methods of questionnaire and interview-based assessments. First, direct observation allows the clinician to objectively assess the behavior, the events which preceded the emission of the behavior, and the consequences which follow from it. Questionnaires may not be specific enough to target specific incidents of problem behavior emission, and interviews may fail to uncover critical variables that were undetected by the client. Second, any of the indirect assessments require the client to remember what occurred, when it occurred, and hypothesize causes for why it occurred. Such issues are prone to distortion and perhaps even bias from the client. The direct observation method removes such artifacts in the assessment process and allows the clinician to independently assess the critical features of the behavioral episode.

Direct observation may take one of a variety of forms, from the clinician physically traveling with the client to observe the performance of the behavior in question, to videotaping of the client at high-risk times such that upon retrieval of the video, an observation may be made and behavioral functions could be deduced. Self-observation of the behavior may occur whereby the client him/herself records the behavior of interest and the variables which surround the episode of emission. Potential limitations to direct observation include the rate in which the behavior may occur and the often secretive nature of many impulse control disorders. Before ruling out direct observations, clinicians should attempt to identify the specific features of the disorder for their individual clients. It may be the case that fire setting is frequent enough that a parent may be able to capture multiple events on a video camera in a room or that hair pulling is exhibited in public, although at low rates. Thus, it may be difficult, but not impossible, to assess such behaviors in their natural environment via direct observation.

Direct observation has great utility for the assessment of impulse control disorders. Yet, unstructured direct observation alone cannot yield causality regarding the functional relations for a given behavior. The clinician can only make hypotheses about what antecedents and consequences appear to be sustaining the behavior. For example, upon witnessing a client steal three times over the course of a week, the clinician notes that his/her client appeared physically out of breath prior to the incident and, upon completing the act of stealing, appeared to be visually calmer. That clinician can deduce that perhaps anxiety reduction is the function of the stealing, yet, without an actual manipulation of environmental variables, which is the essence of an experimental analysis, only hypotheses can be made about behavioral function. Given the nature of many impulse control disorders, direct observation may be the closest approximation to identifying behavioral function that a clinician can attain. Such an approximation can be very useful when treating an impulse control disorder, but should be taken only as a tentative function until behavior change is clearly displayed by the client.

EXPERIMENTAL ANALYSES

There is a general lack of experimental methods for assessing the functions of impulse control disorders. The rarity of the disorders and the ethical and potential dangerous nature of some of the behaviors limit true experimentation. For example, episodes of intermittent explosive aggression may cause harm to the individual or to others. Thus, there are serious ethical concerns regarding the experimental manipulation of such behaviors and situations. There has been some research that addresses the ethical dilemma of experimentally encouraging potentially damaging behavior by having the subject don protective equipment during the functional analysis (Borrero, Vollmer, Wright, Lerman, & Kelley, 2002; Le & Smith, 2002). However, Borrero et al. found that such protective equipment suppressed levels of trichotillomania and head hitting so much that the functions of the behavior could not be determined. (The authors do suggest that such protective equipment could be used as a potential intervention.)

There has been some debate regarding the potential of analogue situations in determining the function of these behaviors. However, it is thought that they may be too different from the location where the behavior takes place to use them to experimentally demonstrate functions of the behavior. There also has been some research on using biofeedback to determine the level of arousal of those with impulse control disorders.

Pyromania

It would be difficult to do a functional analysis of pyromania because of the low frequency and danger of the target behavior. Jackson, Glass, and Hope (1987) attempted to overcome this limitation by reviewing the possible functions of fire setting. These authors identified three antecedent conditions ("psychosocial disadvantage, general dissatisfaction with life and the self, and ineffective social interaction" [p. 176]) as well as the potential classical conditioning pairing of the exciting results of the fire (fire trucks, sirens, etc.) and the stimulation resulting from these consequences with the actual fire. They also emphasized that the infrequent occurrence of the fires may prevent satiation to this stimulation. The authors suggested that simple educational strategies and focusing on the fire-setting behavior itself are not likely to be successful treatment strategies because they discount the influences of antecedents and consequences. They also asserted that punitive approaches may be detrimental because they encourage secretive fire-setting behavior.

Jackson, Glass, and Hope's model suggests that examining social influences and self-control in the fire setter would be beneficial as well as looking at the development of pyromania compared to normal fire play. These authors hoped that their speculative analyses would lead to more empirically driven research. Unfortunately, experimental studies of the function of impulse control disorders are still quite rare.

One such empirical study (Last, Griest, & Kazdin, 1985) looked at heart rate and skin potential as measured by a polygraph on a child who set fires. The study had the child look at slides depicting fire-related stimuli or non-fire-related stimuli. The subject also rated the stimuli on four dimensions (excitation, fearfulness, pleasantness, and ability to attend to slide). Following the presentation of each slide, the subject was asked to write down five words he associated with the slide. He also verbalized his thoughts during the presentation of each slide. The researchers found that the fire stimuli elicited greater skin potential responses that exceeded baseline and nonfire stimuli responses. However, heart rate was lower during the fire stimuli. The word-listing procedure indicated that the fire-related stimuli elicited negative statements such as "it's awful" or "I hate it." The authors suggested that these responses may be due to the subject's history of punishment with fire-setting behavior. Unfortunately, the researchers were not able to implement a treatment based on this assessment because the subject left the facility. However, they did suggest that an exposure based treatment may have been effective in reducing the physical symptoms of arousal in the presence of the fire-related stimuli.

This type of research may be potentially important because it is a way to objectively measure the *DSM-IV* criteria of feelings of building tension, followed by feelings of relief in kleptomania, pyromania, and trichotillomania.

Other Impulse Control Disorders

Perhaps the most commonly studied impulse control disorder is trichotillomania (see Chapter 8 by Miltenberger, Woods, and Himle for a detailed discussion of experimental functional analysis in trichotillomania). Unfortunately, functional analyses of other impulse control disorders are quite rare in the literature. In one example, Keeney, Fisher, Adelinis, and Wilder (2000) conducted a functional analysis of a woman with intermittent explosive disorder and mental retardation. They found that her aggression and self-injurious behavior were maintained by both negative reinforcement (escape from demands) and positive reinforcement via attention. Such experimental analyses of impulse control disorders have clear implications for treatment (as discussed in the following sections), and it is unfortunate that theses examinations are not more prevalent in the literature.

FUNCTION-BASED INTERVENTIONS

The key means by which the clinician should approach intervening on an impulse control disorder is to identify function and then replace the targeted behavior of concern with a more adaptive positive behavior that serves the same function. For example, if it was identified that a child pyromaniac engaged in the behavior of fire setting to gain social attention from his/her parents, perhaps even in the form of negative social attention of being scolded and asked repeatedly "Why are you doing this?" then the clinician should attempt to teach the parents ways in which they can provide the same form of attention to the child for non-fire-setting behaviors. Perhaps with a large number of children, this client gains significant amounts of attention only when setting fires in the home. The challenge for the parents and clinicians is to identify times and behaviors for this child that will yield significant amounts of attention while fire setting yields only the minimal safety precautions. Such an intervention, termed differential reinforcement for alternative or other behaviors, has increased in usage in the behavioral literature. Replacing the reinforcing consequences for the behavior of concern with identical consequences for an alterative behavior is not limited to treatment of impulse control disorders, yet this being said, such an intervention works appropriately within this syndromal classification.

If the clinician considers that functional intervention that rests on the results of the functional analysis is the key to treatment of impulse control disorders, the wide varying topographical nature of this disorder classification becomes a moot point. Instead, the clinician should seek to target function and treat the behavior based on function, and this should be the underpinning clinical philosophy. What will follow is that the kleptomaniac, pathological gambler, and intermittent explosive person all receive similar treatment on one level—that treatment is going to be based on the nature of the maintaining variables which sustain their disorder.

For example, Keeney et al. (2000) used the information obtained in their functional analysis of a woman with intermittent explosive disorder to develop a response cost contingency. The response cost consisted of the women receiving continuous access to highly preferred stimuli (either attention or music) unless she engaged in aggression or self-injurious behavior. Under these procedures, destructive behavior still produced escape, one of the conditions identified by the functional analysis as increasing the woman's destructive behavior, but it also resulted in the loss of music or attention. This response cost-effectively reduced her escape-maintained behavior. A clinician could use a similar treatment for behavior maintained by both negative reinforcement (escape) and positive reinforcement (attention or tangible) whether that behavior was aggression in the case of an individual with intermittent explosive disorder or fire setting in the case of a pyromaniac.

While functional interventions are approached similarly throughout this diagnosis class of impulse control disorders, the clinician should understand that the treatment from one client to the next within a diagnosis will differ based on function. Treating one pathological gambler with anxiety reduction and guided imagery will not necessarily work for another pathological gambler who may be gambling for increased socialization with peers. Treating all persons within a diagnostic class identically not based on function will yield poor clinical significance and weak treatment utility. A clear example of this is the homogeneous treatment approach of Gamblers Anonymous for pathological gamblers, which yields a mere 8% success rate of nongambling after 1 year (Stewart & Brown, 1988).

Petry (2005) presented a very promising functional intervention approach for the treatment of pathological gambling. The approach involves one-on-one therapy between client and clinician, whereby over the course of an eight-week period the client learns to identify the functional relations which may be maintaining his/her problem behavior. Specifically, the therapist prompts the client to identify the antecedents for gambling and when those antecedents occur most often. Interestingly, the therapist also instructs the client to identify antecedents for not gambling, like getting busy at work or having a relative visiting from out of town. Teaching the client that there are in fact triggers for both gambling and nongambling behavior is noted as key to learning that gambling, even for the most addicted person, does not happen all the time. Similar techniques for understanding the consequences of gambling are used, whereby the clinician probes the client to describe the immediate consequences of gambling. Sometimes these consequences may be good, such as the person won some money, or bad, such as when they person lost and could not pay his/her bills. Yet, the clinician also probes for the long-term consequences of the gambling event, and often, if not always, those consequences are negative. Once the gambler discovers the functional relations of why gambling behaviors occur along with what happens down the road after gambling, treatment improvements are often reported (Petry, 2005).

CASE STUDY

The following case study illustrates the need for effective assessment and treatment of an impulse control disorder of a pathological gambler. Excessive gambling can appear on the surface as a problem with money management or a behavioral addiction. In fact, as this case example demonstrates, it can be a problem with self-management of anxiety. Given the heterogeneity of functions across these diagnostic labels, many other interventions may be possible for impulse control disorders.

Pat was a 45-year-old male with a 20-year history of excessive gambling. Pat realized that he had been gambling more than he should occasionally for the past 20 years, but only recently sought out professional treatment. Upon entry into a treatment facility, Pat underwent an intake which consisted of a structured interview and standardized questionnaires. Specifically, Pat was asked to complete the *South Oaks Gambling Screen* (Lesieur & Blume, 1987), the *Yale-Brown Compulsive Scale Modified for Pathological Gambling* (DeCaria et al., 1998), and the MAS (Durand & Crimmins, 1992), which was modified to assess gambling functions (Dixon, 2005). These administrations took approximately 1 hr to complete and were followed with the *Structured Clinical Interview for Pathological Gambling* (SCI-PG; Grant, 2001), which all taken together proceeded to provide a greater understanding of behavior function. The following are excerpts from the initial meeting with the client:

Clinician: Thank you for coming in today Pat and beginning the first step towards the treatment of your excessive gambling. As I recall from our phone conversation last week, you are interested in therapy for the treatment of your problem gambling. Before we begin the treatment process let me ask you to walk through a few questionnaires with me regarding the severity [i.e., as measured by the SOGS], the degree to which you have control of your gambling [i.e., as measured by the Yale-Brown], and possibly the reasons or triggers that make you want to gamble [i.e., as measured by the MAS-revised].

After completion of the questionnaires and the structured interview, this initial session concluded with the therapist stating:

Clinician: I will further explore your responses to these questionnaires and interview by the time we get together and meet next week for our first formal therapy session. The reason why we had you complete these instruments Pat was to get a better understanding of you and your gambling, because the reasons why people gamble are all different. We need to get a clear understanding of why you gamble, and customize the treatment for you individually. Over the next week, I would like you to complete the following form [see Figure 20.1] which will allow you to rate on a scale of 1 to 10 your degree of control you had over your gambling for that day, as well as the following form which allows you to place a checkmark in the box each day you did or did not gamble [see Figure 20.2]. Don't worry about days that you gambled or felt like your gambling was out of control.

DATE: _____

Today it felt like I had

1   2   3   4   5   6   7   8   9   10
NONE   SOME   TOTAL
much control over my gambling.
DATE: _____

Today it felt like I had

1   2   3   4   5   6   7   8   9   10
NONE   SOME   TOTAL
much control over my gambling.
DATE: _____

Today it felt like I had

1   2   3   4   5   6   7   8   9   10
NONE   SOME   TOTAL
much control over my gambling.
DATE: _____

Today it felt like I had

1   2   3   4   5   6   7   8   9   10
NONE   SOME   TOTAL
much control over my gambling.

FIGURE 20.1 Self-recording form for patient with the impulse control disorder of pathological gambling.

DAY	Week 1	Week 2	Week 3	Week 4	Week 5	Week 6	Week 7
Sun	X	X	X				
Mon	X	X	X				
Tues	!!!!	!!!!	X				
Wed	X	X	X				
Thurs	!!!!	!!!!	!!!!				
Fri	X	X	X				
Sat	X	X	X				

Instructions: For each day, place an X in the box for days you did not gamble, and fill the box in with!!!! for days in which you did gamble. Remember, it does not matter how good or bad your sheet looks. Just be honest with yourself.

FIGURE 20.2 Homework form for recording days gambled and not gambled by a client in treatment for pathological gambling.

It does not matter. What matters instead is that you record these things honestly so that we can get a good idea of what is going on with you each day.

Upon the completion of the 2 hr of initial intake, Pat was scheduled for eight weeks of one-on-one intensive behavioral therapy (Petry, 2005) in which he attempted to gain a greater understanding of the functions to which his excessive gambling behavior served and possible means by which his problems with gambling could be resolved. During the initial weeks in therapy, Pat learned a great deal about some of the environmental events that may have been functionally related to his problem gambling behavior. His gambling behavior appeared to be negatively reinforced by escape from stress and the demands at his job or home life. The clinician presented Pat with homework, which gave him the opportunity to self-record his own gambling behavior and the antecedents and the consequences of that behavior (see Figure 20.3). For example, upon his return to therapy in a subsequent week, Pat discussed his homework with the therapist and was surprised to see that most of his gambling occurred on Tuesdays and Thursdays, which were days at the office requiring stressful meetings. Pat also noted that he felt agitated leaving work those days, but that agitation

Date and Time	Triggers for the urge to gamble	What did you do?	How did you feel and what happened after you did this?
Monday 11/14 9pm	Argument with spouse	Went and worked in the barn	More comfortable and at ease
Tuesday 11/15	Meeting at work went bad	Stopped at the casino	Good at first, then upset about losing money
Thursday 11/17	Board room discussion at work	Bought lotto tickets	Felt better about the day. Did not care about work meeting
Tuesday 11/22	Worked 2 hours late	Played blackjack at casino until morning	Forgot about the time and won 20 dollars

FIGURE 20.3   Think hard every time you have the urge to gamble. Record the following information each time you had the urge to gamble this week. Whenever possible, complete these entries IMMEDIATELY after the urge sets in.

quickly subsided upon entry into the local casino or when buying lotto tickets. As Pat noted:

Pat: I have been keeping a pretty good record of when I gamble and what some of the feelings I have following my gambling are. First, it seems like I feel better, and my stress level goes down. However, once I realize that I just lost my money, and that my wife is going to get mad again, my stress level goes back up. Sometimes this vicious circle of stress–gambling–no stress–stress just keeps repeating itself. I don't know what to do, and I don't think I can stop.

Clinician: Well Pat, it appears that you seem to be on the right track with understanding the reasons, or the functions, for what your gambling might be caused by. Let's take a look at your homework. What we can see [from Figures 20.2 and 20.3] is that it appears you are gambling most often on Tuesdays and Thursdays. Is that correct?

Pat:	Yes it is. I guess I really never thought of that before.
Clinician:	Now, if we look at this sheet [see Figure 20.3] we can see that many times that you gambled, you noted that you had something troubling at work, a meeting or something at the end of the day. Is this true?
Pat:	Yes indeed it is. I can't stand my boss John. He drives me crazy with all the demands he puts on me. I feel like I am going to explode. Sometimes I feel he is just like my wife. Always nagging at me. Pat do this, Pat do that. I go crazy.
Clinician:	That is interesting Pat. Many people don't like their boss. In fact mine is kind of demanding too. It gets to me as well. However, what I noticed on this homework sheet here Pat is that you had an urge to gamble the other day when you had a fight with your wife, but in fact you did not gamble did you?
Pat:	Boy that was a bad fight. I really wanted to gamble. I went out to the barn to get my keys to the truck, and noticed there were some baseboards I wanted to stain to complete the bathroom remodel. So, I guess I just got busy staining these boards, and forgot to go to the casino. When I was done staining, I really did not feel like going anymore.
Clinician:	Well Pat, I think we may be on to something here. How did you feel after you were in the barn working on those baseboards?
Pat:	I felt relaxed, and in my own zone. The time went by fast, and I was rocking out to some classic rock songs on the radio I had not heard in a while. See my wife does not like that kind of music. She is more into country western.
Clinician:	OK, great. I have an idea. Do you have other projects at home that need to be done, that you seem not to have time to do?
Pat:	Oh yes. There are many, I just don't seem to finish.
Clinician:	Great, for the next week I want you to think about what these projects are, and each time before you leave for work on Tuesday and Thursday, I want you to write down one of these projects, and carry that piece of paper with you as you walk out the door. I want you to look at that paper, as you drive home, and I want you to put that paper in the barn as soon as you get home. If you think you are going to gamble or want to gamble, fine. But get home first. Can we make this the deal? No matter what, get home and put that paper in the barn. OK?
Pat:	Sounds fine. I don't know what this will do for me. But OK.

As the week went by, Pat found himself spending more time than he had in the past in the barn working on home projects. He was surprised, and reported more control over his gambling (as noted using Figure 20.1). Although he gambled on Thursday, it was after he completed fixing an old toilet and, once at the casino, lost interest quickly and returned home. While meeting with the therapist the following week, Pat had some revelations he shared.

Pat: This week I found myself much more in control of my gambling. I really did not find things so crazy as they usually are.

Clinician: What do you mean Pat?

Pat: Well, I did what you said about putting that piece of paper in the barn on my way home, and you know what, it seemed to distract me from the crap at the office. I mean I surely was pissed off leaving work, but at the same time I really seemed like there was part of me that said screw it I am going home and fixing that back door. I thought, screw these guys, I am going home. When I got there, I started jamming some Lynyrd Skynyrd and work just started to seem like miles away.

Clinician: Let me ask you Pat, did it feel as far away as it does while you gamble? I mean when you started playing the music and working on that door, did you feel better, like you do when you are gambling?

Pat: As a matter of fact, I did. Wow that is crazy! Am I going to get addicted to working in the barn? My wife will surely like that!

Clinician: No, I don't think so. But what I want you to realize Pat is that you seem to be somewhat successfully replacing your gambling behavior with another behavior that has the same outcome for you. You are finding a break from the demands of the day, and those stressful meetings at work, by doing some home projects. In the past work stressed you out, and you gambled. And the stress went away. Now that same stress of work is there, but you found another means to get away from it all. And in fact, it didn't cost you anything like when you gambled.

Pat: Interesting, and I even have extra money now, and perhaps I could use some of it to buy a new table saw.

Clinician: That is great. Well, let's continue to keep track of your progress as the weeks continue.

What followed for Pat was an understanding that his gambling was maintained by the function of removal of anxiety or stress from his workday. Together with his clinician, Pat identified an alternative behavior which served the same function for him as gambling did. See Figure 20.4 for a cumulative display of Pat's gambling during treatment. The clinician was

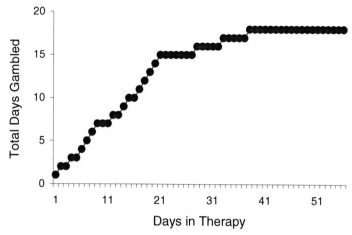

FIGURE 20.4 Data depicting total days gambled during therapy. (Note, no baseline data are displayed here.)

not able to remove the reinforcing functions of gambling. Rather, the clinician helped Pat discriminate and describe the functional relationship between environmental events and his gambling behavior in words and writing (cf. Skinner, 1953). The clinician then instructed Pat to engage in a form of self-management. At times when gambling was more likely, Pat learned to rearrange his environment to make alternate, functionally equivalent forms of behavior more likely than they had been previously. He did this by presenting himself with an antecedent—a written prompt to engage in a hobby—so that he was more likely to engage in these alternate activities that also reduced tension. One might consider gambling and leisure activities are two concurrent operants. Prior to intervention, gambling was a very low effort response with a high probability of negative reinforcement. The probability of leisure was very low. Intervention increased the probability of the second response through prompting, and it competed with problematic gambling (cf. Herrnstein, 1970).

This discovery process did not happen overnight. Pat may need to learn other forms of self-management as he learns to discriminate and describe other problematic behaviors and their antecedents. However, as Pat learns to discriminate stressful behaviors, he learns to redesign his environment again to make alternative healthy behaviors more likely, rather than engage in the historical default of taking a trip to the casino to gamble all his problems away. That gamble often paid off in the short term but never paid off in the long term. With effective functional treatment, Pat hit the jackpot—the road to recovery from an impulse control disorder through Skinnerian self-management.

REFERENCES

American Psychiatric Association. (1980). *Diagnostic and statistical manual of mental disorders* (3rd ed.). Washington, D.C.: Author.

American Psychiatric Association. (1987). *Diagnostic and statistical manual of mental disorders* (3rd ed., *rev.*). Washington, D.C.: Author.

American Psychiatric Association. (1994). *Diagnostic and statistical manual of mental disorders* (4th ed.). Washington, D.C.: Author.

Baylé, F. J., Caci, H., Millet, B., Richa, S., & Olié, J. (2003). Psychopathology and comorbidity of psychiatric disorders in patients with kleptomania. *American Journal of Psychiatry, 160*, 1509–1513.

Black, D. W., & Moyer, T. (1998). Clinical features and psychiatric comorbidity of subjects with pathological gambling behavior. *Psychiatric Services, 49*, 1434–1439.

Borrero, J. C., Vollmer, T. R., Wright, C. S., Lerman, D. C., & Kelley, M. E. (2002). Further evaluation of the role of protective equipment in the functional analysis of self-injurious behavior. *Journal of Applied Behavior Analysis, 35*, 69–72.

Bradford, J., Geller, J., Lesieur, H., Rosenthal, R., & Wise, M. (1994). Impulse control disorders. In T. A. Widiger (Ed.), *DSM-IV sourcebook* (pp. 1007–1032). Washington, D.C.: American Psychiatric Association Press.

Coccaro, E. F., Kavoussi, R. J., Berman, M. E., & Lish, J. D. (1998). Intermittent Explosive Disorder-Revised: Development, reliability, and validity of research criteria. *Comprehensive Psychiatry, 39*, 368–376.

Dannon, P. N. (2002). Kleptomania: An impulse disorder? *International Journal of Psychiatry in Clinical Practice, 6*, 3–7.

DeCaria, C. M., Hollander, E., Begax, T., Schmeidler, J., Wong, C. M., Cartwright, C., et al. (1998). *Reliability and validity of pathological gambling modification of the Yale-Brown Obsessive-Compulsive Scale (PG-YBOCS): Preliminary findings*. Presented at the 12th National Conference on Problem Gambling, Las Vegas, NV.

Dixon, M. R. (2005). *The Motivation and Assessment Scale (MAS)—revised for gambling.* Unpublished manuscript.

Durand, V. M., & Crimmins, D. B. (1988). Identifying the variables maintaining self-injurious behavior. *Journal of Autism and Developmental Disorders, 18*, 99–117.

Grant, J. (2001). Demographic and clinical feature of 131 pathological gamblers. *The Journal of Clinical Psychiatry, 62*, 957–969.

Grant, J. E., & Kim, S. W. (2002) Temperament and early environmental influences in kleptomania. *Comprehensive Psychiatry, 43*, 223–228.

Hall, J. R. (2005). Children playing with fire. *National Fire Protection Association Fire Analysis and Research*. Retrieved on October 12, 2005, from http://www.nfpa.org/assets/files/PDF/ChildrenPlayingExSummary.pdf

Herrnstein, R. J. (1970.) On the law of effect. *Journal of the Experimental Analysis of Behavior, 13*, 234–266.

Hollander, E., Kaplan, A., & Pallanti, S. (2004). Pharmacological treatments. In J. E. Grant & M. N. Potenza (Eds.), *Pathological gambling: A clinical guide to treatment* (pp. 189–206). Washington, D.C.: American Psychiatric Publishing.

Jackson, H. F., Glass, C., & Hope, S. (1987). A functional analysis of recidivistic arson. *British Journal of Clinical Psychology, 26*, 175–185.

Keeny, K. M., Fisher, W. W, Adelinis, J. D., & Wilder, D. A. (2000). The effects of response cost in the treatment of aberrant behavior maintained by negative reinforcement. *Journal of Applied Behavior Analysis, 33,* 255–258.

Kim, S. W., & Grant, J. E. (2001). Personality dimensions in pathological gambling disorder and obsessive-compulsive disorder. *Psychiatry Research, 104,* 205–212.

Kolko, D. J., & Kazdin, A. E. (1988). Prevalence of firesetting and related behaviors among child psychiatric patients. *Journal of Consulting and Clinical Psychology, 56,* 628–630.

Kolko, D. J., & Kazdin, A. E. (1989a). Assessment of dimensions of childhood firesetting among patients and nonpatients: The firesetting risk interview. *Journal of Abnormal Child Psychology, 17,* 157–176.

Kolko, D. J., & Kazdin, A. E. (1989b). The Children's Firesetting Interview with psychiatrically referred and nonreferred children. *Journal of Abnormal Child Psychology, 17,* 609–624.

Last, C. G., Griest, D., & Kazdin, A. E. (1985). Physiological and cognitive assessment of a fire-setting child. *Behavior Modification, 9,* 94–102.

Le, D. D., & Smith, R. G. (2002). Functional analysis of self-injury with and without protective equipment. *Journal of Developmental and Physical Disabilities, 14,* 277–290.

Linden, R. D., Pope, H. G., & Jonas, J. M. (1986). Pathological gambling and major affective disorder: Preliminary findings. *Journal of Clinical Psychiatry, 47,* 201–203.

Lesieur, H. R., & Blume, S. B. (1987). The South Oaks Gambling Screen (SOGS): A new instrument for the identification of pathological gamblers. *American Journal of Psychiatry, 144,* 1184–1188.

Matson, J. L., Bamburg, J. W., Cherry, K. E. & Paclawskyj, T. R. (1999). A validity study on the Questions About Behavioral Function (QABF) scale: Predicting treatment success for self-injury, aggression, and stereotypies. *Research in Developmental Disabilities, 20,* 142–160.

Matson, J. L., & Vollmer, T. R. (1995). *User's guide: Questions About Behavioral Function (QABF).* Baton Rouge, LA: Scientific Publishers.

McElroy, S. L. (1999). Recognition and treatment of DSM-IV intermittent explosive disorder. *Journal of Clinical Psychiatry, 16,* 12–16.

McElroy, S. L., Soutullo, C. A., Beckman, D. A., Taylor, P., & Keck, P. E. (1998). DSM-IV intermittent explosive disorder: A report of 27 cases. *Journal of Clinical Psychiatry, 59,* 203–210.

National Research Council. (1999). *Pathological gambling: A critical review.* Washington, D.C.: National Academy Press.

Olvera, R. L. (2002). Intermittent explosive disorder: Epidemiology, diagnosis and management. *CNS Drugs, 16,* 517–526.

Petry, N. M. (2005). *Pathological gambling: Etiology, comorbidity, and treatment.* Washington, D.C.: American Psychological Association.

Segal, D. L., & Falk, S. B. (1998) Structured interviews and rating scales. In A. S. Bellack & M. Hersen (Eds.), *Behavioral assessment: A practical handbook* (4th ed., pp. 158–178). Needham Heights, MA: Allyn & Bacon.

Shaffer, H. J., Hall, M. N., & Vander Bilt, J. (1999). Estimating the prevalence of disordered gambling behavior in the Unites States and Canada: A research synthesis. *American Journal of Public Health, 89,* 1369–1376.

Skinner, B. F. (1953). *Science and human behavior.* New York: Macmillan.

Stewart, R. M., & Brown, R. I. F. (1988). An outcome study of Gamblers Anonymous. *British Journal of Psychiatry, 152,* 284–288.

Stinchfield, R., Govoni, R., & Firsch, G. R. (2005). DSM-IV diagnostic criteria for pathological gambling: Reliability, validity, and classification accuracy. *American Journal on Addictions, 14,* 73–82.

Yoo, H. J., Williams, D. C., Napolitano, D. A., Peyton, R. T., Baer, D. M., & Schroeder, S. R. (2003). Rate-decreasing effects of the atypical neuroleptic risperidone attenuated by conditions of reinforcement in a woman with mental retardation. *Journal of Applied Behavior Analysis, 36,* 245–248.

21

FUNCTIONAL ASSESSMENT WITH CLINICAL POPULATIONS: CURRENT STATUS AND FUTURE DIRECTIONS

CYNTHIA M. ANDERSON
University of Oregon

In applied psychology, treatment utility is one of the most—if not the most—important criteria by which assessments should be judged. A pre-intervention assessment should result in improved outcomes relative to outcomes likely achieved without the results of the assessment for the target individual or individuals. Unfortunately, the assessment-intervention link has not been well established in many areas of psychology. Even in "behavioral" psychology, where the emphasis is primarily on observable behavior and the relation between behavior and environmental events, the primary goal of assessment is diagnostic classification [e.g., via the *Diagnostic and Statistical Manual of Mental Disorders* (*DSM-IV-TR,* American Psychiatric Association, 2000)]. The one exception to this emphasis on structural diagnosis is to be found in the field of behavior analysis. Behavior analysts have generally eschewed the emphasis on diagnosis based on topography of responses and instead insisted that behavior is best understood by focusing not on what the response looks like, but instead on the context in which the response occurs. Until quite recently, however, this approach was applied primarily to individuals with disabilities exhibiting challenging behavior. Thus, a contribution of this book is that it fills a needed gap in the literature by demonstrating how a focus on environment-behavior relations might lead to more effective interventions for a wide variety of human problems.

In this chapter, the benefits and limitations of nomothetic classification are reviewed, and advantages of a functional approach to assessment are described. Next, overarching themes from chapters in the current book are reviewed, and directions for future research on functional assessment and function-based support are suggested.

SYSTEMS OF CLASSIFICATION

The *DSM-IV-TR* is the prominent classification system used by behavioral health providers, such as psychologists, psychiatrists, and social workers, in the United States. The *DSM* is a nosological system based on the medical model and provides a structural description of the behavior(s) of interest. Because the *DSM-IV-TR* is based on a medical model of pathology, there is a tendency to view a given psychological disorder as an illness or a disease resulting from internal, intraorganismic variables (Hayes & Follette, 1993; Nelson & Hayes, 1986). Of course, this focus on internal, often hypothetical "causes" is antithetical to a behavior analytic approach, which emphasizes the contextual variables that are related to the behavior of interest and are important in the prediction and control of behavior (Skinner, 1953). The medical model also has political implications. Morey, Skinner, and Blashfield (1986) noted that, among the behavioral health professions, psychiatry has maintained a dominance, as is evidenced both economically (psychiatrists tend to earn larger salaries than do other professionals) and legally (psychiatrists are typically the only behavioral health practitioners who have admission and commitment privileges at hospitals). The dominance of psychiatrists, who are trained within the medical field, is justified only if behavior pathology is viewed as an illness that is best treated within the medical model.

BENEFITS AND LIMITATIONS OF DSM CLASSIFICATION

The *DSM* certainly has some utility for behavioral health providers. It provides a structure for communication, information retrieval, and description—three purposes of a nosological system. From a behavior analytic perspective, the *DSM* is useful for these purposes because it provides descriptive information about behaviors that tend to covary. Although many diagnostic criteria traditionally were constructed based on clinical intuition and unsystematic observation, there is an increasing use of literature reviews, statistical analyses, and field trials to form diagnostic categories (Scotti, Ujcich, Weigle, & Holland, 1996). Unfortunately, the degree to which the *DSM* nosology is useful for achieving arguably the most important goals of a classification system—identification causal and manipulable environmental variables—is equivocal (Gresham & Kern, 2004; Hayes & Follette, 1993; Korchin & Schuldberg, 1981). Further, diagnosis based on *DSM* criteria does not affect treatment outcome (Gresham & Gansle, 1992; Hayes, Nelson, & Jarrett, 1987). From a behavioral analytic perspective, the *DSM* does not guide clinicians to efficacious treatment choices precisely because of its emphasis on classification via topography and reliance on the medical model.

In contrast to a structural model of classification, behavior analysis endorses a *functional* classification scheme; the focus is on *why* behavior is occurring, on what environmental variables evoke and maintain the behavior of interest. As a result, a preintervention functional assessment is linked directly to an intervention focusing on manipulating those environmental variables hypothesized based on the functional assessment to be related to the problem of interest. Functional assessment has been used successfully with individuals with disabilities exhibiting challenging behavior since the 1960s, and recent reviews document the relative superiority of function-based versus non-function-based interventions for decreasing challenging behavior (Carr et al., 1999; Dunlap & Childs, 1996; Lane, Umbreit, & Beebe-Frankenberger, 1999; Lee & Miltenberger, 1997; Lennox, Miltenberger, Spengler, & Erfanian, 1988; Scotti et al., 1996). Further, the use of functional assessment with school-aged children has increased dramatically since two recent reauthorizations of the *Individuals with Disabilities Education Act* (1997; 2004), which included language mandating the use of functional assessment in certain situations. Unfortunately, the development and use of methods of functional assessment appropriate for clinical populations has received only scant attention until quite recently. As is illustrated in chapters throughout this book, functional assessment has a large role to play in (a) offering an explanation of behavior pathology that emphasizes the role of the environment in its maintenance and exacerbation and (b) developing effective interventions for a wide variety of clinical problems. The book highlights as well many of the questions that remain to be answered regarding functional assessment and functionally derived interventions.

A FUNCTIONAL APPROACH TO BEHAVIOR PATHOLOGY

As is illustrated throughout the book, a functional approach to behavior disorders is a conceptually parsimonious approach. Regardless of the topography of the behaviors of concern or the variables that resulted in presenting concerns, the focus is the same—identifying *environmental variables,* of which behavior is a function. The utility of evaluating effects of environmental variables on behavior is fairly obvious when problems seem to be clearly related to environmental events—marital discord, for example. When a disorder is due, or is assumed to be due, primarily to nonenvironmental factors, such as genetics, physical structure, or neurochemistry, a functional approach may be eschewed as overly simplistic or as ignoring the most relevant factors—those within the person. As is illustrated time and again by the authors in this text, however, a functional approach often provides a more parsimonious approach and one that leads logically to interventions.

From a behavior analytic perspective, biological and physiological variables often impact current functioning, and understanding of their role is critical for successful intervention. An example is pediatric feeding disorders; Piazza and Addison (Chapter 7, "Function-Based Assessment and Treatment of Pediatric Feeding Disorders") clearly articulate the multiple roles that physical abnormalities as well as medical variables might play in the onset and maintenance of food refusal or selectivity. Biological and physiological variables also may set the occasion for current problems but may no longer interact directly with environmental events. Traumatic brain injury is an example of such an effect. As illustrated by Dixon and Bihler (Chapter 12, "Brain Injury"), an individual who has experienced a brain injury may now respond to environmental stimuli in very different ways than was typical prior to the injury. In cases such as this, the behavior analytic approach acknowledges the role of such variables but does not focus on those variables—which in most cases are unlikely to be alterable. The focus instead is on evaluating current functioning—problem behaviors that are occurring and the presence of any skill deficits—and then identifying environmental factors responsible for the occurrence or nonoccurrence of those target responses. This is illustrated well by Fisher, Drosel, Yury, and Cherup (Chapter 11, "A Contextual Model of Restraint-Free Care for Persons with Dementia") in their discussion of the role of functional assessment with individuals diagnosed with dementia. In most cases there is no intervention that will reduce the dementia, so the focus instead is on decreasing problematic behavior, such as aggression, and on increasing the individual's skill repertoire. Both goals are accomplished via a preintervention functional assessment which identifies variables responsible for the maintenance of problem behavior which can be altered in intervention and identifies as well effective reinforcers which can be used to strengthen adaptive skills.

It should be apparent that, regardless of presenting concerns, emphasis is placed on the role environmental variables play in maintenance and exacerbation of behavior. The result is an explanation of behavior that is similar across populations—although specific variables, such as type of attention that functions as a reinforcer, may vary, the principles remain the same.

LINK BETWEEN ASSESSMENT AND INTERVENTION

The result of a functional assessment is a hypothesis statement identifying environmental variables that likely evoke and maintain the behavior of concern. This approach is thus logically linked to intervention as specific events that might be altered to effect a change in responding are identified. This is in direct contrast to a structural approach to diagnosis, which is linked in no logical way to intervention, as causal mechanisms or mechanisms responsible for maintenance are not identified. In recent years there has been increased emphasis placed on evidence-based approaches for psychiatric disorders (Huppert, Fabbro, & Barlow, 2006), and the result has been substantive growth in research-based approaches for various disorders. A thorough exploration of the limitations of this approach is beyond the scope of this chapter (but see Glasgow, Davidson, Dobkin, Ockene, & Spring, 2006). However two points are especially relevant here.

First, and as pointed out by Neef and Northup (Chapter 5, "Attention Deficit Hyperactivity Disorder"), the vast majority of diagnoses consist of a long list of behavioral characteristics that must be present in some combination. For example, a diagnosis of autism requires that an individual exhibit behaviors that fall into three different categories—restricted pattern of behavior, impaired communication, social skills deficits—with each category consisting of four possible behaviors that might be exhibited. An individual can be diagnosed with autism if he or she exhibits at least two behaviors from each category, resulting in an almost infinite possibility of presentations. Of course, topography is not related to the function of behavior, so even individuals emitting topographically similar responses may require different interventions. Hopko, Hopko, and Lejuez (Chapter 15, "Mood Disorders") illustrate this in discussing the cases of three women, all presenting with identical behaviors categorized as depression following bilateral mastectomies as intervention for breast cancer. One woman's depressed behavior was a result of aversive social interactions with her partner, another's resulted from faulty rules she had created surrounding breast surgery, and the third woman's depressed behavior resulted from losing her job due to extended absence resulting from illness. Given these difficulties, a functional assessment which identifies idiographic casual relations for each client likely will lead to a more efficacious intervention.

A second limitation of the growing emphasis on interventions linked to diagnosis is the way in which evidence is gathered for these interventions. The vast majority of work supporting such interventions is clinical trials research. Such work involves implementing a standard protocol with a group of clients and typically comparing outcomes to a control group. Interventions typically are carried out by well-trained researchers in university clinics, and participants are carefully screened such that individuals who meet criteria for more than one *DSM* diagnosis or who present with other potentially confounding difficulties such as marital distress or substance abuse are excluded (Hoagwood & Johnson, 2003). This is, of course, not typical of most clinical practices. First, the majority of clinicians have not received detailed training in the interventions they implement and receive minimal supervision during intervention. Further, most individuals presenting for treatment in mental health clinics are not experiencing only one problem but instead present with comorbid problems or have stressful life events occurring simultaneously. Finally, because statistical analyses are used to evaluate outcomes, the extent to which the results of a study actually translate into a significant clinical improvement for any participant in the study is not known. As a result of these concerns, a growing number of researchers and practitioners are calling for alternative strategies for advancing the evidence-base for interventions (e.g., Glasgow et al., 2006; Klesges, Dzewaltowski, & Christensen, 2006; Thyer, 1997).

OVERARCHING THEMES AND DIRECTIONS FOR FUTURE RESEARCH

As illustrated throughout the book, the technology for and applicability of functional assessment is growing. Importantly, current work both adds to the existing literature base on functional assessment and also raises important directions for future research. Each of these areas is addressed next.

TECHNOLOGY OF FUNCTIONAL ASSESSMENT

To date, the vast majority of work in the area of functional assessment and functionally derived interventions has been conducted with individuals with significant disabilities emitting severe problem behavior at high frequency. As a result, methods of functional assessment have been developed primarily for that population. Speaking broadly, methods of functional assessment are classified as either indirect (e.g., rating scales) or direct (e.g., ABC observations, functional analysis). These methods can be placed along a continuum of control exerted over environmental events with indirect assessments lying at one end and experimental methods at the other end and descriptive assessments lying along the middle of the continuum.

Direct Functional Assessment

Direct methods of functional assessment are those through which the clinician develops hypotheses about environment-behavior relations by observing the individual engaging in target responses. Functional analysis—a direct method of functional assessment involving experimental manipulation of environmental variables such that causal functional relations can be identified—is considered the gold standard of functional assessment, as it allows for the greatest degree of environmental control. Of existing methods of functional analysis, the analog functional analysis developed by Iwata, Dorsey, Slifer, Bauman, and Richman (1982/1994) and variations thereof are used most often. This method consists of three to four test conditions within which a single antecedent event (e.g., attention deprivation) and a single consequence (e.g., attention delivery) are manipulated systematically. The analog functional analysis is conducted using a single-subject design—typically a multielement design—which allows for demonstration of functional relations. Most often the assessment is conducted by trained practitioners in a clinical setting. The analog functional analysis provides a model for assessing effects of multiple controlling variables on problem behavior, and recent work suggests that the validity of the analog functional analysis might be enhanced by inclusion of idiographic variables directly relevant to the target individual (Call, Wacker, Ringdahl, & Boelter, 2005; Mueller, Sterling-Turner, & Moore, 2005; O'Reilly, Sigafoos, Lancioni, Edrisinha, & Andrews, 2005). For example, Wilder and Wong (Chapter 14, "Schizophrenia and Other Psychotic Disorders") describe several specific modifications necessary when conducting a functional analysis with individuals diagnosed with schizophrenia, including the types of tasks used in a demand condition and the instructions given to clients prior to conducting the assessment.

Inclusion of idiographic variables requires that information be gathered prior to conducting an experimental analysis; one strategy for accomplishing this goal is via a descriptive assessment. Descriptive assessments may be conducted in at least two ways. First, an observer may simply record instances of problem behavior and environmental events that occur within some temporal proximity of the problem behavior using time sampling or interval recording procedures (Bijou, Peterson, & Ault, 1968). Categories of antecedent and consequent events that are temporally associated with problem behavior are identified based on the observations. Unfortunately, because environmental events are not manipulated, functional relations cannot be verified in the ABC assessment. Although specific events may appear to reliably precede or follow problem behavior during the observation, such events may not be directly related to the problem behavior. For example, problem behavior following prompts might be immediately followed by escape. Shortly after the task has been withdrawn, the participant may be provided with attention. If the ABC assessment is used to identify only those events that are temporally contiguous with problem behavior, the attention delivery may not even be recorded. Further, without experimental manipulation, it is difficult to determine which consequence—escape or attention—actually maintains responding in this situation. A second limitation of descriptive assessments is that problem behavior may not be observed. This may occur, for example, if caregivers have restructured the environment such that antecedent variables that reliably evoke problem behavior rarely occur. In such a case, extensive observation may be necessary to develop hypotheses about functional relations.

An alternative to this type of descriptive observation is to conduct a structural analysis (Anderson & Long, 2002; Stichter & Conroy, 2005; Stichter, Lewis, Johnson, & Trussell, 2004). Structural analyses involve manipulating environmental variables—usually identified via an interview—in a systematic manner and recording instances of problem behavior and other environmental events. For example, Anderson and Long (2002) manipulated three antecedent variables—requests to complete academic tasks, termination of preferred activities, and deprivation of attention using a multielement design—with four children exhibiting severe problem behavior. Caregivers conducted the assessment and were instructed to respond to problem behavior as they typically did. Anderson and Long reported differentiated responding across conditions, and interventions based on the assessment were effective for three of the four participants (intervention was not developed for the fourth). Structural analyses may be especially useful in typical settings and with individuals whose problem behavior is under the control of complex stimuli; however, because environmental events are not manipulated systematically, functional control is not demonstrated.

Direct methods of functional assessment are preferred, as they require a lower degree of inference relative to indirect methods; however, direct methods are inappropriate for some settings and with some populations. In particular, direct methods may not be useful if an individual's responding is affected by observation by others (such as covert behavior, behavior that occurs only in the presence of specific individuals), if the setting is not amenable to outside observers (such as an individual's place of employment), or if the behavior of concern occurs at low frequency. In such cases, clinicians must rely on indirect methods of functional assessment.

Indirect Methods of Functional Assessment

Indirect methods of functional assessment require the greatest degree of inference regarding environment-behavior relations, as the clinician does not actually observe the target behavior but instead gathers information via an informant. Although indirect assessments allow one to gather a great deal of information in a relatively brief amount of time, scant research has focused on the reliability, validity, or treatment utility of existing methods. Further, the few studies that have evaluated psychometric properties of indirect measures have produced mixed to poor results (Barton-Arwood, Wehby, Gunter, & Lane, 2003; Floyd, Phaneuf, & Wilczynski, 2005; Fox & Davis, 2005; Newcomer & Lewis, 2004). Further, most structured indirect functional assessment methods were developed for use with individuals with disabilities (e.g., the *Motivation Assessment Scale*, Durand & Crimmins, 1988) and hence may not be appropriate for typically developing individuals exhibiting clinical problems such as those described in the current book. As a result, most clinicians desiring to use a functional approach to assessment and intervention must develop their own indirect functional assessment tools.

Methods of Functional Assessment: Needed Research

Most existing methods of functional assessment were developed for, and have been evaluated with, individuals with disabilities. Although the basic classes of environment-behavior relations (e.g., negative reinforcement, positive reinforcement) remain the same across populations, the development of methods of functional assessment applicable to specific populations likely will increase the utility of functional assessment as a method for identifying environment-behavior relations and developing effective interventions. Two broad areas will need to be addressed regarding features of functional assessment strategies: types of functional assessments conducted and variables included in a functional account of human behavior.

Research to Identify and Validate Methods of Functional Assessment

One broad question to be addressed is whether structured methods of functional assessment are needed or if instead clinicians simply should develop a method specific to the individual(s) they are working with. Although individuals with extensive background in behavior analysis likely are able to accomplish the latter in a competent manner, the extent to which practitioners without such a background could do so is questionable. If we hope to extend the applicability and utility of our science beyond those with extensive training in behavior analysis to include, for example, social workers and master's level therapists, behavior analysts will need to focus on developing and evaluating methods of functional assessment useful for typically developing individuals. For example, structured functional assessment interviews for use with children diagnosed with pediatric feeding disorders likely would be useful for many practitioners, such as occupational therapists, speech therapists, and social workers, who commonly work with these children but do not have a solid background in behavior analysis from which to draw. It also might be useful for researchers to attempt to develop set structural analysis conditions that are useful in different settings and with different populations. For example, perhaps a generic set of antecedent conditions could be developed for individuals diagnosed with schizophrenia presenting with behavioral excesses. The specific stimuli used in each assessment would differ across participants (e.g., the type of preferred stimuli removed), but the general procedural guidelines would remain the same. The need for such research is illustrated by recent calls within the fields of school psychology and special education for methods of functional assessment with documented utility for school-aged children presenting with high-incidence disabilities such as ADHD and conduct problems (Floyd et al., 2005; Fox & Davis, 2005; Scott et al., 2004; Shriver, Anderson, & Proctor, 2001).

Variables Included in a Functional Account of Human Behavior

Most published research on functional assessment focuses on relations between discrete environmental variables—most often actions emitted by another individual—that occur close in time to the target response. As functional assessment technology is applied to highly verbal individuals exhibiting complex problems, research increasingly is needed to address temporal features of stimuli and the role of verbal behavior.

Temporal Relations

Effects of temporal relations between environmental events and responding typically are not considered in a functional assessment, as the focus primarily is on variables that occur in close temporal contiguity to the target response. It is likely, however, that this view of temporal relations will need to be expanded to account for (a) distal relations between events and responses and (b) effects of time delays on the values of reinforcers.

With regard to distal relations, the emphasis on events that occur in close proximity to a response is understandable for at least two reasons. First, extensive basic and applied research demonstrates that consequences that occur close in time to a response typically are more effective than delayed consequences. Second, when an event is distal to a response, its effects may be difficult to judge given that many other events occur between it and the target response. For example, one may speculate that an individual skips an afternoon class as a result of a fight with his or her partner that morning—perhaps the argument served to establish avoiding a tedious lecture as a reinforcer. While this is plausible, multiple other events that occurred during the temporal gap—such as sitting in traffic on the way to school, the fact that the sun is shining for the first time in a month—also may have established avoiding the class as a reinforcer. As behavior analysts focus on complex human behavior, effects of more distal stimuli may need to be taken into account more often within a functional assessment. For example, an illicit drug ingested hours before may function as an establishing operation for current behavior.

Importantly, a handful of studies have established a clear functional relation between temporally distant events and problem behavior (Horner, Day, & Day, 1997; Ray & Watson, 2001; Reed, Dolezal, Cooper-Brown, & Wacker, 2005). For example, Horner et al. demonstrated that functional relations demonstrated via an analog functional analysis varied systematically based on the prior occurrence or nonoccurrence of distal events including sleep deprivation and cancellation of a previously scheduled event. Research is needed first to identify distal events that may affect responding in systematic ways and then to determine whether a clear causal relation between those events and the target behavior can be identified (see Horner et al. for an elegant methodology for accomplishing this task). Research should address as well whether a complete and pragmatic account of behavior can be reached without the inclusion of such distal variables; that is, can an effective intervention be developed if distal events are not included?

A second area to be addressed within functional assessment research is the role that time plays in the valuation and devaluation of consequences. A quantitative model, delay discounting (Mazur, 1987), has been used to demonstrate that the value of a reinforcer systematically is discounted or devalued as a result of its delay to presentation. A substantive body of basic work supports delay discounting with both humans and nonhumans (Madden, Petry, Badger, & Bickel., 1997; Mazur, 1987, 1998; Myerson & Green, 1995; Petry & Casarella, 1999). Within this paradigm, individuals who show increased delay discounting, i.e., choose smaller rewards after a brief delay rather than larger but more delayed rewards, are referred to as more "impulsive" relative to those for whom delayed outcomes hold their value longer. Building on this literature base, and as discussed by Neef and Northup (Chapter 5, "Attention Deficit Hyperactivity Disorder") and Higgins, Heil, and Sigmon (Chapter 13, "A Behavioral Approach to the Treatment of Substance Use Disorders"), several researchers have suggested that this sensitivity to immediate reinforcement is a defining feature of impulsivity exhibited by children diagnosed with ADHD (Schweitzer & Sulzer-Azaroff, 1995; Sonuga-Barke, 1994; Sonuga-Barke, Daley, Thompson, Laver-Bradbury, & Weeks, 2001) and individuals diagnosed with substance abuse (Bickel & Marsch, 2001; Madden et al., 1997; Petry & Casarella, 1999). It seems plausible that temporal delays play a role in other problematic behavior as well including, for example, obesity (e.g., choosing the more immediate reinforcers associated with eating highly caloric food over the delayed reinforcers associated with maintaining a healthy weight), habit disorders (e.g., choosing the immediate sensory reinforcer associated with pulling out eyelashes over the larger, delayed rewards such as having lush eyelashes), and many childhood behavior disorders (e.g., choosing brief "negative" attention as an immediate reinforcer rather than the more delayed "positive" attention and associated consequences that follow appropriate behavior). As of yet, effects of temporal delays on consequences have not been included in methods of functional assessment in any systematic way. Neef and Northup (Chapter 5, "Attention Deficit Hyperactivity Disorder") offer precise, pragmatic strategies for assessing which dimensions (e.g., delay, magnitude, quality) of a reinforcer control behavior; however, this framework has not been extended to other populations. With individuals who abuse substances, for example, basic research has documented that many substance abusers exhibit impulsive responding (see Higgins et al., Chapter 13, "Schizophrenia and Other Psychotic Disorders," for a review), and this work forms part of the conceptual basis for interventions such as the Community Reinforcement plus Vouchers Approach described by Higgins et al. As of yet, however, a systematic method for evaluating whether a given individual exhibits such impulsive behavior as part of a preintervention functional assessment has not been developed.

Even when relatively standard interventions such as Community Reinforcement plus Vouchers exist, the value of determining whether an individual is overly sensitive to temporal delays prior to treatment is fairly obvious. If, for example, a substance-abusing person is not particularly sensitive to temporal delays but responds differentially instead to the amount of reinforcement available, then the typical intervention might not be effective. What might be more effective is to begin with larger rewards for abstaining from substance abuse than otherwise would be used. Research thus is needed to first evaluate the extent to which delay discounting is applicable to other human problems and to next develop procedures for the evaluation of impulsive behavior within a functional assessment.

Role of Verbal Behavior

Most humans are verbal beings and a growing body of human operant (Pilgrim & Galizio, 2000; Tyndall, Roche, & James, 2004) and applied research (de Rose, de Souza, & Hanna, 1996; Dougher, 1998; Leslie, Tierney, Robinson, Keenan, & Watt, 1993; Melchiori, de Souza, & de Rose, 2000) has focused on how our capacity to respond to verbal information as opposed to physical stimuli affects behavior. Two interrelated areas of research are relevant to functional assessment: work on rule-governed behavior and research on relational responding.

Stimulus control can be developed in two ways: via direct contact with differential consequences or through rules (Pierce & Cheney, 2004; Skinner, 1969). Several chapters in the current work discuss effects of rules on subsequent responding with regard to personality disorders (Cuper, Merwin, & Lynch, Chapter 19, "Personality Disorders"), eating disorders (Farmer & Latner, Chapter 18, "Eating Disorders"), anxiety disorders (Friman, Chapter 16, "The Fear Factor: A Functional Perspective on Anxiety"), and mood disorders (Hopko et al., Chapter 15, "Mood Disorders"). If an individual's behavior is wholly or in part a response to rules the individual has created, then a comprehensive functional assessment must address the rules; the goal here would be to identify the rules, the responses that often follow, and the consequences identified within the rule that maintain responding. For example, an individual who meets criteria for depression might tell himself, "If I stay in bed, no more bad things will happen to me today." If the man follows the rule, the behavior (staying in bed) is negatively reinforced by the contingency specified in the rule (avoiding "bad things"). Research suggests that rule-governed behavior is relatively insensitive to direct contact with contingencies (Hayes, Brownstein, Haas, & Greenway, 1986; Joyce & Chase, 1990; Shimoff, Catania, & Matthews, 1981), so even if the man had gotten out of bed on a given day and nothing bad had happened, this rule likely would continue to control his behavior on future days. When behavior is under the control of rules that are in opposition to actual contingencies, an intervention will need to address this control either directly or indirectly to be effective. Research thus is needed to (a) identify strategies for evaluating rule-governed behavior within a functional assessment and (b) assess whether strategies that address control by rules as part of an intervention lead to more positive outcomes than would be attained without their use.

Closely related to rule-governed behavior is a growing body of work beginning with Sidman (1994) that documents a so far uniquely human ability to respond to relations among stimuli that were not taught directly. Of derived relations, stimulus equivalence has generated the most research to date, but, as reviewed by Friman in Chapter 16, "The Fear Factor: A Functional Perspective on Anxiety," other relations may be derived as well. What makes equivalence and other relations so relevant to clinical problems is that stimulus functions transfer from one stimulus to another when such relations are established (Dougher & Markham, 1994; Greenway, Dougher, & Wulfert, 1996; Hayes, Hayes, Sato, & Ono, 1994; Lyddy, Barnes-Holmes, & Hampson, 2001; Markham & Markham, 2002). These findings render us better able to understand complex patterns of responding that seem unrelated to a history of direct contact with environmental stimuli. Relational responding explains, for example, why simply reading about an event (e.g., the reunion of two long-separated lovers) evokes emotions similar to those experienced if we witnessed the event or if it happened to us. Without relational responding as a framework, such responding is difficult to understand, as the words written on a page are arbitrary stimuli—they bear no formal resemblance to the events they represent.

As illustrated by Friman in Chapter 16, "The Fear Factor: A Functional Perspective on Anxiety," relational responding is directly relevant to an understanding of seemingly complex anxiety disorders. A growing body of research demonstrates the relevance of relational responding to multiple other human problems (Greco, Blackledge, Coyne, & Ehrenreich, 2005; Hayes et al., 2004). What is needed at this point are guidelines for including the assessment of relational responding within a functional assessment. As reviewed above, most methods of functional assessment focus on immediate and observable contingencies for responding; such methods may be less applicable to behavior under the control of complex arbitrary relations among stimuli.

APPLICABILITY OF FUNCTIONAL ASSESSMENT

Functional assessment has a strong empirical base, both from the voluminous basic and applied research documenting effects of environmental variables on responding and from a substantive body of research documenting effective interventions based on a pretreatment functional assessment. To date, however, the vast majority of research has been conducted with individuals with significant disabilities exhibiting severe challenging behavior. What is needed at this point is research documenting the utility of functional assessment and, hence, functionally derived interventions across settings and populations.

Behavior analysts need to first demonstrate that functional assessment is applicable and useful for the broad spectrum of human problems. In addition, we must demonstrate that interventions based on a functional assessment are not just effective but that they are *more effective* than what would otherwise have been used. Accomplishing this in a convincing manner likely will require a broad definition of effectiveness. One dimension, of course, is a reduction in the problematic behavior of interest. Equally important, however, is that the intervention results in increases in important, prosocial behaviors and outcomes, such as improved grades, more friends, a more enjoyable job. We also must determine whether functionally derived interventions result in more rapid change and whether the results are more durable over time and generalizable across settings.

SUMMARY

In conclusion, a functional approach to human problems has many advantages relative to the more prevalent approach—classification based on topography of responses. A functional approach provides a parsimonious framework for conceptualizing human behavior—one that is applicable to all behavior exhibited by humans and nonhumans. As importantly, a functional approach is linked directly to development of interventions, as a functional assessment identifies environmental variables that might be manipulated to effect a change in behavior. Throughout this book, authors illustrate the conceptual and empirical basis for functional assessment with various populations. They illustrate as well that the application of functional assessment to clinical populations is an area ripe for both basic and applied research.

REFERENCES

American Psychiatric Association. (2000). *Diagnostic and Statistical Manual of Mental Disorders (4th ed.)*. Washington, DC: Author.

Anderson, C. M., & Long, E. S. (2002). Use of a structured descriptive assessment methodology to identify variables affecting problem behavior. *Journal of Applied Behavior Analysis, 35*(2), 137–154.

Barton-Arwood, S. M., Wehby, J. H., Gunter, P. L., & Lane, K. L. (2003). Functional behavior assessment rating scales: Intrarater reliability with students with emotional or behavioral disorders. *Behavioral Disorders, 28*, 386–400.

Bickel, W. K. & Marsch, L. A. (2001). Toward a behavioral economic understanding of drug dependence: Delay discounting processes. *Addiction, 96*, 73–86.

Bijou, W. W., Peterson, R. F., & Ault, M. H. (1968). A method to integrate descriptive and experimental field studies at the level of data and empirical concepts. *Journal of Applied Behavior Analysis, 1*, 175–191.

Call, N. A., Wacker, D. P., Ringdahl, J. E., & Boelter, E. W. (2005). Combined antecedent variables as motivating operations within functional analyses. *Journal of Applied Behavior Analysis, 38*, 385–389.

Carr, E. G., Horner, R. H., Turnbull, A. P., Marquis, J. G., McLaughlin, D. M., McAtee, M. L., et al. (1999). *Positive behavior support for people with developmental disabilities: A research synthesis*. Washington, DC: American Association on Mental Retardation.

de Rose, J. C., de Souza, D. G., & Hanna, E. S. (1996). Teaching reading and spelling: Exclusion and stimulus equivalence. *Journal of Applied Behavior Analysis, 29*, 451–469.

Dougher, M. J. (1998). Stimulus equivalence and the untrained acquisition of stimulus functions. *Behavior Therapy, 2*, 577–591.

Dougher, M. J., & Markham, M. R. (1994). Stimulus equivalence, functional equivalence and the transfer of function. In S. C. Hayes, L. J. Hayes, M. Sato, & K. Ono (Eds.), *Behavior analysis of language and cognition* (pp. 71–90). Reno, NV: Context Press.

Dunlap, G., & Childs, K. E. (1996). Intervention research in emotional and behavioral disorders: An analysis of studies from 1980–1993. *Behavioral Disorders, 21*, 125–136.

Durand, V., & Crimmins, D. B. (1988). Identifying the variables maintaining self-injurious behavior. *Journal of Autism and Developmental Disorders, 18*, 99–117.

Floyd, R. G., Phaneuf, R. L., & Wilczynski, S. M. (2005). Measurement properties of indirect assessment methods for functional behavioral assessment: A review of research. *School Psychology Review, 34*, 58–73.

Fox, J., & Davis, C. (2005). Functional behavior assessment in schools: Current research findings and future directions. *Journal of Behavioral Education, 14*, 1–4.

Glasgow, R. E., Davidson, K. W., Dobkin, P. L., Ockene, J., & Spring, B. (2006). Practical behavioral trials to advance evidence-based behavioral medicine. *Annals of Behavioral Medicine, 34*, 5–13.

Greco, L. A., Blackledge, J. T., Coyne, L. W., & Ehrenreich, J. (2005). Integrating acceptance and mindfulness into treatments for child and adolescent anxiety disorders: Acceptance and commitment therapy as an example. In S. Orsillo & L. Roemer (Eds.), *Acceptance and mindfulness-based approaches to anxiety: Conceptualization and treatment* (pp. 301–324). New York: Springer.

Greenway, D. E., Dougher, M. J., & Wulfert, E. (1996). Transfer of consequential functions via stimulus equivalence: Generalization to different testing contexts. *Psychological Record, 46*, 131–143.

Gresham, F. M., & Gansle, K. A. (1992). Misguided assumptions of DSM-III-R: Implications for school psychological practice. *School Psychology Quarterly, 7*, 79–95.

Gresham, F. M., & Kern, L. (2004). Internalizing behavior problems in children and adolescents. In R. B. Rutherford, M. M. Quinn, & S. R. Mathur (Eds.), *Handbook of research in emotional and behavioral disorders* (pp. 262–281). New York: Guilford Press.

Hayes, S. C., Brownstein, A. J., Haas, J. R., & Greenway, D. E. (1986). Instructions, multiple schedules, and extinction: Distinguishing rule-governed behavior from schedule controlled behavior. *Journal of the Experimental Analysis of Behavior, 46,* 137–147.

Hayes, S. C., & Follette, W. C. (1993). The challenge faced by behavioral assessment. *European Journal of Psychological Assessment, 9,* 182–188.

Hayes, S. C., Hayes, L. J., Sato, M., & Ono, K. (Eds.). (1994). *Behavior analysis of language and cognition.* Reno, NV: Context Press.

Hayes, S. C., Nelson, R. O., & Jarrett, R. B. (1987). The treatment utility of assessment: A functional approach to evaluating assessment quality. *American Psychologist, 42,* 963–974.

Hayes, S. C., Strosahl, K., Wilson, K. G., Bissett, R. T., Pistorello, J., Toarmino, D., et al. (2004). Measuring experiential avoidance: A preliminary test of a working model. *Psychological Record, 54,* 553–578.

Hoagwood, K., & Johnson, J. (2003). School psychology: A public health framework. I. From evidence-based practices to evidence-based policies. *Journal of School Psychology, 4,* 3–21.

Horner, R. H., Day, H., & Day, J. R. (1997). Using neutralizing routines to reduce problem behaviors. *Journal of Applied Behavior Analysis, 30,* 601–614.

Huppert, J. D., Fabbro, A., & Barlow, D. (2006). Evidence-based practice and psychological treatments. In C. D. Goodheart, A. E. Kazdin, & R. J. Sternberg (Eds.), *Evidence-based psychotherapy: Where practice and research meet* (pp. 131–152). Washington, DC: American Psychological Association.

Individuals with Disabilities Education Act Amendments of 1997, 20 U.S.C. § 1400 (1997).

Individuals with Disabilities Education Improvement Act, 20 U.S.C. § 1400 (2004).

Iwata, B. A., Dorsey, M. F., Slifer, K. J., Bauman, K. E., & Richman, G. S. (1994). Toward a functional analysis of self-injury. *Journal of Applied Behavior Analysis, 27,* 197–209. (Reprinted from *Analysis and Intervention in Developmental Disabilities, 2,* 3–20, 1982).

Joyce, J. H., & Chase, P. N. (1990). Effects of response variability on the sensitivity of rule-governed behavior. *Journal of the Experimental Analysis of Behavior, 54,* 251–262.

Klesges, L. M., Dzewaltowski, D. A., & Christensen, A. J. (2006). Are we creating relevant behavioral medicine research? Show me the evidence! Comment. *Annals of Behavioral Medicine, 31,* 3–4.

Korchin, S. J., & Schuldberg, D. (1981). The future of clinical assessment. *American Psychologist, 36,* 1147–1158.

Lane, K. L., Umbreit, J., & Beebe-Frankenberger, M. E. (1999). Functional assessment research on students with or at risk for EBD: 1990 to the present. *Journal of Positive Behavior Interventions, 1,* 101–111.

Lee, M. I., & Miltenberger, R. G. (1997). Functional assessment and binge eating: A review of the literature and suggestions for future research. *Behavior Modification, 21,* 159–171.

Lennox, D. B., Miltenberger, R. G., Spengler, P., & Erfanian, N. (1988). Decelerative treatment practices with persons who have mental retardation: A review of five years of the literature. *American Journal on Mental Retardation, 92,* 492–501.

Leslie, J. C., Tierney, K. J., Robinson, C., Keenan, M., & Watt, A. (1993). Differences between clinically anxious and non-anxious subjects in a stimulus equivalence training task involving threat words. *Psychological Record, 43,* 153–161.

Lyddy, F., Barnes-Holmes, D., & Hampson, P. J. (2001). Transfer of sequence function via equivalence in a connectionist network. *Psychological Record, 51,* 409–428.

Madden, G. J., Petry, N. M., Badger, G. J., and Bickel, W. K. (1997). Impulsive and self-control choices in opioid-dependent patients and non-drug-using control participants: Drug and monetary rewards. *Experimental and Clinical Psychopharmacology, 5,* 256–262.

Markham, R. G., & Markham, M. R. (2002). On the role of covarying functions in stimulus class formation and transfer of function. *Journal of the Experimental Analysis of Behavior, 78,* 509–524.

Mazur, J. E. (1987). An adjusting procedure for studying delayed reinforcement. In M. L. Commons, J. E. Mazur, J. A. Nevin, & H. Rachlin (Eds.), *Quantitative analyses of behavior: Vol. 5. The effect of delay and intervening events on reinforcement value* (pp. 55–73). Hillsdale, NJ: Erlbaum.

Mazur, J. E. (1998). Choice with delayed and probabilistic reinforcers: Effects of prereinforcer and postreinforcer stimulus. *Journal of the Experimental Analysis of Behavior, 70,* 253–265.

Melchiori, L. E., de Souza, D. G., & de Rose, J. C. (2000). Reading, equivalence, and recombination of units: A replication with students with different learning histories. *Journal of Applied Behavior Analysis, 33,* 97–100.

Morey, L. C., Skinner, H. A., & Blashfield, R. K. (1986). Trends in the classification of abnormal behavior. In A. R. Ciminero, K. S. Calhoun, & H. E. Adams (Eds.), *Handbook of behavioral assessment* (2nd ed., pp. 47–75). New York: Wiley.

Mueller, M. M., Sterling-Turner, H. E., & Moore, J. W. (2005). Towards developing a classroom-based functional analysis condition to assess escape-to-attention as a variable maintaining problem behavior. *School Psychology Review, 34,* 425–431.

Myerson, J., & Green, L. (1995). Discounting of delayed rewards: Models of individual choice. *Journal of the Experimental Analysis of Behavior, 64,* 263–276.

Nelson, R., & Hayes, S. (1986). The nature of behavioral assessment. In R. Nelson & S. Hayes (Eds.), *Conceptual foundations of behavioral assessment* (pp. 3–41). New York: Guilford.

Newcomer, L. L., & Lewis, T. J. (2004). Functional behavioral assessment: An investigation of assessment reliability and effectiveness of function-based interventions. *Journal of Emotional and Behavioral Disorders, 12,* 168–181.

O'Reilly, M., Sigafoos, J., Lancioni, G., Edrisinha, C., & Andrews, A. (2005). An examination of the effects of a classroom activity schedule on levels of self-injury and engagement for a child with severe autism. *Journal of Autism and Developmental Disorders, 35,* 305–311.

Petry, N. M., & Casarella, T. (1999). Excessive discounting of delayed rewards in substance abusers with gambling problems. *Drug and Alcohol Dependence, 56,* 25–32.

Pierce, W. D., & Cheney, C. D. (2004). *Behavior analysis and learning* (3rd ed.). Mahway, NJ: Lawrence Erlbaum Associates.

Pilgrim, C., & Galizio, M. (2000). Stimulus equivalence and units of analysis. In J. C. Leslie & D. Blackman (Eds.), *Issues in experimental and applied analyses of human behavior* (pp. 111–126). Reno, NV: Context Press.

Ray, K. P., & Watson, T. (2001). Analysis of the effects of temporally distant events on school behavior. *School Psychology Quarterly, 16,* 324–342.

Reed, G. K., Dolezal, D. N., Cooper-Brown, L. J., & Wacker, D. P. (2005). The effects of sleep disruption on the treatment of a feeding disorder. *Journal of Applied Behavior Analysis, 38,* 243–245.

Schweitzer, J. B., & Sulzer-Azaroff, B. (1995). Self-control in boys with attention deficit hyperactivity disorder: Effects of added stimulation and time. *Journal of Child Psychology and Psychiatry, 36,* 671–686.

Scott, T. M., Bucalos, A., Liaupsin, C., Nelson, C., Jolivette, K., DeShea, L., et al. (2004). Using functional behavior assessment in general education settings: Making a case for effectiveness and efficiency. *Behavioral Disorders, 29*, 189–201.

Scotti, J. R., Ujcich, K. J., Weigle, K. L., & Holland, C. M. (1996). Interventions with challenging behavior of persons with developmental disabilities: A review of current research practices. *Journal of the Association for Persons with Severe Handicaps, 21*, 123–134.

Shimoff, E., Catania, A. C., & Matthews, B. A. (1981). Uninstructed human responding: Sensitivity of low rate performance to schedule contingencies. *Journal of the Experimental Analysis of Behavior, 36*, 207–220.

Shriver, M. D., Anderson, C. M., & Proctor, B. (2001). Evaluating the validity of functional behavior assessment. *School Psychology Review, 30*, 180–192.

Sidman, M. (1994). *Equivalence relations and behavior: A research story.* Boston: Authors Cooperative.

Skinner, B. F. (1953). *Science and human behavior.* New York: Macmillan.

Skinner, B. F. (1969). *Contingencies of reinforcement: A theoretical analysis.* New York: Appleton-Century-Crofts.

Sonuga-Barke, E. J. (1994). On dysfunction and function in psychological theories of childhood disorder. *Journal of Child Psychology and Psychiatry, 35*, 801–815.

Sonuga-Barke, E. J., Daley, D., Thompson, M., Laver-Bradbury, C., & Weeks, A. (2001). Parent-based therapies for preschool attention-deficit/hyperactivity disorder: A randomized, controlled trial with a community sample. *Journal of the American Academy of Child & Adolescent Psychiatry, 40*, 402–408.

Stichter, J. P., & Conroy, M. A. (2005). Using structural analysis in natural settings: A responsive functional assessment strategy. *Journal of Behavioral Education, 14*, 19–34.

Stichter, J. P., Lewis, T. J., Johnson, N., & Trussell, R. (2004). Toward a structural assessment: Analyzing the merits of an assessment tool for a student with E/BD. *Assessment for Effective Intervention, 30*, 25–40.

Thyer, B. A. (1997). Professor Higgins' dilemma: Eliza Doolittle grows up—A review of *Sourcebook of Psychological Treatment Manuals for Adult Disorders,* edited by Vincent Van Hasselt and Michel Hersen. *Journal of Applied Behavior Analysis, 30*, 731–734.

Tyndall, I. T., Roche, B., & James, J. E. (2004). The relation between stimulus function and equivalence class formation. *Journal of the Experimental Analysis of Behavior, 81*, 257–266.

ABOUT THE CONTRIBUTORS

Laura R. Addison's research interests include pediatric feeding disorders and functional assessment of speech in children with autism and other developmental disabilities using a Skinnerean framework to conceptualize language development.

Cynthia M. Anderson is Associate Professor, School Psychology, University of Oregon, and also is a researcher in Educational and Community Supports at University of Oregon. Her research focuses on functional assessment and function-based supports for children exhibiting challenging behavior, pediatric feeding disorders, and systems-change in schools.

Holly Bihler is currently a graduate student in the Behavior Analysis and Therapy Masters program at Southern Illinois University in Carbondale where she was the recipient of a graduate fellowship. She is actively involved in brain injury research as well as a member of the gambling laboratory at Southern Illinois University. Holly received her BA in 2003 from Western Michigan University.

Stacey M. Cherup received a bachelor's degree in psychology from Western Michigan University in 2004. She is currently pursuing a doctorate in clinical psychology at the University of Nevada, Reno. Past research has included evaluating preference assessment modalities with persons with dementia, as well as interventions to increase engagement in pleasurable activities. Her current research interests focus on behavioral management of disruptive behaviors exhibited by persons with dementia in long-term care facilities, as well as staff training in behavioral interventions. She is currently employed at the Nevada Caregiver Support Center in Reno.

Prudence Cuper is a doctoral student in the Clinical Psychology Department at Duke University. Her interests include dialectical behavior therapy and the treatment of personality disorders, and she works as a research assistant at the Cognitive Behavioral Research and Treatment program at the Duke University Medical Center.

Robert Didden is an Associate Professor in Developmental Disabilities at the Department of Special Education of the Radboud University in Nijmegen, The Netherlands. He also works as a psychologist at the Hanzeborg Centre for people with mild mental retardation with behavioral and psychiatric disorders in Zutphen, The Netherlands. His research and clinical interests include applied behavior analysis, especially as applied to behavioral and mental health problems, and instructional procedures for children and adults with developmental disabilities. He serves as an associate editor for *Perceptual and Motor Skills* and the Dutch-language journal *Nederlands Tijdschrift voor de Zorg aan Verstandelijk Gehandicapten.*

Mark Dixon is an Associate Professor in the Behavior Analysis and Therapy (BAT) program, University of Southern Illinois, Carbondale. He has authored 50 journal articles, 5 book chapters, 2 books, and over 175 papers and presentations in a variety of areas including choice and self-control, verbal behavior, gambling, organizational effectiveness, computer programming for psychological research, and developmental disabilities. Dr. Dixon also is the Director of the Behavioral Consultation Group, a service project of the BAT program designed to place graduate students in human service agencies as behavior analysts or organizational consultants. Some of his current research projects include examining the variables involved in maintaining or terminating gambling behavior and the designing of effective behavioral interventions for persons with traumatic brain injuries. He currently serves as the president of the ABA Special Interest Group: Behaviorists Interested in Gambling.

Erica Doran is currently a graduate student in the Learning Processes program at Queens College and the City University of New York's Graduate Center. She is actively involved in research into numerous behavioral processes, particularly the formation of equivalence classes. She received her BA in Psychology in 1988 from George Washington University and received her JD in 1991 from Fordham University School of Law.

Claudia Drossel received a doctorate in experimental psychology from Temple University in 2004. She is currently pursuing a doctorate in clinical psychology at the University of Nevada, Reno, and is the Associate Director of the Nevada Caregiver Support Center, a grant-funded agency that provides behavioral health services to persons with dementia and their families. Her research focuses on functional assessments of cognitive and behavioral deficits in dementia populations and on the implication of those assessments for the design of effective interventions. She is also interested in the development and application of the functional analytic approach and treatment dissemination.

Richard F. Farmer is a psychologist and researcher currently affiliated with Oregon Research Institute in Eugene, Oregon. After earning a PhD in clinical psychology from the University of North Carolina at Greensboro in 1993, he went on to serve as an Assistant and Associate Professor of Psychology at Idaho State University, and then as a Senior Lecturer in

Psychology at the University of Canterbury, New Zealand. His main areas of research and clinical interest include behavioral assessment and therapy, personality disorders, eating disorders, impulsivity and disinhibition, and experiential avoidance.

Jane E. Fisher completed a doctorate in clinical psychology at Indiana University, Bloomington. She is a Professor of Clinical Psychology and former Director of Clinical Training at the University of Nevada, Reno. Dr. Fisher's research interests include behavioral health and aging, applied behavior analysis, and the integration of evidence-based behavioral healthcare in primary and long-term care settings. Her research on the development of restraint-free interventions for elderly persons with cognitive disorders is funded by the National Institute on Aging.

Patrick C. Friman is the Director of Outpatient Behavioral Pediatrics and Family Services at Father Flanagan's Boys' Home and an Adjunct Professor of Pediatrics at the University of Nebraska School of Medicine. His research and clinical interests include habit disorders, incontinence, child behavior problems at home and school, and a broad range of topics organized under the general category known as behavioral pediatrics. He currently is the editor of the *Journal of Applied Behavior Analysis*, serves on the editorial board of eight other journals, and is the author or coauthor of more than 150 journal articles and book chapters.

Sarah Heil is Research Assistant Professor of Psychology and Research Assistant Professor of Psychiatry. Her research interests involve women's health and substance abuse, characterizing nicotine withdrawal and other factors related to smoking during pregnancy, as well as testing and disseminating voucher-based interventions to promote smoking cessation and relapse prevention during pregnancy and postpartum. A second interest is in testing pharmacotherapies to treat heroin and opiate dependence during pregnancy and improving treatment outcomes for both mothers and babies.

Stephen Higgins is Professor of Psychology and Professor of Psychiatry, Co-Director, Substance Abuse Treatment Center; Co-Director, Human Behavioral Pharmacology Laboratory; and Co-Director, Behavioral Pharmacology of Drug Dependence Training program at the University of Vermont. He has published extensively in the areas of drug abuse, cocaine dependence, behavior therapy, and applied behavior analysis. His research interests include behavioral pharmacology, treatment of cocaine dependence, and behavioral and cardiac effects of stimulant-alcohol combinations. He has published extensively in the areas of contingency management and treatment of cocaine and tobacco abuse and addiction and has received extensive grant support for this work with the National Institute of Drug Addiction.

Michael B. Himle is a doctoral student in clinical psychology at the University of Wisconsin, Milwaukee. He is the Coordinator of the UW-

Milwaukee Tic Disorders and Trichotillomania Specialty Clinics and has over 5 years of experience working with children and adults with tic disorders, trichotillomania, and body-focused repetitive behaviors. He has published several empirical articles and book chapters on the assessment and treatment of these problems. His current research interests focus primarily on understanding the role of reinforcement and contextual variables in the expression and suppression of tics.

Derek R. Hopko is Assistant Professor of Clinical Psychology at The University of Tennessee. He graduated from West Virginia University and completed his residency and postdoctoral training at the University of Texas Medical School. His research and clinical interests focus on the behavioral assessment and treatment of mood and anxiety disorders. Dr. Hopko has strong interests in health psychology and conducts behavioral treatment outcome research with cancer patients diagnosed with clinical depression. He is a recipient of grant funding from the National Institute of Mental Health (NIMH), has approximately 55 peer-reviewed publications, and serves on the editorial board of five journals.

Sandra D. Hopko is an employee of the Cariten Assist Employee Assistance program in Knoxville, Tennessee. She conducts her therapy largely within the framework of behavior analytic philosophy, and her areas of clinical specialization include mood and anxiety disorders as well as substance abuse problems. In spite of Sandra's focus on clinical work, she has published several empirical and conceptual manuscripts and is the codeveloper of the *Brief Behavioral Activation Treatment for Depression*.

Marianne Jackson is a doctoral student in the Behavior Analysis program at the University of Nevada, Reno. She received her bachelor's degree in psychology from the University of Strathclyde, Scotland, and began working with autistic children in the field of applied behavior analysis. At the University of Nevada, Reno, she has been, for 3 years, the Assistant Director of the PATH program serving adults with intellectual disabilities. Her master's thesis investigated the role of discrimination ability and equivalence relations. She is a Behavioral Consultant to the Washoe County School District and serves on the Applied Behavior Analysis Executive Council as student representative.

Taylor Johnson is currently a graduate student in the Behavior Analysis and Therapy Masters program at Southern Illinois University in Carbondale, where she is the recipient of a graduate fellowship. She is actively involved in brain injury research as well as a member of the gambling laboratory at Southern Illinois University. Taylor received her BA in English with a minor in Psychology in 2003 from the University of Utah.

Craig H. Kennedy is a Professor of Special Education and Pediatrics at Vanderbilt University and directs the Vanderbilt Kennedy Center Behavior Analysis Clinic. His interests focus on aggressive behavior and their precursors, including stereotypy. His current research analyzes the precurrent and current causes of aggression, emphasizing the interaction of behavioral processes, brain circuits, and regulatory genes.

Janet D. Latner is in the Department of Psychology at the University of Hawaii at Manoa. Her research interests are in the area of eating disorders and obesity, including identifying and changing the antecedents to binge eating, and addressing the appetitive deficits and cognitive dysfunctions that may maintain and perpetuate binge eating in individuals with bulimia nervosa and binge eating disorder. She is also interested in identifying and alleviating the stigmatization and psychosocial burden of obesity, and in improving the long-term outcome of obesity treatment through the use of self-help and continuing care treatment.

Carl W. Lejuez is an Associate Professor of Clinical Psychology at the University of Maryland. He graduated from West Virginia University and completed his residency at Brown University. His research focuses on the translation of basic behavioral research into clinical practice, with a specific focus on mood, anxiety, and substance use disorders. He is a recipient of the American Psychological Association Division of Experimental Psychology Young Investigator Award, has published over 70 peer-reviewed articles and book chapters, and is a grant recipient from the National Institute of Mental Health, including a recent R01 grant that examines the efficacy of behavioral activation treatment for chronic smokers.

Thomas R. Lynch is Assistant Professor in the Departments of Psychology and Psychiatry at Duke University and is the Director of the Duke Cognitive Behavioral Research and Treatment program. He received his PhD from Kent State University and did postdoctoral training at Duke University. He has over 55 published peer-reviewed articles and book chapters and has received five research grants from the National Institutes of Health, including a career development award. He currently is the Principal Investigator on two NIDA R01 grants: A multisite study of dialectical behavior therapy for borderline personality disorder with opiate dependence and a virtual reality study examining novel methods to enhance cue exposure treatment for cocaine addiction.

Michael Marroquin is a doctoral student at the Queens College and the Graduate School and University Center, City University of New York (CUNY), Learning Processes psychology subprogram specializing in developmental disabilities. He currently works as an applied behavior analysis therapist for children with autism spectrum disorder and teaches undergraduate psychology courses at Queens College (CUNY). He is a student member of the American Psychological Association and the student program representative to the Association for Behavior Analysis.

Rhonda M. Merwin is a clinical psychology intern in the Cognitive Behavioral Research and Treatment program at Duke University Medical Center and a doctoral student at the University of Mississippi. She has expertise in Acceptance and Commitment Therapy and has published empirical work on Relational Frame Theory. She received a fellowship award to complete some of this work while at the University of Mississippi. She is interested in acceptance-based interventions, verbal processes, and functional contextualistic approaches to self-relevant thoughts.

Raymond G. Miltenberger is the Jordan A. Engberg Professor of Psychology at North Dakota State University where he has taught since 1985. He is also an adjunct professor of neuroscience at the University of North Dakota School of Medicine. He received his bachelor's degree in psychology from Wabash College in 1978 and his PhD in clinical psychology from Western Michigan University in 1985 after completing an internship in behavioral pediatrics and developmental disabilities at the Kennedy Institute of the Johns Hopkins University School of Medicine. Dr. Miltenberger's research interests are in the areas of children's safety skills, repetitive behavior disorders, and the analysis and treatment of problem behaviors in children and individuals with developmental disabilities.

Nancy A. Neef is a Professor of Special Education at The Ohio State University. Her current research interests are in the area of behavior analysis and attention deficit hyperactivity disorder.

John Northup is currently an Associate Professor in School Psychology at the University of Iowa. He received his PhD from the University of Iowa, completed a postdoctoral fellowship at the Kennedy Krieger Institute, John Hopkins University School of Medicine, and was previously an Associate Professor at Louisiana State University. His research interests are in the areas of the assessment and treatment of the disruptive behavior disorders. He is currently conducting research on the development of functional analysis and assessment procedures for typically developing children and in the area of evaluations of medication such as Ritalin in the classroom and drug-behavior interactions.

Cathleen Piazza is the Director of the Pediatric Feeding Disorders program at the Marcus Institute, Atlanta, Georgia, and an Associate Professor of Psychiatry and Behavioral Sciences at the Johns Hopkins University School of Medicine. She received her doctorate from Tulane University and completed her predoctoral internship and a postdoctoral fellowship at the Kennedy Institute and the Johns Hopkins University School of Medicine. After her training, Dr. Piazza continued as a faculty member at the Kennedy Krieger Institute, where she served as the director of the Severe Behavior Unit, the chief psychologist of the Neurobehavioral Unit, and the director of the Pediatric Behavioral Sleep Clinic. Dr.

Piazza is a former associate editor and is currently on the editorial board of the *Journal of Applied Behavior Analysis*. In 2002, Dr. Piazza was named a Woman of Distinction by the Chron's and Colitis Association and also was identified as the most productive female researcher in the areas of behavior analysis and behavior therapy in the 1990s.

Joseph Plaud is a clinical psychologist whose graduate training was primarily focused on behavioral assessment and therapy. He received his BA in psychology in 1987 from Clark University in Worcester, Massachusetts. He received his PhD in clinical psychology from the University of Maine in 1993, after completing his clinical internship at the University of Mississippi and Jackson Department of Veterans Affairs Medical Centers. Dr. Plaud joined the clinical psychology faculty at the University of North Dakota in 1993 until 1998, where he was involved in the training of clinical and experimental psychology graduate students, as well as pursuing his teaching and research activities in behavior analysis, behavior modification and therapy, and behavioral assessment. Dr. Plaud is also interested in the philosophical and historical foundations of psychology, with particular interests in the theoretical underpinnings of behaviorism and behavior therapy, and the accurate dissemination of behavior analysis in public forums. Dr. Plaud has published and lectured widely in the assessment and treatment of sexual offenders. He is a Visiting Scholar of Human Development at Brown University, the director of Applied Behavioral Consultants, Inc., and editor-in-chief of the *Journal of Sexual Offender Civil Commitment: Science and the Law.*

Stacey Sigmon is Research Assistant Professor of Psychology and Research Assistant Professor of Psychiatry at The University of Vermont. Her research interests are in behavioral pharmacology and substance abuse treatment. As part of a 4-year NIDA grant, she is developing and programmatically evaluating a combined behavioral-pharmacological treatment for prescription opioid abuse. She is also conducting a study investigating the efficacy of a voucher-based smoking cessation intervention in methadone-maintained smokers. She is the director of The Chittenden Center, which is Vermont's first and only methadone clinic and currently treats 200 opioid-dependent patients.

Peter Sturmey, PhD is Professor of Psychology at Queens College and the Graduate Center, City University of New York. He has published over 100 peer-reviewed articles and numerous book chapters. He is on the editorial board of several journals, including *Research in Developmental Disabilities* and *Research in Autism Spectrum Disorders*. Previous books include *Functional Analysis in Clinical Psychology* (1996, Wiley UK), *Offenders with Developmental Disabilities* (with Lindsay & Taylor, 2004, Wiley, UK), *Mood Disorders in People with Mental Retardation* (2005, NADD Press), and *Autism Spectrum Disorders: Applied Behavior Analysis, Evidence and Practice* (with Fitzer, In press, Proed Inc.).

His main research interests are developmental disabilities and applied behavior analysis.

Robert G. Wahler is Professor of Psychology and Director of the Clinical Psychology program at the University of Tennessee, Knoxville. His primary clinical practice and research is focused on disruptive children and how parents can alter the course of these youngsters' maladaptive behavior. His work in the consulting room and in the laboratory has followed principles of applied behavior analysis since 1964.

John Ward-Horner is a doctoral student in the Learning Processes psychology subprogram at Queens College and The Graduate Center of the City University of New York, where he is the recipient of a graduate fellowship. His research interest includes parent and staff training of discrete-trial teaching, and the generalization of discrete-trial teaching skills. He currently works as a behavior therapist with children with autism and also teaches an undergraduate experimental psychology class.

David A. Wilder is an Associate Professor in the School of Psychology at the Florida Institute of Technology. He completed a postdoctoral fellowship at the Johns Hopkins University School of Medicine and worked as an assistant professor of psychology at the University of the Pacific in Stockton, California. His research interests include functional assessment and intervention in a variety of populations, including children with behavior problems and adults with schizophrenia and organizational behavior management.

W. Larry Williams has been at the University of Nevada, Reno, since 1995 where he is an Associate Professor and currently the Associate Chair of the Psychology Department. From 1978 to 1984 he helped design, taught in, and chaired the first masters' program in special education in Brazil and Latin America. From 1985 to 1994 he was director of several behavioral services in Toronto, Canada . His interests are in discrimination processes in persons with developmental disabilities, organizational behavior management in human service settings, assessment and clinical interventions in the developmentally disabled population, and behavioral consultation.

Stephen E. Wong is an Associate Professor in the School of Social Work at Florida International University. He has worked as a research associate and research psychologist with the Department of Psychiatry of the University of California at Los Angeles, and as an Assistant Professor in the School of Social Service Administration at the University of Chicago. Dr. Wong's research interests include the treatment of severe mental disorders; the teaching of social and independent living skills; single-subject experimental designs; and the ideological, political, and economic forces shaping mental health services.

Douglas W. Woods is Associate Professor of Psychology at the University of Wisconsin, Milwaukee. His research interests include assessing and treating tic disorders, trichotillomania, and other OCD-spectrum disorders in children and adults, behavior therapy as applied to childhood feeding problems, oppositional and conduct problems, and social rejection.

Craig Yury received a bachelor's degree from the University of Manitoba and is currently a doctoral candidate in clinical psychology at the University of Nevada, Reno. His research interests center on the development and evaluation of psychological treatments and assessment procedures for the geriatric population. He has conducted several research projects examining the effects of environmental support for maintaining behavior in persons with dementia. Currently, he is developing an instrument designed to improve the scope of the assessment of behavioral functioning for use in dementia research.

INDEX